Manual of Transportation Engineering Studies

H. Douglas Robertson, Ph.D., P.E., Editor
Joseph E. Hummer, Ph.D, P.E., Assistant Editor
Donna C. Nelson, Ph.D, P.E., Assistant Editor

Institute of Transportation Engineers

Prentice Hall Englewood Cliffs, NJ 07632

Library of Congress Cataloging-in-Publication Data

Manual of transportation engineering studies / H. Douglas Robertson, editor.
p. cm.
"Institute of Transportation Engineers."
Includes bibliographical references and index.
ISBN 0-13-097569-9
1. Traffic engineering. 2. Transportation engineering.
I. Robertson, H. Douglas. II. Institute of Transportation Engineers.
HE333.M28 1994
629.04--dc20
93-13519
CIP

Acquisitions Editor: Bill Zobrist
Production Editor: Joe Scordato
Copy Editor: Barbara Zeiders
Cover Designer: H. J. Salzbach
Buyers: Linda Behrens, Dave Dickey

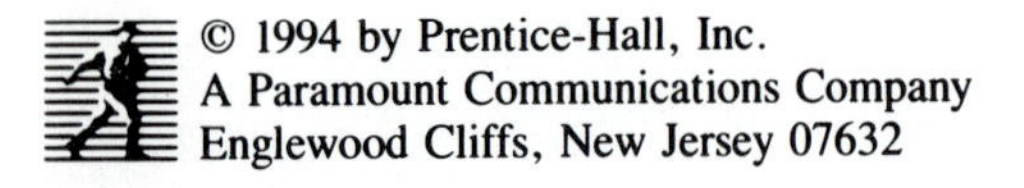

A Paramount Communications Company
Englewood Cliffs, New Jersey 07632

Printed in the United States of America

10 9 8 7 6 5 4 3 2 1

ISBN 0-13-097569-9

ISBN 0-13-097569-9

Prentice-Hall International (UK) Limited, *London*
Prentice-Hall of Australia Pty. Limited, *Sydney*
Prentice-Hall Canada Inc., *Toronto*
Prentice-Hall Hispanoamericana, S.A., *Mexico*
Prentice-Hall of India Private Limited, *New Delhi*
Prentice-Hall of Japan, Inc., *Tokyo*
Simon & Schuster Asia Pte. Ltd., *Singapore*
Editora Prentice-Hall do Brasil, Ltda., *Rio de Janeiro*

The Institute of Transportation Engineers (ITE) is made up of more than 11,000 transportation engineers and planners in over 70 countries. These transportation professionals are responsible for the safe, efficient and environmentally compatible movement of people and goods on streets, highways, and transit systems. For more than 60 years the Institute has been providing transportation professionals with programs and resources to help them meet those responsibilities. Institute programs and resources include handbooks, technical reports, a monthly journal, professional development seminars, local, regional, and international meetings, and other forums for the exchange of opinions, ideas, techniques, and research.

For current information on Institute's programs, please contact:

INSTITUTE OF TRANSPORTATION ENGINEERS
525 School Street, S. W., Suite 410
Washington, DC 20024-2729 USA
Telephone: (202) 554-8050
Facsimile: (202) 863-5486

OTHER INSTITUTE OF TRANSPORTATION ENGINEERS
BOOKS AVAILABLE FROM PRENTICE HALL

Residential Street Design and Traffic Control
Transportation and Traffic Engineering Handbook, 2nd Edition
Transportation and Land Development
Manual of Traffic Signal Design, 2nd edition
Traffic Engineering Handbook, 4th edition
Traffic Signal Installation and Maintenance Manual
Transportation Planning Handbook

Contents

Foreword

This first edition of the *Manual of Transportation Engineering Studies* is an updated and expanded version to the *Manual of Traffic Engineering Studies*, 4th Edition, by Box and Oppenlander.

The Institute's original intent was to update the 1976 4th edition and produce a 5th edition. As the outline was developed, we concluded that technology had advanced significantly in the area of data collection since 1976. It was then decided to rewrite, rather than update existing chapters. In addition, several new chapter topics were identified, expanding the scope beyond traffic engineering studies, to transportation engineering.

This publication is designed to aid transportation professionals and communities to study their transportation problems in a structured manner, following procedures accepted by the profession.

The *Manual of Transportation Engineering Studies* should be used in conjunction with the Institute's *Traffic Engineering Handbook, Transportation Planning Handbook* and other resources in order to prove most useful.

As we release the first edition of the *Manual of Transportation Engineering Studies,* I hope that it will continue to be improved in later editions. Your comments and suggestions are earnestly solicited as a means towards making those improvements.

H. Douglas Robertson
Editor

Acknowledgments

This publication is the result of many hours of concerted effort by the individuals listed below and on the following page. The Institute is appreciative of their dedicated service to the profession.

Work on the book began in 1990 with the confirmation of the editing and author teams. Working closely with ITE Headquarters staff and members of the Technical Review Committee, an outline was established and authors were assigned to chapters. Each chapter draft was reviewed by the members of the review committee who had expertise in that subject. Chapter authors incorporated the reviewers' comments. Final chapter drafts were reviewed by the editing team.

My role as editor was greatly enhanced by my two assistant editors, Joseph E. Hummer and Donna C. Nelson, who also served with me as principal authors. Appreciation also goes to the remainder of the author team, L. Ellis King and Marsha D. Anderson. The authors extend their thanks to the members of the Technical Review Committee for the expertise and helpful insight they provided. We greatly acknowledge Thomas W. Brahms and Juan M. Morales of the ITE Headquarters staff for their patience, administrative support and coordination efforts with the many people involved in this project. We also acknowledge the Prentice Hall professionals for their endeavors.

H. Douglas Robertson
Editor

Technical Review Committee

BOWMAN, Brian L.
BOYD, Ian C.
BRAY, Norman E.
BURKE, Michael L.
CALLOW, John F.
DALGLEISH, Michael J.
DAVIS, John A.
EILTS, Leonard G.
FRUIN, John J.
HAGEN, Lawrence T.
HAMBURG, John
HARKEY, David L.
KAUFMAN, David C.
KINZEL, Chris D.
KOEPKE, Frank J.
LIEBERMAN, William
LYLES, Richard W.
MALINSKY, Grant F.
SAITO, Mitsuru
STRONG, Dennis W.
SULLIVAN, Edward C.
TIMMERMAN, Henry E.
WARREN, Harvey L.
WOODS, Donald L.
ZEGEER, Charles V.

1 Introduction

H. Douglas Robertson, Ph.D., P.E.
Joseph E. Hummer, Ph.D., P.E.
Donna C. Nelson, Ph.D., P.E.

Motor vehicles continue to be the principal means of transportation in the United States. Despite the problems of congestion, delay, parking, pollution, and safety, motor vehicles are expected to dominate surface transportation for the foreseeable future. Solutions to transportation problems and improvements to transportation facilities and services can reasonably be developed after the magnitude, location, and extent of the problems or the need for improvements are well understood. Such understanding comes from factual information gathered in an unbiased, objective manner and analyzed to present a clear, concise picture of the nature of the problem and the impact it is having on the existing transportation system. In a similar manner, the need for transportation improvements can be defined and evaluated.

PURPOSE OF THE MANUAL

This manual is designed to guide and assist traffic and transportation engineers, technicians, and those assigned traffic engineering responsibilities. It also serves as a reference for practicing engineers and engineering students. The primary focus of the manual is on ''how to conduct'' transportation engineering studies in the field. Each chapter introduces the type of study and describes the methods of data collection, the types of equipment used, the personnel and level of training needed, the amount of data required, the procedures to follow, and the techniques available to reduce and analyze the data. Applications of the collected data or information are discussed only briefly. The focus is on planning the study, preparing for field data collection, executing the data collection plan, and reducing and analyzing of the data. Guidelines for both oral and written presentation of study results are offered.

ORGANIZATION OF THE MANUAL

Each chapter is written so that it will stand alone without the need for extensive cross-referencing. Information that is applicable to several types of studies is presented in the appendices. Such topics include general study design, questionnaire design, fundamental statistical analysis, and presentation techniques. An appendix containing typical forms useful in transportation studies is also included. The chapters are generally presented beginning with basic studies and moving to complex or specialized studies. Data from the fundamental studies, such as traffic volume, speed, delay, and inventory, may also be required as part of the more complex or specialized studies, such as studies of traffic impact, accidents, or conflicts, or observance studies.

GENERAL TIPS FOR CONDUCTING FIELD STUDIES

This section contains general instructions and tips that are common to all the transportation engineering studies covered in this manual. These common elements include how to choose the right study (or none at all), how to practice the data collection method, how to prepare on the day of the study, how to avoid disasters during the study, and how to enhance data collector safety. The information in this section is vital for those without experience in transportation engineering studies and should be reviewed occasionally by those with experience to make sure that essentials are not being overlooked.

Choosing a Study Method

Field studies are expensive and should not be conducted without considering the alternatives. The following questions are relevant when deciding whether to use a field study, and which type of study is appropriate:

- What analysis method will be used to solve the problem being faced? (Do not proceed until an analysis method is selected.)
- What input data are needed for the analysis method?
- Are there acceptable values from previous work that can be used as input data? (If ''yes'' for all inputs, do not use a field study.)
- Are data available that can be manipulated to become acceptable as input data? For example, if turning movement counts are needed, are estimates from available link volume data and network geometry acceptable? (If ''yes,'' do not use a field study.)
- Are field study techniques available that will provide the input data needed? (If ''no,'' do not use a field study.)
- Are the time, money, personnel, and other resources needed to conduct the field study available? (If ''no,'' do not use a field study until the resources become available.)
- Is there more than one field study technique that will provide the needed input data with available resources? (If ''yes,'' use the most cost-efficient study technique.)

In summary, whether to conduct a field study, and the choice of a particular study method, are driven by the needs of the question that must be answered and the analysis that is planned. At any time before or during the field study, the engineer has the option of canceling the study if there is a change in the conditions that led to the choice of study method.

Practicing the Study Technique

After study sites, times, personnel, and equipment are selected and arranged, it is usually necessary to practice the study technique. Practice may not be necessary for field personnel who have experience conducting the study of interest, but it is essential for inexperienced personnel. The practice session is typically scheduled for the day before or several days before the field study is to be conducted, under conditions similar to the most extreme expected during the study. For instance, if a day-long study is planned, practice during the peak hours will be beneficial. Data collectors should practice under the direct supervision of the engineer for a short time, so that obvious mistakes can be corrected. The engineer and the data collector should also record data independently for a short period of time; a comparison of those data will reveal less obvious errors. The site, time, personnel, equipment, method, and data collection form should be scrutinized by the engineer during and after the practice session, and the critical comments of data collectors should also be sought. If extensive changes in the study technique are made as a result of a practice session, it may be a good idea to schedule another practice session to test those changes.

Immediate Preparations for the Study

Too often, inefficient data collection or data collector discomfort result from inadequate preparation for the study. The checklist in Figure 1-1 provides a means for data collectors to make sure that they have not forgotten anything. Not all of the items on the checklist are needed for every study. However, the best policy, especially for studies at remote sites, is to take even marginal items from the checklist to the site. Unused items will not cause much trouble sitting in an automobile trunk, and the consequences of forgetting an item may be large. Checklist item 1 is the most important thing a data collector can do to avoid failure prior to the study: Equipment must be calibrated and checked from input, through storage, to output.

Conducting the Study

On a study day, there are two main actions required of the engineer (if she is not also collecting data). First, the engineer must monitor the data collectors to make sure that they are using agreed-upon procedures and are not falsifying data. An unannounced visit to the site, or even the promise of one, is usually enough to ensure an honest effort by the data collectors. Data are often reviewed after the study for any unusual pattern that might indicate falsification as well. Second, the engineer must stay close to the telephone in case the data collectors need to confer with her. Any number of unusual occurrences can confront data collectors unfamiliar with the analysis planned or unaware of alternate courses of action. Data collectors should be encouraged to call and place the responsibility for a decision with the engineer.

Data collectors must arrive at least 15 minutes early (for most studies) at the site in order to assess conditions, distribute equipment, record crucial "header" information, assume positions, and begin at the scheduled time. The "header" information must appear on each form or data record and includes items such as (but not limited to) site name, date, time, observer name, and weather condition. The sooner that header information is recorded, the better. Data collectors waiting until the end of the day to record header information on a stack of forms are taking an unnecessary risk. Data collectors should note any unusual occurrences in the transportation system that could affect the

Before going to the site, do not forget . . .

____ 1. To check the data collection equipment. Make sure that it records, stores, and/or produces output as you require. Make sure that the equipment is calibrated properly.
____ 2. To label the equipment as needed (e.g., turning movement counters can be labeled with the approaches being watched).
____ 3. To bring the data collection equipment. Also, bring spares of equipment such as stopwatches that are small and unreliable.
____ 4. An accurate watch set to the correct time.
____ 5. The correct sizes of spare batteries for all the equipment.
____ 6. An abundance of forms. Also bring a clean copy of the form from which more copies can be made if needed.
____ 7. Paper for taking notes.
____ 8. Plenty of pens.
____ 9. Clipboards or other writing surfaces.
____ 10. A letter from the landowner (if private property will be used) and/or from a responsible agent of the highway authority giving permission to collect data. Contact names and telephone numbers are sometimes adequate.
____ 11. A few business cards of the engineer supervising the study. A contact name and telephone number are sometimes adequate.
____ 12. A short, simple answer to the question, ''What are you doing here?''
____ 13. The telephone number where the supervising engineer can be reached on the day of the study, in case questions arise.
____ 14. A map showing the site or directions to the site, if it is unfamiliar.
____ 15. Folding chairs.
____ 16. For a long study, a cooler or insulated container with beverages.
____ 17. A hat, sun visor, or sunglasses.
____ 18. Sunburn protection.
____ 19. Extra cold-weather clothes, such as a sweatshirt and gloves.
____ 20. Extra warm-weather clothes, such as a T-shirt and shorts.

Figure 1-1 Field study preparation checklist.

data being collected. Any deviations from accepted collection procedure should also be noted and probably should be cleared first by a responsible lead observer or engineer. Data should always be recorded in ink to prevent fading, smudging, and erasing.

Data collectors on public or private property will often be asked, in tones ranging from polite to threatening, ''What are you doing here?'' Data collectors should be taught a short, standard response to this question that will satisfy most members of the public without distracting too long from the data collection task. A calm, professional approach and a referral to the supervising engineer are usually enough to diffuse even very suspicious inquiries.

Data Collector Safety

The first responsibility of the data collector during the field study is to maintain his personal safety, the safety of the other data collectors, and the safety of the traveling public. Traffic accidents are the primary safety threats during transportation studies. To avoid traffic accidents, data collectors should follow these commonsense defenses:

- Stay as far from the traveled way as possible.
- Stay alert for errant vehicles.
- Wear a fluorescent orange vest if working near the traveled way.
- Do not interfere with existing traffic patterns and distract drivers as little as possible.
- Use standard traffic control devices, if applicable, to inform drivers of a closed lane, closed shoulder, activity near the traveled way, or other substantially changed driving conditions.
- Allow the driver of a vehicle to collect data only if that does not detract from his ability to drive.
- When in a vehicle collecting data in the traffic stream, keep seat belts buckled and do not block the vision of or distract the driver.

Crime is also a threat to data collector safety during field studies. The best defense for data collectors when criminal behavior threatens is usually to abandon the study and leave the area. Safety from crime can be enhanced by the following measures:

- Minimize nighttime data collection.
- Collect data in teams of at least two persons who remain in sight of each other at all times.
- Alert the local police when a data collection effort is under way.
- Position data collection personnel in automobiles.
- Avoid the overt display of moderate- or high-priced equipment.
- Keep personnel in the office aware of the data collection schedule.

Other threats, from lightning to stray dogs, arise occasionally during studies. As with threats from crime, the wisest strategy for many of these other threats is to abandon the study and leave the area.

Summary

Safe, efficient, and effective data collection requires skill, attention to detail, and common sense. The importance of "good" data cannot be overstated. Important conclusions are drawn from field data that form the basis for decisions that affect the expenditure of large amounts of money and can have a significant effect on the safety of the public at large. Data collection demands the same level of professionalism as any other task undertaken by an engineer or engineering technician.

2 Volume Studies

H. Douglas Robertson, Ph.D., P.E.
Joseph E. Hummer, Ph.D., P.E.

INTRODUCTION

Engineers often use counts of the number of vehicles or pedestrians passing a point, entering an intersection, or using a particular facility such as a travel lane, crosswalk, or sidewalk. Counts are usually samples of actual volumes, although continuous counting is sometimes performed for certain situations or circumstances. Sampling periods may range from a few minutes to a month or more. The length of the sampling period is a function of the type of count being taken and the use to which the volume data will be put (Box and Oppenlander, 1976).

In this chapter the focus is on the common methods for counting traffic in the field; how volume data are sampled, expanded, and analyzed; and how count programs are established. Brief descriptions of specific studies are presented along with references containing more detail. Examples throughout the chapter show field data collection and summary forms. Appendix G presents summary forms suitable for copying.

METHODS OF COUNTING

The two basic methods of counting traffic are manual observation and mechanical or automatic recording.

Manual Observation

Purpose and Application

Many types of counts require classifications that are obtained more easily and accurately with trained observers. Examples include vehicle occupancy, pedestrians, turning movements, and vehicle classifications. The latter may also be obtained by automatic means.

Other reasons for conducting manual counts are time and resources. Practical applications often require less than 10 hours of data at any given location. Thus the effort and expense to set up and remove automated equipment is not justified. While there will always be a need for manual counting, advances in technology may soon shift the present balance between the use of manual and automatic methods.

Equipment

Tally Sheets. The simplest means of conducting manual counts is to record each observed vehicle with a tick mark on a prepared field form. Figure 2-1 shows a field sheet for a vehicle turning movement count. The form allows for whatever classifications may be desired. A watch or stopwatch is required to cue the observer to the desired count interval, and a new form is used at the start of each interval. The raw counts are tallied, summarized, or keyed into a computer upon return to the office.

Mechanical Count Boards. Mechanical count boards consist of various combinations of accumulating counters mounted on a board to facilitate the type of count being made. Typical counters are accumulating pushbutton devices with three to five registers. To conduct a four-way intersection count, four counters would be positioned on each side of a board to represent each approach to the intersection. If only vehicle turning movements were desired, each counter would have three registers (for left, through, and right). If pedestrians were to also be included, four or five registers would make up each counter.

Many configurations of registers and counters are possible. Figure 2-2 shows examples of mechanical count boards. Each button represents a different stratification of vehicle or pedestrian being counted. The observer pushes the correct button each time a vehicle or pedestrian passes. The limited number of buttons that can reasonably be presented to an observer usually restricts the number of vehicle classes that can be counted on mechanical boards. Tally sheets and electronic count boards (see below) are somewhat less restrictive. A watch or stopwatch is required to cue the observer to the desired count interval. When the end of an interval is reached, the observer reads the counter, records the data on a field form, and resets the counter to zero. The observers or analysts summarize the data from the field forms or key them into a computer upon return to the office. Figure 2-3 illustrates a field summary form for an intersection turning movement count.

Electronic Count Boards. Battery-operated, hand-held, electronic count boards are currently the most common device to aid in the collection of traffic count data. They operate in a fashion similar to that of mechanical count boards, with a few important differences. They are lighter weight, more compact, and easier to handle. They contain an internal clock that separates the data by whatever interval is chosen, so no field forms are needed. Since field forms are often mislabeled or unlabeled, easily lost, and difficult to handle in rainy, cold, or windy weather, electronic counting boards have an advantage over tally sheets and mechanical counting boards. Most important, they preclude the need for manual data reduction and summary. Data may be transferred directly from the field to a computer in the office via a modem or dumped into the computer upon return from the field. Regardless of the transfer means, the data are summarized, analyzed, and the

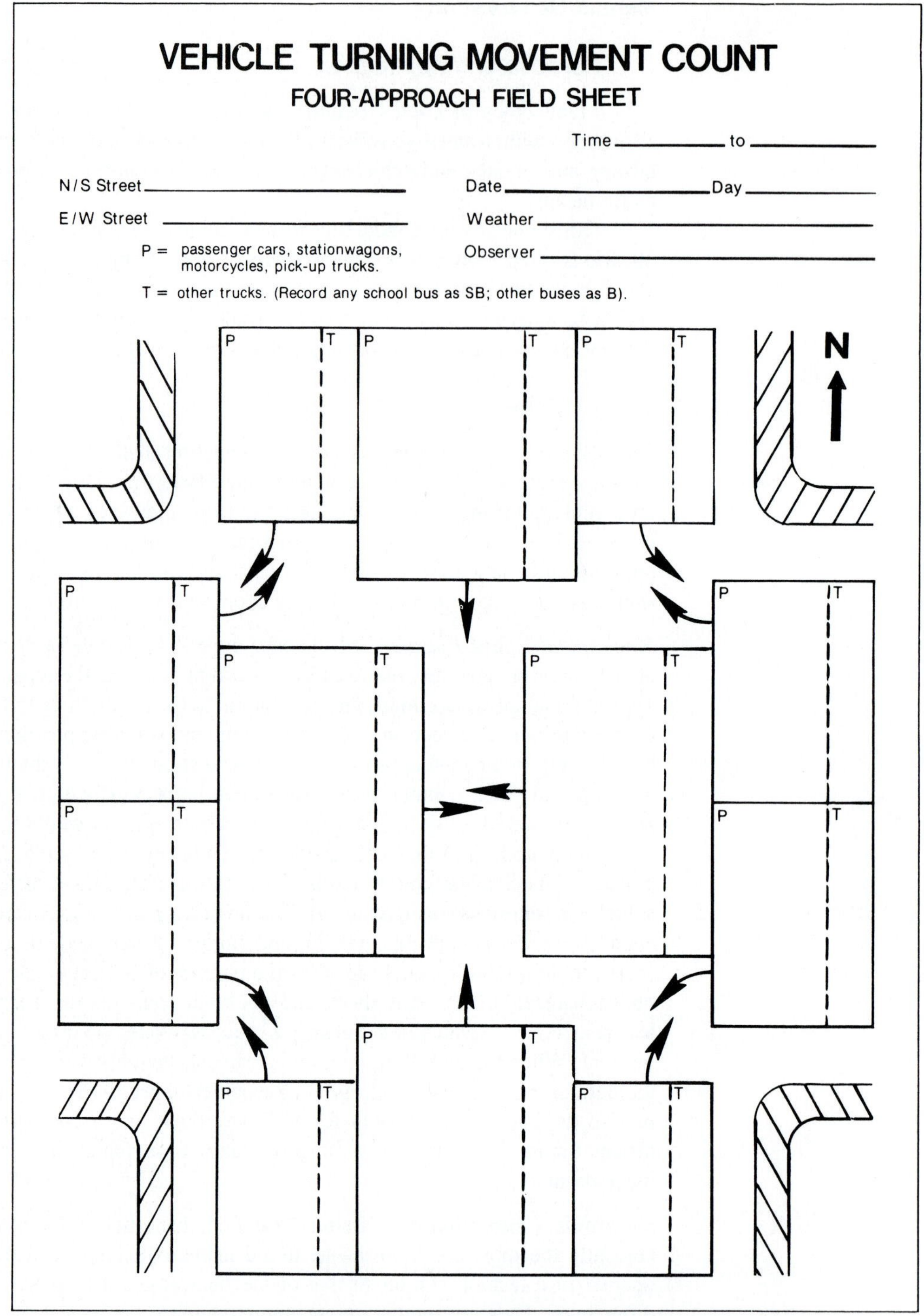
VEHICLE TURNING MOVEMENT COUNT

FOUR-APPROACH FIELD SHEET

Time ________ to ________

N/S Street ________ Date ________ Day ________

E/W Street ________ Weather ________

P = passenger cars, stationwagons, motorcycles, pick-up trucks. Observer ________

T = other trucks. (Record any school bus as SB; other buses as B).

Figure 2-1 Example of an intersection field tally sheet. *Source:* Box and Oppenlander, 1976, p. 21.

results displayed in a selected presentation format by means of computer software. This eliminates the data reduction step required with tally sheets and mechanical count boards.

Many electronic count boards are capable of handling several types of common traffic studies, including turning movement, classification, gap, stop delay, saturation flow rate, stop sign delay, spot speed, and travel-time studies. For agencies requiring more than occasional manual traffic counts, the electronic count board or hand-held com-

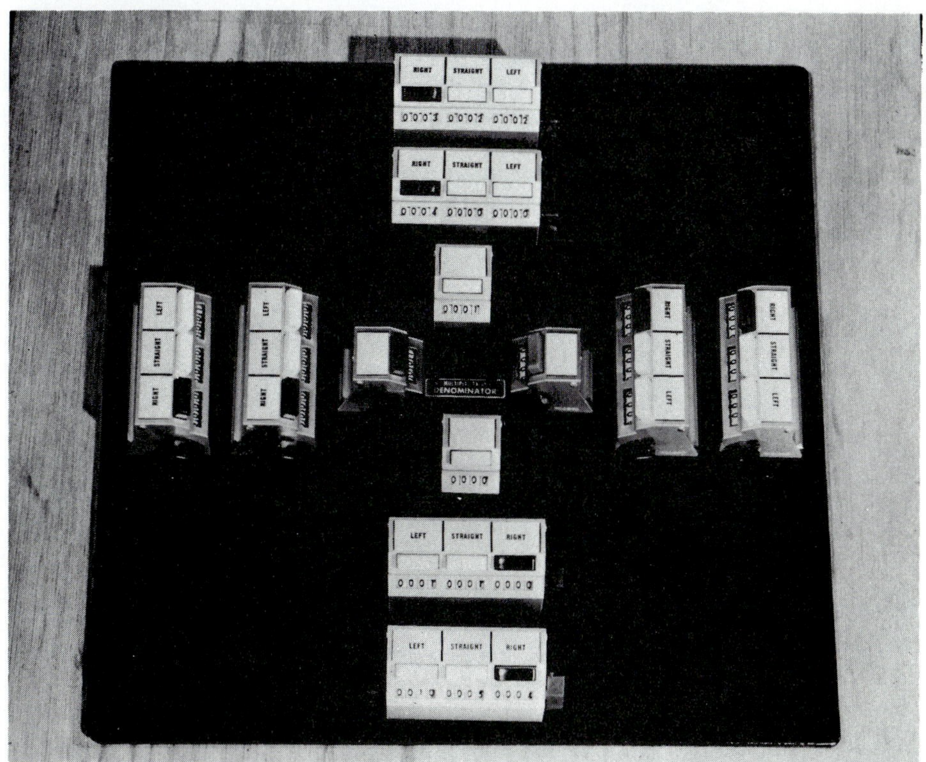

Figure 2-2 Examples of mechanical count boards. *Source:* Box and Oppenlander, 1976, p. 27.

TABULAR SUMMARY OF VEHICLE COUNTS

Observer ______ Date ______ Day ______ City ______

INTERSECTION OF ______ AND ______

R = Right turn
S = Straight
L = Left turn

TIME BEGINS	from NORTH				from SOUTH				TOTAL North South	from EAST				from WEST				TOTAL East West	TOTAL ALL
	R	S	L	Total	R	S	L	Total		R	S	L	Total	R	S	L	Total		

Figure 2-3 Example field summary form for an intersection turning movement count. *Source:* Box and Oppenlander, 1976, p. 29.

puter is a cost-effective, laborsaving tool. Figure 2-4 illustrates a typical electronic count board. The window on the counter displays data, menus of commands, and messages about the status of the counter. Many current electronic count boards provide a shift key for special functions such as recording particular vehicle classes. However, observers are more efficient and accurate using boards that do not require pressing a shift key simultaneously with another key for each vehicle class (Bonsall et al., 1988).

Personnel Required

Manual traffic counting requires trained observers. They must be relieved periodically to avoid fatigue and degraded performance. Breaks of 10 to 15 minutes should be scheduled at least every two hours. If the data collection period is more than 8 hours, breaks of 30 to 45 minutes should be allowed every four hours. The size of the data collection team depends on the length of the counting period, the type of count being performed, the number of lanes or crosswalks being observed, and the volume level of traffic. One observer can easily count turning movements at a four-way, low-volume, signalized intersection with one-lane approaches as long as special classifications and/or vehicle occupancy are not required. As any or all of the foregoing variables increase, the complexity of the counting task increases and additional observers will be needed.

Duties may be divided among observers in various ways. At a signalized intersection, one observer may record the north and west approaches while the other observer watches the south and east approaches. In that way only one approach is moving for each observer at any given time. Another way to divide duties is for one observer to record occupancy or certain classes of vehicles, while the other observer counts total volumes. At complex sites, individual lanes, crosswalks, or classifications may be assigned to individual observers. Also at complex sites, one observer may have the sole job of relieving the other observers on a rotating schedule basis.

Figure 2-4 Electronic count boards used for traffic counting. *Source:* JAMAR Sales Company, Inc.

Field Procedures

Preparation. An accurate and reliable manual traffic count begins in the office. A locally developed checklist is a valuable aid, even to experienced teams, to ensure that all preparations for the field study have been completed before the team arrives at the site to be counted. Figure 1-1 is a general checklist that can serve as the beginning of a locally developed checklist. Preparations should start with a review of the purpose of and type of count to be performed, the count period and time intervals required, and any information known about the site (i.e., geometric layout, volume levels by time of day, signal timing, etc.). This information will help determine the type of equipment to be used, the field procedures to follow, and the number of observers required.

The selection of equipment will dictate the types of data forms needed, if any. Header information should be filled in to the extent possible in the office, and the forms should be arranged in the order in which they will be used by each observer in the field. The checklist should also include equipment items, such as pens, batteries, stopwatches, and blank videotapes, as appropriate. Having to return to the office to retrieve forgotten items may delay the start of the study or cause it to be postponed. An inadequate number of forms to complete the study could also invalidate the study, resulting in wasted resources.

Equipment must operate properly to ensure accurate counting. Good counting boards have firm keys that provide the observer with tactile and audio confirmation when a key has been pressed successfully. Units with "soft" keys should be repaired or discarded. Each observer must make sure that his or her counter adds one and only one to the total when a key is pressed and that clearing the count resets the total to zero. An office review of the procedures to be followed and a check of the proper operation of all equipment complete the preparation stage of the study.

Observer Location. Observers must position themselves where they can most clearly view the traffic they are counting. Observers must avoid vantage points blocked regularly by trucks, buses, parked cars, or other features. They should be located well away from the edge of the travel way, both as a personal safety precaution and to avoid distracting drivers. A position above the level of the street and clear of obstructions usually affords the best vantage point. If several observers are counting at the same site, they must maintain visual contact with one another and be able to communicate so as to coordinate their activities.

Protection from the elements is also an important consideration for the observer. Proper clothing to suit prevailing weather conditions is paramount. Safety vests should be worn if the observer is near traffic at any time. Observers may count from inside vehicles as long as their view is unobstructed. Sitting in an automobile is safer and more comfortable during inclement weather than sitting outside. Observers should park the vehicle in a legal space that is close enough to the intersection so that they can see the farthest lane of their assigned approaches. Parking on private property is sometimes convenient, but observers must first obtain the property owner's permission. While sitting outside, observers may use chairs to prevent fatigue and umbrellas for protection from the sun, as long as these devices are not distracting to drivers. A sign indicating that a traffic count is under way usually satisfies driver curiosity about observers they can view from the roadway.

Data Recording. The key to successful traffic counts lies in keeping the data organized and labeled correctly. Counts may produce a large number of data forms. Each form must be clearly labeled with such information as the count location, observer's name, time of study, and conditions under which the counts are made. The form itself should clearly indicate the movements, classifications, and time intervals.

The observer must concentrate his or her attention on accurately recording each count in the proper place or with the proper button. Special care must be taken with electronic counting boards to ensure that they are properly oriented to the geographic and geometric layout of the intersection. When two or more observers are working together, time intervals must be maintained and coordinated accurately. Observers should also look for and note on their forms or in a log any temporary traffic events, such as accidents or maintenance activities, that may lead to unusual traffic counts.

When mechanical count boards are used, provisions must be made to record the accumulated counts and reset the counters at the end of each interval. Two procedures may be used to accomplish this without significant error. They are referred to as the short-break and alternating count procedures (McShane and Roess, 1990). The *short-break procedure* is to take a 1- to 3-minute break at the end of, but included in, each count interval. The observers use this time to record their counts and reset their counters. The volume for the counting period is estimated by extrapolating the actual count over the short-break portion of the count interval using the following:

$$V' = V \times CF \tag{2-1}$$

$$CF = \frac{CP}{CP - SB} \tag{2-2}$$

where

CF = count expansion factor
CP = counting period, minutes
SB = short break, minutes
V' = adjusted count, vehicles
V = actual count, vehicles

In the *alternating count procedure*, the observer performs a full count every other interval, resting on alternate intervals. The volumes occurring during the missing count intervals are estimated by interpolation. Table 2-1 provides an example where the two procedures are combined. Remember, the longer the gaps between actual counts, the more susceptible extrapolations and interpolations are to error. With either the short-break or alternating count procedure, it is important that all observers are in communication with each other so they can start and end simultaneously. It is also important that all data forms are clearly marked with the procedure used.

TABLE 2-1
Example of the Short-Break and Alternating Count Procedures Combined

	4-Min Counts			*Adjusted Counts*[a]	
Counting Period	*Mvt. A*	*Mvt. B*	*CF*	*Mvt. A*	*Mvt. B*
5:00–5:05 P.M.	100	—	5/4	125	b
5:05–5:10	—	90	5/4	128	113
5:10–5:15	104	—	5/4	130	117
5:15–5:20	—	96	5/4	132	120
5:20–5:25	106	—	5/4	133	122
5:25–5:30	—	99	5/4	131	124
5:30–5:35	102	—	5/4	128	128
5:35–5:40	—	105	5/4	127	131
5:40–5:45	100	—	5/4	125	127
5:45–5:50	—	98	5/4	b	123

[a]Rounded to nearest vehicle.
[b]End count may be extrapolated but cannot be interpolated.
Source: McShane and Roess, 1990, p. 88.

Automatic Counts

Purpose and Application

There are many applications of volume data that do not require complex classifications. A simple count of vehicles is sufficient. Often, these types of counts are needed for extended periods of time (i.e., days, weeks, or even months). The use of observers for such purposes would be cost prohibitive. Automatic counting provides the means of gathering large amounts of volume data at a reasonable expenditure of time and resources. The main disadvantage of automatic counts relative to manual counts is equipment reliability. With current equipment, there is a high chance that detection or recording equipment will fail during a given study due to malfunctions, vandalism, large vehicle interference, weather, or numerous other threats. Analysts must prepare contingency plans to use in case of equipment failure.

Advances in technology are producing equipment capable of recording more than just simple vehicle counts. Sophisticated vehicle classification counts are now possible. Turning movement counts are being recorded automatically by connecting count recorders to vehicle detectors installed at signalized intersections. Advances are also being made with visual recording and image sensing systems. Transponders on-board the vehicle that can furnish data to roadside readers are under development. In summary, the balance between manual and automatic counting is changing.

Equipment

There are many types and models of automatic volume data collection equipment. This equipment consists of two basic components: a data recorder, and sensors to detect the presence of vehicles and/or pedestrians. The recording component is essentially the same for both portable and permanent applications. The types of sensors used may differ. Some equipment also has the capability of communicating the collected data to a central facility for processing. The *ITE Journal* is a good source of advertisements for the latest in automatic data collection equipment.

Space does not permit a discussion of the wide range of equipment currently available today. Besides, advancing technology is causing continuous changes that would quickly render such a discussion obsolete. Automatic counters are employed in two principal ways: as portable counters or as permanent counters. In the sections that follow both categories of automatic counters are discussed in a generic sense. Differences in sensing devices are noted. In addition, the discussion includes a brief description of visual automatic counters.

Portable Counters. Manual observation is a form of portable counting. A team of observers is dispatched to a specific location to gather data for a limited period. Upon completion of that task, the team moves to another location. Portable counters serve in the same temporary manner but through the use of automatic counting equipment. Another difference is that the period of data collection is usually longer for automatic than for manual counts. Portable counters are most commonly used for 24-hour counts but may be deployed for periods of days or weeks. They may also be used in situations where safety precludes the use of observers (e.g., in tunnels, on bridges, or in adverse weather).

Portable count recorders range from simple accumulating counters to microcomputer-driven classification counters. As mentioned above, many of the recording devices may be used in either a portable or a permanent configuration. The method of sensing (or detecting) the vehicles counted represents the principal difference between portable and permanent counting equipment. Portable counters generally use pneumatic road tubes, piezoelectric strips, tape switches, or temporary induction loop detectors. Technicians place these sensors across the travel lanes being counted and connect them to the recording component.

Permanent Counters. Agencies establish permanent count stations where they desire long-term, continuous counts (e.g., 24 hours a day, 365 days a year). The volumes collected at these stations are usually part of an areawide program to monitor traffic characteristics and trends over time. The same recording component may be used as with portable counters. The sensors, however, are usually more permanent in nature. The most common type of permanent detector in use today is the induction loop, which is installed in the pavement. Other forms of permanent detectors use radar, sound, microwaves, and infrared light.

Videotape. Observers can count volumes by viewing a videotape recorded during the time periods of interest. Observers can record their counts with an electronic count board, with tick marks on a tally sheet, with a mechanical counter, or directly into a computer. A special computer program is necessary with the latter method. With adequate light conditions and a good vantage point, one camera can capture all turning movements at a typical intersection. A digital clock in the video image is an excellent way to note the end of intervals. However, a notation by the camera operator or the camera operator speaking into the microphone are adequate substitutes for the clock.

The major advantages of using videotape for turning movement counts are the accuracy possible and the ability to use the tape for other studies. Greater accuracy is possible with videotape than with any other common method of counting volume because observers can view the tape repeatedly. The first observation of a tape is likely to be less accurate than manual observation in the field. However, observers viewing a tape many times are likely to converge on the "true" count. Also, the engineer in the office can view the tape to answer questions about the classification of certain vehicles or the impact of unusual events during the count period. Observers can view the tape at slow motion for additional scrutiny. The ability to conduct other studies with a videotape, including the intersection and driveway studies discussed in Chapter 5, can lead to very efficient uses of the labor of field crews. Usually, other studies require a higher-quality videotape than a turning movement count, so engineers should consider the needs of the other studies first when specifying videotape use.

Even where lighting and vantage points allow videotaping, the disadvantages of using videotape for volume counts usually outweigh the advantages. Most agencies do not need absolutely accurate counts and do not collect other data from the tape. Videotape cameras and players are relatively expensive. In addition, setting up the camera, recording, and viewing the tape typically require more labor than other methods (Bonsall et al., 1988). Do not count volumes with videotape unless great accuracy is needed or the recording will be reused.

Video imaging systems remove the need for a human being to view the tape and record observations. Instead, a computer scans the image automatically and notes changes in the image that fit the profile for the event of interest. In recent years, video imaging system cost has fallen and quality has risen so that 80 to 90% accuracy during the day or at night is routine. Video imaging systems may soon be a common way to count volumes. Work is currently under way in the United States and elsewhere to refine image sensing and processing systems that would enhance the automated extraction of traffic data, to include vehicle and pedestrian volumes, from video images [see, e.g., Michalopoulos (1990) and Vieren et al. (1991)].

Personnel Required

The only personnel required for automatic counts are those needed to install and recover the equipment. Crew sizes of two or three are usually sufficient to deploy most portable counting equipment. The recording component can be handled by one person; however, one or two persons will be needed to install the road tubes or tape switches, while an additional person watches for traffic. Recovery of the equipment can usually be performed by one or two persons. The installation of permanent counters with induction loop detectors may require a larger crew and the closure of travel lanes.

Field Procedures

Preparation. Field work should never be undertaken without proper preparation in the office. A locally prepared checklist is an invaluable aid even for the most routine task. The purpose of the count will drive the type of equipment to be used and the deployment procedures to follow. All equipment should be checked to see that it is functioning properly. An ample supply of accessories (i.e., nails, clamps, tape, adhesive, chains, locks, batteries) and all necessary tools should be on hand.

Selecting the Count Location. Agencies decide in the office on which street or highway the count will be made and the general location (midblock or intersection) where the counters will be placed. These decisions depend on the type of study being performed. The exact location of the count recorder and sensors is usually determined in the field. Guidelines for the deployment of automatic counters include:

- Do not place sensors across marked or unmarked parking lanes, where a parked vehicle could activate the sensor continuously.
- Avoid placing sensors on pavement expansion joints, sharp pavement edges, or curves.
- Deploy sensors at right angles to the traffic flow.
- For directional counts, keep at least 1 foot of space between the sensor and the centerline of the roadway.
- Fasten the sensor securely to the pavement with nails, clamps, tape, and/or adhesives made specifically for this purpose. Loose sensors will prevent the collection of data and may pose a hazard to motorist and pedestrians.
- At intersections or near driveways, place sensors where double counting of turning vehicles can be avoided.
- Locate the count recorder near a sign post or tree and secure it with a lock and chain, or place it in a locked signal control cabinet to prevent vandalism.
- Keep the cable or tube that connects the sensor to the recorder as short as possible.
- Record sensor placement by noting the physical location on a condition diagram sketch.
- Use a test vehicle to ensure that bidirectional counters are recording the proper direction.
- Set the count interval to ensure that totals will occur on the hour to make the data compatible with other counts.
- Note the time that counter operation begins.
- Check the installation periodically to ensure that it is in place and functioning properly. In cold-climate states, agencies should check sensors whenever it snows to ensure that snow plows have not removed the sensors from the road.

Installation and Retrieval. The primary concern during installation and retrieval operations is the safety of the field crew. The crew's vehicle should be clearly visible to traffic and should be parked away from the traveled way. All crew members should wear reflective clothing at all times. Deployments and recoveries should be accomplished during periods of low traffic volume and good visibility. If nighttime operations are necessary, the crew should employ extra safety measures (i.e., lights, cones, warning signs). Anytime that crew members must enter the roadway, at least one crew member should have the sole duty of watching for traffic and warning the rest of the crew. Details of installation techniques are varied and in many cases product specific. Information of this nature is generally available from the product manufacturer. Police assistance may be required to ensure the safety of the crew and the public.

DATA REDUCTION AND ANALYSIS

Following collection, raw data must be placed in a form suitable for analysis. This reduction usually consists of converting tally marks to numbers, summarizing the data by calculating subtotals and totals, and arranging the data in a format for performing analyses. The analysis may range from a simple extraction of descriptive information to a sophisticated statistical treatment of the data. The analysis will depend on the type of study being conducted.

Manual versus Automatic Data Reduction

Data collected by observers using tally marks or count boards must be reduced to a form suitable for analysis. Tally marks are counted for each time interval and classification and

From Main St. South						*From 1st Ave. West*					
Left		*Through*		*Right*		*Left*		*Through*		*Right*	
P	*T*	*P*	*T*	*P*	*T*	*P*	*T*	*P*	*T*	*P*	*T*
20	9	533	73	37	20	44	9	400	40	53	32
From Main St. North						*From 1st Ave. East*					
Left		*Through*		*Right*		*Left*		*Through*		*Right*	
P	*T*	*P*	*T*	*P*	*T*	*P*	*T*	*P*	*T*	*P*	*T*
19	1	518	14	85	5	58	2	299	61	70	5

Figure 2-5 Peak-hour volume summary table. *Source:* McShane, W. R. and Roess, R. P., *Traffic Engineering,* Prentice Hall, Englewood Cliffs, NJ, 1990, p. 94.

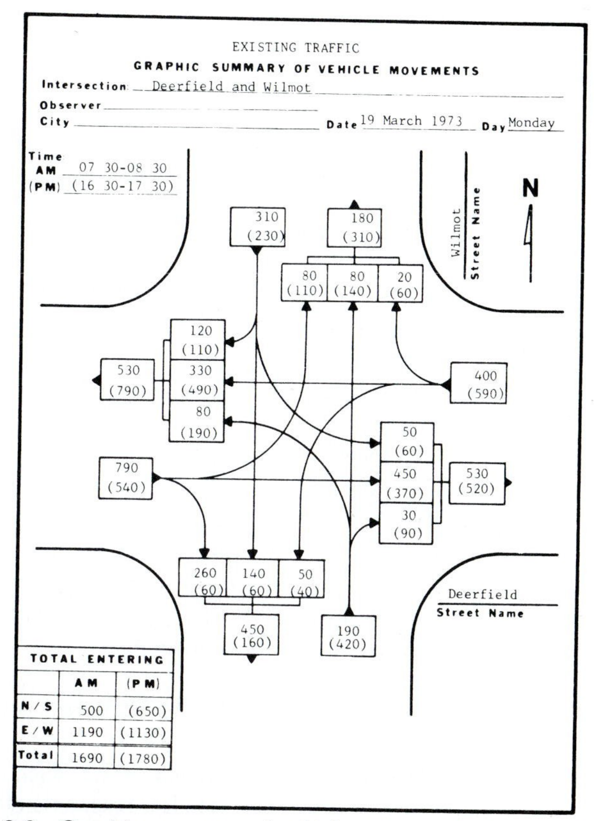

Figure 2-6 Graphic summary of vehicle movements. *Source:* Box and Oppenlander, 1976, p. 30.

the counts are entered on summary sheets as shown in Figure 2-3 or 2-5. Analysts can summarize data from count boards in a similar manner. Summary data may also be displayed in a graphic format as shown in Figure 2-6 for a four-leg intersection. In addition to the counts for each movement, totals for each approach and the intersection as a whole are shown for both peak hours.

Continuous count recorders (mechanical counters) record data either on registers or on paper tape. These data must be reduced and tabulated as described above. Figure 2-7 shows a sample form for summarizing a 24-hour weekly count. Such reductions are

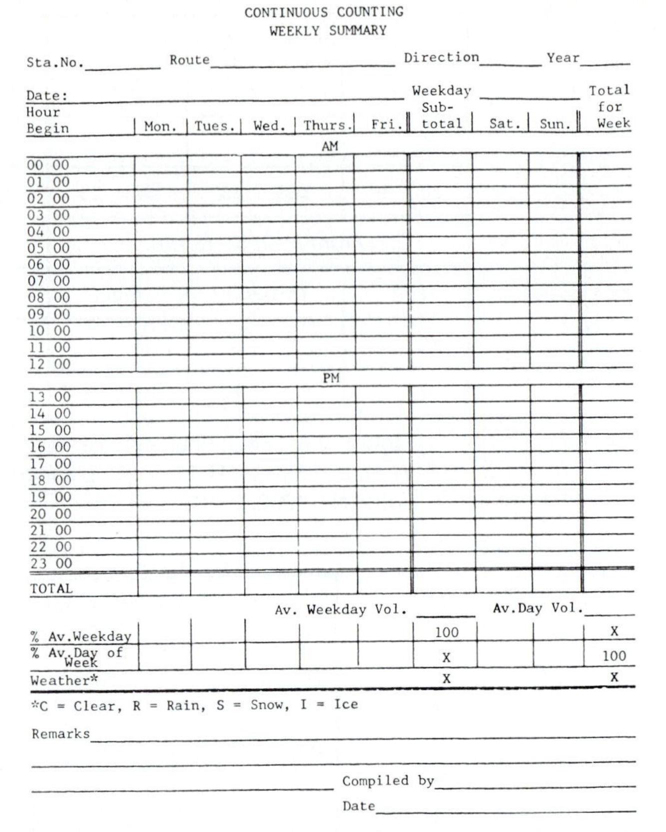

CONTINUOUS COUNTING
WEEKLY SUMMARY

Sta.No._______ Route___________________ Direction_______ Year______

Date: ________________________________ Weekday __________

Hour Begin	Mon.	Tues.	Wed.	Thurs.	Fri.	Weekday Sub-total	Sat.	Sun.	Total for Week
AM									
00 00									
01 00									
02 00									
03 00									
04 00									
05 00									
06 00									
07 00									
08 00									
09 00									
10 00									
11 00									
12 00									
PM									
13 00									
14 00									
15 00									
16 00									
17 00									
18 00									
19 00									
20 00									
21 00									
22 00									
23 00									
TOTAL									

Av. Weekday Vol. ______ Av.Day Vol.______

	Mon.	Tues.	Wed.	Thurs.	Fri.	Weekday Sub-total	Sat.	Sun.	Total for Week
% Av.Weekday						100			X
% Av.Day of Week						X			100
Weather*						X			X

*C = Clear, R = Rain, S = Snow, I = Ice

Remarks____

Compiled by________________

Date________________

Figure 2-7 Continuous counting weekly summary. *Source:* Box and Oppenlander, 1976, p. 30.

labor intensive and time consuming and analysts must check the result carefully to ensure accuracy.

As mentioned previously, computer-driven counters do not require manual data reduction. Count data are stored in summary form in the count recorder's memory and then "dumped" into the computer either in the office or from the field via a modem. Computer software accomplishes the reductions and produces summaries of the data and desired calculated values (e.g., average daily traffic, peak-hour factors, and percent turns). These tools save analyst time and often eliminate errors. Figure 2-8 shows an example of a computer-generated count summary.

Converting Axle Counts to Vehicle Counts

Counters driven by single-point sensors such as pneumatic road tubes, tape switches, or loops record *axles,* not vehicles. Each axle crossing the sensor causes a pulse that is recorded. Since not all vehicles have two axles, the axle count cannot simply be divided by 2 but must be adjusted to reflect the proportion of vehicles with more than two axles in the traffic stream. To adjust the axle count, analysts use a short classification count of the number of vehicles with different numbers of axles. An average number of axles is calculated from the sample, which is assumed to represent the entire count. The number and duration of sample counts needed depends on the variation of vehicle mix over the course of a day or week. If the proportion of vehicle types remains relatively constant over the count period, one sample 1 to 2 hours long should be sufficient. If the count lasts a week or more, samples should be taken during a weekday and a weekend.

The example shown in Table 2-2 illustrates how to use the results of a sample classification count. The 2-hour sample is used to adjust a count of 8500 axles on a two-lane

PAGE: 1
FILE: None

Major St.: Park Rd
Minor St.: Intersection 2
Weather : Clear

Primary Movements: Vehicles

DATE: 9/07/88

Time Begin	From North RT	From North THRU	From North LT	From East RT	From East THRU	From East LT	From South RT	From South THRU	From South LT	From West RT	From West THRU	From West LT	Vehicle Total
6:20	3	20	0	0	0	0	0	55	2	0	0	1	81
6:35	0	30	0	0	0	0	0	80	2	0	0	0	112
6:40	4	22	0	0	0	0	0	75	2	2	0	1	106
6:45	9	47	0	0	0	0	0	77	7	2	0	3	145
6:50	7	57	0	0	0	0	0	72	5	4	0	2	147
6:55	9	67	0	0	0	0	0	79	8	4	0	2	169
HR TOTAL	52	243	0	0	0	0	0	438	26	12	0	9	760
7:00 AM	12	65	0	0	0	0	0	72	11	10	0	2	172
7:05	15	68	0	0	0	0	0	72	7	5	0	13	180
7:10	18	62	0	0	0	0	0	87	23	13	0	4	207
7:15	10	50	0	0	0	0	0	111	12	6	0	11	200
7:20	6	48	0	0	0	0	0	113	3	4	0	9	192
7:25	4	33	0	0	0	0	0	78	1	8	0	15	139
7:30	0	41	0	0	0	0	0	106	1	0	0	1	149
7:35	0	40	0	0	0	0	0	114	0	0	0	0	154
7:40	1	43	0	0	0	0	0	100	1	1	0	1	147
7:45	2	46	0	0	0	0	0	32	0	0	0	2	122
7:50	1	48	0	0	0	0	0	97	1	0	0	1	148
7:55	0	51	0	0	0	0	0	37	0	2	0	1	141
HR TOTAL	69	575	0	0	0	0	0	1119	60	49	0	59	1951
DAY TOTAL	101	938	0	0	0	0	0	1557	86	61	0	68	2711

PEAK PERIOD ANALYSIS FOR THE PERIOD: 6:30 AM–8:00 AM

DIRECTION FROM	START PEAK HOUR	PEAK HR FACTOR	VOLUMES Right	VOLUMES Thru	VOLUMES Left	VOLUMES Total	PERCENTS Right	PERCENTS Thru	PERCENTS Left
North	6:45 AM	0.71	91	621	0	712	13	87	0
East	6:45 AM	0.00	0	0	0	0	0	0	0
South	6:55 AM	0.80	0	1111	68	1179	0	94	6
West	6:30 AM	0.43	58	0	62	120	48	0	52
Entire Intersection									
North	6:30 AM	0.71	91	621	0	712	13	87	0
East		0.00	0	0	0	0	0	0	0
South		0.79	0	1081	79	1160	0	93	7
West		0.43	57	0	62	119	48	0	52

Figure 2-8 Computer output for an intersection turning movement count.

TABLE 2-2
Example of Conversion of Axle Counts to Vehicle Counts

Results of Two-Hour Classification Counts	
Classification	*Number of Vehicles*
2-axle	80
3-axle	10
4-axle	5
5-axle	5

Conversion Computations		
Classification	*Number of Vehicles*	*Number of Axles*
2-axle	× 80	= 160
3-axle	× 10	= 30
4-axle	× 5	= 20
5-axle	× 5	= 25
	100	235

Average number of axles/veh = 235/100 = 2.35
24-hour vehicle count = 8500/2.35 = 3617 vpd

Source: McShane and Roess, 1990, p. 90.

rural highway during a 24-hour period to obtain an estimate of the total number of vehicles per day (McShane and Roess, 1990). Automatic counters are available that will perform classification counts, thus precluding the need for adjusting axle counts.

Count Periods

The counting period selected for a given location depends on the planned use of the data and the methods available for collecting the data. The count period should be representative of the time of day, day of week, or month of year that is of interest in the study. For example, Mondays and Fridays are usually not typical weekdays. Engineers and planners rarely need turning movement counts, vehicle classifications, or pedestrian counts from nights, Sundays, or holidays. Saturday counts are sometimes needed for shopping areas. The count period should avoid special events and adverse weather unless the purpose is to study such phenomena. Count periods may range from an hour to a year. Manual counts are usually for periods less than 1 day. Typical count periods for turning movements, sample counts, vehicle classifications, and pedestrians include:

- 2 hours; peak period
- 4 hours; morning and afternoon peak periods
- 6 hours; morning, midday, and afternoon peak periods
- 12 hours; daytime (e.g., 7:00 A.M.–7:00 P.M.)

Count intervals are typically 5 or 15 minutes. For capacity analysis purposes, 15-minute counts are adequate. If a peak-hour factor is sought, 5-minute counts are preferable. Electronic counting boards provide totals automatically for any interval. With a computer program to analyze these counts, shorter intervals require no greater observer or analyst effort than longer intervals and provide more detailed results. The choice of an appropriate count interval depends primarily on the needs of the analysis.

Automatic counts are usually taken for a minimum of 24 hours. They may extend for 7 days, a month, or even a year. The interval most commonly used is 1 hour. Smaller intervals may be desired for certain purposes. However, smaller intervals require greater computer storage space or reduction time.

Sample Counts and Count Expansions

All counts are samples. Even permanent count stations represent a sample of specific locations among the many locations in a given area. Count periods are also a sample of the overall long-term traffic flow. Time and resources do not permit the continuous counting of every intersection and unique roadway section on all existing streets and highways. Consequently, sample counts are taken over shorter time periods at specific locations. These counts are then adjusted and/or expanded to produce estimates of the expected traffic flow at that or similar locations.

Short Counts

The short-break and alternating count procedures were described earlier in the chapter and involve alternating count and rest periods. Interpolation is used to "fill in" the rest periods. In a similar fashion, hourly counts may be obtained by expanding short counts of 5, 6, 10, 12, 15, 20, or 30 minutes by use of an appropriate multiplier (e.g., 4 times a 15-minute count provides an hourly estimate). The level of accuracy of such expansions increases with the length of the sample count. If traffic flow is relatively constant over the hour, a short sample count will produce a reliable estimate of hourly flow. If flow is highly variable during the hour, a longer sampling period will be needed to obtain a good estimate. Short counts of subhourly flow allow one team of observers to cover several locations in the time it would take to perform a 1-hour count at a single location.

Small Network Counts

Short counts may be expanded by use of a *control station.* If a number of sample counts are needed in a relatively small area, analysts select one location representative of the area streets to be sampled. It is important that the control station service the same type of street and variations of traffic being sampled on the other streets. The control station is counted continuously during the entire sampling period using the same count interval (e.g., 15 minutes) as on the sampled streets. The counts taken at the sampling locations are called *coverage counts.* Both the coverage and control counts are taken at midblock to avoid the complexity of turning movements. Each link or street segment to be sampled should be counted at least once during the sampling period. The counts may be made manually or with automatic counters.

The control count data establishes the volume variation pattern for the entire sampling period. The pattern is quantified by calculating, for the control count data, the proportion of total sampling period volume occurring during each count interval. Assuming that this pattern applies to all of the sampled locations in the study area, the full sampling period volume for a coverage count location is obtained by dividing the sample count by the control count proportion for the corresponding count interval. Figure 2-9 shows an illustration of the procedure. In the example, a control station was used and four coverage counts were made over a 2-hour sampling period. The count interval was 15 minutes beginning at each quarter hour. McShane and Roess (1990) provide a more extensive discussion of this count expansion procedure with examples.

Volume Data Presentations

Traffic volume data may be portrayed in a number of ways. The selection of the presentation method depends on the planned use of the data and the audience that will view the data. Analysts most often depict volume data in summary tables or in one of several graphical forms. Graphs and bar charts are suitable for illustrating traffic volumes over time. Figure 2-10 provides examples of bar charts showing monthly, daily, and hourly traffic variations. Peak periods of flow are readily discernible. Pie charts are useful to

Control Station

Time	*Count*	*Proportion of Period Total*
7:00–7:15 A.M.	720	720/6864 = 0.105
7:15–7:30	776	776/6864 = 0.113
7:30–7:45	837	837/6864 = 0.122
7:45–8:00	951	951/6864 = 0.138
8:00–8:15	1022	1022/6864 = 0.149
8:15–8:30	986	986/6864 = 0.144
8:30–8:45	874	874/6864 = 0.127
8:45–9:00	698	698/6864 = 0.102
Total = 6864 vehicles		

Expansion of Coverage Counts

Location	*Time*	*Count*	*2-Hour Volume (Est.)*
1	7:15–7:30 A.M.	784	784/0.113 = 6938
2	7:45–8:00	1192	1192/0.138 = 8638
3	8:15–8:30	863	863/0.144 = 5993
4	8:45–9:00	532	532/0.102 = 5216

Figure 2-9 Count expansion example.

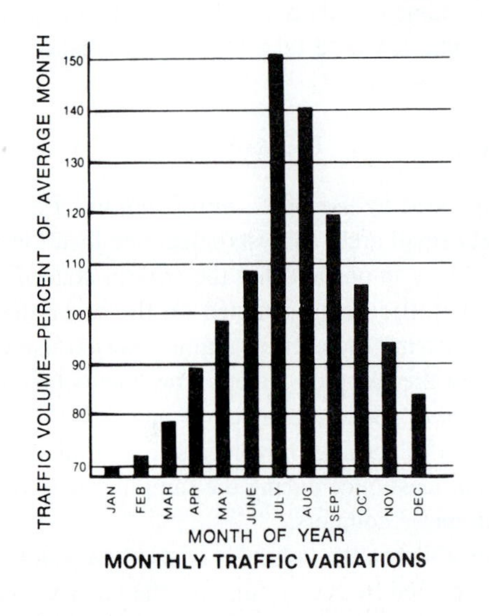

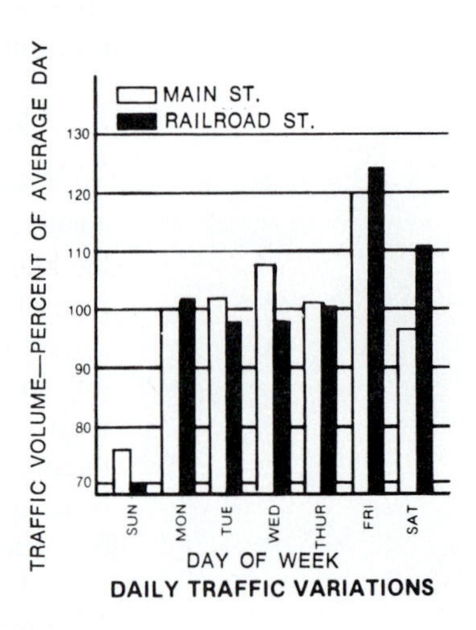

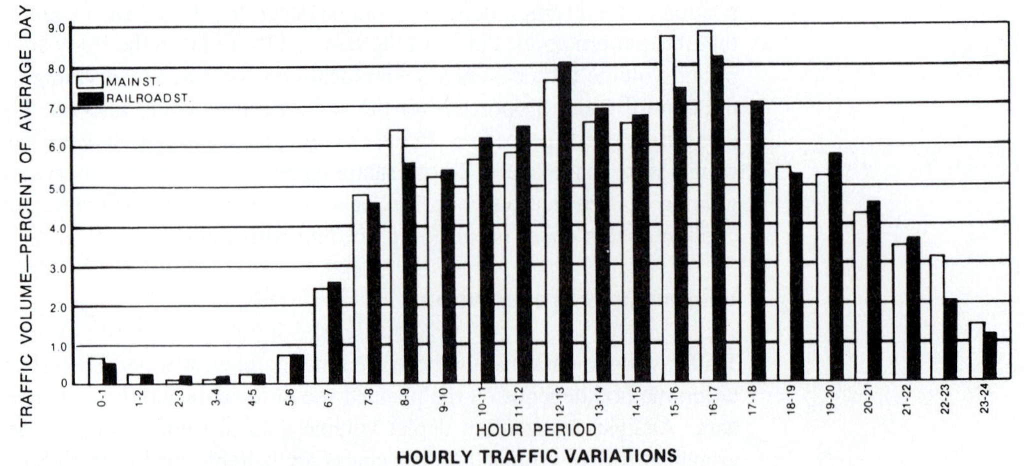

Figure 2-10 Bar charts of monthly, daily, and hourly traffic volumes. *Source:* Box and Oppenlander, 1976, p. 190.

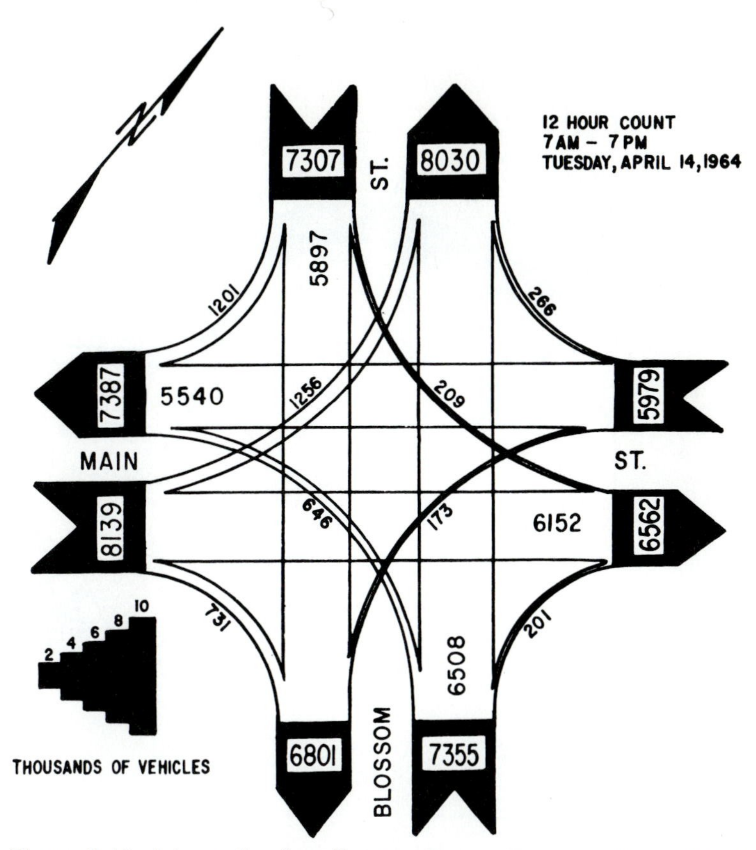

Figure 2-11 Intersection flow diagram. *Source: Transportation and Traffic Engineering Handbook,* 1st ed., Institute of Transportation Engineers, Washington, DC, 1976, p. 410.

show proportions of volumes by type of vehicle. An intersection graphic summary, such as the one shown in Figure 2-6, presents a picture of how much traffic flows through the intersection for a specified period.

An intersection flow diagram (Figure 2-11) provides a similar picture. The width of the flow band indicates the level of traffic volume. Traffic flows may be depicted in the same way on a route map (Figure 2-12). The thickness of the bands provides a visualization of the relative volumes found on the streets of a highway network. Perhaps the most common means of presenting volume data for an area or system of streets is a traffic count map. Figure 2-13 shows a map with total volumes in vehicles per day (vpd) in both directions for a city street system.

SPECIFIC COUNTING STUDIES

Traffic volume counting is not always a simple, straightforward task. Some types of volume studies are complex and difficult to perform. They require special preparations and observer training. In this section the issues that the data collector must confront to conduct the most common of these complex studies are discussed.

Intersection Counts

The most complex location in a traffic system is the intersection. Each approach has up to four possible turning movements: U, left, through, and right (although in most studies, U-turns are included with left turns). Many applications require a separate count of

FLOW MAP

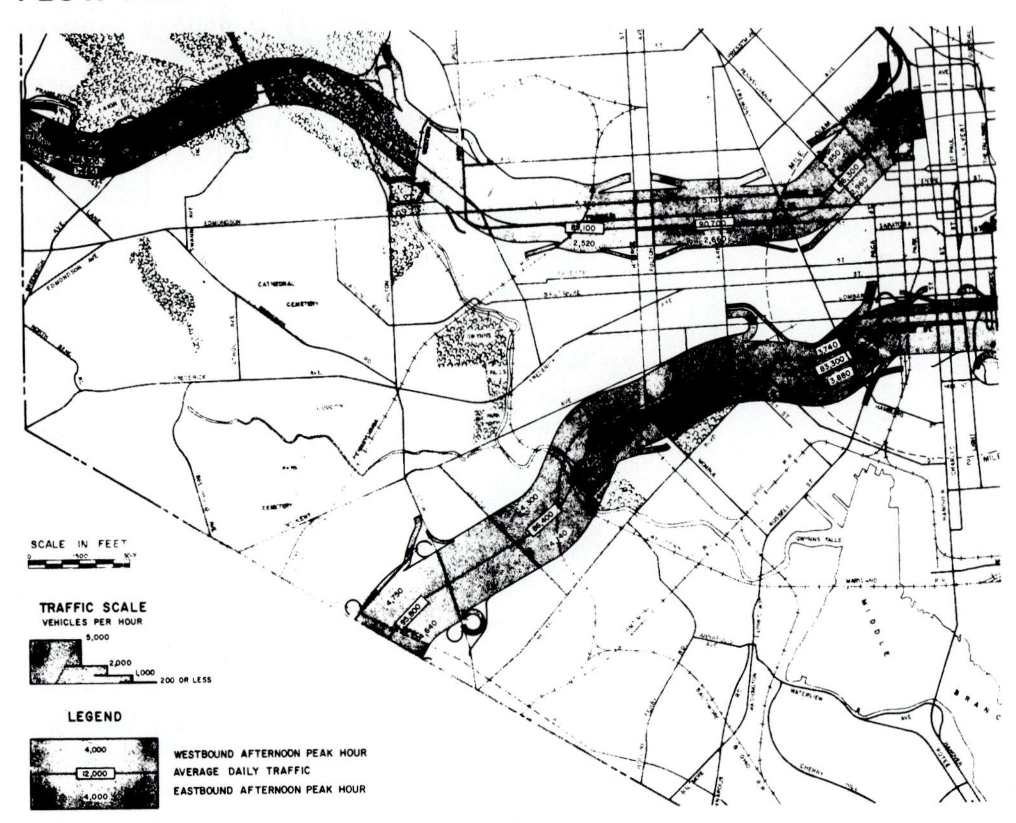

Figure 2-12 Traffic flow map. *Source:* Box and Oppenlander, 1976, p. 38.

COUNT MAP

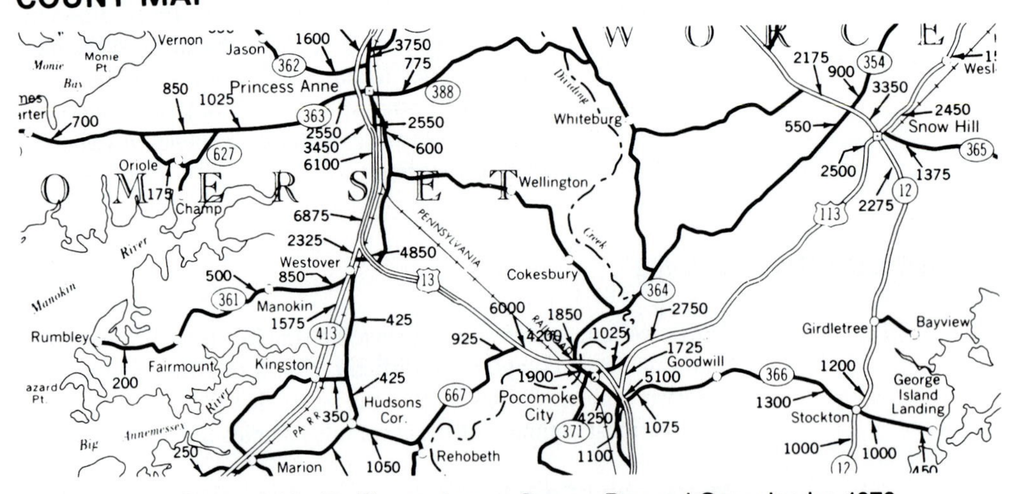

Figure 2-13 Traffic count map. *Source:* Box and Oppenlander, 1976, p. 38.

buses, passenger cars, and trucks. At a four-leg intersection, an observer could be faced with recording 48 separate data elements during each sampling period. Except for very light traffic conditions, intersection counts require multiple observers. If many vehicle classes are to be examined at a busy intersection with several simultaneous movements, each observer must be able to record data for two or three lanes. Simplified methods of identifying vehicle classes are sometimes desirable. For example, one could classify all motor vehicles with two to four tires as automobiles and all motor vehicles with six or more tires as trucks. The classification scheme must be well understood by all observers before the beginning of the count.

Signalized Intersections

Since all approaches do not have the right-of-way simultaneously, an observer may alternate counting movements in two directions (e.g., eastbound and southbound) as the signal phase changes. Some models of electronic count boards are equipped with a separate box attached by cable to the main box to allow one record to be made while two observers each count two approaches. Counts at signalized intersections are complicated because one or more movements occur during each phase, each signal cycle contains two or more phases, and the green time for each phase often is not equal to the other phases. In order not to bias the count toward any particular set of movements, the count interval must be an even multiple of the signal cycle length. If the short-break method is being used, both the count interval and the short-break interval must be even multiples of the cycle length.

Actuated signals further complicate counting, because both the cycle lengths and the green times vary from cycle to cycle. One rule of thumb is to select counting intervals that will include at least five cycles, using the maximum cycle length to determine the interval. Assuming that the signal controller is responding to demand, the counts should be representative of the demand, despite the variations in timing (McShane and Roess, 1990).

Arrival versus Departure Volumes

Intersection volume counts are usually recorded as vehicles cross the stop bar and enter the intersection. This point is chosen so that turning movements can be observed accurately. If the intersection becomes saturated (demand exceeds capacity), queues will develop that may require more than a single cycle to dissipate. When this occurs the departure counts do not reflect the demand volumes. In these cases, arrival volumes should be recorded.

Arrival volumes are not easy to observe. The queues are constantly changing and may extend beyond the line of vision of the observer. Additional observers are normally needed to count queue lengths while the primary observers count departure volumes. For greatest accuracy, the queue should be counted every cycle. However, the queue may be counted at the end of each counting interval. All queue counts should be made at the beginning of a red phase.

Table 2-3 illustrates how to estimate arrival volumes by observing departures and queue lengths. One can calculate the arrival count for each interval by adding the net change in queue length to the observed departure count. Note from the example that while the total departure and arrival volumes are the same, the distribution of volumes across counting intervals is different. This procedure estimates arrival volume for the approach. Turning movements for these arrival volumes may be obtained by assuming that arrivals follow the same distribution of left, right, and through movements as the departure volumes (McShane and Roess, 1990).

TABLE 2-3
Example of Estimating Arrival Volumes from Departure Counts

Time Period	Total Departure Count (veh)	Queue Length (veh)	Arrival Volume (veh)
4:00–4:15 P.M.	50	0	50
4:15–4:30	55	0	55
4:30–4:45	62	5	62 + 5 = 67
4:45–5:00	65	10	65 + 10 − 5 = 70
5:15–5:30	60	12	60 + 12 − 10 = 62
5:30–5:45	60	5	60 + 5 − 12 = 53
5:45–6:00	62	0	62 − 5 = 57
	55	0	55
	469		469

Source: McShane and Roess, 1990, p. 93.

Pedestrian Counts

Pedestrian counts are usually taken at intersection crosswalks, at midblock crossings, or along sidewalks or walking paths. These data are used for traffic signal and crosswalk warrant studies, for capacity analysis, in accident studies, for site impact analysis, and in other planning applications. In Chapter 13 procedures for performing pedestrian volume counts are described.

Cordon Counts

Agencies make a cordon count by encircling an area such as a central business district (CBD) or other major activity center with an imaginary boundary and counting vehicles and pedestrians at all of the points where streets cross the cordon. Observers classify each vehicle by type, direction of travel, and occupancy, and typically use 15- to 60-minute intervals. The counts show the amount of traffic entering or leaving and enable an estimation of the vehicle and person accumulations within the area.

Agencies use cordon counts most commonly as part of an origin–destination (O-D) survey as a basis for expanding interview data. The counts are taken in conjunction with the interviews. O-D studies are described in Chapter 7. Cordon counts may also be taken for trend analysis purposes. For this application, agencies count one weekday each year, during a month with an average daily traffic that is close to the annual average daily traffic. The counts are made at the same time each year.

Count stations are located on the cordon boundary at midblock locations. Agencies can keep the number of stations to a minimum by taking advantage of natural or human-made barriers. Alleys and very low-volume streets may be ignored, if the aggregate loss is less than 3 to 4% of the total. Counts should be made on the same day. However, if the agency maintains a set of control stations, cordon counts made on different days can be adjusted using the control station data. Short counts should *not* be used, because the distribution of traffic at each crossing location is critical to the determination of accumulation. Counts of transit passengers should be available from the local transit agency or can be made using the methods presented in Chapter 17. Figure 2-14 shows typical field and summary sheets for a cordon count.

Vehicle accumulations within the cordon are found by summing the entering and leaving counts at all count stations by time interval. The counts usually begin when the street system is at its lowest flow. Agencies can count the number of vehicles parked on-street and off-street to estimate the number of vehicles inside the cordon when the study begins. Table 2-4 illustrates the procedure for computing accumulation. Figure 2-15 illustrates a method of displaying cordon count data summaries.

CORDON COUNT

PASSENGER VEHICLES,
TRUCKS, AND MISCELLANEOUS VEHICLES
FIELD SHEET

TRAFFIC________ BOUND ON______________STREET

WEATHER__________ FROM_____ TO_____

¼ HOUR STARTING	NUMBERS OF		
	PASSENGER VEHICLES INCLUDING TAXICABS	TRUCKS	OTHER VEH. EXCL. BUSES & STR. CARS
—			
—			

DATE____________ RECORDER____________

CORDON COUNT

PEDESTRIANS — FIELD SHEET

TRAFFIC ON_____ SIDE OF______________STREET

WEATHER__________ HRS. FROM_____ TO.____

¼ HOUR STARTING	INBOUND	OUTBOUND
—		
—		
—		

DATE____________ RECORDER____________

CORDON COUNT

PEDESTRIANS SUMMARY SHEET

DATE____________ PEDESTRIANS ON____________STREET
WEATHER__________ ____________STREET
____________STREET
COMPILED BY____________ ____________STREET

¼ HOUR STARTING	PEDESTRIANS ON																							
	ST.						ST.						ST.						ST.					
	SIDE		SIDE		TOTAL		SIDE		SIDE		TOTAL		SIDE		SIDE		TOTAL		SIDE		SIDE		TOTAL	
	IN	OUT	IN	OUT	IN	OUT	IN	OUT	IN	OUT	IN	OUT	IN	OUT	IN	OUT	IN	OUT	IN	OUT	IN	OUT	IN	OUT
TOTAL																								
AVERAGE																								
PEAK HOUR																								

CORDON COUNT

PASSENGER VEHICLES,
TRUCKS, AND MISCELLANEOUS VEHICLES
SUMMARY SHEET

DATE____________ TRAFFIC ON__________STREET
WEATHER_______ COMPILED BY____________

¼ HOUR STARTING	NUMBER INBOUND					NUMBER OUTBOUND					TOTAL IN PLUS OUT
	PASSENGER VEHICLES	TRUCKS	TRANSIT	MISC.	TOTAL	PASSENGER VEHICLES	TRUCKS	TRANSIT	MISC.	TOTAL	
TOTAL											
AVERAGE											
PEAK HOUR											

CORDON COUNT

ALL PERSONS SUMMARY SHEET

DATE________ TRAFFIC ON__________ STREET
WEATHER________ COMPILED BY__________

¼ HOUR STARTING	PERSONS INBOUND VIA					TOTAL INBOUND	PERSONS OUTBOUND VIA					TOTAL OUTBOUND	TOTAL IN PLUS OUT
	PASSENGER VEHICLE	TRUCK	MISC VEHICLE	TRANSIT VEHICLE	WALKING		PASSENGER VEHICLE	TRUCK	MISC VEHICLE	TRANSIT VEHICLE	WALKING		
TOTAL													
AVERAGE													
PEAK HOUR													

Figure 2-14 Typical field and summary sheets for a cordon count.
Source: Box and Oppenlander, 1976, p. 40.

TABLE 2-4
Accumulation Computations for a Cordon Count

Time	Vehicles Entering	Vehicles Leaving	Accumulation (veh)
4:00–5:00 A.M.	—	—	250
5:00–6:00	100	20	330
6:00–7:00	150	40	440
7:00–8:00	200	40	600
8:00–9:00	290	80	810
9:00–10:00	350	120	1040
10:00–11:00	340	200	1180
11:00–12:00 NOON	350	350	1180
12:00–1:00 P.M.	260	300	1140
1:00–2:00	200	380	960
2:00–3:00	180	420	720
3:00–4:00	100	350	470
4:00–5:00	120	320	270

Source: McShane and Roess, 1990, p. 103.

Screen Line Counts

Screen line counts are made to record travel from one area to another. The *screen line* is some form of natural or human-made barrier with a limited number of crossing points. Analysts use screen line counts to check and adjust the results of O-D studies. They may also be used to detect trends or long-term changes in land use, commercial activity, and travel patterns.

For screen line counts, hourly intervals for a 12- to 24-hour period are usually used on a weekday. Using several counting periods that are 1 week or more apart will preclude any unusual conditions occurring on a given day from causing a bias in the data. Classification counts may also be desirable when conducting a screen line count. Upon completion of the count, the screen-line crossings (hourly or total) are compared to the crossings predicted by the home-interview study. The result is then used to adjust the model that predicts O-D flows. Figure 2-16 illustrates the layout of a screen line and the display of screen line and O-D survey data.

Areawide Counts

All states maintain ongoing count programs on state highways for planning, estimating vehicle miles of travel, tracking volume trends, and conducting traffic engineering analyses. Another objective of these programs is to estimate AADT at coverage count locations. Cities and counties may also have similar programs for roads and streets in their jurisdictions.

Control Counts

Daily and seasonal (monthly) volume variation patterns are established and monitored using control counts in an areawide program. Counts are made either continuously or periodically throughout the year. The most useful counts are made at permanent-count stations, which operate 24 hours a day, 365 days a year. Control-count stations supplement the data obtained from permanent-count stations. Typically, at major control-count stations, a 7-day continuous count is made during each month of the year. At minor control-count stations, a 5-day weekday continuous count is made every other month. There are generally twice as many minor as major control-count stations. Control counts

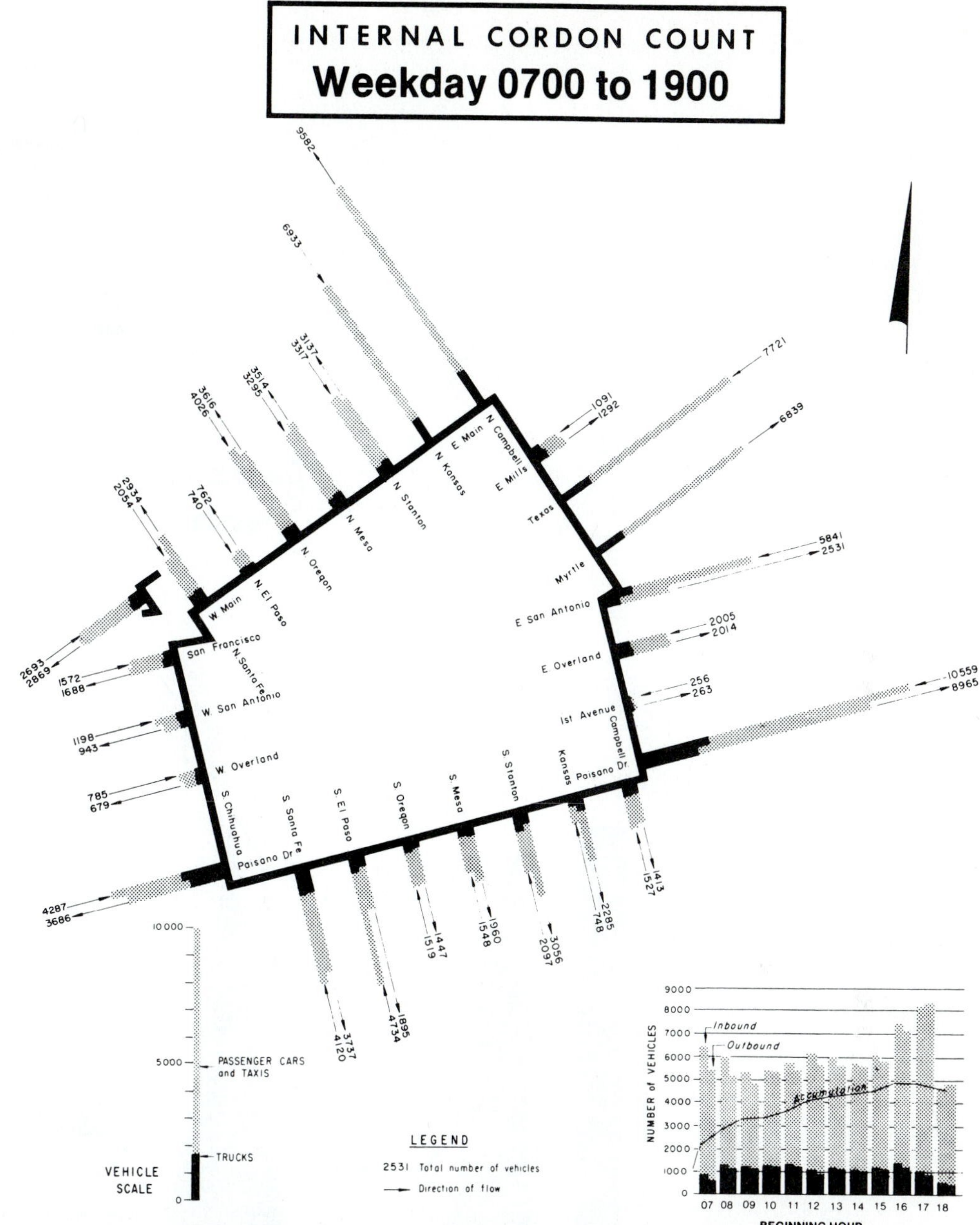

Figure 2-15 Cordon count summary data display. *Source:* Box and Oppenlander, 1976, p. 41.

in urban areas may be referred to as *key counts.* For every 20 to 25 coverage stations, there should be a permanent or control-count station.

Daily and Seasonal Factors

Permanent-count stations provide the most accurate source of data for computing daily and seasonal adjustment factors. The first step in such a computation is to find the average volume for each day of the week over the entire year. The average of this 7-day profile is the ADT of a typical week. The daily adjustment factor is found by dividing

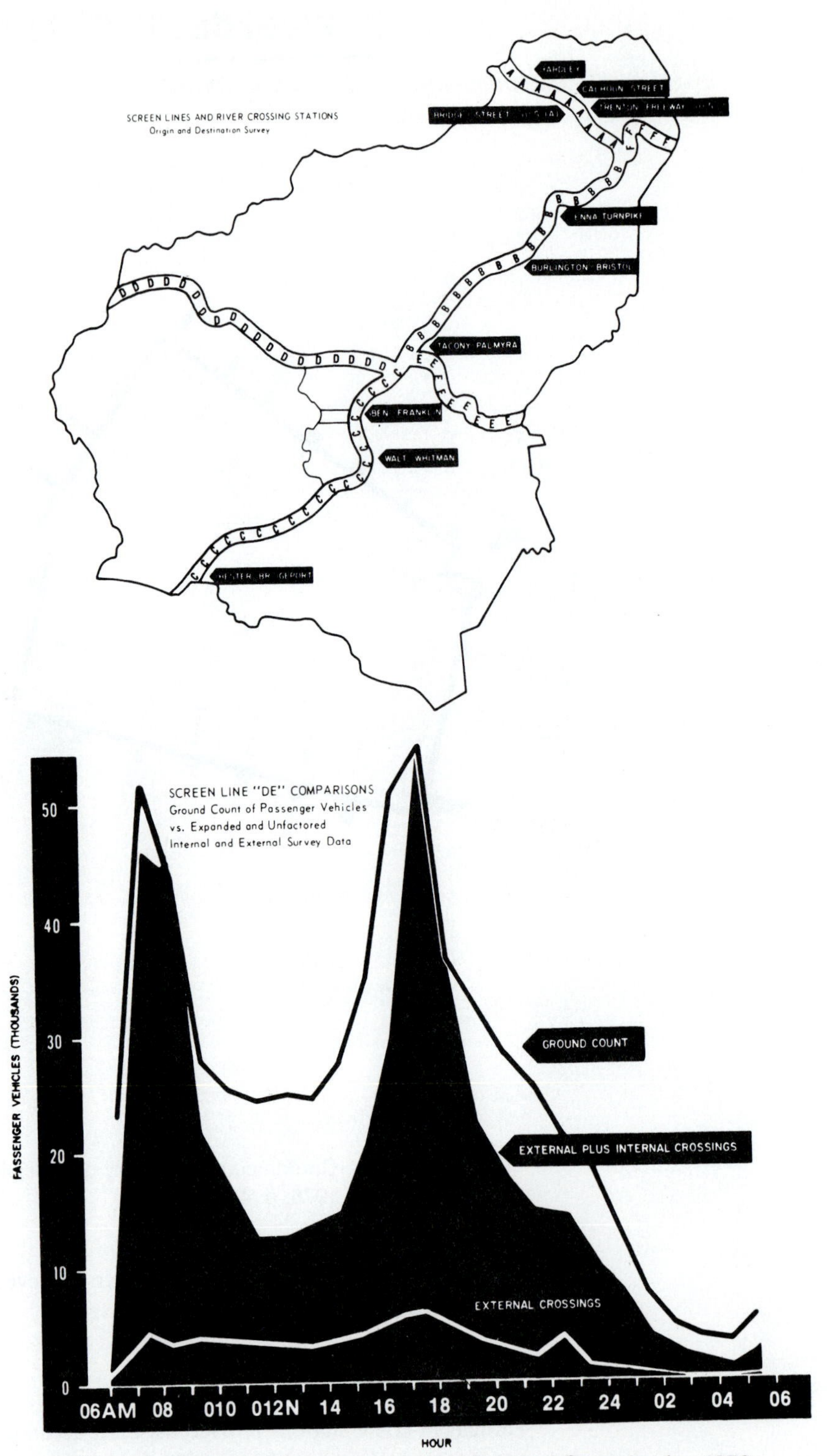

Figure 2-16 Screen line study. *Source:* Box and Oppenlander, 1976, p. 43.

TABLE 2-5
Illustrative Computation of Daily Variation Factors

Day	*Average Yearly Volume for Day (vpd)*	*Daily Factor*
Monday	1332	1429/1332 = 1.07
Tuesday	1275	1429/1275 = 1.12
Wednesday	1289	1429/1289 = 1.11
Thursday	1300	1429/1300 = 1.10
Friday	1406	1429/1406 = 1.02
Saturday	1588	1429/1588 = 0.90
Sunday	1820	1429/1820 = 0.80
Total = 10,000 veh		
ADT = 1429 vpd		

Source: McShane and Roess, 1990, p. 100.

TABLE 2-6
Illustrative Computation of Monthly Variation Factors

Month	*Total Traffic (veh)*	*ADT for Month (vpd)*	*Monthly Factors (AADT/ADT)*
January	19,840	/31 = 640	797/640 = 1.25
February	16,660	/28 = 595	797/595 = 1.34
March	21,235	/31 = 685	797/685 = 1.16
April	24,300	/30 = 810	797/810 = 0.98
May	25,885	/31 = 835	797/835 = 0.95
June	26,280	/30 = 876	797/876 = 0.91
July	27,652	/31 = 892	797/892 = 0.89
August	30,008	/31 = 968	797/968 = 0.82
September	28,620	/30 = 954	797/954 = 0.84
October	26,350	/31 = 850	797/850 = 0.94
November	22,290	/30 = 743	797/743 = 1.07
December	21,731	/31 = 701	797/701 = 1.14
Total = 290,851			
AADT = 290,851/365 = 797			

Source: McShane and Roess, 1990, p. 100.

the ADT by the average volume for each day of the week. Table 2-5 illustrates the computation of daily adjustment factors.

The computation of seasonal or monthly variation factors follows a similar procedure. The ADT for each month is the monthly volume from the permanent-count station divided by the number of days in the month. The AADT is then computed and divided by the ADT for each month to arrive at the monthly adjustment factors. Table 2-6 illustrates the computation of monthly variation factors. Daily and seasonal factors can be computed in a similar way from control-count data. Since control counts are samples rather than continuous counts, the margin for error is greater. However, carefully planned control counts will produce reliable estimates. For further discussion, see McShane and Roess (1990).

Coverage Counts

To adjust a 24-hour coverage count for a given location to an estimate of AADT, multiply the count by the appropriate daily and monthly variation factors. The factors used should be from a permanent- or control-count station location similar in geometry and traffic characteristics to the location of the coverage count.

SUMMARY

In this chapter the various methods used to perform traffic volume counts have been described. Issues related to data reduction and analysis and presented specific types of counting studies have been discussed. For further information and more details on traffic volume studies, refer to McShane and Roess (1990) and Pignataro (1973).

REFERENCES

BONSALL, P. W., F. GHARI-SAREMI, M. R. TIGHT, AND N. W. MARIER (1988). The performance of handheld data-capture devices in traffic and transport surveys, *Traffic Engineering and Control,* Vol. 29, No. 1.

BOX, P. C., AND J. C. OPPENLANDER (1976). *Manual of Traffic Engineering Studies,* 4th ed., Institute of Transportation Engineers, Washington, DC, p. 13.

MCSHANE, W. R., AND R. P. ROESS (1990). *Traffic Engineering,* Prentice Hall, Englewood Cliffs, NJ, pp. 84–106.

MICHALOPOULOS, P. G. (1990). Automated extraction of traffic parameters through video image processing, *Proceedings of the ITE Annual Meeting,* Orlando, FL, pp. 33–38.

PIGNATARO, L. J. (1973). *Traffic Engineering: Theory and Practice,* Prentice Hall, Englewood Cliffs, NJ, pp.143–163.

VIEREN, C., J. P. DEPARIS, P. BONNET, AND J. G. POSTAIRE (1991). Dynamic scene modeling for automatic traffic data extraction, *Journal of Transportation Engineering,* Vol. 117, No. 1, pp. 47–56.

3

Spot Speed Studies

H. Douglas Robertson, Ph.D., P.E.

INTRODUCTION

Speed is an important transportation consideration, because highway users relate speed to economics, safety, time, comfort, and convenience. Spot speed studies are designed to measure speeds at specific locations under the traffic and environmental conditions prevailing at the time of the study. There are two types of average speed measures that express the rate of movement or speed of a vehicle. The first, time–mean speed, is covered in this chapter. The second, space–mean speed, is addressed in Chapter 4.

Applications

Speed is a basic measure of traffic performance. Thus spot speed data have a number of applications, including:

1. Determining traffic operation and control parameters
    a. Speed limits
    b. Advisory speeds
    c. Critical intersection approach speeds
    d. No-passing zone markings
    e. School routes, zones, and crossings
    f. Traffic sign locations
    g. Location and timing of traffic signals

2. Establishing highway design elements
   a. Horizontal and vertical curvature
   b. Superelevation
   c. Gradient and length of grade
   d. Length of speed change lanes
   e. Length of passing and no-passing zones
   f. Sight distances
3. Analyzing highway capacity
4. Assessing highway safety
   a. Hazardous or problem location identification
   b. Accident analysis
   c. Complaint investigation
5. Monitoring speed trends
6. Measuring effectiveness of controls or programs
   a. Traffic control devices
   b. Operational changes (e.g., speed limits)
   c. Enforcement programs

Definitions

- *Speed* is the rate of movement of a vehicle in distance per unit of time. Common units are miles per hour (mph), feet per second (fps), kilometers per hour (kph), and meters per second (mps).
- *Spot speed* is the instantaneous measure of speed at a specific location on a roadway.
- *Time–mean speed* is the arithmetic mean or average of several spot speed measurements. It is the sum of the measured spot speeds divided by the number of measurements.
- *Space–mean speed* is another type of average speed. It is the length of a segment divided by the mean travel time of several vehicles or trips over the segment. This measure is addressed in Chapter 4.
- *Median spot speed* is the middle value in a series of spot speeds that have been ranked in order of magnitude.
- *Modal spot speed* is the value that occurs most frequently in a sample of spot speed measurements. *i*th *percentile spot speed* is that value at or below which *i* percent of the spot speeds occur. For example, the 85th percentile speed is that speed at or below which 85% of the total observed values fall in a sample of measured spot speeds.
- *Pace* is the specified increment of spot speed, usually 10 mph, that includes the greatest number of speed measurements.
- *Standard deviation* is a commonly used measure of the spread of individual speeds around the mean. It is the square root of the sum of squares of the deviations of the individual spot speeds from the mean divided by the number of measurements less one.

DATA COLLECTION

Speed data are collected by one of two general approaches: indirect and direct measurements. Indirect measurements provide an estimate of speed, because they are actually measurements of time for a vehicle to travel a known distance between two points. By employing a short "trap," such as two inductive loops spaced 6 feet apart, the difference between the indirectly measured space speeds and the desired spot speeds becomes insignificant. Direct measurements of speed are made using the Doppler principle (i.e., radar).

Two basic methods of data collection are the individual vehicle selection method and the all-vehicle sampling method. Both methods can use direct measurement or indirect measurement. Each method is discussed below in terms of study purpose, location, time period, personnel and equipment, sample size, and field procedures. The most common ways to measure speed in detail are presented in this chapter, and references are provided for users with an interest in the less frequently used ways to measure speed.

Safety

As with all field studies, safety is the paramount consideration in conducting spot speed studies. The measurement of speeds involves workers in the proximity of the roadway, whether it is to install and recover detection devices from the roadway or to operate data collection equipment on the roadside. Workers must use care and vigilence at all times while working near the roadway. Workers should park their vehicles off the traveled way, wear reflective vests, and act in a manner that does not distract motorists or influence their driving behavior. Workers should place and recover detection devices in the roadway under low-volume conditions. Standard lane closure procedures and warning devices should be used if extended time in the roadway is required. Study teams may need police assistance to direct traffic during the deployment and recovery of equipment.

Individual Vehicle Selection Method

Analysts use this method when the study purpose can be satisfied with a relatively small sample of spot speeds taken over relatively short time periods. The objectives of such studies are usually very specific and limited in scope to certain types of locations, time periods, and conditions. Examples of such applications include measuring the effectiveness of a traffic control device, spot checking the effect of speed enforcement, or establishing the location of a traffic sign.

Location, Time, and Conditions of the Study

Selecting the spot to use the individual vehicle selection method; the time period over which to collect the data; and the roadway, traffic, and weather conditions under which to conduct the study is generally a matter of common sense. The objective and scope of the study dictate the specific location for collecting the data, the time of day and day of week, and the conditions under which the speed data are desired. If approach speeds to an intersection are the sample of interest, speed measurements should be taken upstream on the approach just before the point that traffic begins to decelerate for a possible stop at the intersection. If speeds are being sampled as part of a nighttime accident study, the data should be collected during the hours of darkness. If wet pavement is a factor of interest in the study, analysts should measure speeds when it is raining. If the study team needs free-flow speeds, they should conduct the study during off-peak time periods.

Personnel and Equipment

The individual vehicle selection method may use observers with a speed trap, but is generally carried out using the Doppler principle, direct measurement technique. While radar is currently the principal means for direct speed measurement, researchers are making advances in speed detection through the use of infrared and laser technologies.

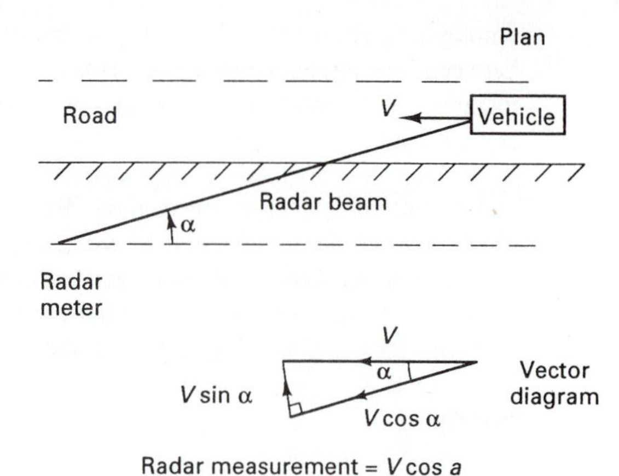

Figure 3-1 Angle of incidence of the radar beam to traffic. *Source:* Taylor, M. A. P. and Young, W., *Traffic Analysis: New Technology and New Solutions,* Hargreen Publishing Company, North Melbourne, Victoria 3051, Australia, 1988, p. 155.

Radar Meters. The most commonly used device for directly measuring speeds is the radar gun. This device may be hand-held, mounted in a vehicle, or mounted on a tripod. The device transmits a continuous beam of high-frequency microwaves toward a moving target vehicle. These waves reflect off the vehicle back to the unit. The change in frequency between the transmitted and reflected waves is proportional to the speed of the target vehicle relative to the speed of the radar unit. If the radar unit is fixed (not moving), it does not matter if the target vehicle is traveling toward the unit or away from the unit.

The strength of the reflected microwave signal decreases with increasing distances. Effective ranges vary from a few hundred feet to 2 miles. The transmitted beam becomes wider as it travels out from the unit, so it may cover more than one traffic lane. Most units allow for adjustments in range and beam width. Check the manufacturer's specifications when selecting the device to ensure that it will meet your study requirements.

The accuracy of radar units is affected by two errors, roundoff and angle. Units typically display the measured speed in digital form rounded down to the nearest whole unit of speed. For example, a reading of 55 mph would mean that this estimate was actually between 55 and 56 mph.

Angle error occurs because the angle of incidence of the radar beam to the travel direction of the target vehicle produces a reading on the unit that is less than the actual speed. As shown in Figure 3-1, the measurement is a function of the cosine of the incidence angle. While in law enforcement this error provides a margin in favor of the target vehicle, other applications of spot speed data may require a correction to the reading to assure a proper level of accuracy. Table 3-1 illustrates the effect of the angle error on true speed. Because of the absolute nature of these two error sources, the relative error decreases as speed increases. Some radar units have a built-in correction for angle error based on preset angles of incidence (Taylor and Young, 1988).

A radar unit is easily operated by one person. Operators can write down the digital readings displayed on the unit or can record the readings verbally on a tape recorder and transfer them later to paper or computer in the office for analysis. If traffic is heavy or the sampling strategy is complex, two observers may be needed: one to call out the readings for the vehicles of interest and one to record the speeds on an appropriate form.

Analysts must recognize the bias associated with using radar units. Many vehicles are equipped with radar detectors that warn drivers that a radar unit is operating in their vicinity. Drivers typically slow when warned by a detector, which affects the speeds observed in the traffic stream. Many of these detectors are visible to the observer as the vehicle passes, which allows that observation to be discounted.

TABLE 3-1
Radar: True Speed and Cosine Error

	Measured Speed at True Speed of:					
Angle (deg)	*30 mph*	*40 mph*	*50 mph*	*55 mph*	*60 mph*	*70 mph*
0	30	40	50	55	60	70
1	29.99	39.99	49.99	54.99	59.99	69.99
3	29.96	39.94	49.93	54.92	59.92	69.90
5	29.89	39.85	49.81	54.79	59.77	69.73
10	29.54	39.39	49.24	54.16	59.09	68.94
15	28.98	38.64	48.30	53.12	57.94	67.61
20	28.19	37.59	46.99	51.68	56.38	65.78
30	25.98	34.64	43.30	47.63	51.96	60.62
45	21.21	28.28	35.36	38.89	42.43	49.50
60	15.00	20.00	25.00	27.50	30.00	35.00
90	00.00	00.00	00.00	00.00	00.00	00.00

Source: Lunenfeld and McDade, 1983.

Manual Speed Traps. Spot speeds may be estimated by manually measuring the time it takes a vehicle to travel between two defined points on the roadway a known distance apart. This technique, commonly referred to as a speed trap, is a "low-technology" approach to speed data collection. With the ready availability of radar units, as described in the preceding section, and the automatic speed traps discussed in later sections, manual speed traps are seldom used. Information about this technique and the procedures for using it are presented by Box and Oppenlander (1976), Lunenfeld and McDade (1983), and Pignataro (1973).

Videotaping. Photography has been used by researchers in performing speed studies for many years. More recently, videotaping has replaced film and time-lapse photography as the medium of choice. Videotaping is simply recording vehicles that pass through a manual speed trap and extracting speed data from the tape. The usefulness of this method for research lies in the expanded amount of information recorded on the tape and the ability to analyze and reanalyze speed and other types of data as well. The technique applies primarily when analysts require total information about the scene over relatively short periods of time (Taylor et al., 1989). Effective use of videotaping requires a video camcorder equipped with a character generator or similar device for recording a time stamp and a variable-speed or frame-step playback unit. These features are not present in the majority of readily available video equipment.

Sample-Size Requirements

Analysts who must collect a sufficient number of spot speed observations should be made to allow statistical analysis of the study results. A minimum sample size can be determined for a desired degree of statistical accuracy by using equation (3-1) to calculate the number of speeds to be measured, when mean speed is the statistic of interest.

$$N = \left(S\frac{K}{E}\right)^2 \tag{3-1}$$

where

N = minimum number of measured speeds
S = estimated sample standard deviation, mph
K = constant corresponding to the desired confidence level
E = permitted error in the average speed estimate, mph

TABLE 3-2
Standard Deviations of Spot Speeds for Sample-Size Determination

		Average Standard Deviation	
Traffic Areas	*Highway Type*	*mph*	*kph*
Rural	Two-lane	5.3	8.5
	Four-lane	4.2	6.8
Intermediate	Two-lane	5.3	8.5
	Four-lane	5.3	8.5
Urban	Two-lane	4.8	7.7
	Four-lane	4.9	7.9
	Rounded value:	5.0	8.0

Source: Box and Oppenlander, 1976, p. 80.

TABLE 3-3
Constant Corresponding to Level of Confidence

Constant, K	*Confidence Level (%)*
1.00	68.3
1.50	86.6
1.64	90.0
1.96	95.0
2.00	95.5
2.50	98.8
2.58	99.0
3.00	99.7

Source: Box and Oppenlander, 1976, p. 81.

Analysts can estimate S for this equation from previous speed studies under similar conditions of study or from speed monitoring data at a nearby location. In the absence of these data, Table 3-2 presents estimated values of average standard deviations as a function of traffic area and highway type. For the greatest accuracy, analysts can conduct the study, calculate the actual standard deviation of the data, and check to see if the sample size was adequate. If not, additional data would have to be collected under the same conditions as the first study. Another technique is to use a calculator to update continuously a running total, average speed, and standard deviation. When the standard deviation becomes stable, an adequate sample size has been obtained.

Confidence level is the probability that the difference between the calculated mean speed from the sample and the true average speed at the study location is less than the permitted error. This concept is discussed in greater detail in Appendix C. The constant K that is used in equation (3-1) corresponds to confidence-level values as shown in Table 3-3. The permitted error, E, in equation (3-1) reflects the precision required in estimating the mean speed. This parameter is an absolute tolerance and is expressed as plus and minus a specified value. Typical permitted errors range from $\pm$ 1.0 to $\pm$ 5.0 mph.

If the statistic of interest is some percentile speed, such as the 85th percentile, equation (3-2) is appropriate for determining the sample size required.

$$N = \frac{S^2K^2(2 + U^2)}{2E^2} \tag{3-2}$$

TABLE 3-4
Constant Corresponding to Percentile Speed

Constant, U	*Percentile Speed*
0.00	50th
1.04	15th or 85th
1.48	7th or 93rd
1.64	5th or 95th

where N, S, K, and E are as defined for equation (3-1) and U is the constant corresponding to the desired percentile speed. Table 3-4 presents constants corresponding to percentile speeds. As a general rule, the minimum sample size should never be less than 30 measured spot speeds.

Field Procedures for Radar Spot Speed Studies

Successful spot speed data collection depends on how well two aspects of the study are conducted. Poor treatment of these issues can adversely affect the accuracy of the measurements and/or bias the results. The first issue involves the configuration of the site for data collection, and the second pertains to how individual vehicles are selected for measurement.

Layout of Site. The positioning of the radar unit is constrained by three considerations:

1. Capabilities of the radar unit
2. Minimizing the angle of incidence
3. Concealing the unit from the view of motorists

The capabilities of radar units vary considerably. Units must be set up and operated in accordance with the manufacturer's specifications and instructions. As discussed earlier, the larger the angle of incidence between the radar beam and the direction of travel of the target vehicle, the larger the cosine error. An angle of less than 15 degrees keeps the error under 2 mph.

Concealment of the radar unit and operators will prevent motorist distraction (a safety concern) and reaction (a potential source of bias). The equipment and crew may be concealed by vegetation or roadside structures, or they may simply be located out of view of target vehicles. Roadway maintenance vehicles, which motorists expect to see along the roadside, may be used to conceal the unit and crew. This approach will not work in states where the police also use maintenance vehicles to conceal radar units. Figure 3-2 shows several ways to position the radar unit (FHWA, 1980). Concealment will not defeat radar detectors.

Selection of Target Vehicles. The guiding principle is to select target vehicles randomly that represent the population of vehicles under study. Thus analysts must clearly define the study population (i.e., free-flow vehicles, large trucks, platoon leaders, all vehicles, etc.). Once the population is defined, study teams can adopt a selection strategy to provide a random sample of that population. With all strategies, teams must exercise caution to avoid selecting vehicles with radar detectors.

Except for studies conducted under low-volume conditions, it will be impossible to obtain a radar measurement of every vehicle. For example, vehicles may mask other vehicles from the radar beam. Large vehicles return a stronger signal than do small vehicles, thus overriding the smaller vehicle speed. Vehicles from the opposite direction or in a different lane from that under study may override the measurement of the target vehicle.

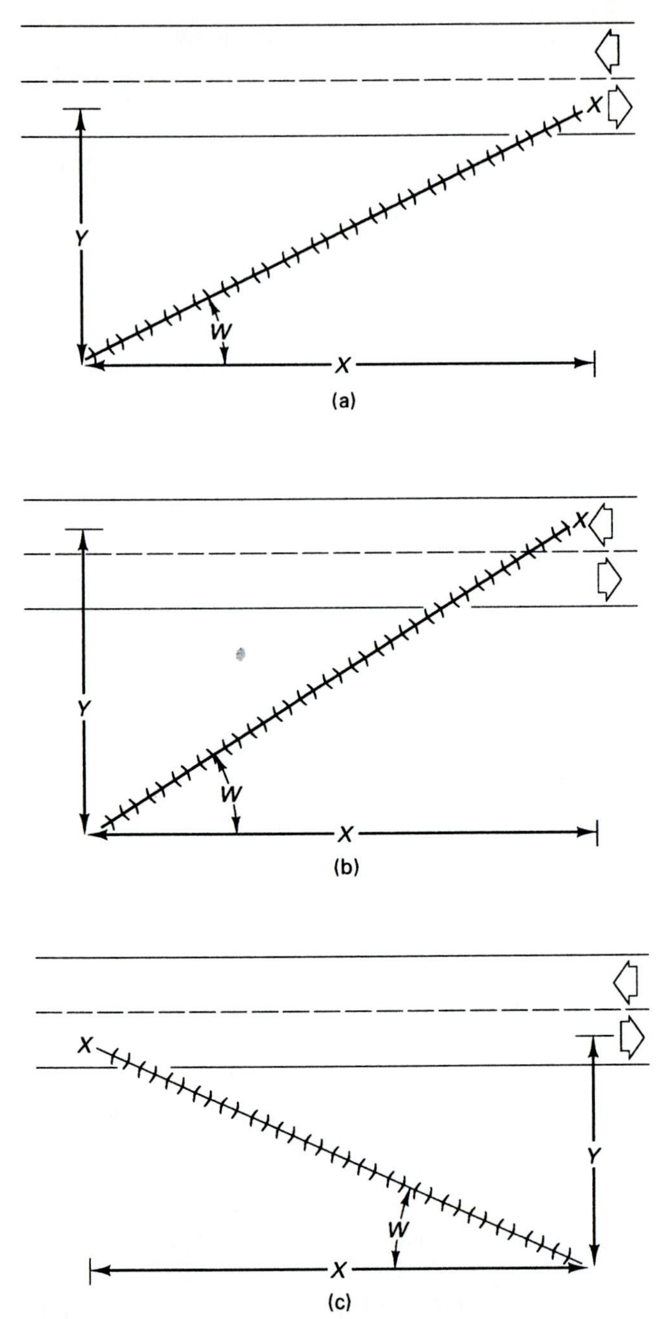

Figure 3-2 Radar unit setups. (a) Method A; (b) Method B; (c) Method C; (d) Method D. *Source:* FHWA, 1980.

Observers have a natural tendency to record those vehicles that "stand out" in some way, such as fast vehicles, slow vehicles, trucks, or platoon leaders. A procedure that controls for this bias is to select every third, fifth, tenth, or *n*th vehicle. Take care, however, that the *n*th vehicle is not controlled by some external effect. For example, every tenth vehicle on an arterial street could be a platoon leader if a coordinated traffic signal system establishes a regular pattern of traffic flow.

Documentation. The final layout of the data collection site should be fully described in any report of speed data. Observers should make an accurate sketch of the site showing the number of lanes, the position of the radar unit, and the *x* and *y* dimensions as shown in Figure 3-2. The *x* and *y* dimensions permit calculation of the angle of incidence so that

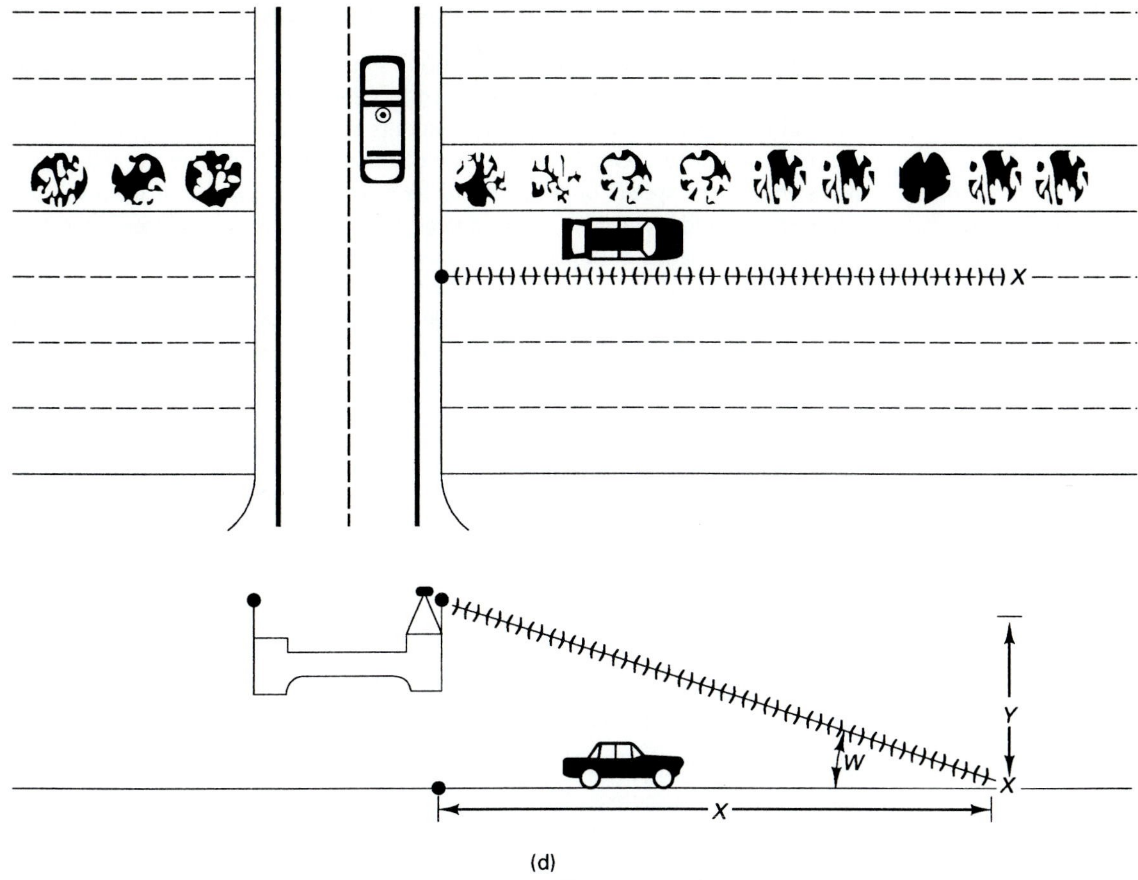

(d)

Figure 3-2 Continued.

a cosine error correction may be applied, if desired. Observers should record the start time, end time, any downtime, and the conditions prevailing during the study. Photographs of the layout may also prove useful.

Calibration. The radar manufacturer's recommended calibration tests should be made before the start and again at the end of data collection. The results should be included in any report of speed data. It is also wise to make an initial test in the office to ensure that the radar unit is operable before traveling to the site.

Field Procedures for Manual Speed Traps and Videotaping

As indicated in a preceding section, analysts seldom use manual speed traps and videotaping. Readers can find information about the procedures for conducting spot speed studies using manual speed traps or videotaping in Box and Oppenlander (1976), Lunenfeld and McDade (1983), McShane and Roess (1990), Pignataro (1973), and Taylor et al. (1989).

All-Vehicle Sampling

Analysts use this method when the purpose of the study requires or can be accommodated by measuring the spot speeds of all vehicles passing a point for a sample of time periods. The objectives of such studies are usually more general than studies using the individual vehicle selection method, but may also be specific and somewhat limited in scope with respect to certain types of locations, time periods, and conditions. Examples of such applications include monitoring speed trends, assessing highway safety, or establishing speed limits.

Location, Time, and Conditions of the Study

As with individual vehicle selection, selecting the spot to take speed measurements; the time period over which to collect the data; and the roadway, traffic, and weather conditions under which to conduct the spot speed study is generally a matter of common sense. The objective and scope of the study also dictate the selection of these items. If average speeds on a section of freeway are the sample of interest, speed measurements should be taken at the midpoint of a typical section. If speeds are being sampled to determine the 85th percentile speed for use in establishing a speed limit, the data should be collected over one or more 24-hour periods on several typical days. If free-flow versus platoon speeds are key to addressing a research question, the collection technique must be capable of distinguishing these two groups from all other vehicles.

Personnel and Equipment

The all-vehicle sampling method utilizes automatic data collection equipment such as detectors placed in the travel lanes that serve as input devices to recorders located at the site. Data from the recorders may be downloaded with a computer in the field or by telephone modem to a computer in the office. Significant advances have been made in the use of both permanent and portable detectors, recorders, and computers. Computers are now capable of sensing different types of vehicles, recording travel times over traps, calculating speeds, classifying vehicles, and storing large quantities of data. These advances permit analysts to study large samples of vehicles over long time periods. In the following sections the principal techniques and equipment in use today are described.

Detectors. The most commonly used devices for measuring speed are induction loop and tape switch detectors. These devices are normally deployed in pairs. Two loops or tape switches placed a measured distance apart form a speed trap that measures the time it takes the vehicle to travel from one detector to the next. Speed monitoring stations generally operate with permanent loop installations. Agencies place these loops in saw cuts in the pavement, sealed for protection from the environment, in the same manner that loop detectors are installed on approaches to signalized intersections.

Figure 3-3 illustrates typical permanent loop configurations for measuring speeds and for counting traffic. Figure 3-4 shows two possible data collection configurations of permanent loops and temporary tape switches to record speed and vehicle classification. In Figure 3-4(a), the loops measure vehicle presence and speed, and the tape switch counts axles. In Figure 3-4(b), the tape switches detect axles and speed, and the loop detects vehicle presence. Tape switches may be installed temporarily with tape, glue, or nails (Harkey et al., 1991). Study teams can tape portable loops to the roadway. A more practical and durable approach, however, is to mount the loop on a rubber mat. These mats may be rolled for storage and transport, are relatively easy to handle, and can be installed and recovered quickly. Nails and/or strong adhesive tape will hold the mats in place. Mats may be purchased commercially or fabricated locally. Harkey et al. (1987) present details for the construction and deployment of inductive loop mats.

To measure speeds on multilane roads, study teams must place detectors with separate inputs to the recorder in each lane. Figure 3-5 shows the detector lane placement for simultaneous measurement of speeds in two opposing directions. Notice that the detector mats are offset so that the lead wires are far enough apart to prevent crosstalk or interference, which can cause false readings (Harkey et al., 1989).

Automatic Recorders/Computers. A number of vehicle classification and recording counters capable of calculating speeds are available. These recorders are capable of storing large amounts of individual vehicle data or even larger amounts of classification data. Many recorders can operate continuously for several days. Data are downloaded from the recorder via a laptop computer or portable floppy disk drive in the field or via telephone modem to a computer in the office.

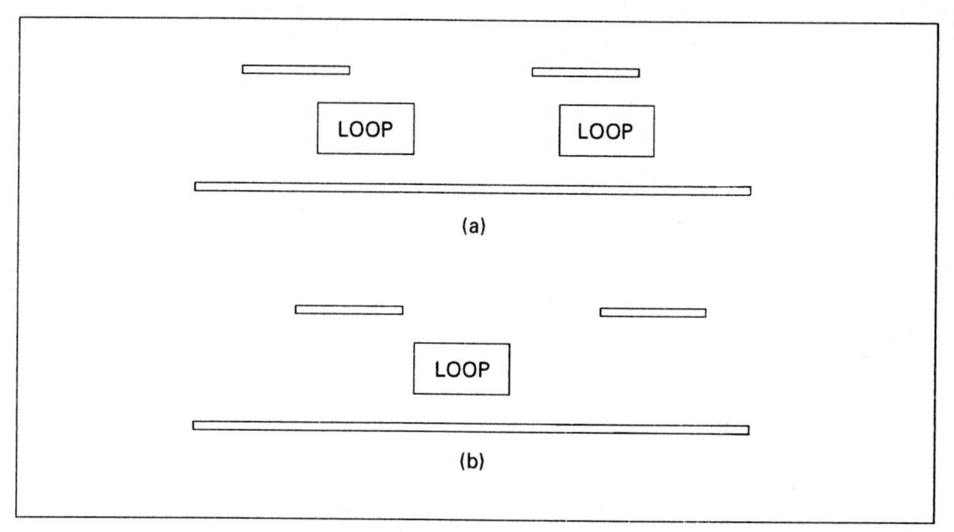

Figure 3-3 Types of permanent loop configurations. (a) Permanent speed monitoring station. (b) Permanent vehicle counting station. *Source:* Harkey, D. L., Davis, S. E., and Stewart, J. R., "Safety Impacts of Different Speed Limits for Cars and Trucks: Data Collection Plan", The Scientex Corporation, Charlotte, NC, Feb. 1991, pp. 20–21.

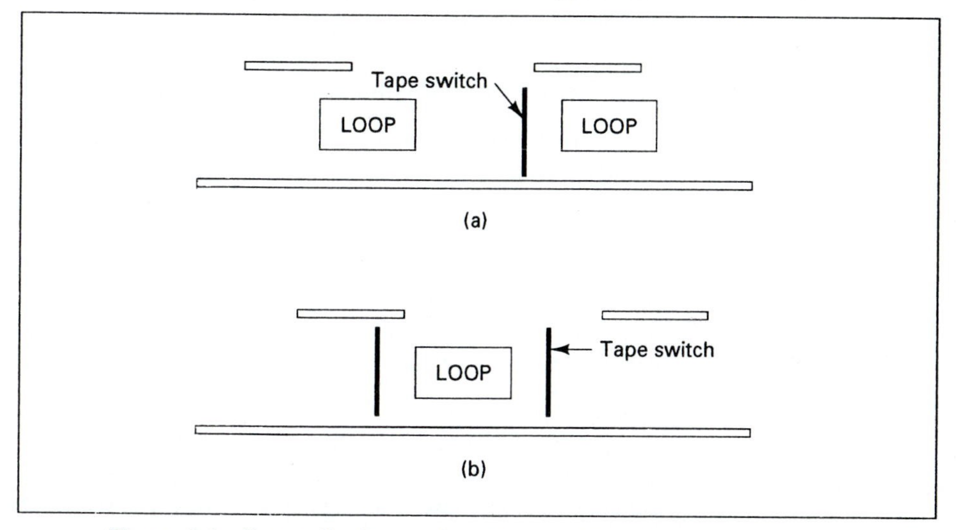

Figure 3-4 Data collection configurations. (a) Two loops used with one tape switch. (b) One loop used with two tape switches. *Source:* Harkey, Davis, and Stewart, 1991, pp. 20–21.

Personnel. One of the advantages of automatic speed data collection is that personnel are needed only during installation and recovery of the data collection equipment. Once installed, the equipment can operate unattended for several days. Agencies normally use two or three people to deploy and recover detectors and recorders. Safety dictates that one person never attempt this task alone. During installation and recovery, the crew and its vehicle should be highly visible to motorists. The crew should use cones and warning signs. One person should watch for traffic anytime that a crew member is in the roadway.

Sample-Size Requirements

With the all-vehicle sampling method, obtaining an adequate sample is seldom a problem because deployments are usually made for at least a 24-hour period. Sample-size requirements may be estimated in the same manner as described earlier for the individual

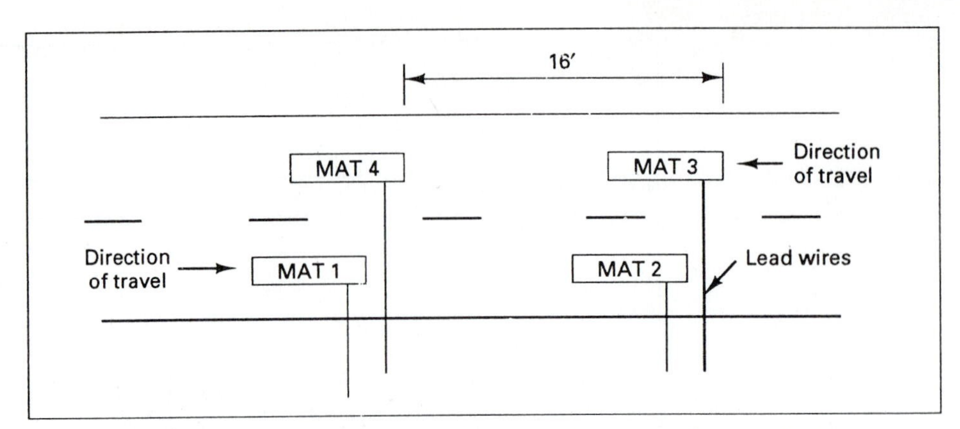

Figure 3-5 Lane placement of inductive loop mats. *Source:* Harkey, Davis, Robertson and Stewart, 1989.

vehicle selection method. Analysts seeking samples of certain types of vehicles may need to collect a large total sample to ensure that the types of interest are adequately represented. Since the automatic data collection system captures every vehicle (except in the case of a malfunction), the important sampling issue analysts need to address is the time period when the data collected will be representative of the desired study conditions. The answer depends on the purpose of the study, the resources available, and common sense.

Field Procedures for Automatic Spot Speed Studies

Successful spot speed studies using automatic data collection equipment depend on the operational reliability of the detectors and recorders, the physical installation of the detectors and lead wiring, and the calibration and quality control measures employed. External factors that can affect data collection include the temperature, weather, level of traffic volume, mix of vehicle types, and the environment (i.e., dirt, dust, debris, at the data collection site). Study procedures begin in the office with coordination preparations and operational checks of equipment. In the field, the principal tasks are deployment, calibration, recovery, and documentation.

Office Preparations. The first task is to coordinate all data collection activities with appropriate state and local officials, including transportation, traffic, and law enforcement agencies. These agencies need to be informed and analysts need to ensure that their studies do not interfere with ongoing activities. The second task is to brief the field team on the data collection plan to ensure that data are collected at the proper place in the desired manner for the required time period. The third task involves the team's preparations. All tools, supplies, and equipment should be assembled and inspected. Checklists may prove helpful, and the general checklist in Figure 1-1 may be a good starting point. Each piece of equipment should be tested to see that it is functioning properly. This precaution can prevent costly trips back to the office to replace missing supplies or malfunctioning equipment.

Deployment. Safety is the first consideration during deployment. It is preferable to close each lane to traffic while work is under way. The detectors should be prepared on the roadside to minimize the time each lane is closed. Workers then place each set of loop mats and/or tape switches as prescribed in the data collection plan. The 3-foot by 6-foot mats are usually placed 10 feet apart. The accuracy of loop and tape switch spacing is critical and the crew must be careful. After placing, the lead wires are connected to the recorder and the loops or switches are checked for proper functioning. After any needed repairs are made, the crew can secure the detectors to the pavement. The field crew should have some leeway to select the exact position to deploy the data collection system.

The crew can avoid broken pavement and locate the recorder near some fixed object to which it can be secured.

Calibration. With the detectors and recorders in place, the next step is to check the accuracy of the equipment in measuring the axle counts, axle spacings, and speeds of the traffic stream. The crew can check axle spacing against the known spacing of the data collection vehicle. Axle counts are made visually by the team observing several passing vehicle types, and those counts are compared to the readings on the recorder. The crew can adjust the recorder until the spacing is correct within one-tenth of a foot and the axle counts match. A calibrated radar gun is used to measure vehicle speeds, which are compared to the speeds of those same vehicles monitored by the data collection equipment. A typical tolerance limit is ± 1.0 mph. Again, the crew can adjust the recorder until this tolerance is met. It is advisable to check the functions and accuracy of the equipment at least once during every 24-hour data collection period.

Recovery. After the data collection period, the recorded data are checked for accuracy by computing the ratio of usable vehicle data to total vehicle data. To compute this ratio, compare the total vehicle count to the usable vehicle count. A usable vehicle is a vehicle with both a length and a speed. Sometimes a vehicle will be sensed by only one of the loops or switches in the pair. This occurs when the vehicle path does not cross both sensors or when one of the sensors fails. Such vehicles are stored in the total vehicle count but not in the usable vehicle count. The ratio of usable to total should exceed 0.75. If this criterion is not met, the team should collect data during another period.

Crews recover data collection equipment by reversing the process they used to deploy it. Again, safety should be the principal concern. If the equipment is to be deployed at another site, the data should be transferred from the recorder either at the site or via a telephone modem to the office. If the site is being monitored for an extended period of time, the data should be ''dumped'' periodically.

Documentation. The final layout of the data collection site should be fully described in any report of speed data. The crew should make an accurate sketch of the site, showing the number of lanes, the position of the detectors, and the location of the recorders. The crew should record the start time, end time, any downtime, and the conditions prevailing during the study. Equipment malfunctions and repairs and the results of calibration and accuracy checks should be recorded. Photographs of the layout may also prove useful. Appendix G contains a set of sample data collection forms for automatic spot speed data collection.

DATA REDUCTION AND ANALYSIS

A typical spot speed study analysis is comprised of three parts. Data reduction is the first part and is simply the arrangement of the measured speeds, or ''raw'' data, into a convenient tabular or graphical form. The second part is the calculation and presentation of descriptive statistics, which illustrate the collection of speed data by means of a few representative values or variables. The third part of a typical analysis is statistical inference, which permits the development of statistical estimates and the testing of statistical hypotheses. Appendix C presents detailed information on field study analyses and uses many examples from spot speed data. Therefore, the discussion in this section is brief, and the reader is referred to Appendix C for greater detail.

Data Reduction

Both the individual vehicle selection method and the all-vehicle sampling method produce measured values of individual vehicle speed. The speeds are usually arrayed in tabular form in the sequence they are recorded. To illustrate, Table 3-5 contains data from

TABLE 3-5
Speed Data Obtained on a Rural Highway

Car No.	Speed (mph)	Car No.	Speed (mph)	Car No.	Speed (mph)	Car No.	Speed (mph)
1	35.1	23	46.1	45	47.8	67	56.0
2	44.0	24	54.2	46	47.1	68	49.1
3	45.8	25	52.3	47	34.8	69	49.2
4	44.3	26	57.3	48	52.4	70	56.4
5	36.3	27	46.8	49	49.1	71	48.5
6	54.0	28	57.8	50	37.1	72	45.4
7	42.1	29	36.8	51	65.0	73	48.6
8	50.1	30	55.8	52	49.5	74	52.0
9	51.8	31	43.3	53	52.2	75	49.8
10	50.8	32	55.3	54	48.4	76	63.4
11	38.3	33	39.0	55	42.8	77	60.1
12	44.6	34	53.7	56	49.5	78	48.8
13	45.2	35	40.8	57	48.6	79	52.1
14	41.1	36	54.5	58	41.2	80	48.7
15	55.1	37	51.6	59	48.0	81	61.8
16	50.2	38	51.7	60	58.0	82	56.6
17	54.3	39	50.3	61	49.0	83	48.2
18	45.4	40	59.8	62	41.8	84	62.1
19	55.2	41	40.3	63	48.3	85	53.3
20	45.7	42	55.1	64	45.9	86	53.4
21	54.1	43	45.0	65	44.7		
22	54.0	44	48.3	66	49.5		

Source: Garber and Hoel, 1988.

TABLE 3-6
Frequency Distribution Table for Set of Speed Data

Speed Class (mph)	Class Midvalue, u_i	Class Frequency (Number of Observations in Class), f_i	$f_i u_i$	Percentage of Observations in Class	Cumulative Percentage of All Observations	$f(u_i - \bar{u})^2$
34–35.9	35.0	2	70	2.3	2.30	420.5
36–37.9	37.0	3	111	3.5	5.80	468.75
38–39.9	39.0	2	78	2.3	8.10	220.50
40–41.9	41.0	5	205	5.8	13.90	361.25
42–43.9	43.0	3	129	3.5	17.40	126.75
44–45.9	45.0	11	495	12.8	30.20	222.75
46–47.9	47.0	4	188	4.7	34.90	25.00
48–49.9	49.0	18	882	21.0	55.90	9.0
50–51.9	51.0	7	357	8.1	64.0	15.75
52–53.9	53.0	8	424	9.3	73.3	98.00
54–55.9	55.0	11	605	12.8	86.1	332.75
56–57.9	57.0	5	285	5.8	91.9	281.25
58–59.9	59.0	2	118	2.3	94.2	180.50
60–61.9	61.0	2	122	2.3	96.5	264.50
62–63.9	63.0	2	126	2.3	98.8	364.50
64–65.9	65.0	1	65	1.2	100.0	240.25
		86	4260			3632.00

Source: Garber and Hoel, 1988.

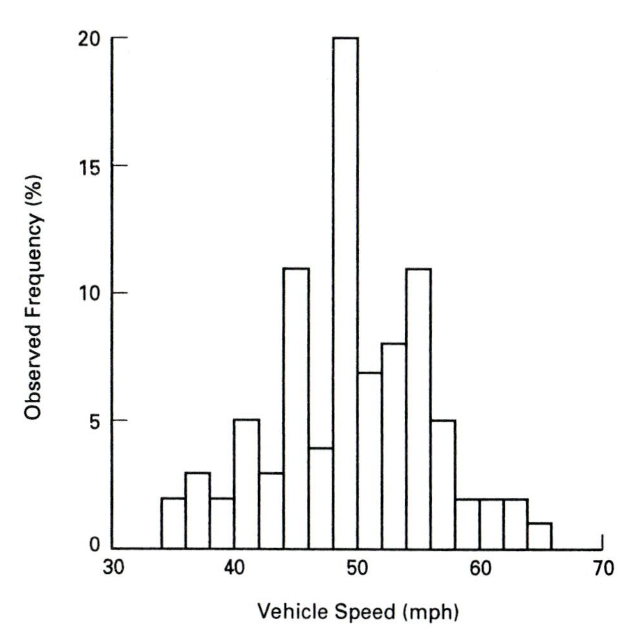

Figure 3-6 Histogram of observed vehicles' speeds. *Source:* Reprinted by permission from page 94 of *Traffic and Highway Engineering* by Garber, N.J. and Hoel, L.A.; copyright © 1988 by West Publishing Company. All rights reserved.

a speed study conducted in rural Virginia that used the individual vehicle selection method. Table 3-6 shows a frequency distribution table for the data in Table 3-5 in which the speeds have been grouped into 16 classes with an interval of 2 mph. Class frequencies, the percentage of observations in each class, and the cumulative percentage of all observations per class are also shown.

Analysts can display speed data in several graphical forms. Figure 3-6 is a histogram of the classified frequency data in Table 3-6. A frequency distribution curve may be plotted from the percentage of observations column of Table 3-6, as shown in Figure 3-7. Similarly, analysts can form the S-shaped cumulative distribution curves shown in Figure 3-8 by plotting the cumulative percentages of all observations versus speed. Data from automatic speed data recorders can also be presented in tabular form, either as raw data (Figure 3-9) or in summary form (Figure 3-10). The all-vehicle sampling method usually yields a large quantity of data, so that computer software is necessary for reduction and analysis. Several software programs are available that provide analysis and graphing of speed data (Center for Microcomputers in Transportation, 1992).

Descriptive and Inferential Statistics

Several variables that describe a collection of speed observations may be calculated from both the raw and classified data. The most common variables include the mean, mode, median, standard deviation, pace, and percentile. The pace is defined as the 10-mph range that contains the greatest percentage of observations. Some of these variables may be obtained directly from the frequency and cumulative distribution curves. The calculations may also be made manually or through the use of computer software. Most of the software that agencies use with data recorders is capable of producing these variables.

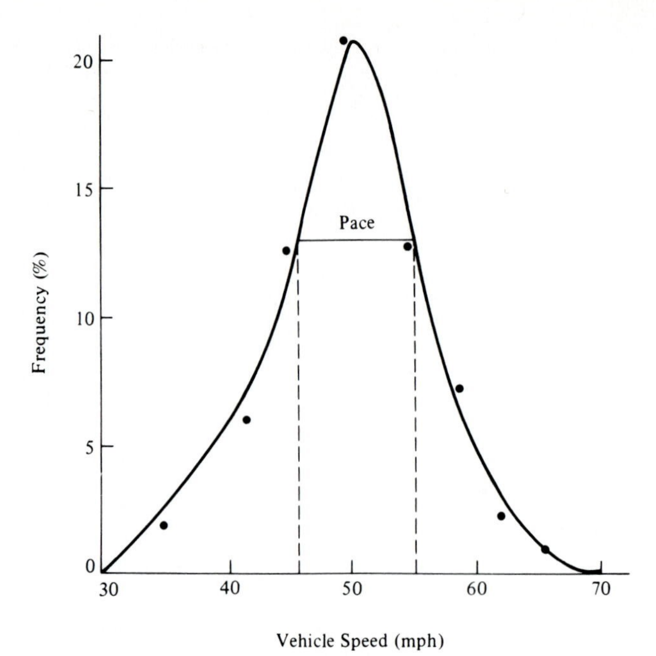

Figure 3-7 Frequency distribution. *Source:* Garber and Hoel, 1988.

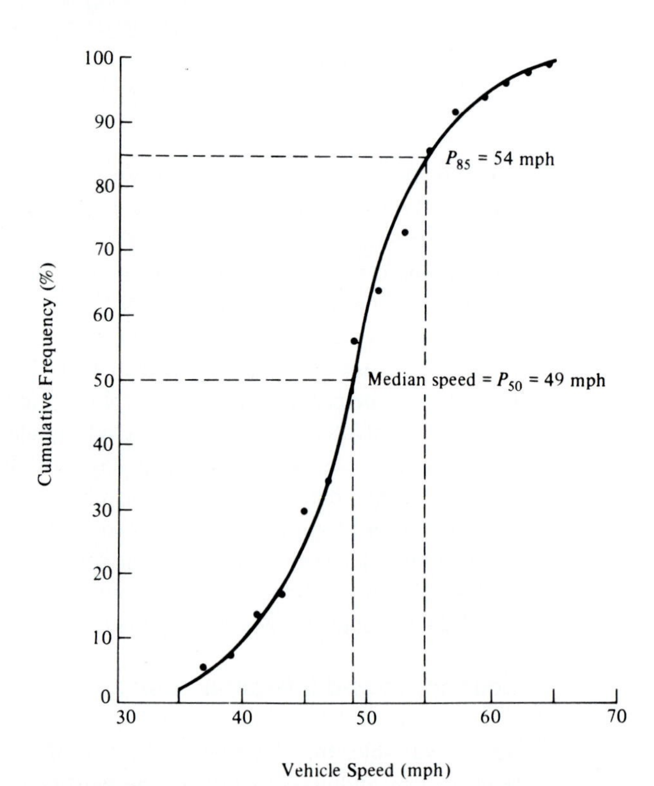

Figure 3-8 Cumulative distribution. *Source:* Garber and Hoel, 1988.

Analysts can draw inferences from testing these variables statistically. For example, analysts can determine if significant differences exist in mean spot speeds for different traffic or roadway conditions. Also, analysts can determine if the speed distributions for data samples from two different locations or time periods are the same or different. Appendix C provides details on statistical tests and describes how to perform the necessary calculations.

```
File OPENED at 11/20/90 16:48
STORAGE: Raw
SITE: 177MP63.5
INFO#1: RURAL NC
INFO#2: WITH RURAL VA 177MP19.3
2 Active Lanes. Date Format = MM/DD/YY. Unit Type = 1

3: LANE USED TRIG = LOOP  AXLE = TPSW  PRES = LOOP
   Sensor Spacing = 16.0' Loop2 Length = 6.0'
   INFO:___________ 1 Record Interval:
   INTERVAL#1 Start = 00:00  Length = 00:15

4: LANE USED TRIG = LOOP  AXLE = TPSW  PRES = LOOP
   Sensor Spacing = 16.00' Loop2 Length = 6.0'
   INFO:___________ 1 Record Interval:
   INTERVAL#1 Start = 00:00  Length = 00:15

3: 16:48:22 54MPH 2 Axles, 9.7'
3: 16:48:24 57MPH 7 Axles, 14.9' 4.5' 28.5' 4.2' 8.6' 20.6'
3: 16:48:26 56MPH 7 Axles, 16.7' 4.3' 26.5' 3.9' 8.0' 19.6'
3: 16:48:28 61MPH 2 Axles, 9.3'
4: 16:48:46 64MPH 2 Axles, 8.9'
4: 16:48:47 61MPH 2 Axles, 8.5'
4: 16:48:51 59MPH 2 Axles, 8.5'
3: 16:48:51 59MPH 5 Axles, 16.6' 4.3' 31.4' 3.9'
3: 16:49:24 59MPH 2 Axles, 11.1'
4: 16:50:19 55MPH 2 Axles, 10.5'
3: 16:50:25 59MPH 2 Axles, 9.0'
```

Figure 3-9 Example of a raw data file. *Source:* Harkey, Davis, and Stewart, 1991.

SUMMARY TABLE FOR SITE 10619 : LARGE URBAN | 40-50 | MULTILANE

DATE (YYMMDD)...... 870819 - 870820
BEGIN TIME (HHMM).. 1200
SPEED LIMIT (mph).. 45

LANE(S)............ 1 2 3 4
END TIME (HHMM).... 1200
FREE FLOW HEADWAY.. 4.0
PACE (mph)......... 10

						FREE FLOW																	ALL VEH				NON-FREE FLOW		
									PERCENTILES							>SPEED LIMIT +				PACE							% sHEADWAY		
TIME	COUNT	%1	%2	%3	%4	%	MEAN	S.D.	5%	15%	30%	50%	70%	85%	95%	>SL	> 5	> 10	> 15	LL	UL	%	MEAN	85%	>SL	s1.0	s2.0	s3.0
100	65	20	80	0	0	86	47.1	7.1	35.1	38.7	45.8	48.4	50.1	52.8	56.1	71	38	7	4	44	53	67	46.7	52.8	67	6	9	12
200	30	13	80	7	0	90	50.9	5.6	41.3	44.8	46.3	52.0	54.5	56.6	58.6	81	56	26	4	46	55	59	50.6	56.8	76	10	10	10
300	27	15	78	4	4	89	49.5	4.6	41.4	44.5	46.3	49.3	51.7	54.9	57.2	88	46	13	0	45	54	75	49.3	54.8	81	7	11	11
400	16	31	63	6	0	75	49.3	6.4	34.8	42.2	45.5	49.9	50.9	53.5	58.0	83	50	17	8	43	52	75	49.9	55.1	87	25	25	25
500	47	11	81	4	4	89	49.0	6.3	40.1	43.0	45.2	48.7	52.3	54.6	59.9	71	45	12	7	43	52	59	49.3	54.6	74	6	6	9
600	87	13	83	2	2	95	48.2	6.4	37.6	40.6	44.8	48.1	51.0	54.2	58.8	70	43	12	5	44	53	60	48.2	54.4	71	3	5	5
700	285	13	84	2	1	87	48.5	5.9	38.8	43.4	45.8	48.4	50.9	53.7	57.9	73	41	10	2	42	51	68	48.6	53.7	73	1	5	11
800	580	8	91	1	1	73	49.0	7.1	40.7	44.6	46.9	49.2	51.8	53.9	58.0	81	45	12	3	44	53	73	48.8	53.8	79	4	12	21
900	540	8	88	2	2	75	48.2	6.0	39.4	43.4	46.1	48.4	50.9	53.3	57.0	75	37	9	3	44	53	72	48.1	52.9	75	3	12	19
1000	399	8	89	3	1	78	47.4	7.1	35.4	41.1	45.4	48.3	50.3	53.5	56.7	71	33	10	1	44	53	65	47.2	53.6	72	4	11	16
1100	433	9	87	3	1	81	47.2	6.3	36.3	41.1	44.7	47.5	50.5	53.6	56.6	66	33	9	1	44	53	63	47.3	53.7	66	3	8	15
1200	474	10	85	2	2	81	46.7	7.1	35.8	40.9	44.1	46.8	50.0	52.8	56.7	65	30	8	3	42	51	63	46.4	52.7	61	3	11	16
1300	575	7	89	2	2	82	47.6	5.6	38.8	42.5	44.8	47.5	50.0	52.9	56.6	70	31	8	2	42	51	69	47.6	52.9	71	2	9	15
1400	519	7	90	3	1	83	47.5	6.6	38.1	41.9	45.3	47.6	50.1	52.9	56.4	70	32	8	2	43	52	66	47.3	52.9	69	1	8	12
1500	511	8	89	2	1	80	48.2	5.5	39.9	42.9	45.3	48.1	50.9	53.5	56.9	73	35	10	3	43	52	69	48.3	53.6	73	3	10	15
1600	661	8	87	3	1	76	48.0	6.1	38.8	42.6	45.4	48.4	50.9	53.5	55.7	73	37	9	3	43	52	67	47.7	52.7	71	2	11	19
1700	783	8	90	1	1	72	48.9	6.2	41.0	44.7	46.8	49.1	51.7	53.8	57.5	83	41	10	2	45	54	73	48.7	53.7	82	3	12	20
1800	875	7	91	1	1	69	48.7	6.5	40.1	44.0	46.8	49.1	51.0	53.7	57.7	79	42	10	3	44	53	72	48.6	53.6	79	2	12	22
1900	708	7	91	1	1	74	48.4	5.9	40.1	43.4	46.0	48.4	50.9	53.6	57.6	76	39	9	2	43	52	71	48.1	53.5	75	2	12	19
2000	541	8	90	2	0	80	48.0	5.8	38.3	42.8	45.5	48.2	50.8	53.7	57.2	73	33	10	1	43	52	67	48.0	53.6	74	2	8	13
2100	482	10	89	1	0	85	45.8	6.1	35.7	40.5	43.4	46.2	48.8	51.0	55.6	59	23	5	1	42	51	67	45.8	50.9	59	1	6	10
2200	421	11	87	1	0	87	45.2	6.0	36.2	40.1	43.4	45.5	47.8	50.0	53.6	57	15	2	1	41	50	71	45.2	50.0	57	1	5	8
2300	230	14	83	2	0	91	46.3	6.1	35.6	41.5	44.0	46.8	49.2	51.8	54.7	64	26	4	0	42	51	69	46.3	51.8	64	3	7	8
2400	149	13	85	1	1	93	45.9	6.0	34.8	41.3	42.9	46.1	48.4	50.2	55.7	59	17	7	1	40	49	71	45.9	50.2	59	1	3	7

						FREE FLOW																	ALL VEH				NON-FREE FLOW		
									PERCENTILES							>SPEED LIMIT +				PACE							% sHEADWAY		
LENGTH	COUNT	%1	%2	%3	%4	%	MEAN	S.D.	5%	15%	30%	50%	70%	85%	95%	>SL	> 5	> 10	> 15	LL	UL	%	MEAN	85%	>SL	s1.0	s2.0	s3.0
<10	826					70	46.4	11.1	19.0	40.6	45.3	48.2	51.7	54.6	58.9	70	39	13	3	44	53	58	46.0	54.6	70	13	20	26
<20	8346					79	47.9	5.7	38.8	42.7	45.4	48.2	50.8	53.5	56.7	72	34	8	1	42	51	68	47.9	52.9	72	1	8	14
TRK>20	266					83	45.9	7.2	35.7	40.3	43.3	46.2	49.3	51.8	56.5	55	26	6	1	42	51	65	45.9	51.8	57	0	5	10
ALL	9438	9	88	2	1	78	47.7	6.3	38.3	42.6	45.4	48.2	50.8	53.5	56.8	71	34	8	2	42	51	67	47.7	53.0	72	2	9	15

Figure 3-10 Sample output produced by the FHWA speed analysis program.
Source: Harkey, Davis, Robertson, and Stewart, 1989.

REFERENCES

BOX, P. C., AND J. C. OPPENLANDER (1976). *Manual of Traffic Engineering Studies,* 4th ed., Institute of Transportation Engineers, Washington, DC, pp. 78–91.

CENTER FOR MICROCOMPUTERS IN TRANSPORTATION (1992). *McTrans Catalog*, University of Florida, Gainesville, FL, June, p. 22.

FHWA (1980). *Speed Monitoring Program Procedural Manual*, Federal Highway Administration, Washington, DC, May.

GARBER, N. J., AND L. A. HOEL (1988). *Traffic and Highway Engineering,* West Publishing, St. Paul, MN, pp. 84–97.

HARKEY, D. L., S. E. DAVIS, J. R. STEWART, AND H. D. ROBERTSON (1987). *Assessment of Current Speed Zoning Criteria: Pilot Study Report,* Federal Highway Administration, Washington, DC, October, pp. 11–20, 49–55.

HARKEY, D. L., S. E. DAVIS, H. D. ROBERTSON, AND J. R. STEWART (1989). *Assessment of Current Speed Zoning Criteria: Final Report,* FHWA/RD-89-161, Federal Highway Administration, Washington, DC, June, pp. 8–12.

HARKEY, D. L., S. E. DAVIS, AND J. R. STEWART (1991). *Safety Impacts of Different Speed Limits for Cars and Trucks: Data Collection Plan,* The Scientex Corporation, Charlotte, NC, February, pp. 17–22.

LUNENFELD, H., AND J. D. MCDADE (1983). *Traffic Performance Data Collection,* FHWA-TO-83-2, Federal Highway Administration, Washington, DC, June 6, Unit 4.

MCSHANE, W. R. AND R. P. ROESS (1990). *Traffic Engineering,* Prentice Hall, Englewood Cliffs, NJ, pp. 134–139.

PIGNATARO, L. J.(1973). *Traffic Engineering Theory and Practice,* Prentice Hall, Englewood Cliffs, NJ, pp. 116–142.

TAYLOR, M. A. P., AND W. YOUNG (1988). *Traffic Analysis: New Technology and New Solutions,* Hargreen Publishing, North Melbourne, Victoria, Australia, pp. 150–157.

TAYLOR, M. A. P., W. YOUNG, AND R. G. THOMPSON (1989). *Headway and Speed Data Acquisition Using Video,* Transportation Research Record 1225, Transportation Research Board, Washington, DC, pp. 130–139.

4

Travel-Time and Delay Studies

H. Douglas Robertson, Ph.D., P.E.

INTRODUCTION

Travel time and delay are two of the principal measures of highway system performance used by traffic engineers, planners, and analysts. Vehicle speed is directly related to travel time and delay and is also used to evaluate traffic and highway systems. There are two types of average speed: time–mean speed (or mean spot speed) and space–mean speed (or mean travel speed). Measuring time–mean speed is described in Chapter 3. The measurement of space–mean speed is discussed in this chapter.

Travel time varies inversely with travel speed. A travel-time study provides data on the amount of time it takes to traverse a section of street or highway. These data, combined with the length of the section under study, produce mean travel speed. Travel-time and delay studies are conducted when the sources and amounts of delay occurring within the section are also noted. This chapter treats the measurement of delay along a roadway segment. Intersection delay studies are addressed in Chapter 5.

Applications

Engineers and planners use data from travel-time and delay studies in a number of tasks, including:

- Determining the efficiency of a route with respect to its ability to carry traffic and relative to other routes through the use of sufficiency ratings or congestion indices
- Providing input to capacity analysis of roadway segments
- Identifying problem locations as indicated by delay
- Evaluating the effectiveness of traffic operation improvements
- Providing input to transportation planning models, trip assignment models, and route-diversion models
- Providing input to economic analyses of alternatives
- Generating travel-time contour maps
- Providing input to studies that evaluate trends in efficiency and level of service over time

Definitions

- *Travel time* is the time taken by a vehicle to traverse a given segment of street or highway.
- *Running time* is the time a vehicle is actually in motion (or moving faster than a predesignated speed) while traversing a given segment of street or highway.
- *Speed* is the rate of movement of a vehicle in distance per unit time.
- *Travel speed* is the distance traveled divided by the travel time.
- *Running speed* is the distance traveled divided by the running time.
- *Space–mean speed* or *mean travel speed* is the segment distance divided by the mean travel time over the segment of several vehicles or of a single vehicle completing several trips.
- *Mean running speed* is the segment distance divided by the mean running time over the segment of several vehicles or of a single vehicle completing several trips.
- *Delay* is the time lost by a vehicle due to causes beyond the control of the driver.
- *Operational delay* is the component of delay caused by the presence and interference of other traffic. This type of delay may be in the form of side friction, where other traffic interferes with the traffic stream (e.g., parking maneuvers) or in the form of internal friction, where the interference takes place within the traffic stream (e.g., encountering a reduction in capacity).
- *Fixed delay* is the component of delay caused by traffic control devices that is independent of traffic volume and operational delay.
- *Stopped-time delay* is that part of delay when the vehicle is not moving (or moving slower than a predesignated speed).
- *Travel-time delay* is the difference between the actual travel time and the travel time based on a vehicle traversing the study segment at an average speed equal to that for uncongested traffic flow on the segment.

The definitions of stopped delay and travel-time delay given here differ slightly from those given in Chapter 5, due to the orientation of Chapter 5 exclusively to studies at intersections, whereas studies of longer roadway segments are discussed in this chapter.

METHODS OF STUDY

Travel-time and delay studies may be conducted using the average vehicle, moving vehicle, license plate, direct observation, or interview method. The first two methods require test vehicles, while the other methods do not. The choice of method depends on the purpose of the study; the type of roadway segment under study; the length of the segment; the time of day of interest; and the personnel, equipment, and resources available. The two most common methods (i.e., average vehicle and moving vehicle) are presented in detail, whereas the other methods are described only briefly. Appendix G contains data forms that are suitable for copying.

AVERAGE VEHICLE METHOD

The average vehicle method measures travel time; running time; distance traveled; and the type, location, duration, and cause of traffic delays along the study route. The data are recorded as the test vehicle traverses the study route. From these data, travel speed, space–mean speed, and running speed may be calculated. This method is applicable to any type of route, but is most widely used on arterial streets with at-grade intersections. The recommended minimum total length of route to be studied is 1 mile. Engineers may conduct areawide travel time and delay surveys on the major arterials leading to and from the central business district (CBD) and display the results as time–contour maps.

Time of Study

Agencies usually study travel time and delay during the peak hours in the directions of heaviest traffic flow. It may also be desirable to compare travel times, speeds, and delays between peak and off-peak periods or between sets of other conditions. Some of these other conditions include good versus adverse weather and commuter versus special event traffic.

Personnel and Equipment

The average vehicle method requires a test car and the means to record time and distance. These can be recorded manually or automatically.

Manual Data Collection

Manual data collection requires a driver and observer/recorder, two stopwatches, and data collection forms. The distances between control points and the length of the total route may be obtained from accurate, drawn-to-scale plans or maps or from the vehicle odometer.

Automatic Data Collection

Several varieties of hand-held and laptop computers with accompanying software are available to perform travel-time and delay studies (University of Florida Transportation Research Center, 1990; Center for Microcomputers in Transportation, 1992; HMR Communications, 1990; RWD Engineering, 1991). The computer is connected to a distance measuring instrument (DMI) that is attached to the test car. Hand or foot switches enable the driver to operate the recording equipment safely negating the need for an observer/recorder. Less sophisticated equipment, such as the traffic analyzer and tacograph, are suitable for travel-time and delay studies if laptop computers are not available.

TABLE 4-1

Approximate Minimum Sample-Size Requirements for Travel-Time and Delay Studies with Confidence Level of 95.0%

Average Range in Running Speed (mph)	*Minimum Number of Runs for a Permitted Error of:*				
	1.0 mph	*2.0 mph*	*3.0 mph*	*4.0 mph*	*5.0 mph*
2.5	4	2	2	2	2
5.0	8	4	3	2	2
10.0	21	8	5	4	3
15.0	38	14	8	6	5
20.0	59	21	12	8	6

Average Range in Running Speed (kph)	*Minimum Number of Runs for a Permitted Error of:*				
	2.0 kph	*3.5 kph*	*5.0 kph*	*6.5 kph*	*8.0 kph*
5.0	4	3	2	2	2
10.0	8	4	3	3	2
15.0	14	7	5	3	3
20.0	21	9	6	5	4
25.0	28	13	8	6	5
30.0	38	16	10	7	6

Source: Adapted from Box and Oppenlander, 1976, p. 96.

Sample-Size Requirements

The purpose of the travel-time and delay study dictates the size of sample (in terms of number of test runs) required. The following are suggested ranges of permitted errors in the estimate of the mean travel speed related to study purpose:

- Transportation planning and highway needs studies: ± 3.0 to ± 5.0 mph.
- Traffic operations, trend analysis, and economic evaluations ± 2.0 to ± 4.0 mph.
- Before-and-after studies ± 1.0 to ± 3.0 mph.

Engineers can apply these suggestions to similar studies as well.

Table 4-1 suggests numbers of test runs to perform a travel-time and delay study. These values have a confidence level of 95% and are based on desired permitted error (as described above) and the average range in running speed (as described below). Running speed is used because it is a more stable measure than travel speed. The analyst can determine the required sample size after a minimum of two test runs. Four initial test runs are recommended. Upon completion of the initial test runs, calculate the running speeds for each run in sequence. Compute the absolute differences between each pair of sequential runs (i.e., between first and second, second and third, etc.). Sum the differences and divide by the number of differences calculated to arrive at the average range in running speed, as shown in equation (4-1).

$$R = \frac{\Sigma A}{N - 1} \qquad (4\text{-}1)$$

where

R = average range in running speed

ΣA = sum of calculated speed differences

N = number of completed test runs

Select the approximate minimum sample size from Table 4-1 for the computed average range in running speed and the desired error permitted. If the required number of test runs is greater than the initial number of test runs, make additional runs under the same conditions as the initial runs to complete the sample required. Determine a sample size for each direction of travel and for each set of traffic and/or environmental conditions of interest. For example, assume that four runs were made to obtain an average range in running speed of 10 mph and that the desired permitted error is ± 3 mph. Table 4-1 indicates that 5 is the minimum number of runs required, assuming a confidence level of 95%. Therefore, one additional run would have to be made to achieve the required sample size.

The observer(s) must be sensitive to changes in the traffic or environmental conditions. The sample number of runs represents a single set of conditions. For example, speeds will probably vary during a peak period. Therefore, it may be necessary to conduct separate travel-time and delay studies for different portions of the peak period.

Field Procedure

Before test runs begin, observers select the start point, end point, and control point locations along the route where they will record time measures. On arterial and other types of surface streets, these locations are usually at major intersections or other easily identifiable control points. The choice of the near curb, far curb, or center of the intersection as the control point should be consistent throughout the study route.

Test Car Technique

The driver of the test vehicle proceeds along the study route in accordance with one of the following techniques:

1. *Average-car technique:* test vehicle travels according to the driver's judgment of the average speed of the traffic stream.
2. *Floating-car technique:* driver "floats" with the traffic by attempting to safely pass as many vehicles as pass the test vehicle.
3. *Maximum-car technique:* test vehicle is driven at the posted speed limit unless impeded by actual traffic conditions or safety considerations.

The selection of test car technique is based on the purpose of the study and the study team's judgment of the technique that best reflects the traffic stream being investigated. Most study teams prefer the average-car technique.

Manual Data Collection

Position the test car a short distance upstream of the begin point. "Zero" both stopwatches and complete the header information on the data forms. Figure 4-1 shows a sample field sheet. Review the duties of the driver and observer. Make a couple of dry runs to measure the distances between checkpoints and to rehearse the procedure. If the rehearsal runs go smoothly, count them as test runs. Test runs should begin promptly at the beginning of the desired study period so as to complete the required sample of runs before conditions along the route change. If the required sample cannot be obtained in a

TRAVEL-TIME AND DELAY STUDY
AVERAGE VEHICLE METHOD
FIELD SHEET

DATE __________ WEATHER __________________ TRIP NO. ______________

ROUTE ____________________ DIRECTION ______________________________

TRIP STARTED AT ______________ AT ______________________________
(LOCATION) MILAGE)

TRIP ENDED AT ______________ AT ______________________________
(LOCATION) (MILEAGE)

CONTROL POINTS		STOPS OR SLOWS		
LOCATION	TIME	LOCATION	SEC Delay	

TRIP LENGTH	TRIP TIME	TRAVEL SPEED
RUNNING TIME	STOPPED TIME	RUNNING SPEED

SYMBOLS OF DELAY CAUSE: S-TRAFFIC SIGNALS SS-STOP SIGN LT-LEFT TURNS

PK-PARKED CARS DP-DOUBLE PARKING T-GENERAL

PED-PEDESTRIANS BP-BUS PASSENGERS LOADING OR UNLOADING

COMMENTS ____

RECORDER ______________

Figure 4-1 Sample travel-time and delay study field sheet: average-vehicle method. *Source:* Adapted from Box, P. C. and Oppenlander, J. C., *Manual of Traffic Engineering Studies,* 4th Ed., Institute of Transportation Engineers, Washington, DC, 1976.

single period, complete the remaining runs on another day under the same conditions. As the test vehicle passes the begin point, the driver starts the first stopwatch. The test car proceeds through the study route according to the driving technique selected. The observer records time readings from the first stopwatch as the vehicle passes each control point. When the test vehicle stops or is forced to travel slowly (i.e., 5 mph or less), the observer uses the second stopwatch to measure the amount of delay and notes the location, duration, and cause of each delay on the field sheet.

As the test vehicle passes the end point of the study route, the driver reads the first stopwatch, and the observer notes the total time of the run on the field sheet. The test vehicle then returns to the begin point, the driver and observer reset the stopwatches, the observer prepares a new field sheet, and the next run begins. This procedure is repeated until the required number of sample runs is reached or until conditions that could affect the study change. If the reverse direction is also being studied, the same procedure governs the return trip with the data recorded on a separate field sheet.

Automatic Data Collection

Position the test car a short distance upstream of the begin point. Turn on and initialize the data recording equipment. Calibrate the distance measuring instrument (DMI) before arriving at the study site. Train the driver (and observer, if used) to operate the automatic equipment before starting a study. Safety is a primary consideration in conducting this type of study with a driver only. Make a dry run and enter into the data recorder the location of the begin, end, and control points for the route under study. Begin the test runs promptly at the start of the desired study period, and complete the required number of runs before conditions change. If the required sample cannot be obtained in a single period, complete the remaining runs on another day under the same conditions.

As the test vehicle passes the begin point, the driver activates the data recorder. The test car proceeds through the study route according to the driving technique selected. As each control point is passed, the driver activates the appropriate switch. When the test vehicle stops or is forced to travel slowly (the DMI will automatically sense this), the driver uses the appropriate switch to enter the cause of each delay. The DMI automatically records the location and duration of the delay. Upon reaching the end point, the driver activates the appropriate switch to note that fact. The test vehicle then returns to the begin point, the driver resets the data recorder, and the next run begins. This procedure is repeated until the required number of sample runs is reached or until conditions that could affect the study change.

Data Analysis and Summary of Results

To study travel-time data, analysts convert the time and distance measures to space–mean speeds. The average range in running speed is computed, as described above, to help determine the sample size required. The travel speed for each test vehicle run is calculated using the following equation:

$$S = 3600 \frac{D}{T} \tag{4-2}$$

where

S = travel speed, mph
D = length of study route or section, miles
T = travel time, seconds

The equation for space–mean speed or mean travel speed is

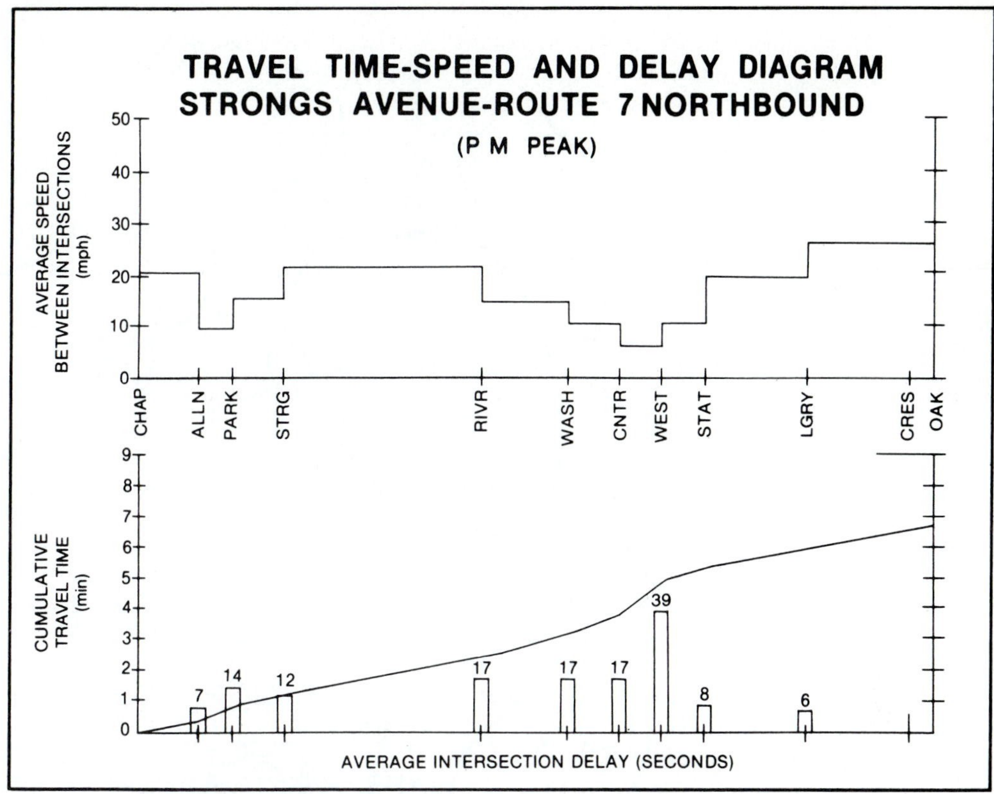

Figure 4-2 Travel time-speed, and delay summary along a route. *Source:* Box and Oppenlander, 1976, p. 104.

$$S = \frac{3600ND}{\Sigma T} \tag{4-3}$$

where

S = space–mean speed or mean travel speed, mph
D = length of study route or section, miles
ΣT = sum of travel times for all test runs, seconds
N = number of test runs

Analysts calculate running speed and mean running speed using equations (4-2) and (4-3), except that they use running speed in place of travel speed and running time in place of travel time. Mean speed may be calculated for each section of the study route in addition to the mean speed of the total route.

Since the study team measures delay directly, they can summarize the delays of each type (operational, fixed, stopped, and total) for each section and for each run. Mean delays are calculated by dividing the sums of delays by the number of runs. Each of the resulting speed and delay measures may be treated to statistical analyses, such as those described in Appendix C.

Graphical summaries are useful in displaying the various travel-speed and delay measures. Figure 4-2 is an example. For areawide studies, time contours overlaid on a map of the city's streets (as shown in Figure 4-3) offer a practical graphical display of travel times. Computer software that accompanies automatic data collection equipment offers a powerful range of analysis and data summary capabilities. Figures 4-4 and 4-5 present examples of these capabilities.

Courtesy: Baltimore Metropolitan Area Transportation Study, by Wilbur Smith and Associates

Figure 4-3 Area-wide time contours for a city. *Source:* Box and Oppenlander, 1976, p. 105.

MOVING VEHICLE METHOD

The moving vehicle method provides estimates of hourly volume, average travel-time, and space–mean speed as a vehicle makes round trips through the section being investigated. These estimates are obtained by measuring travel time, opposing traffic, overtaking traffic, and passed traffic. Analysts apply this method to any type of route, but most often on freeways or on uniform sections of arterial streets with one mile or more between signals. The method is applicable only on two-way routes where opposing traffic is visible at all times. The test vehicle must also be able to turn around at each end of the test sections (Garber and Hoel, 1988).

Date/Time of Report: 7/31/90; 10:03

Time/Space Trajectory for Sample Run

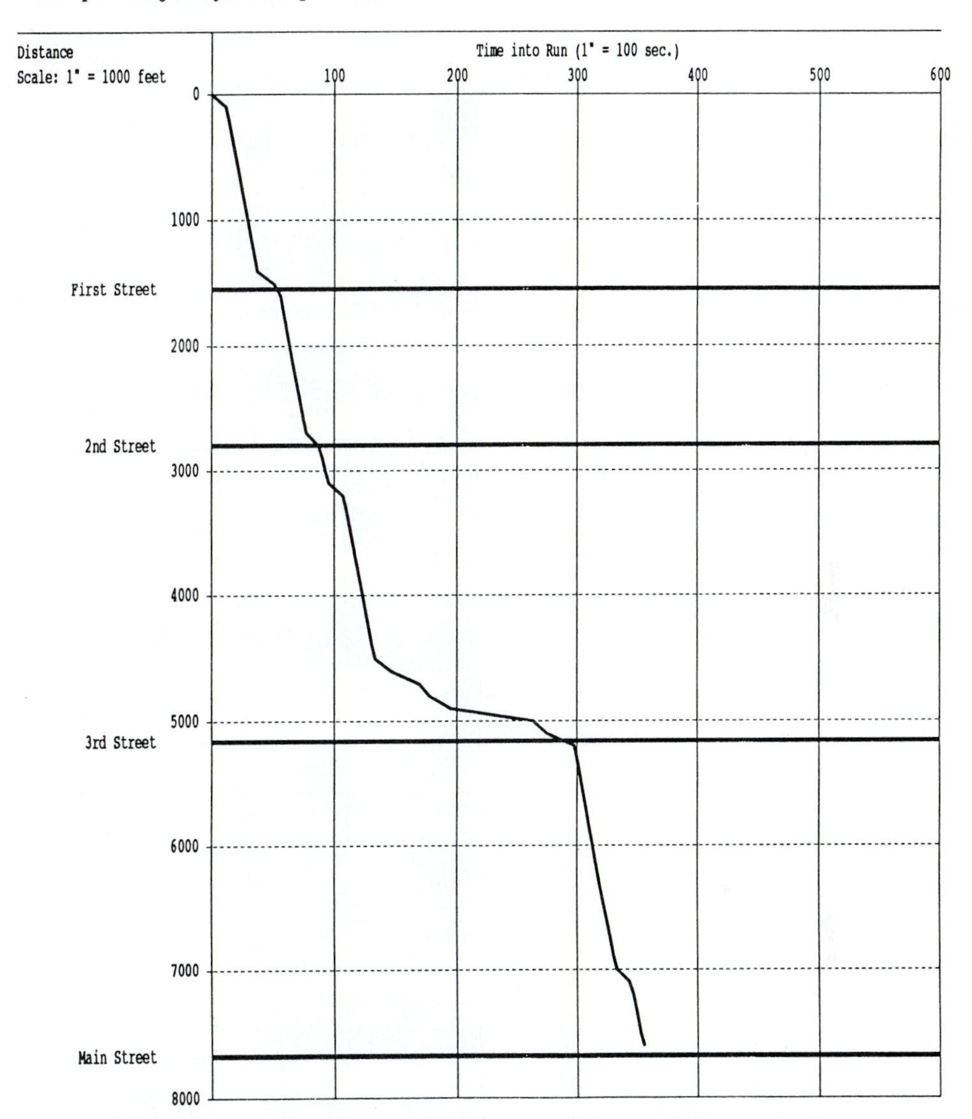

Figure 4-4 Time/space trajectories for sample run. *Source:* "PC-Travel Sample Reports," JAMAR Sales Company, Inc., Ivyland, PA, 1990.

Personnel and Equipment

The moving vehicle method requires a test car and the means to record time, distance, and traffic volume. On multilane facilities with heavy traffic, study teams may use videotape to record the runs. The team will then reduce the data in the office.

Manual Data Collection

A driver and observer/recorder, a stopwatch, hand-held counters, and data collection forms are required. The study team can scale the length of the test section from accurate plans or maps or can obtain the length (with less accuracy) from the vehicle odometer.

Date/Time of Report: 7/31/90; 10:03
Speed Profile for Run: 1. **Pulse Data File:** fryburg.pls
Run Title: Test Arterial With Multiple Delays

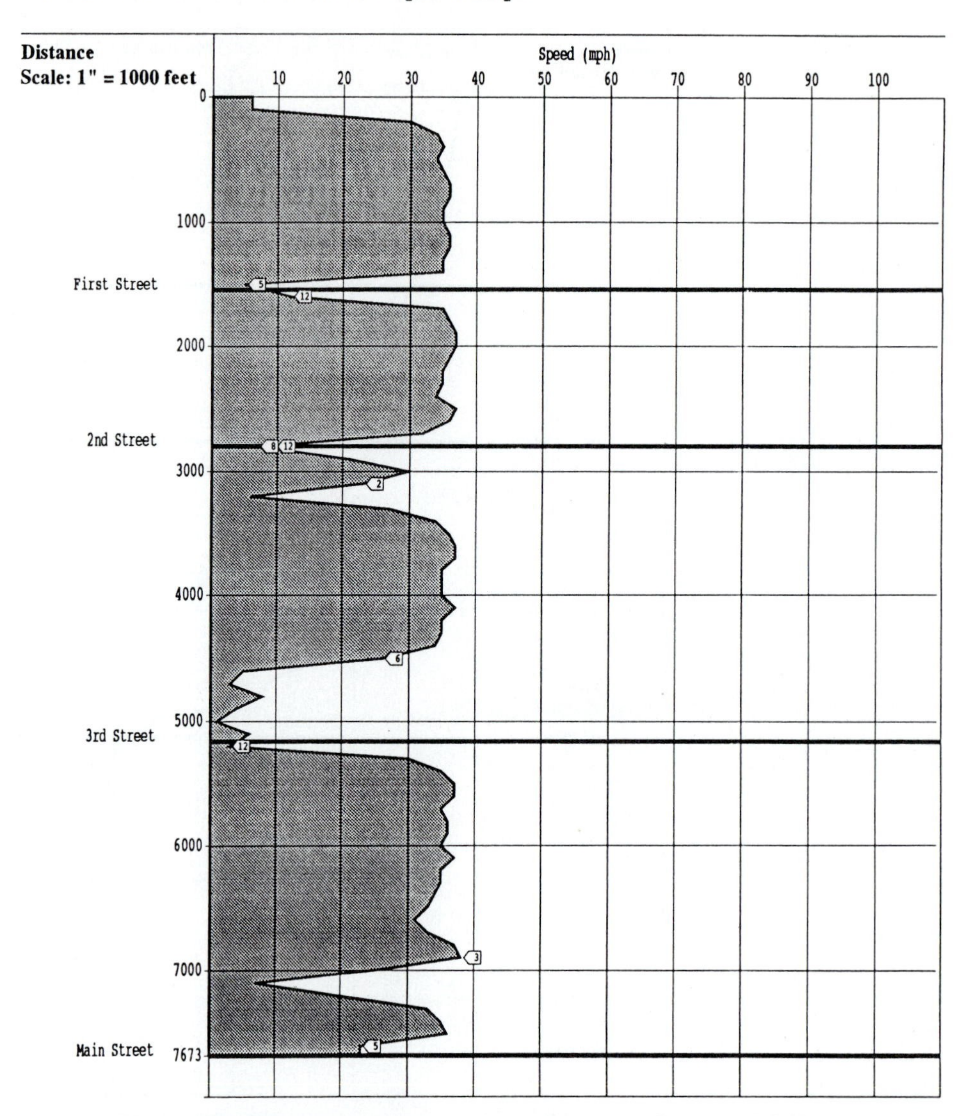

Figure 4-5 Sample test arterial with multiple runs. *Source:* "PC-Travel Reports," 1990.

Automatic Data Collection

Engineers have used specially designed data recording hardware in the past to perform travel-time and delay studies using the moving vehicle method (Robertson and Courage, 1973). At present, there is no readily available hardware or software specifically designed for the moving vehicle method. Hand-held or laptop computers could, however, be configured and software developed to serve this purpose. Because of safety concerns, automation would not eliminate the need for an observer/recorder.

Sample-Size Requirements

Sample-size refers to the number of test runs made in each direction along the test section. Analysts can obtain reliable, unbiased estimates of volumes and travel-times with a minimum of six test runs in each direction under comparable conditions (Mortimer, 1957).

TRAVEL-TIME AND DELAY STUDY
MOVING VEHICLE METHOD
FIELD SHEET

ROUTE_______________ DATE_______________
START POINT_______________ END POINT_______________
WEATHER_______________

RUN	START TIME	FINISH TIME	TRAVEL TIME	VEHICLES MET	VEHICLES O.TAKING	VEHICLES PASSED
__BOUND						
1						
2						
3						
4						
5						
6						
7						
8						
TOTAL						
AVERAGE						
__BOUND						
1						
2						
3						
4						
5						
6						
7						
8						
TOTAL						
AVERAGE						

COMMENTS _______________

RECORDER(S)_______________

Figure 4-6 Sample travel-time and delay study field sheet: moving vehicle method.

Field Procedure

Before test runs begin, the start-point and end-point locations are selected along the route to define the test section. On freeways, these locations are at interchanges that permit a fast turnaround. On arterial and other types of surface streets, these locations are usually at major intersections where U-turns are legal or where a reasonably quick turnaround is possible.

Test Car Technique

The driver of the test car proceeds along the study section with traffic at a comfortable and safe speed. To compensate for the inability of the test car to travel at exactly the average travel speed of the traffic stream, the number of vehicles passed is subtracted from the number of vehicles overtaken by the test car while traveling in the direction of interest.

Manual Data Collection

The test car begins a short distance upstream of the begin point. The driver zeros the stopwatch and the observer completes the header information on the data forms. Figure 4-6 shows a sample field sheet. The driver and observer review their duties. A couple of dry runs are made to measure the length of the section and to rehearse the procedure. If the rehearsal runs go smoothly, the analyst can count them as test runs. One round trip through the test section constitutes a test run. Begin test runs promptly at the beginning of the desired study period so as to complete the required sample of runs before conditions along the route change. If a single period does not provide the required sample, complete the remaining runs on another day under the same conditions.

As the test vehicle passes the begin point, the driver starts the stopwatch. The test car proceeds through the study section in the direction opposing the direction of traffic being studied. The observer counts and records the opposing traffic met during the trip. As the test car passes the end point of the study section, the driver reads the stopwatch and the observer notes the time of the trip on the field sheet. The test car turns around and travels the same section in the opposite direction. The driver starts the stopwatch to measure the time of the trip, and the observer counts the number of vehicles passed by the test car and the number of vehicles overtaken by the test car. The team repeats this procedure until they reach the required number of sample runs (round trips) or until conditions that could affect the study change.

Both directions may be studied simultaneously. In this situation, the driver counts the vehicles passed and overtaken, while the observer counts the vehicles traveling in the opposite direction. The team must ensure that the counts are properly recorded on the hand-held counters. If traffic is heavy in the opposing direction, it may be necessary to weight the count with a multiplier factor (e.g., one count equals five vehicles). On freeways with more than three lanes in each direction, the team may need a second observer. Safety is a primary consideration in conducting these studies. If traffic is heavy, use additional observers so that the driver can concentrate on the driving task.

Automatic Data Collection

The procedure is the same as for manual data collection, except that the observer uses a hand-held or laptop computer to record time and traffic count data. The procedure requires the same number of personnel. Automated data reduction saves time and money during the analysis.

Data Analysis and Summary of Results

Methodology

Consider the diagram in Figure 4-7. The subscripts n and s refer to the direction the test car is traveling when the item is measured. The test car starts at point A and travels south to B. The observer counts all the vehicles met from the opposite direction (M_s). The time of the southbound trip (T_s) is recorded, the driver turns around at B, proceeds north to A, and records the time of the northbound trip (T_n). This constitutes a round trip or one run. The observer counts any vehicle that is passed by the test car (P_n) and any vehicle that passes the test car (O_n) during the northbound trip.

Hourly Volume

During this round trip, the test car is essentially measuring the number of vehicles that will pass the starting point, A, in the time it takes the test car to make a round trip between A and B. The vehicles met (M_s) will pass point A before the test car can return

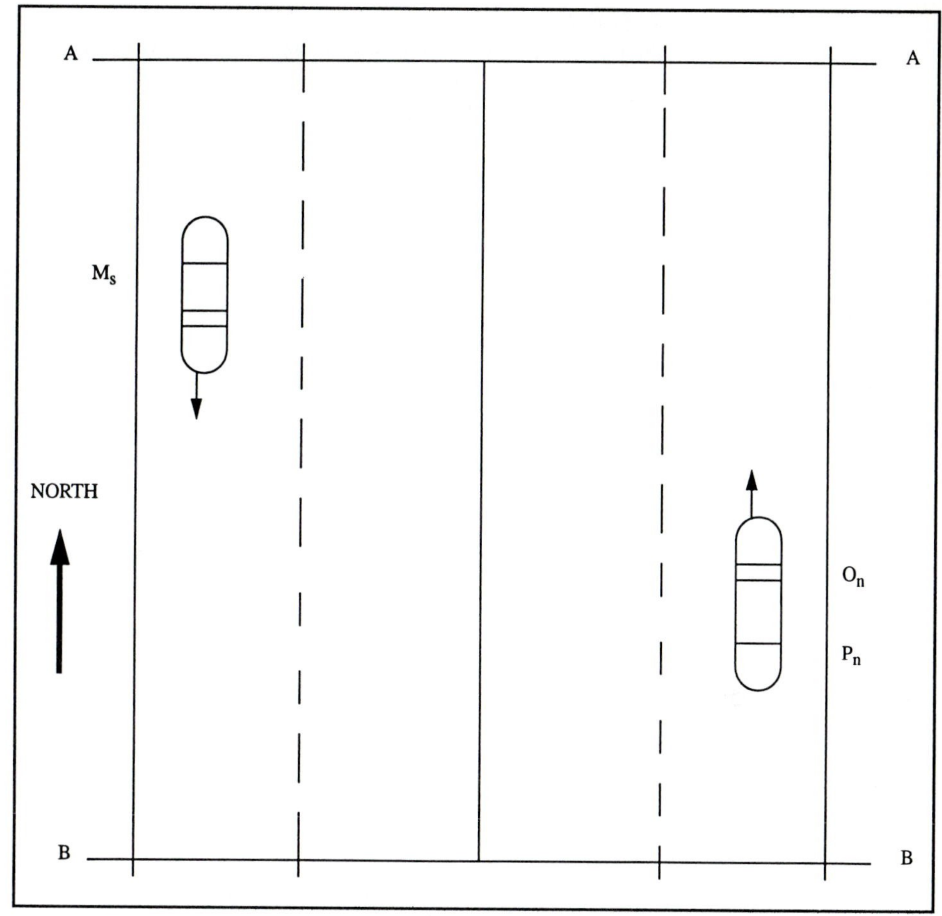

Figure 4-7 Moving vehicle method variables. *Source:* Adapted from Pignataro, L. J., *Traffic Engineering: Theory and Practice,* Prentice Hall, Englewood Cliffs, NJ, 1973.

from B to A. Those vehicles overtaking the test car (O_n) minus those passed by the test car (P_n) will compensate for the test car not traveling at exactly the average speed of the northbound traffic. Therefore, the volume past point A, in the northbound direction, in the time it takes the test car to make a round trip, is $M_s + O_n - P_n$. The formula for northbound volume, then, is

$$V_n = 60\left(\frac{M_s + O_n - P_n}{T_n + T_s}\right) \tag{4-4}$$

where

Vn = volume per hour, northbound
M_s = opposing count of vehicles met when the test car was traveling south
O_n = number of vehicles overtaking the test car as it traveled north
P_n = number of vehicles passed by the test car as it traveled north
T_n = travel time when traveling north, minutes
T_s = travel time when traveling south, minutes

The formula for southbound volume is the same except all subscripts are reversed. If the test car turned instantaneously at B, the volume count would be exact. However, some time is lost during turning which introduces a small error, so the analyst should

TABLE 4-2
Illustrative Example Data for the Moving Vehicle Method

Northbound Trips	T_n (min)	M_n	O_n	P_n
1N	2.65	85	1	0
2N	2.70	83	3	2
3N	2.35	77	0	2
4N	3.00	85	2	0
5N	2.42	90	1	1
6N	2.54	84	2	1
Total	15.66	504	9	6
Average	2.61	84.0	1.5	1.0
Southbound Trips	T_s (min)	M_s	O_s	P_s
1S	2.33	112	2	0
2S	2.30	113	0	2
3S	2.71	119	0	0
4S	2.16	120	1	1
5S	2.54	105	0	2
6S	2.48	100	0	1
Total	14.52	669	3	6
Average	2.42	111.5	0.5	1.0

Source: Adapted from Pignataro, 1973, Table 7-4.

consider the result of Equation (4-4) an estimate. Since a single run may not be representative of average conditions, several runs are made and the analyst uses an average (Pignataro, 1973).

Average Travel Time

The average travel time for one-directional flow is calculated using the following formula:

$$T_n = T_n - \frac{60(O_n - P_n)}{V_n} \tag{4-5}$$

where T_n is the average travel time of all traffic northbound, and the other variables are as defined for equation (4-4). As with equation (4-4), the southbound volume is the same except all subscripts are reversed. The quantity $(O_n - P_n)$ represents a correction factor to account for the fact that the test car may not have traveled at the average speed.

Space–Mean Speed

Analysts can calculate the space–mean speed for one directional flow using:

$$S_n = \frac{60\ d}{T_n} \tag{4-6}$$

where S_n is the space–mean speed northbound, in mph, and d is the length of test section, in miles.

EXAMPLE 4-1

Table 4-2 shows data for a 0.75-mile length of test section. From Equation (4-4), the hourly volumes in the northbound and southbound directions are, respectively,

$$V_n = 60\left(\frac{111.5 + 1.5 - 1.0}{2.61 + 2.42}\right)$$
$$= 1336 \text{ vehicles per hour}$$

$$V_s = 60\left(\frac{84.0 + 0.5 - 1.0}{2.42 + 2.61}\right)$$
$$= 996 \text{ vehicles per hour}$$

From equation (4-5), the average travel times for the northbound and southbound directions are, respectively,

$$T_n = 2.61 - \frac{60(1.5 - 1.0)}{1336}$$
$$= 2.59 \text{ minutes}$$
$$T_s = 2.42 - \frac{60(0.5 - 1.0)}{996}$$
$$= 2.45 \text{ minutes}$$

From equation (4-6), the space–mean speeds for the northbound and southbound directions are, respectively,

$$S_n = 60\left(\frac{0.75}{2.59}\right)$$
$$= 17.4 \text{ mph}$$
$$S_s = 60\left(\frac{0.75}{2.45}\right)$$
$$= 18.4 \text{ mph}$$

LICENSE PLATE METHOD

This method produces travel time only, from which average travel speed may be calculated once the distance between observation points is measured. A test vehicle is not required. Observers position themselves at the entrance and exit to the test section and at other major intersections along the route. As vehicles pass the observers at each location, the observers record the last three or four digits of the license tag along with the time from a stopwatch. A tape recorder is useful to avoid missing vehicles. The tag numbers and times are then matched in the office, either manually or by computer, to obtain travel times. The matching is laborious if done manually. A sample size of 50 license number matches usually provides adequate accuracy (Box and Oppenlander, 1976). Computer software for matching license plate numbers is available (Center for Microcomputers in Transportation, 1992).

Toll-road cards provide a variation of the license plate method to measure travel times over a facility between toll booths. Time of arrival at each booth must be imprinted on the card. Knowing the distance between booths will allow the calculation of overall travel speeds. With the deployment of electronic toll collection systems, the measurement of vehicle travel times will be fully automated (McShane and Roess, 1990).

DIRECT OBSERVATION METHOD

Observers at an elevated vantage point can measure travel time directly between two points a known distance apart. The method requires good visibility and is not suitable for sections greater than 1/2 mile in length.

INTERVIEW METHOD

Selected individuals who are willing to cooperate may provide a satisfactory sample from which to obtain travel times and delays without the use of a test vehicle or observers. These persons are asked to record their start and end times for designated routes. They also record the times and durations of delay. This is a variation on the average car method, except that in place of a single test vehicle, there are multiple test vehicles. Employees who drive on the job, truck drivers, and taxi drivers often make good subjects. This method is useful when a large amount of data is needed in a short time (Pignataro, 1973). The interview method requires some training and equipment (e.g., stopwatches) for the subjects. The reliability of the results may not equal that of methods that employ better-trained data collectors. Refer to Appendix B for additional information.

REFERENCES

Box, P. C., and J. C. Oppenlander (1976). *Manual of Traffic Engineering Studies,* 4th ed., Institute of Transportation Engineers, Washington, DC, pp. 93–105.

Center for Microcomputers in Transportation (1992). *McTrans Catalog*, University of Florida Transportation Research Center, Gainsville, FL, pp. 22,41.

Garber, N.J., and L. A. Hoel (1988). *Traffic and Highway Engineering,* West Publishing, St. Paul, MN, pp. 118–125.

HMR Communications (1990). *Highway Instrumentation Products Catalog,* HMR, West Newton, PA, pp. 7–11.

McShane, W. R., and R. P. Roess (1990). *Traffic Engineering,* Prentice Hall, Englewood Cliffs, NJ, p.140.

Mortimer, (1957). *Moving Vehicle Method of Estimating Traffic Volumes and Speeds,* Highway Research Board Bulletin 156, Highway Research Board, Washington, DC.

Pignataro, L. J. (1973). *Traffic Engineering Theory and Practice,* Prentice Hall, Englewood Cliffs, NJ, pp. 106–110.

Robertson, H. D., and K. G. Courage (1973). *Traffic Signal Studies Using a Digital Tape Recorder,* Traffic Research Report 615-6, Civil and Coastal Engineering Department, University of Florida, Gainesville, FL, June.

RWD Engineering (1991). *PC-Travel,* JAMAR Sales Company, Inc., Ivyland, PA.

University of Florida Transportation Research Center (1990). *A System for Evaluating Urban Traffic Congestion with Moving Vehicle Studies, User's Guide,* University of Florida, Gainesville, FL, February.

5

Intersection and Driveway Studies

Joseph E. Hummer, Ph.D., P.E.

INTRODUCTION

Intersection and driveway studies are among the most common studies in transportation engineering. In particular, many agencies routinely count turning movements and study intersection delay. Other intersection and driveway studies include queue length, saturation flow and lost time, gap and gap acceptance, and intersection sight distance studies. Engineers perform these studies less often but find them useful for special locations or when calibrating basic relationships. Engineers use the results of intersection and driveway studies to determine whether signals are warranted and to determine intersection capacity, traffic signal timing, site development impacts, safe speeds, and other important parameters.

The basic data needs from intersection and driveway studies have changed very little through the years. However, the equipment used to conduct the studies has changed greatly in recent years and is likely to keep changing in the coming years. In this chapter the basic procedure for each study is described together with the equipment that is most commonly used by agencies in North America for delay, queue length, saturation flow and lost time, gap and gap acceptance, and intersection sight distance studies. The procedures for intersection turning movement counts were detailed in Chapter 2. Following the presentation of a study's basic procedure, other equipment and methods are described.

INTERSECTION DELAY

Intersection delay studies are very common. Intersection delay data have many uses, including measurement of the quality of traffic flow and evaluation of the need for traffic signals. Engineers can estimate intersection delay with equations or simulation models. However, the inputs to the equations and models can be extensive, and the results can be erroneous, so field studies are often used at operating intersections. One of the major problems with intersection delay studies is the definition of delay. There are many types of delay, and using terms casually can lead to error. The following are among the most useful terms describing delay at intersections (Buehler et al., 1976).

- *Travel-time delay* is the difference between the time a vehicle passes a point downstream of the intersection where it has regained normal speed and the time it would have passed that point had it been able to continue through the intersection at its approach speed.
- *Stopped-time delay* (or simply *stopped delay*) is the time a vehicle is standing still while waiting in line in the approach to an intersection.
- *Time-in-queue delay* is the difference between the time a vehicle joins the rear of a queue and the time the vehicle clears the intersection.
- *Total delay* is the difference between a vehicle's traversal time through an intersection and the time that vehicle would have required if allowed to travel at its desired speed.

Engineers use stopped delay most often because it is the easiest to measure and because the *Highway Capacity Manual* (TRB, 1985) bases its definition of signalized intersection level of service on stopped delay, among other reasons. For a given traffic stream the types of delay defined above are related, but not strongly. For example, stopped delay multiplied by 1.3 provides a rough approximation of total delay for typical midday uncongested conditions at signalized intersections. However, simply lengthening the red signal period alters that factor substantially (Teply, 1989).

Manual Procedure for Measuring Stopped Delay

The most common method used to measure stopped delay in the field requires observers to record a count at a certain time interval of the number of vehicles stopped on the intersection approach of interest. The analyst assumes that each vehicle counted was stopped for an entire interval. The total count of stopped vehicles during all intervals multiplied by the length of the time interval provides the stopped delay estimate. Dividing this delay estimate by the number of vehicles departing the approach provides an estimate of stopped delay per vehicle that is useful for highway capacity calculations and other purposes. Estimates of stopped delay using this manual procedure are acceptable for many applications. However, when delays are short or volumes are low, analysts should use these results with extreme caution (Teply, 1989).

Figure 5-1 shows, with sample data, a form for recording the number of stopped vehicles at the end of each interval. Appendix G provides a blank form suitable for copying. The form is useful when the time increment is evenly divisible into 60 seconds. A form with two columns—one for time and one for number of stopped vehicles—is convenient for time increments that are not evenly divisible into 60 seconds. Observers should complete the rows and/or columns for "time" before data collection so that they know when to record a count. The form in Figure 5-1 provides the simple equations for estimating stopped delay per vehicle from raw data.

The personnel needed depend on the equipment used and the volume of traffic at the intersection being studied. A single observer can accurately record the number of

INTERSECTION STOPPED - DELAY FIELD SHEET

Intersection GRAND RIVER AT ABBOTT

Time 4:30pm To 4:45pm

Date 10/8/91 Day Tuesday

Weather Clear, 70°

Observer(s) CB

Street Grand River

East Bound Traffic Lane Center, Through

Number of Stopped Vehicles; V_s

Min \ Sec	0	15	30	45				
0	1	3	18	9				
1	0	6	7	8				
2	9	1	8	8				
3	9	6	0	8				
4	7	6	13	1				
5	6	6	1	4				
6	9	0	4	2				
7	0	0	3	4				
8	5	4	0	1				
9	4	9	16	8				
10	3	8	7	1				
11	0	1	5	7				
12	6	0	4	9				
13	7	7	5	3				
14	0	6	9	1				
Totals	66	63	100	74				

$\Sigma V_s =$ 303

Volume, V = 149

Stopped Delay $= \frac{\Sigma V_s \times I}{V} =$ 30.5 sec/veh

Figure 5-1 Form for recording the number of stopped vehicles and estimating stopped delay. *Source:* Transportation Research Board, "Highway Capacity Manual," *Special Report 209,* National Research Council, Washington, DC, 1985, p. 9–83

stopped vehicles on two lanes with moderate length queues (up to 25 vehicles per lane) or on one lane with long queues (Reilly et al., 1976). However, a single observer working with moderate or long queues needs an audio signal of the end of an interval that is discernible above the roar of the traffic. If a watch that provides a loud, perodical audio signal is not available, observers can use a tape recording with a signal after every time increment. Observers must check the tape player before a study and several times during a study to ensure that low battery levels are not slowing the tape. Where long queues exist or if no audio signal is possible, the study needs two observers. One observer can then watch the time, signal when an interval has ended, and record the count of stopped vehicles while the second observer concentrates exclusively on counting. Analysts need a turning movement count, or a classification of vehicles entering the intersection as having stopped or having not stopped, to estimate stopped delay per vehicle. Therefore, additional observers to conduct those counts may be necessary.

Stopped vehicle observers should position themselves near the right shoulder or on the right sidewalk at the approximate midpoint of the maximum queue. Observers must have a clear view of the lanes they are observing and should move around as necessary to avoid being blocked by buses or trucks. Observers can sit in automobiles in locations where blocked vision is not a problem. Park the automobile in a legal space well out of the travel lanes.

Observers should count vehicles with locked wheels as being stopped. Observers should also count vehicles creeping forward as being stopped if they are moving at less than a slow walking speed, which is about 3 mph (Buehler et al., 1976). Reilly et al. (1976) offered a more precise definition of a stopped vehicle. Observers count a creeping vehicle as *stopped* if:

1. It had been stopped.
2. It is, at the time of observation, creeping forward in a queue that is not discharging.
3. The gap between it and the vehicle in front of it is less than three car lengths (50 feet).

Observers would not consider a creeping vehicle as stopped if it violates one of these three conditions. The situation when the signal has just turned green and the queue has just begun to discharge is the most difficult to record accurately. The observer should take a quick glance and note which vehicles are at the ends of the queue before counting the vehicles between the ends. A recording interval between 10 and 20 seconds is typical. This allows observers to make accurate counts while minimizing error in the stopped delay estimate. Past manuals recommended the choice of an interval such as 13 seconds that was not evenly divisible into the signal cycle length. However, current practice is to use any convenient interval because the error introduced by an even increment is negligible.

Observers should record at least 60 intervals. Estimates of stopped delay during peak periods are most useful. Stopped delay estimates will vary widely within short times, especially when peak periods begin and end, so engineers should interpolate between time periods with extreme caution. Light rain should not affect traffic volumes or queuing behavior on commuter routes during peak periods, so observers can study stopped delay if they can keep their forms dry. Do not conduct stopped delay studies in weather that affects normal volumes or driving behavior. Observers need to note on each data form all the usual "header" information, including locations, times, and weather conditions.

Analysts need a measure of traffic volume to calculate stopped delay per vehicle. Sometimes it is convenient to count the vehicles proceeding through the intersection without stopping and the vehicles that stop. The sum of these quantities provides the denominator for the "stopped delay per vehicle" statistic, while the percentage of vehicles stopping is itself a useful indicator of intersection performance. A single observer located near the intersection can usually count vehicles stopping and not stopping on an approach with moderate volume using a mechanical or electronic counting board. Observers should count vehicles that stop as they stop, and count vehicles that do not stop as they pass through the intersection. Observers of stopped delay and counters of stopping and nonstopping vehicles must communicate (usually with hand signals) to assure that observations start and end simultaneously.

Stopped Delay Studies with Other Equipment

Observers can collect stopped delay data with electronic counting boards such as the board shown in Chapter 2. Agencies can purchase boards with templates that show which keys to press. Observers can select items from the board's menu to ensure that keystrokes are correctly interpreted as stopped delay data. The advantages of using electronic counters are that they produce an audible (although sometimes not loud enough) tone at the time an interval expires and that they reduce data automatically. If the traffic volumes are low, an observer may be able to use a single board to record stopped delay data and stopping and nonstopping traffic volumes.

Several studies have used videotape with an on-screen clock to measure stopped delay. Videotape provides a permanent record of the study period that may be used for further review of the stopped delay data or for other studies. Videotape may also reduce the number of field personnel needed for a stopped delay study. However, video recordings frequently suffer from poor lighting conditions and vantage points. Long queues are especially difficult to capture. In addition, producing an estimate of stopped delay from videotape requires much labor in the office.

Other Delay Studies

Buehler et al. (1976) argued that time-in-queue delay (TIQD) is more closely related to travel-time delay than stopped delay and is fairly easy to measure. They recommended a procedure for determining TIQD in which an observer records the number of queued vehicles after each 10- to 20-second interval. The observer notes the number of vehicles in the queue regardless of whether the first few vehicles have started into the intersection. TIQD is then the total number of vehicles recorded over all intervals multiplied by the time interval length. Figure 5-2 shows how one would estimate TIQD and stopped delay using a 10-second interval for a single signal cycle and illustrates why TIQD is greater than stopped delay. Buehler et al. note that observers should not just concentrate on the back of the queue. Observers must account for vehicles that leave or join the middle of the queue, and not all queued vehicles will clear the intersection during a single green phase.

Engineers can measure travel-time delay using an electronic counting board or laptop computer and some special computer programs, as described by Teply and Evans (1989). They developed a method in which observers strike keys as vehicles pass a point well upstream of the intersection, as vehicles pass the stop bar, and as the green interval begins and ends. The computer algorithm reconstructs the time–space trajectories of individual vehicles. Teply and Evans describe the method as practical and effective. They note that several agencies have used the method to study the effectiveness of signal progression when stopped delay and TIQD would have been poor measures.

Total delay is generally more difficult to estimate from field measurements than is stopped delay or TIQD. The test car or license plate methods described in Chapter 4 are appropriate for total delay at an intersection if the only significant delay-producing element between the checkpoints is the intersection. Test car or license plate methods require estimates of mean desired speeds on the highways of interest.

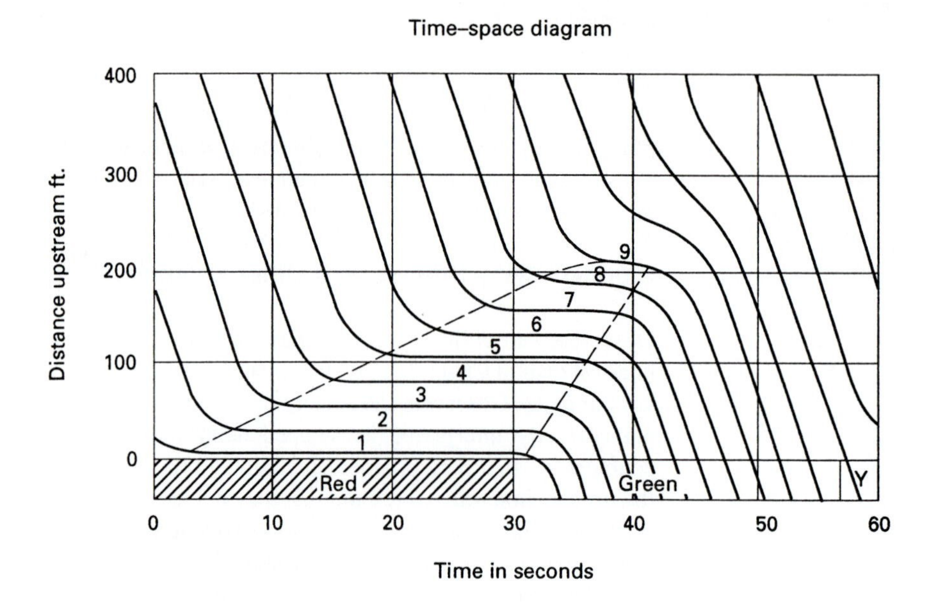

Sampling time ⇒ (sec)	10	20	30	40	50	60	Delay estimate, sec.
Stopped Delay	2	5	7	2	0	0	160
TIQD	2	5	7	9	0	0	230

Figure 5-2 Estimating stopped delay and TIQD. *Source:* Buehler, M. G., T. J. Hicks, and D. S. Berry, "Measuring Delay by Sampling Queue Backup," *Transportation Research Record 615,* Transportation Research Board, National Research Council, Washington, DC, 1976.

QUEUE LENGTH

Queue-length studies have several important applications. Queue-length data can help determine the length of storage lane needed or can provide a useful measure of traffic signal efficiency. Queue-length studies for these purposes are conducted by a method that is very similar to the manual method measuring stopped delay discussed above. Observers count the number of vehicles in a standing or slowly moving queue at designated time intervals. Observers can make notations in the field, or can count from photographs or videotapes. At signalized intersections observers record counts at the start of the green interval and the end of the yellow interval. Counts at unsignalized intersections are usually made at equal intervals of 30 seconds or 1 minute (Box, 1982).

An engineer can investigate the feasibility of a proposed driveway location on an intersection approach using a slightly different queue length study method (Box, 1982). In this case, observers record the amount of time that the queue blocks the proposed driveway location. Observers can use several stopwatches for multiple locations. One would typically conduct this type of queue-length study during the peak hour of the driveway and/or the intersection approach. Dividing the blockage time for a proposed driveway location by the total study time produces the percentage of time the location is blocked, which is a very useful measure.

SATURATION FLOW AND LOST TIME

Saturation flow and lost time are two of the basic building blocks of traffic engineering. Engineers use these measures to time signals and estimate intersection capacity. *Saturation flow* is the number of vehicles that can pass a given point on a highway in a given period of time. In studies of intersections, engineers focus on the flow past the stop bar in a lane in an hour of uninterrupted green signal (also termed the ''ideal'' saturation flow). *Lost time* is the unused portion of the signal cycle. There are two significant components to lost time for each signal phase:

1. *Startup lost time* occurs between the time the green signal begins and the queue begins moving efficiently.
2. *Clearance lost time* occurs between the time the last vehicle crosses the stop bar and the next signal phase begins.

Many agencies use standard constant values for saturation flow and lost time in analyses. However, saturation flow and lost time vary significantly between intersections and between different times of day. To avoid errors caused by inappropriate use of a standard value, some agencies measure saturation flow and lost time directly before performing other analyses. More often, agencies sample saturation flow and lost time periodically at several sites and calibrate their equations based on those samples. The procedures for measuring saturation flow and lost time are described in this section. The procedures are relatively simple and one can use a variety of equipment to perform them. Because the need for good estimates of saturation flow and lost time has been emphasized recently, more agencies are likely to perform these studies in the future.

Studying Saturation Flow with a Stopwatch

An observer generally collects saturation flow data at intersections with a stopwatch. Figure 5-3 shows a form that is useful for collecting saturation flow observations with a stopwatch. Appendix G contains a blank form suitable for copying. The observer starts the watch when the rear axle of the fourth vehicle in a queue that had been stationary while waiting for the green signal crosses the stop bar. The observer stops the watch when the rear axle of the seventh, eighth, ninth, or tenth vehicle in the queue (whichever was the last vehicle in the stopped queue at the instant the signal turned green) crosses the stop bar. For example, suppose that the stopped queue is eight vehicles long at the instant the signal turns green. The observer would start the watch for vehicle 4, stop the watch for vehicle 8, and enter the elapsed time in the ''eighth vehicle'' column of the form in Figure 5-3. The observer cannot record a measurement if the queue is less than seven vehicles long when the signal turns green because short queues provide unstable data. If the queue is more than 10 vehicles long, the observer stops the watch at the tenth vehicle. Ten vehicles is a convenient maximum that decreases the chances of error due to the effects of spillback or due to vehicles stopping for the red signal. Observers must ignore vehicles joining the queue after the green signal appears. One observer records saturation flow data for one lane at a time. Saturation flow rates estimated for a lane usually apply to adjacent lanes of the same type on the same approach. One observer with a clear view of adjacent approaches can alternately record data from a lane on each if the approaches use different parts of the signal cycle.

The factors that affect saturation flow rates are grade, lane width, intersection location (central business district versus other), type of lane, and presence of adjacent parking lanes (TRB, 1985). Therefore, the engineer must carefully select approaches to measure saturation flow to ensure an unbiased result. Do not use a saturation flow estimate from a steep approach to analyze a flat approach, for instance. Heavy vehicles also affect saturation flow rates, so observers should not record data if a heavy vehicle is in

FIELD SHEET - SATURATION FLOW STUDY

Location: Sagamore Parkway at Yeager Road
West Bound Traffic Lane: Right Through
Date: 4/2/92 Thurs Time: 4:00pm-6:00pm City: West Lafayette
Observers: MH Weather: Cloudy 50°
Grade: 1 % Lane Width: 12 ft Area: Suburban Other:

Obs No.	Time (seconds) between 4th vehicle and ...			
	7th veh.	8th veh.	9th veh.	10th veh.
1		8.2		
2	4.8			
3				11.4
4				11.2
5			10.1	
6	6.3			
7		6.3		
8			9.1	
9	5.3			
10				10.6
11		7.9		
12	6.1			
13				10.5
14		7.4		
15				11.3
16				11.4
17		7.0		
18	6.0			
19			9.2	
20	5.8			

Obs No.	Time (seconds) between 4th vehicle and ...			
	7th veh.	8th veh.	9th veh.	10th veh.
21				11.5
22				11.1
23	5.2			
24		8.3		
25	5.4			
26			9.6	
27			9.0	
28				10.6
29	5.8			
30				10.2
31	12.3			
32				
33				
34				
35				
36				
37				
38				
39				
40				
Column Sums:	57.0	45.1	47.0	109.8
	(a)	(b)	(c)	(d)

$$\text{Mean Saturation Flow (vph)} = \frac{3600 * \text{Total Number of Observations}}{\frac{(a)}{3} + \frac{(b)}{4} + \frac{(c)}{5} + \frac{(d)}{6}}$$

Figure 5-3 Saturation flow data collection form.

TABLE 5-1
Values of Z for Equation (5-1)

Percent Confidence	*Z*
90	1.64
95	1.96
99	2.58
99.5	2.81

one of the first seven positions in the queue. If a heavy vehicle is in position 8, the observer can record the time between the fourth and seventh vehicles, and so on. Also, do not record data during a signal phase in which traffic flow is interrupted by buses, by left-turning traffic waiting for opposing traffic to clear, or by right-turning traffic waiting for pedestrians to clear. Engineers can calculate interrupted saturation flow from ideal saturation flow by the methods of the *Highway Capacity Manual* (TRB, 1985). The procedure for studying saturation flow in an exclusive left-turn or right-turn lane with a protected signal phase is the same as the basic procedure for a through lane.

For agencies that have difficulty finding sites unaffected by the factors mentioned above, McShane and Roess (1990, p. 494) suggest a procedure for estimating ideal saturation flow from measurements at nonideal sites. The engineer can solve the saturation flow equation in the *Highway Capacity Manual* for the ideal saturation flow given the measured nonideal saturation flow and the standard adjustment factors for the nonideal conditions. Time of day, weather, and events that affect driver populations or behavior also affect saturation flow. Measure ideal saturation flows during peak hours, in dry weather, and during times when no special events are affecting drivers. It may be difficult to collect saturation flow data during nonpeak hours in any case, due to small queues.

Observers need to stand near the intersection where they can clearly see the lane(s) being observed, the stop bar, and the signal indication. However, observers also need to see 200 feet or so upstream from the stop bar (to the tenth vehicle in the queue) to check which vehicles are stopped when the signal turns green and to note the position of any heavy vehicles. Observers usually have to walk around on the shoulder or the sidewalk to avoid having their vision blocked by trucks and buses and to get the best possible view of each queue. The comfort and inconspicuousness of the observers usually have to be compromised during a saturation flow study.

One can calculate desirable sample sizes for a saturation flow study from a standard sample size equation. Usually, engineers have some knowledge of the precision of the saturation flow estimate they desire. For instance, an engineer may not want the mean estimated saturation flow rate to differ from the true saturation flow rate by more than d vehicles per hour. She can find the necessary sample size n by

$$n = \left(Z \frac{s}{d}\right)^2 \tag{5-1}$$

where Z is a constant from the standard normal distribution corresponding to a certain confidence level (see Table 5-1) and s is an estimate of the standard deviation of the population of saturation flow rates. A typical value for s is 140 vehicles per hour (ITE Technical Committee 5P-5, 1991). If the engineer is willing to use this typical standard deviation and wants an estimated mean saturation flow rate within 50 vehicles per hour of the true rate with 95% confidence, she would have to observe $n = [1.96(140/50)]^2 = 30$ valid queues. A peak period at a moderately busy intersection usually produces at least 30 valid queues.

Once the data have been collected on the form in Figure 5-3, one can calculate the mean saturation flow rate using the equation on the bottom of the form. Basically, a mean saturation flow rate is estimated by calculating an average number of seconds consumed per vehicle (i.e., headway) and converting that into a number of vehicles per hour. For the sample data shown in Figure 5-3, mean saturation flow in vehicles per hour, SF, is estimated from the equation on the bottom of the form as follows:

$$\begin{aligned} SF &= \frac{3600n}{a/3 + b/4 + c/5 + d/6} \\ &= \frac{3600 \times 31}{57.0/3 + 45.1/4 + 47.0/5 + 109.8/6} \\ &= 1925 \end{aligned} \qquad (5\text{-}2)$$

In equation (5-2), n is the total number of observations, and a, b, c, and d are the sums of the data in the seventh, eighth, ninth, and tenth vehicle columns, respectively, in seconds.

Saturation Flow Studies with Other Equipment

Observers can sometimes collect saturation flow data with equipment other than a stopwatch, such as a laptop computer, audiotape, and videotape. These other technologies have several advantages, including greater accuracy, instant creation of a computer file (for the laptop), and creation of a permanent record that is available for other studies (for audio- and videotape). For nonresearch studies engineers can rarely justify the extra time and expense of these other methods, however. The laptop and audiotape methods require special computer programs to be written or acquired. The laptop method requires observers to press a certain key when the fourth vehicle crosses the stop bar and other keys when the seventh, eighth, ninth, or tenth vehicle crosses. The program records the times those keys are pressed and performs the calculations. The audiotape method requires an observer at the intersection to speak into a tape recorder when the vehicles of interest cross the stop bar (Shanteau, 1988). Back at the office, a data collector plays the tape at the same speed while pressing the appropriate microcomputer key for each audio cue. A program similar to that on the laptop is needed to record the times and perform calculations. The videotape method requires a clear vantage point and good light conditions. In the office, a technician must stop the videotape and record the time on the on-screen clock as the vehicles of interest cross the stop bar.

Lost-Time Studies

Lost time is more difficult to study than saturation flow for several reasons. First, lost times are short, so accurate measurements require quick reflexes. Second, observers can measure clearance lost time only during completely saturated green phases. Finally, many of the variables that affect saturation flow affect lost time, plus others, including signal head position and lens size. The analyst must be careful when applying a lost-time estimate from one lane to other lanes, approaches, or intersections. Observers record lost-time data with stopwatches, with a laptop computer, with audiotape and a microcomputer back at the office, or with videotape that has an on-screen clock. Since engineers need an estimate of saturation flow to compute startup lost time (as described below), they often gather data for the two studies simultaneously.

The major uncertainty in lost-time studies is where to establish the reference point for timing. In other words, "Where is the vehicle considered to be in the intersection?" Previous studies have used (1) the front or rear tires as they crossed the position that had been occupied by the front tires of the first vehicle in the queue, (2) the stop bar, (3) the crosswalk line, (4) the extension of the curb line of the intersecting street, or (5) other points. Berry (1976) showed that these different reference points dramatically affected the startup lost-time estimate (almost 3 seconds different in some cases). For capacity analysis, Berry recommended recording the time when the front bumpers of the vehicles crossed the extension of the nearside curb line of the intersecting street.

Observers measure clearance lost times directly at the end of a saturated green phase. They record the time when the last vehicle through during the phase crosses the reference point and the time when the signal turns green for the next phase. The difference between these times is the clearance lost time. Observers need to find a location where they can observe both the reference point for timing and the signal indication for the next phase.

The data needed to compute startup lost time are the times when the green signal begins and when the third vehicle in a standing queue passes the reference point. Most studies use the third vehicle because that is the last vehicle that commonly experiences any measurable lost time. Engineers can compute startup lost time for a phase by computing the difference between the two recorded times and subtracting three times the average headway for the lane (in seconds per vehicle) found during the saturation flow study. The result from this calculation can be below zero for a particular phase. In that case, engineers assume a value of zero when calculating statistical parameters based on the results.

GAPS AND GAP ACCEPTANCE

Studies of gaps and gap acceptance are useful in evaluating the capacity and level of service of unsignalized intersections, driveway openings, and unprotected left turns, among other applications. Gap studies only measure characteristics of the major traffic stream, and require a knowledge of the behavior of the minor traffic stream to be useful. For example, if an engineer knows that minor-street drivers need a gap of 10 seconds or more to cross a major street, he can collect the number of gaps of 10 seconds or more on the major street and make useful predictions. Gap acceptance studies measure both the distribution of gaps on the main street and the behavior of minor-street drivers (i.e., whether they move into the major street when presented with various gaps). Like saturation flow studies, engineers study gap acceptance at intersections where general capacity models and assumptions are inappropriate or to calibrate general capacity models for local conditions.

Important definitions for gap and gap acceptance studies include:

- *Gap:* time elapsed between the rear bumper of one vehicle and the front bumper of the following vehicle passing a given point.
- *Headway:* time elapsed between the front bumper of one vehicle and the front bumper of the following vehicle passing a given point. Headway is thus always larger than a gap.
- *Lag:* time elapsed between the arrival of a minor-street vehicle ready to move into the major street and the arrival of the front bumper of the next vehicle in the major traffic stream. The distinction between a gap and a lag is critical, because drivers may react differently to them.
- *Accepted gap or accepted lag:* gap or lag that a minor-street vehicle uses to move into the major street.
- *Rejected gap or rejected lag:* gap or lag that a minor-street vehicle at the stop bar does not accept.

Finally, the critical gap is the minimum size gap that a particular driver will accept. Most studies using gaps to examine capacity assume that each driver has a constant critical gap. These studies attempt to determine the distribution of critical gaps in the driver population.

Gap Studies

Observers usually collect gap data using electronic counting boards. When a vehicle in the major traffic stream crosses a reference point at the intersection of interest, the observer presses a key and the board records the time elapsed since the last time the key was pressed. Many electronic counting boards record gap data by grouping the gaps into "bins" with intervals of 2 seconds. The results will then consist of the number of gaps between zero and 2 seconds, 2 and 4 seconds, and so on. Two-second intervals are coarse but acceptable for most gap studies, but larger intervals are generally not useful. With no other data to collect simultaneously, one observer should have no problem collecting gap data for a multilane major street.

Observers can collect gap data during weather that does not affect normal traffic volumes. Buses and trucks usually do not block the view at unsignalized intersections, so usually, observers can sit in an automobile. The size of gaps in a traffic stream depends on the traffic volume, so one must sample gaps during each period of interest that has a volume different from those of adjacent periods. The mean gap has only marginal meaning in analyses using gap data. Statistics that describe the shape of the gap distribution, such as percentiles, are more useful.

One may also collect gap data using certain types of automatic vehicle detectors, laptop computers, audiotapes in combination with microcomputers in the office (as described above for saturation flow studies), videotapes, or stopwatches. With automatic detectors, engineers must ensure that only the lanes of interest are being measured. Laptop computers and audiotapes with microcomputers require special computer programs, as described earlier for saturation flow studies.

Gap Acceptance Studies

Gap acceptance studies are more difficult to conduct than gap studies. A gap acceptance study still requires data on the gaps presented in the major traffic stream. In addition, observers must categorize each data point as an accepted gap, a rejected gap, an untested gap (there was no minor street vehicle present), an accepted lag, or a rejected lag.

The simplest procedure for collecting gap acceptance data with typical agency equipment requires an observer with an audiotape recorder in the field and a technician with a computer in the office (adapted from Shanteau, 1988). The observer would say "major" in the recorder whenever a major-street vehicle passed the reference point and "minor" whenever a minor-street vehicle appeared at the stop bar (not at the back of a queue). Between the appearances of major and minor vehicles, the observer would also say whether the gap or lag was "accepted," "rejected," or "untested." The observer should record "header" information at the beginning of the tape and periodically during the session should record the time in case the technician in the office loses his place. The technician in the office listens to the tape at the same speed at which it was recorded and strikes a preprogrammed key immediately upon hearing one of the five words (a different key for each word). The computer automatically marks the time of each "major" or "minor" keystroke. A relatively simple program determines whether a gap or lag was observed, computes the gap or lag time, and summarizes the distributions of accepted and rejected gaps and lags. The tape recorder must run at a constant speed, the computer clock must be accurate, and the technician must manually edit the data (removing discrepancies by replaying the tape) before the analysis.

Engineers may require more information from the gap acceptance study besides that described above. For example, the gap acceptance behavior of truck drivers may be of interest. Engineers may alter the audiotape procedure described above to include such variations. Observers prefer headset microphones for this type of work (Bonsall et al., 1988). Acceptable headset microphones are not expensive. Observers need good visibility to the reference point but also need to be inconspicuous to avoid influencing driver behavior. It is advantageous for many reasons, including inconspicuousness, for the observer to sit in an automobile during a gap acceptance study.

As with gap studies, data collected in 2-second bins are adequate for most gap acceptance studies. Ramsey and Routledge (1973) suggest that 2-second bins require a sample of 200 acceptances, and 1-second bins require a sample of 500 acceptances (with a somewhat higher-quality result for the 1-second bins). Observers can also collect gap acceptance data with laptop computers at the intersection or with videotape that has an on-screen clock. At intersections with low volumes, two observers with a watch and a form can usually collect gap acceptance data successfully.

Estimating the Critical Gap Distribution

Many previous references have recommended the calculation of an approximate mean critical gap from a gap acceptance data set. There are several problems with this type of analysis, including the fact that the result can be inaccurate (see Hewitt, 1985), and the result is not very useful. By contrast, Hewitt showed that the analysis method of Ramsey and Routledge (1973) was considerably more accurate. The Ramsey and Routledge method uses the same data to estimate efficiently the entire distribution of critical gaps. Additional advantages of the Ramsey and Routledge method are that it does not require any assumptions about the distribution of critical gaps in the driver population, as some analysis methods do, and that engineers can apply it to gap or lag data. Figure 5-4 shows that the method is easy to use by hand or to program on a computer spreadsheet.

INTERSECTION SIGHT DISTANCE

Sight distances on approaches are critical to safe intersection operations. AASHTO (1990) provides recommendations for necessary sight distances at intersections. Sight distances can also be measured in the field. If the measured sight distance is found to be lower than the recommended sight distance, agencies should consider removing sight obstacles, reducing approach speeds, changing traffic control devices, or taking other actions. Figure 5-5 shows intersection sight distance parameters. Usually, two observers on foot with measuring tapes or wheels record intersection sight distance data. The observers communicate with portable radios or hand signals. Workers can check sight distances at low-volume and low-speed intersections with no control (AASHTO case I) or yield signs (AASHTO case II) with a simple sight angle board. Figure 5-6 shows how to construct such a board, and Figure 5-7 shows how to use it.

Engineers can also check sight distances at intersections with yield signs (case II) using a simple, indirect procedure. First, workers measure distances a and b as shown in Figure 5-5. Then the engineer obtains the stopping sight distance of the major road, d_a, from its design speed and Table 5-2. Finally, the engineer solves for the sight distance on the minor street, d_b, using the equation (AASHTO, 1990, p. 744)

$$d_b = \frac{ad_a}{d_a - b} \tag{5-3}$$

If the speed corresponding to d_b from Table 5-2 is lower than the current design speed of the minor road, the sight distance is inadequate.

To measure sight distances at intersections with stop signs (case III), one observer should stand on the minor approach $D + 10$ feet from the curb line of the major street. Currently, AASHTO (1990) recommends a value of 10 feet for D. The other observer then proceeds along the main street and measures distances d_1 and d_2 as shown in Figure 5-5 when the two observers are at the limit of seeing each other. Observers should not stand in active travel lanes while making sightings or recording measurements. Observers should simulate current AASHTO specifications of an eye height (for the observer on the minor street) of 3.5 feet and an object height (for the observer on the major street) of 4.25 feet.

Assume that the following gap acceptance data set with a 2-second interval was collected:

Gap size (sec):	1	3	5	7	9	Total
Number of acceptances:	0	5	15	25	5	50
Number of rejections:	60	45	30	15	0	150
Acceptances + rejections:	60	50	45	40	5	200

The Ramsey and Routledge method begins with conversion of the acceptances + rejections data above to a percentage and entrance into column 1 of Table A below.

Table A

Column No.: *Critical Gap (sec):*	*1* *0*	*2* *2*	*3* *4*	*4* *6*	*5* *8*
Accepted Gap (sec)					
1	30				
3	25				
5	22.5				
7	20				
9	2.5				
Total (%)	100				

The percentages in column 1 of Table A represent the distribution of accepted gaps that would be observed if all drivers had a critical gap of 0 seconds. Next, the remainder of Table A is completed; if all drivers had critical gaps of 2 seconds, the first entry into column 2 would be 0% because none of them would accept a 1-second gap, the second entry would be 25/(100 − 30) = 35.7%, the third entry would be 22.5/(100 − 30) = 32.1%, etc. The completed Table A would be:

Completed Table A

Column No.: *Critical Gap (sec):*	*1* *0*	*2* *2*	*3* *4*	*4* *6*	*5* *8*
Accepted Gap (sec)					
1	30	0	0	0	0
3	25	35.7	0	0	0
5	22.5	32.1	50	0	0
7	20	28.6	44.4	88.9	0
9	2.5	3.6	5.6	11.1	100
Total (%)	100	100	100	100	100

Table B (below), containing the number of drivers with each critical gap size that accept a gap of a given size, is then completed. First, column 5 of Table B is filled out with the number of accepted gaps from the raw data. Next, the first entry in column 1 can be made; since 5 acceptances of a 3-second gap were recorded, all of those drivers must have had critical gaps of 2 seconds. Next, the total of column 1 can be derived from this first entry and Table A; since 35.7% of all drivers with a critical gap of 2 seconds saw and accepted a gap of 3 seconds, the total of column 1 is 5 × 100/35.7 = 14.0. Then, the rest of column 1 can be completed by using the percentages from Table A on the total in column 1, and Table B looks like this:

Figure 5-4 Analysis of gap acceptance data using the Ramsey and Routledge (1973) method.

Table B

Column No.: *Critical Gap (sec):*	*1* 2	*2* 4	*3* 6	*4* 8	*5* *Total*
Accepted Gap (sec)					
3	5				5
5	4.5				15
7	4.0				25
9	0.5				5
Total	14.0				50

The first entry in column 2 is 0, since drivers with critical gaps of 4 seconds do not accept gaps of 3 seconds. The second entry must be the total number of accepted gaps of 5 seconds minus the number of drivers with critical gaps of 2 seconds who accepted gaps of 5 seconds, or 15 − 4.5 = 10.5. Next, the total of column 2 can be derived just as the total for column 1 was derived once an entry was known: 10.5 × 100/50 = 21.0. The remainder of column 2 is completed like the remainder of column 1 was once the total was known. Columns 3 and 4 are completed just like column 2. The completed Table B is shown below.

Completed Table B

Column No.: *Critical Gap (sec):*	*1* 2	*2* 4	*3* 6	*4* 8	*5* *Total*
Accepted Gap (sec)					
3	5	0	0	0	5
5	4.5	10.5	0	0	15
7	4.0	9.3	11.7	0	25
9	0.5	1.2	1.5	1.8	5
Total	14.0	21.0	13.2	1.8	50

The total row of Table B is the distribution of critical gaps among the 50 drivers in the sample and can be made into percentages. The procedure can use intervals other than 2 seconds, can analyze lag rather than gap data, and can examine any sizes of gaps and lags.

Figure 5-4 (continued)

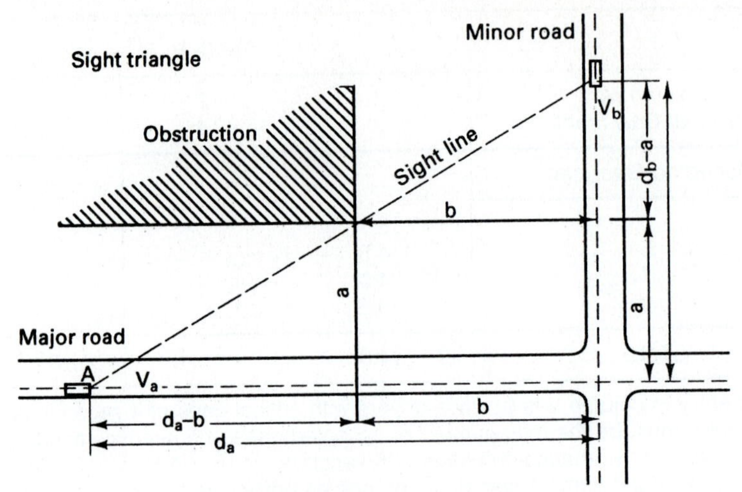

Case I and II
No control or yield control on minor road

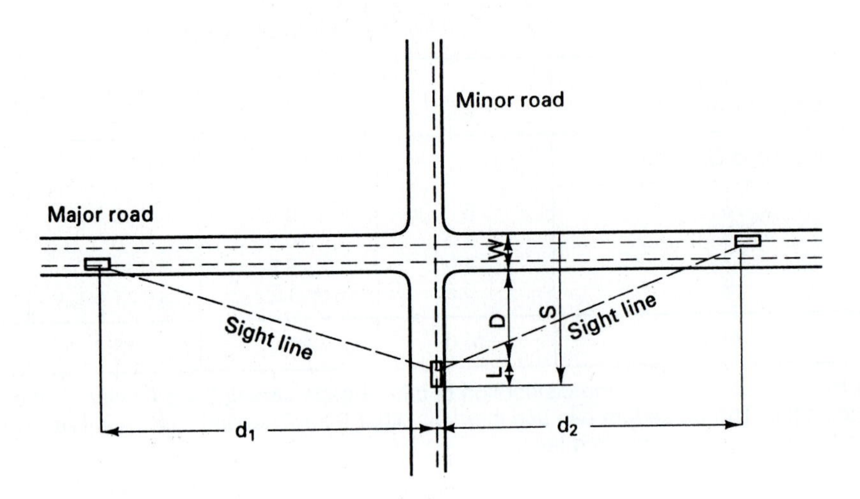

Case III
Stop control on minor road

Figure 5-5 Intersection sight distance terms and measurements. *Source:* AASHTO, "A Policy on Geometric Design of Highways and Streets," 1990.

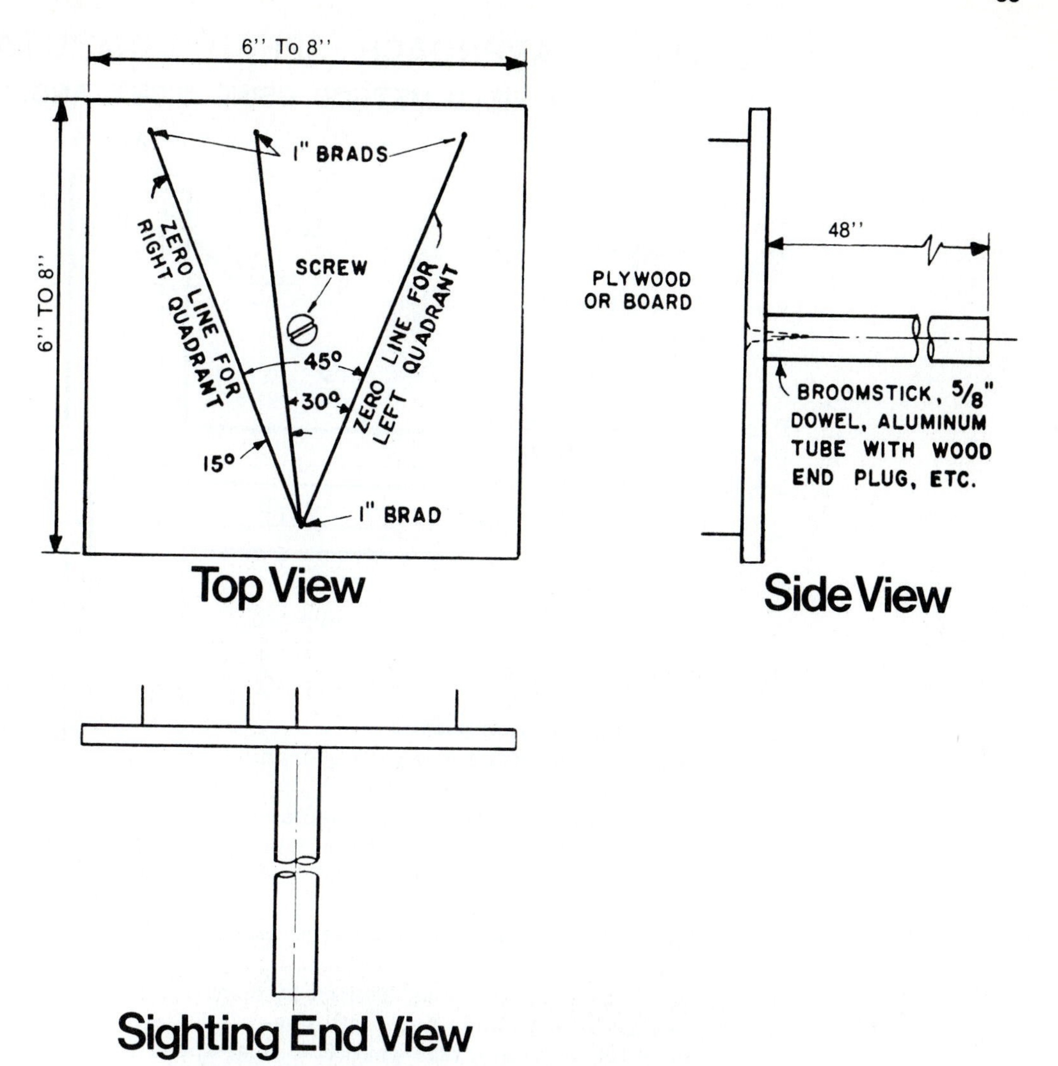

Figure 5-6 Construction of a sight angle board to estimate safe approach speeds at low-speed case I or case II intersections. *Source:* Box and Oppenlander, 1976.

TABLE 5-2
Stopping Sight Distances Corresponding to Various Design Speeds

Design Speed (mph)	*Stopping Sight Distance (ft)*
20	125
25	150
30	200
35	250
40	325
45	400
50	475
55	550
60	650
65	725
70	850

Source: AASHTO, 1990.

CRITICAL APPROACH SPEED COMPUTATION
(SIMPLIFIED FIELD METHOD USING SIGHT ANGLES)

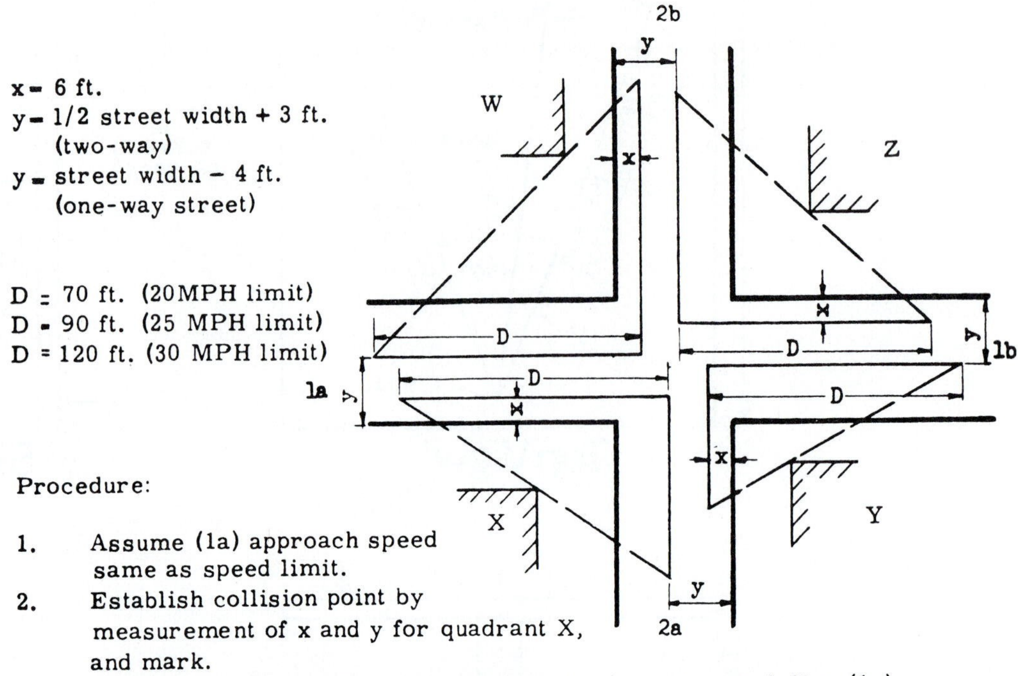

Procedure:

1. Assume (1a) approach speed same as speed limit.
2. Establish collision point by measurement of x and y for quadrant X, and mark.
3. Measure distance D for assumed speed, along approach line (1a) (x distance from curb), and mark point.
4. Set up sight angle board, facing toward collision point and with zero angle line along approach line (1a)
5. Estimate relative location of approach line (2a) at distance y from curb.
6. Sight maximum angle as limited by obstruction X.
7. Classify speed range from table below. This is the maximum safe speed for approach (2a) when (1a) is at speed limit. The reverse is also true (if (2a) approaches at speed limit, above value is maximum safe speed for (1a)).
8. Repeat angle measurement for quadrant W, using proper approach lines (y distance from curb (1a) x distance from curb (2b). Check table for safe approach speed (2b).
9. Measure out distance D on approach (1b) and follow above steps for quadrants Y and Z.

APPROACH SPEED LIMIT	CLASS			
	I (Over 45°)	II (30° to 45°)	III (15° to 30°)	IV (Less than 15°
20 MPH	20 MPH & Over	13 to 19 MPH	5 to 12 MPH	0 to 4 MPH
25 MPH	25 MPH & Over	16 to 24 MPH	6 to 15 MPH	0 to 5 MPH
30 MPH	30 MPH & Over	20 to 29 MPH	10 to 19 MPH	0 to 9 MPH

Figure 5-7 Use of a sight angle board to estimate safe approach speeds at low-speed case I or case II intersections. *Source:* Box and Oppenlander, 1976.

REFERENCES

AASHTO (1990). *A Policy on Geometric Design of Highways and Streets,* American Association of State Highway Transportation Officials, Washington, DC.

BERRY, D. S. (1976). *Discussion: Relationship of Signal Design to Discharge Headway, Approach Capacity, and Delay,* Transportation Research Record 615, Transportation Research Board, Washington, DC.

BONSALL, P. W., F. GHAHRI-SAREMI, M. R. TIGHT, AND N. W. MARIER (1988). The performance of handheld data-capture devices in traffic and transport surveys, *Traffic Engineering and Control,* Vol. 29, No. 1.

BOX, P. C. (1982). Traffic studies, in *Transportation and Traffic Engineering Handbook,* 2nd ed., W. Homburger, L. Keefer, and W. R. McGrath, eds., Institute of Transportation Engineers and Prentice Hall, Englewood Cliffs, NJ.

BOX, P. C., AND J. C. OPPENLANDER (1976), *Manual of Traffic Engineering Studies,* 4th ed., Institute of Transportation Engineers, Washington, DC.

BUEHLER, M. G., T. J. HICKS, AND D. S. BERRY (1976). *Measuring Delay by Sampling Queue Backup,* Transportation Research Record 615, Transportation Research Board, Washington, DC.

HEWITT, R. H. (1985). A comparison between some methods of measuring critical gaps, *Traffic Engineering and Control,* Vol. 26, No. 1.

ITE TECHNICAL COMMITTEE 5P-5 (1991). *Capacities of Multiple Left Turn Lanes,* Draft report of Subcommittee on Geometric Factors, Institute of Transportation Engineers, Washington, DC, July.

MCSHANE, W. R., AND R. P. ROESS (1990). *Traffic Engineering,* Prentice Hall, Englewood Cliffs, NJ.

RAMSEY, J. B. H., AND I. W. ROUTLEDGE (1973). A new approach to analysis of gap acceptance times, *Traffic Engineering Control,* Vol. 15, No. 7.

REILLY, W. R., C. C. GARDNER, AND J. H. KELL (1976). *A Technique for Measurement of Delay at Intersections,* Vol. 3, *User's Manual,* FHWA-RD-76-137, Federal Highway Administration, Washington, DC, September.

SHANTEAU, R. M. (1988). Using cumulative curves to measure saturation flow and lost time, *ITE Journal,* Vol. 58, No. 10.

TEPLY, S. (1989). *Accuracy of Delay Surveys at Signalized Intersections,* Transportation Research Record 1225, Transportation Research Board, Washington, DC.

TEPLY, S., AND G. D. EVANS (1989). *Evaluation of the Quality of Signal Progression by Delay Distributions,* Transportation Research Record 1225, Transportation Research Board, Washington, DC.

TRB (1985). *Highway Capacity Manual,* Special Report 209, Transportation Research Board, Washington, DC.

6

Inventories

H. Douglas Robertson, Ph.D., P.E.

INTRODUCTION

An inventory is a catalog, listing, accounting, record, or display of factual information that describes existing conditions. Agencies store information in manual or automated (computer or video) databases, from which they generate meaningful presentations for various analyses. The type of information included in an inventory is dictated by the purpose for which the information will be used. Some inventory data remain constant, while other data change rapidly over time and must be updated frequently. To be useful, inventories must be accessible and contain relevant data elements. Common types of inventory data include:

- Regulations, laws, and ordinances
- Roadway and roadside
- Intersection
- Traffic control devices (i.e., signs, signals, and markings)
- Parking
- Lighting
- Transit routes
- Traffic generators

A general procedure for conducting street and highway inventories is presented in this chapter. Discussions pertaining to performing other types of inventories can be found in the chapters devoted to those subject areas. Additional information on inventories is available in the *Transportation Planning Handbook* (Kell, 1991) and the *Traffic Engineering Handbook* (Pline, 1992).

A Word of Caution

Inventories can be useful and productive tools for the traffic engineer and planner. However, inventories may also prove to be costly, unreliable, difficult to maintain, and cumbersome to access. A successful inventory has a clearly stated purpose, provides useful and needed information, is easy to access and extract data from, and can be kept current at a reasonable level of effort and cost. If these criteria are not met, it is likely that the money and effort expended to produce the initial inventory has been wasted.

PURPOSE OF THE INVENTORY

Before an agency conducts an inventory, it must address several important questions, including:

- How will the inventory be used?
- What specific information and/or data will serve the purpose of the inventory?
- Can those data be obtained more effectively by means other than an inventory?
- Does the information already exist in another form?

Agencies must identify the intended uses of inventory information so that they can collect relevant data in a suitable form. For example, if the inventory is to be used to schedule maintenance, a set of data elements should be identified that relate to condition and service life. On the other hand, if the inventory is to be used to track maintenance costs, elements related to time, labor, equipment, and materials will be required. Table 6-1 lists some typical uses of inventory data.

Of course, the inventory may serve more than one purpose. Agencies must avoid collecting data elements simply because they are there and might be useful in the future. If after careful study an agency cannot identify a use for a data element, the element is not needed. Many inventory efforts fail because they attempt to collect too much information.

Choice of Data Elements

The choice of data elements is crucial to the success of the inventory. Data elements are often chosen on the basis that having those data in the inventory will preclude a visit to

TABLE 6-1
Typical Uses of Inventory Data

Illustrate street classifications
Locate traffic control devices
Specify condition or service life of devices
Schedule maintenance
Manage costs
Depict application of laws and ordinances
Assist in evaluation of traffic operations
Provide baseline conditions for use in other studies
Locate traffic generators

the field to obtain the information at a later time. Agencies must weigh this savings against (1) how frequently and quickly they will need the data element, and (2) whether a field trip would be necessary anyway to collect or verify other data. Occasional field trips may be more cost-effective than maintaining certain data elements in an inventory.

Inventory data collection and maintenance are costly. Although computers are a valuable tool for constructing inventories, they can also turn an overambitious inventory into a pile of unused computer printouts. The fundamental rule is: *Keep it simple and to the point.* Elements that are critical and are used frequently should be inventoried. Collect other data elements only ''as needed.''

STRUCTURE OF THE INVENTORY

Inventories are usually unique to the agency or jurisdiction that conducts them. While two inventories may have common purposes and common data elements, the structure of the inventories will almost always be different because of differences in the size and layout of the street system, the equipment and resources available to conduct the inventory, and local policies.

Means of Recording and Displaying Inventories

Agencies can record and display inventories in a number of ways. Hard-copy forms, cards, maps, graphs, and tables are the traditional means of recording inventory data. Any of these means may suit the purposes of a given agency, particularly the smaller ones. Increasingly, agencies use computer files and data bases instead of hard-copy forms and cards. From these files, analysts can easily sort the data and produce listings, tables, graphs, and computer-generated maps. Figures 6-1 through 6-6 illustrate various ways of presenting inventory data.

Inventory Location Reference Systems

Almost all transportation inventories use some geographical location system. Next to selecting the data elements, the system chosen to locate the data elements is the most important ingredient in an inventory. Location systems range from simple references using known points (i.e., the speed limit on Main Street from Oak to Pine is 35 mph) to more complex direction and distance schemes. An example of a more complex scheme is when each traffic warning sign may be located by direction and distance from a reference point, such as a milepost or intersection (i.e., a ''Signal Ahead'' sign is located on the south side of Maple Street, 375 feet west of the intersection with River Road). The development and increasing availability of geographic information systems (GISs) has the potential to revolutionize the way that easy-to-access data bases locate and store inventory data elements.

In manual systems, data are filed alphabetically by major street and then directionally by link (section) and node (intersection) along the major street. Locations by street name work relatively well in a small hard-copy system, but agencies often convert the names to numeric or alphanumeric codes when entering them into a computer database system. Computer software packages designed to manage inventory databases are available commercially (Center for Microcomputers in Transportation, 1992).

Numeric or alphanumeric codes allow the computer to search for and/or sort data elements quickly. The codes relate to the street network using a link–node system (McShane and Roess, 1990). Figure 6-7 illustrates part of such a system. In the sample link–node coding system shown in Figure 6-7, each street has a unique two-digit number. Thus each intersection is identified by combining the east-west (EW) street code with the north-south (NS) street code (always in that order). For example, node (intersection) *A* is

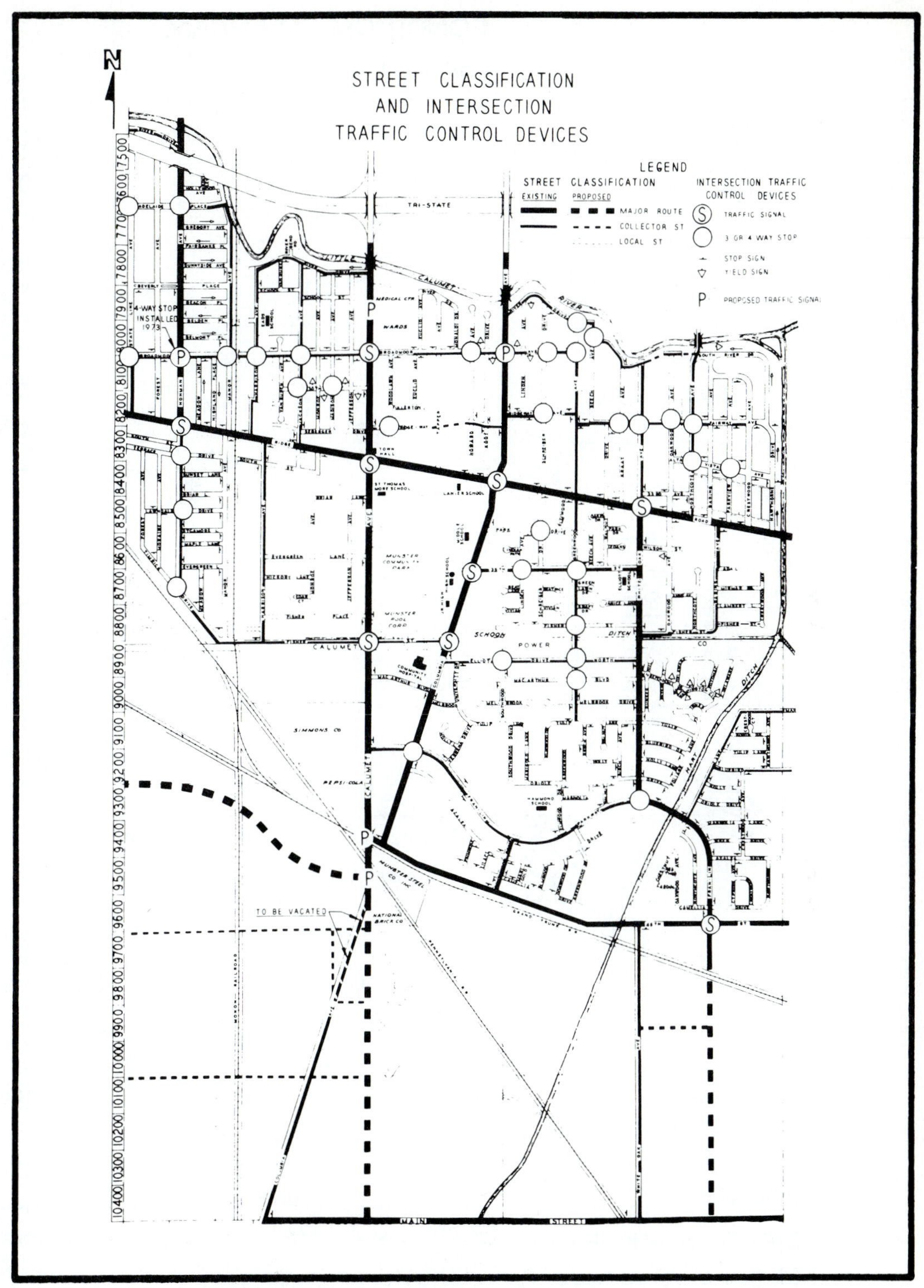

Figure 6-1 Street classification map. *Source:* Box and Oppenlander, 1976.

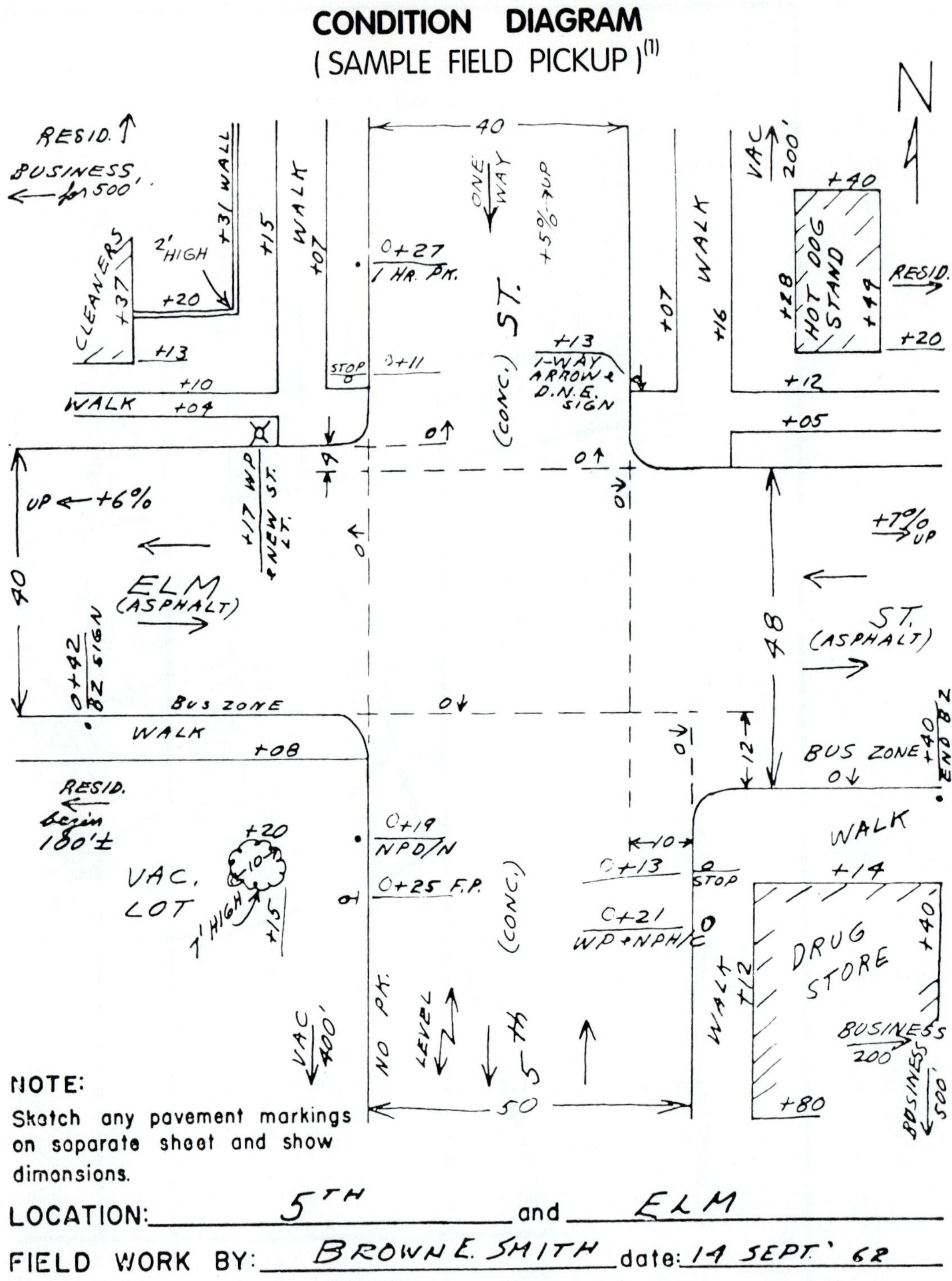

Figure 6-2 Intersection condition diagram. *Source:* Box and Oppenlander, 1976.

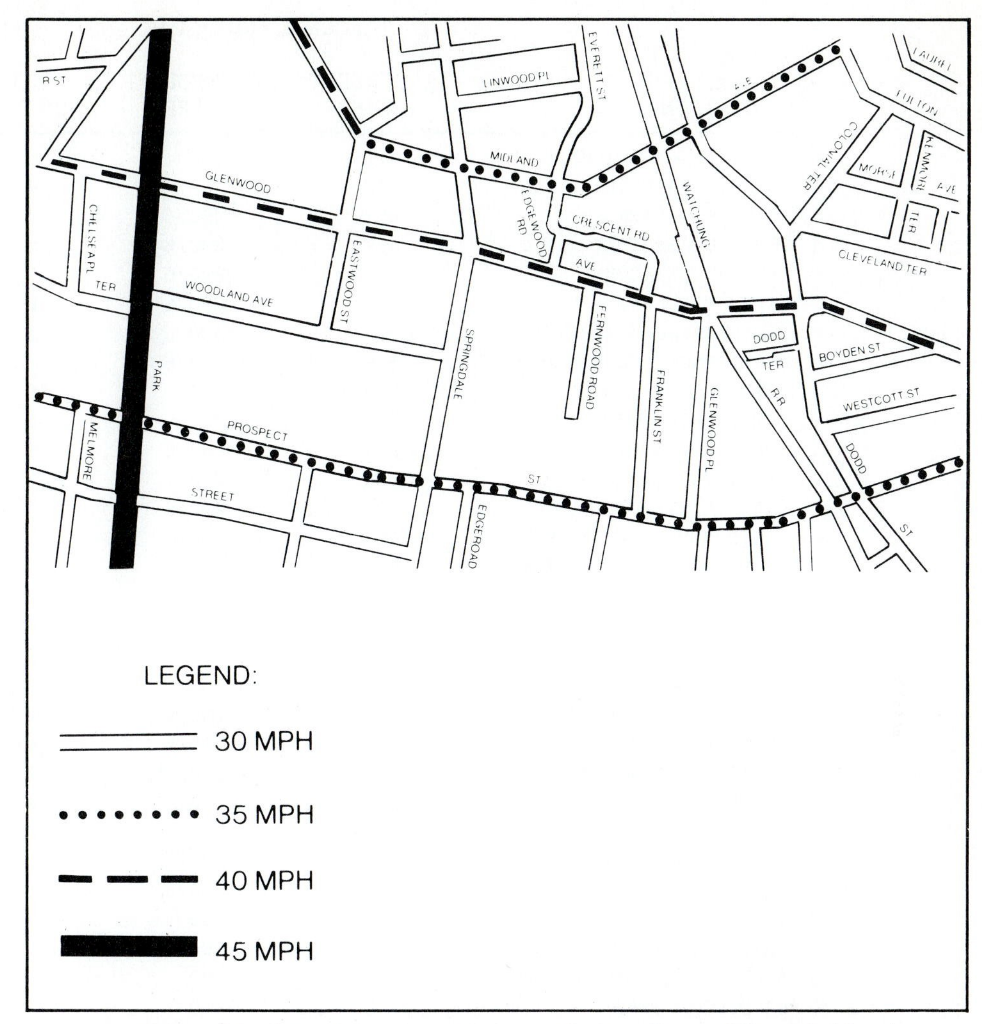

Figure 6-3 Speed zone inventory map. *Source:* Box and Oppenlander, 1976.

coded 3467. Node *B* (a five-leg intersection) is coded 3369, using only the codes of the two major streets. A separate data element called *number of approaches* identifies the intersection as having five legs. Such an element also identifies three-leg intersections, such as node 3468.

Links are identified by the codes for the nodes at each end of the link. For example, link C is 33663466. If it is desirable to identify the link by direction, the order of the node codes would indicate the direction. For example, link *C* northbound would be 34663366, and link *C* southbound would be 33663466. A link may span several intersections (e.g., the span of three EW links along street 35 would be coded 35663569).

Access to the Inventory

The importance of access to inventory data has already been mentioned. The means of accessing data and information must be a primary consideration when establishing any inventory. The method of storage and access must be tailored to the frequency and type of use to be made of the data.

TRACONEX INSTALLATIONS

CONTROLLER LOCATION	INTERSECT NUMBER	CONTROLLER TYPE	NUMBER OF PHASES	DATE INST
1. FAIRVIEW & PROVIDENCE	510	190	8	03-09-77
2. COLONY & FAIRVIEW	538		4	03-23-77
3. FAIRVIEW & PARK	657	290/6	4 + P	04-07-77
4. CENTRAL & EASTWAY	457	290/6	8	04-25-77
5. CENTRAL & KILBORNE	458	290/6	6	05-03-77
6. PARK & WOODLAWN	648	290/6	8	05-18-77
7. INDEPENDENCE & ALBEMARLE	442	390	5	05-31-77
8. PROVIDENCE & SHARON AMITY	509	290/6	8	06-10-77
9. TRYON & SUGAR CREEK	212	290	8	06-13-77
10. TRYON & REMOUNT	665	290	5	06-24-77
11. SOUTH BLVD & WOODLAWN	621	290/6	8	07-12-77
12. EASTWAY & PLAZA	239	290	8	07-15-77
13. SOUTH BLVD & TYVOLA	625	290/6	8	08-24-77
14. SOUTH BLVD & ARCHDALE	628	290/6	8	08-29-77
15. BARCLAY DOWNS & RUNNEYMEDE	542	290/6	5 + P	08-31-77
16. IND BLVD & WESTCHESTER	439	390	5	09-09-77
17. BEATTIES FORD/5TH & TRADE	813	290	4 + P	09-12-77
18. EASTWAY & TRYON	214	190	3	09-22-77
19. SHARON & WENDOVER	521	290/6	4	09-27-77
20. SHARON & RUNNEYMEDE	522	290/6	4	09-27-77
21. SHARON & MORRISON	527	290/6	4	09-23-77
22. EASTWAY & SHAMROCK	249	190/PRE	7	10-10-77
23. GRAHAM & DALTON	225	190	4	10-17-77
24. RAMA & SARDIS	494	290	3 + P	10-21-77
25. RANDOLPH & BILLINGSLEY	487	290/6	3	10-25-77
26. ALBEMARLE & CENTRAL	465	290/6	2	10-28-77
27. IND BLVD & WOODLAND	440	390	5	10-31-77
28. TRYON & CRAIGHEAD	211	390	5	11-03-77
29. ALBEMARLE & SHARON AMITY	444	290	5	11-07-77
30. RANDOLPH & SARDIS	493	290	3	11-09-77
31. DILWORTH & ROMANY	640	290	2	11-14-77

Figure 6-4 Signalized intersection listing.

Storage

Inventories are stored either manually or in a computer. Manual systems may be in the form of maps, card files, file folders, bound listings, microfiche, photologs, or videotapes. These systems are kept in filing cabinets or on bookshelves. They work reasonably well with smaller inventories, but larger systems become unwieldy and consume space. Computer inventories may be stored on hard drives, floppy disks, magnetic tapes, or CD-ROMs. Computer systems consume less space than manual systems do, so they are better suited to large inventory systems.

Retrieval

Nearly all inventory systems are stored by location (i.e., intersection or street segment). Therefore, one must know the location of interest to access inventory data on that location. In some manual systems, all inventory data for a given location may be kept in the same file. Other manual systems store data first by type of inventory and second by location. Computer data systems may be organized in a similar fashion. These systems are usually capable of searching and sorting data by data elements other than location. For example, if there is a need to examine the installation dates of all stop signs in the

TRAFFIC IMPROVEMENT ASSOCIATION OF OAKLAND COUNTY 07/19/76 PAGE 4

SIGN NUMBER	MAIN STREET	SIGN LOCATION DIST.	REF. X-ST.	SIDE OF STRT.	DIR. SIGN FACING	SIGN CODE	SIGN SIZE	DATE	EXPCT LIFE	SIGN CONDITION	VISIBILITY	SUPPORT TYPE	POST CONDITION	STD/INSTD
10000003772300	AMY	39 E	ORTONVILLE	N	E	R1-1	030X30	*04 19 ,76	5 YR	FAIR	GOOD	UPST	GOOD	NSHD
10000003854100	ANDOVER	80 N	EXETER	E	S	R7-1	012X18	*05 9 ,76	5 YR	FAIR	ROTD	UPST	GOOD	NSHD
10000003854200	ANDOVER	80 N	EXETER	E	S	R7-2	012X18	*05 9 ,76	5 YR	FAIR	ROTD	UPST	GOOD	NSHD
10000003854900	ANDOVER	30 S	LONG LK	E	S	R1-1	030X30	*05 9 ,76	0 YR	REPL	GOOD	UPST	GOOD	NSHD
10000003855000	ANDOVER	131 S	LONG LK	W	N	R7-2	012X18	*05 9 ,76	5 YR	FAIR	GOOD	UPST	GOOD	NSHD
05000003900400	ANDREWS	29 E	BOGIE LK RD	N	E	R1-1	024X24	*04 22 ,76	5 YR	FAIR	GOOD	UPST	GOOD	NSD
21000003444600	ANGELENE	37 E	CLINTONVILLE	N	E	R1-1	030X30	*04 9 ,76	0 YR	REPL	GOOD	UPST	GOOD	NSD
10000003859100	APPLE	20 N	KIRKLAND	W	N	R1-1	024X24	*05 9 ,76	5 YR	FAIR	POBS	UPST	GOOD	NSHD
21000001139800	AQUARINA	29 N	WALTON	W	N	R1-1	030X30	*04 5 ,76	5 YR	FAIR	GOOD	UPST	GOOD	NSD
03000003387100	ARDMOOR	1278 N	BURNINGBUSH	E	S	R2-1-25	024X30	*03 19 ,76	0 YR	REPL	GOOD	UPST	GOOD	STD
46000003483300	ARDMORE	35 S	LASHER	E	S	R1-1	030X30	*03 22 ,76	0 YR	REPL	GOOD	UPST	REPL	NSH
02000001085700	ARIZONA CT	43 W	ARIZONA	S	W	R1-2	030X30	*03 30 ,76	0 YR	REPL	GOOD	UPST	GOOD	NSD
15000003019400	ARMSTRONG	71 S	SATURN	ISLD	N	R4-7A	024X30	*04 8 ,76	0 YR	REPL	GOOD	UPST	GOOD	NSH
15000003019500	ARMSTRONG	29 N	WALDON	W	N	R1-1	030X30	*04 8 ,76	0 YR	REPL	GOOD	UPST	REPL	NSD
15000003676900	ARMSTRONG	71 S	SATURN	ISLD	N	R4-7A	024X30	*04 8 ,76	0 YR	REPL	GOOD	UPST	GOOD	NSH
15000003677000	ARMSTRONG	29 N	WALDON	W	N	R1-1	030X30	*04 8 ,76	0 YR	REPL	GOOD	UPST	REPL	NSD
01000003741000	ARMY	455 W	DEVONSHIRE	S	W	R7-1	012X18	*05 20 ,76	0 YR	REPL	GOOD	UPST	GOOD	NSD
01000003741100	ARMY	164 W	DEVONSHIRE	S	W	R7-1	012X18	*05 20 ,76	5 YR	FAIR	GOOD	UPST	REPL	NSD
01000003741400	ARMY	129 E	KINGSTON	S	W	R7-1	012X18	*05 20 ,76	5 YR	FAIR	GOOD	UPST	GOOD	NSD
01000003741600	ARMY	1021 E	KINGSTON	S	W	R7-1	012X18	*05 20 ,76	5 YR	FAIR	GOOD	UPST	GOOD	STD
01000003741700	ARMY	795 W	ROCHESTER	S	W	R7-1	012X18	*05 20 ,76	5 YR	FAIR	GOOD	UPST	GOOD	NSD
01000003743000	ARMY	133 W	DEVONSHIRE	N	E	R7-1	012X18	*05 20 ,76	0 YR	REPL	GOOD	UPST	GOOD	NSD
22000002984500	ARROWHEAD	1410 N	DEERFOOT	E	S	R1-1-RR	030X30	*03 19 ,76	0 YR	REPL	GOOD	UPST	REPL	NSHD
23000003198900	ARTDALE	28 S	BISCAYNE	E	S	R1-2	030X30	*04 28 ,76	5 YR	FAIR	GOOD	UPST	GOOD	NSH

Figure 6-5 Computerized sign inventory printout. *Source:* Cunard, R. A., "Maintenance Management of Street and Highway Signs," *National Cooperative Highway Research Program Synthesis of Practice 157,* Transportation Research Board, Washington, DC, September, 1990, p. 40.

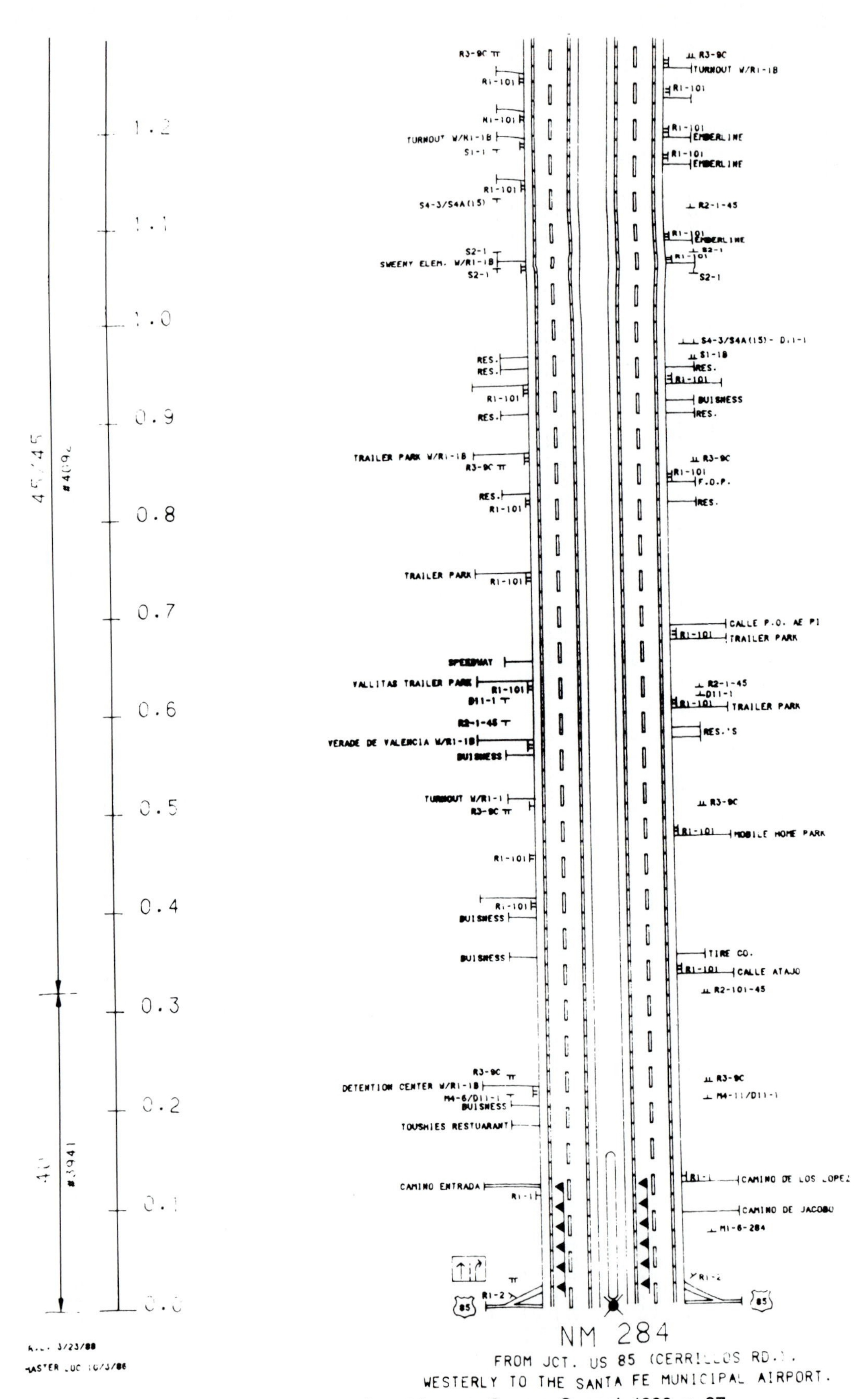

Figure 6-6 CAD system sign strip map. *Source:* Cunard, 1990, p. 37.

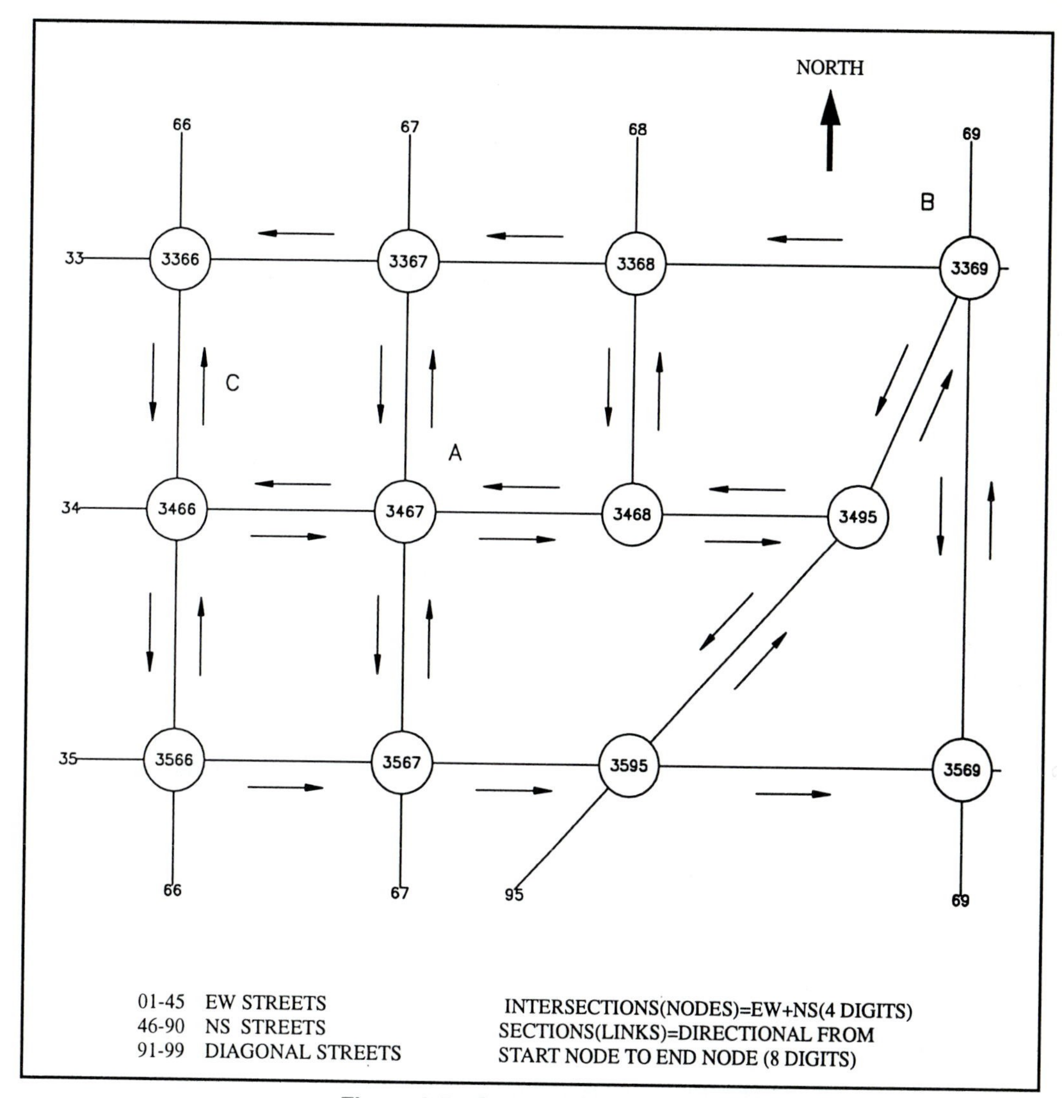

Figure 6-7 Sample link-node coding system.

system for maintenance purposes, each intersection in the manual file would have to be checked to see if stop signs were present. In the computer system, the computer seeks out stop sign locations. Additionally, the computer could sort and retrieve only those intersections with stop sign installation dates prior to a given cutoff date, thus saving both time and effort.

The computer can also generate reports or listings containing only the information desired and therefore avoids the need for manual extraction of data. This capability applies to periodic replacement of devices, spare part inventory control, stock replacement and ordering, dispatch of work crews, maintenance history files, and possible liability defense. In addition, work orders may be generated automatically. Agencies typically design work order formats to enhance inventory updating upon completion of the work.

ESTABLISHING AN INVENTORY

The following is a suggested step-by-step procedure for setting up an initial inventory. Not all of the steps apply to every type of inventory, and thus, some steps may be omitted. Careful thought and planning before starting data collection should lead to a useful and cost-effective inventory.

TABLE 6-2
Typical Street or Highway Segment Data Elements

Unique link identification by name or number
Classification of street or highway
Number of lanes by direction
Width of lanes
Parking conditions
Speed limit
Pavement markings by type of material and condition
Signs by type, location and condition
Bus stop locations
Transit routes
Lighting by type and location
Driveway entrances by location
Adjacent land use
Length of the segment
Date of last information update

Step 1: Determine the Purpose of the Inventory

The purpose(s) of the inventory will determine the applicable steps to be followed and should result in an inventory that best meets the agency's needs at the least expenditure of resources. Review Table 6-1 for typical uses of inventory data.

Step 2: Select the Data Elements to Be Collected

Remember the cautions mentioned earlier about collecting the data elements that will be used and omitting those that are just "nice to have." Table 6-2 illustrates the type of data elements that may be collected for a street or highway segment. Typical data elements for other types of inventories are described elsewhere in this manual.

Step 3: Select a Data Collection Technique

Data collection techniques refer to the methods used to record inventory data elements. Observers may record the elements directly and in the field, or they may film or videotape the elements in the field and record the data in the office. Once a technique is selected, agencies can identify and obtain the equipment needed to record the inventory data elements. The equipment may range from a map, data forms, pencils, and tape recorder to a specially equipped vehicle with cameras or video recorders, laptop computers, and distance-measuring devices. The choice of technique is a function of available budget, time, and personnel. *NCHRP Synthesis of Highway Practice 157* presents a detailed discussion of data collection techniques for sign inventories (Cunard, 1990). The cost-effectiveness of alternative sign inventory procedures is discussed in a report by the Federal Highway Administration (Datta et al., 1985).

Step 4: Prepare a Data Collection Plan

The data collection plan should specify how the agency will collect the inventory data. The plan should contain in detail the equipment needed, the time during which data will be gathered, exactly how each element is to be measured and recorded, how to handle unusual circumstances, and the rules to follow that will ensure consistency in measurements and data recording. If agencies collect data manually, they may design forms or program laptop computers so that observers can record data in the field directly. These actions preclude the data reduction steps required in the office if film or videotape is used. Figure 6-8 shows a sample of a sign inventory data collection form.

City ☐
County ☐ of________ Quadrant________
Military Base ☐
City, Street or County Road________
Direction________

SIGN INVENTORY SHEET

SHEET______ OF______

DATE________

INVENTORY CHIEF________

HELPER________

A	B	C	D		E	F	G	H	I	J	K	L	M	N	O	P
			SIZE								POST					
ODOMETER READING, OR INTERSECTING ROAD/STREET	TYPE OF SIGN	SIGN CODE NUMBER (UNIFORM MANUAL OR STATE MANUAL)	HOR. (IN.)	VERT. (IN.)	TYPE OF FACE	TYPE OF BACKING	CONDITION—DAY	CONDITION—NIGHT	VISIBILITY	NO. IN ASSEMBLY	NO.	TYPE	CONDITION/ PLACEMENT	NON-UNIFORM	ADDITION NEEDED	REMARKS AND DESCRIPTION

Figure 6-8 Sample sign inventory data collection form. *Source:* Cunard, 1990, p. 30.

Step 5: Collect the Data

Agencies should schedule data collection during good weather to enhance accuracy and completeness. It is possible, however, to conduct inventories, particularly manual studies, during less than ideal weather conditions when other more weather-sensitive data collection is impossible. The data collection team should use a checklist such as Figure 1-1 to ensure that they have all of the necessary equipment and materials prior to leaving the office. The team should note any unusual circumstances that arise during data collection and discuss them with the supervisor upon return from the field. Although it is possible to conduct field inventories with one person, the preferable practice is to work in teams of two or more. Such teams allow for quality control as data are recorded and provide for greater safety in that one person can drive while another observes and records.

Step 6: Construct the Database and Data Displays

The means of recording data in the field dictates the specific tasks needed to place the data in a usable format. If agencies record inventory data elements manually, they must transfer the data from the collection forms to maps or tables or computer files (see Figure 6-9). If the data are recorded directly in a computer or other automated device in the field, computer software can handle the data reduction and formatting tasks with relative ease. The software should contain systematic error-check routines to improve the quality of the data. Agencies need to check their computers and software with small samples of data to ensure that the reduction and formatting is free of problems before processing large batches of data.

Inventory data can be displayed in a number of ways, as indicated previously. Maps may either be drafted by hand or drawn using a CAD system. Computer database management software packages are available for both microcomputers and mainframes (Sabria et al., 1985; Center for Microcomputers in Transportation, 1992; FHWA, 1987). These packages manipulate the data efficiently to serve multiple purposes and create numerous formats for displaying the data. Visual displays in the form of photologs, videologs, and aerial photos represent other forms by which inventory data may be presented (Baker, 1982; APWA, 1986; ASCE, 1985).

MAINTAINING AN INVENTORY

Keeping an inventory current is critical to its usefulness. As an inventory becomes outdated, its utility deteriorates rapidly. The success of an inventory is directly proportional to an agency's commitment to maintaining it.

Software That Aids in Inventory Updating

Numerous computer software packages are available to aid in both establishing and maintaining traffic and highway inventories. Traffic maintenance software focuses on pavement markings, signs, signals, and street lighting. Highway maintenance software encompasses a road inventory system of highway and bridge segments. Typical data elements in highway maintenance software include pavement and shoulder type, width, and condition; median type and width; right-of-way width; parking; curbs; turn lanes; traffic volumes; drainage; land use characteristics; and curve and grade statistics (Center for Microcomputers in Transportation, 1992).

SIGN INVENTORY CODING FORM

Sioux City, Iowa

Sheet ________ of ________

Coded by: ______________________

Checked by: ______________________

					Sign Location			Lateral Position				Sign									Post					Date:		
														Size														
Card Number	Sector	Inv. Dir.	Sign Number	Inventory Route	Distance	Reference Cross Street	Direction Facing	Distance	Side of Roadway	Number in Assembly	MUTCD Number	Shape	Color	Horizontal	Vertical	Condition	Reflectivity	Visibility	Height	Backing Mat.	Type	Condition	Remarks/Special Sign Message	Sign Change	Reason for Change	Month	Day	Year
1	2 3	4	5 6 7 8 9	10 11 12 13 14 15	16 17 18 19	20 21 22 23 24 25	26	27 28	29	30	31 32 33 34 35 36 37 38 39 40 41 42 43	44	45	46 47	48 49	50	51	52	53 54	55	56	57	58 59 60 61 62 63 64 65 66 67 68 69 70 71 72	73	74	75 76	77 78	79 80

Figure 6-9 Sample data entry form. *Source:* Cunard, 1990, p. 31.

Frequency of Updates

The frequency with which agencies must update inventories depends on how prone the data elements are to change. The issue is complicated by elements with differing service lives, damage inflicted by accidents and incidents, and replacement necessitated by changes external to the facility itself. Such changes may consist of land use growth and development, safety and/or operational improvements, and shifts in travel patterns and demands.

There are two basic approaches to inventory updating. Perhaps the most prudent and cost-effective method is continuous updating. The simple concept is that once an agency establishes an inventory, it is promptly updated each time there is a change, whether that change is an addition, deletion, modification, or replacement. This approach sounds easy and very logical but requires disciplined attention to detail to implement properly. Computer software has made this approach truly feasible. Not only can agencies enter updates quickly and easily, but the software enables fast and convenient examinations of the database to check the accuracy of its contents and ensure that changes are being recorded promptly. Continuous updating is most appropriate for inventories that are used frequently and thus require a greater degree of accuracy.

The second basic method is periodic updating. As the name indicates, the agency conducts a partial or complete field inventory at specified intervals and records changes that have occurred since the last update. The interval length depends on the level of accuracy required and the frequency of changes experienced by the inventory in question. Periodic updating is most appropriate for inventories that have a low usage level, experience relatively few changes, and require a lower degree of accuracy.

REFERENCES

ASCE (1985). Photomaps inventory traffic control devices, *Civil Engineering,* American Society of Civil Engineers, New York, October.

APWA (1986). Sign inventory uses aerial photos, consultant, *APWA Reporter,* Vol. 53, No. 11.

BAKER, W. T. (1982). *Photologging,* NCHRP Synthesis of Highway Practice 94, Transportation Research Board, National Research Council, Washington, DC, November.

CENTER FOR MICROCOMPUTERS IN TRANSPORTATION (1992). *McTrans Catalog,* University of Florida Transportation Research Center, Gainesville, FL, June, pp. 15, 35–36, 41, 47.

CUNARD, R. A. (1990). *Maintenance Management of Street and Highway Signs,* NCHRP Synthesis of Highway Practice 157, Transportation Research Board, National Research Council, Washington, DC, September, pp. 28–43.

DATTA, T. K., K. H. TSUCHIYAMA, AND K. S. OPIELA (1985). Cost-Effective Inventory Procedures for Highway Inventory Data, DTFH61-83-C-00043, U.S. Department of Transportation, Federal Highway Administration, Washington, DC, June.

FWHA (1987). *Sign Management System Users Guide,* U.S. Department of Transportation, Federal Highway Administration, Washington, DC.

KELL, J. H. (1991). *Transportation Planning Handbook, Prentice Hall, Englewood Cliffs, NJ,* Chapter 11.

MCSHANE, W. R. AND R. P. ROESS (1990). *Traffic Engineering,* Prentice Hall, Englewood Cliffs, NJ, pp. 75–79.

PLINE, J. L. (1992). *Traffic Engineering Handbook,* Prentice Hall, Englewood Cliffs, NJ, pp. 59–60.

SABRIA, F., B. BONNER, AND E. SULLIVAN (1985). *Evaluation of Traffic Control Device Inventory Programs for Microcomputers,* UCB-ITS-TD-85-2, Institute of Transportation Studies, University of California, Berkeley, CA, May.

7

Transportation Planning Data

Donna C. Nelson, Ph.D., P.E.

INTRODUCTION

Transportation planning is a complex process that involves the evaluation and selection of highway or transit facilities to serve present and future demand. Studies are conducted to gather information for transportation modeling efforts or to evaluate the potential impacts of specific programs and projects. The data needed for these studies usually includes information about population and economic activity, travel characteristics, and transportation facilities. The data are often organized and analyzed to:

1. Identify the scale of present system inadequacies
2. Provide basis to forecast future land use and travel
3. Derive population, land use, and travel relationships
4. Calibrate travel demand models

The data needed for a study vary according to the scope and nature of project, past planning activities, cost of data collection, and so on. Generally, the transportation planning process requires that data from a wide variety of sources be complied, organized, and analyzed. Data collection efforts can include most of the major operational surveys described in this book, including volume, speed, travel time, transit, goods movement studies and inventories of land use, traffic control, geometrics, and other characteristics. Refer to those chapters for detailed discussions of the design and implementation of these studies. In addition, Appendix B describes the process of designing, testing, and implementing surveys. Detailed explanations of transportation modeling processes are beyond the scope of this book; refer to the ITE's *Transportation Planning Handbook* for further information on this topic (ITE, 1992).

TYPES OF STUDIES

Planning studies may be categorized into three general time horizons: short, medium, and long term (Stopher and Meyburg, 1975). *Short-term* or *project planning activities* focus on select projects that can be implemented within a 1- to 3-year period. The short-term planning process includes systems monitoring (collection of performance data), identification of system deficiencies, formulation of process goals and objectives; identification of strategies, development of performance criteria, evaluation of strategies, and ranking of strategies using a cost-effectiveness approach. The end product of the process is often a list of recommended projects or programs designed to provide better management of existing facilities by making them more efficient (ASCE, 1986). Examples of planning projects include:

- Planning for selected groups (e.g., the elderly or handicapped)
- Transportation/land use interaction (e.g., trip generation at shopping centers, transportation impacts of a proposed development)
- Goods movement studies
- Residential development projects
- Bikeway planning
- Energy contingency planning
- Impact studies

The transportation plan for each metropolitan area must include a transportation system management element to provide for the short-term transportation needs of the urbanized area. This plan provides for the more efficient use of existing transportation resources and providing for the movement of people and goods.

Medium-term plans usually focus on efforts such as transportation system management (TSM) studies, air quality management planning, and corridor studies. There appears to be no distinct division between short-, medium-, and long-term planning efforts. Rather, those projects commonly categorized as medium term extend beyond the time span and narrow focus of short-term plans but are more specialized and immediate than the long-term planning efforts described below. TSM elements identify traffic engineering, public transportation regulatory, pricing, management, and operational-type improvements to the transportation system.

Long range plans project transportation needs of an area and identify the projects to be constructed over a 20-year period to meet these needs. The plans tend to be capital intensive with relatively long implementation times. Although the problem-solving process is similar for the long- and short-term planning processes, the long-range planning process requires more sophisticated methods for forecasting future transportation demand than are required for short-term planning projects.

DEFINING STUDY AREAS

Most planning studies begin with delineation of the survey or study area. The exterior limit of the survey is called the *cordon line*. For a comprehensive urban survey, this study area may include the entire urbanized area plus a portion of the outer fringes where future growth is expected. For local area studies, the cordon area is established to encompasses the area of interest and to minimize data collection effort. Internal zones can also be defined to permit data to be summarized for a reasonably small area and add more detail to the information on trip movements within the study area.

Establishing the Cordon

The cordon is, in effect, the boundary around the study area for which trip movements are required. The purpose of the study suggests the extents of the survey area. The survey objectives and the constraints on the study should be considered in establishing the exact cordon line. The cordon line may follow physical boundaries to movement (such as hills or rivers) and major transportation facilities (such as freeways or rail lines).

Establishing Traffic Analysis Zones

Once the outer boundaries or limits of the study are set, the study area can be divided into traffic analysis zones (TAZs). There are no strict guidelines on the size and/or number of zones. Too many zones complicate the analysis; too few zones may give an unrealistic grouping of trip ends and may not allow realistic trip routings. The size of zones is determined by the survey area size, population density, desired data items, study purposes, and modeling techniques to be employed.

Study zones can be established in a number of ways. They are generally selected to include homogeneous socioeconomic characteristics, minimize intrazonal trips, and employ physical, political, and historical boundaries where possible. If possible, zone systems should generate and attract approximately equal trips, households, population, or area. Census tracts or census enumeration districts are often considered because population data are easily available by this geographic designation. Census block groups are the smallest geographic areas for which census data are readily available. The zones are usually smaller in the more dense downtown area and larger in sparsely populated outlying areas. For a small survey, the area surrounding a single transport route may be divided into only a few zones; a large metropolitan area might be divided into several hundred zones. A good rule of thumb is that the driving time across the zone should not exceed 3 to 5 minutes. Figure 7-1 shows zones for an O-D survey conducted in one U.S. city.

INVENTORIES

Once the scope and scale of the planning effort have been determined and the boundaries of the study area established, the usual next step is to assemble the data required. The information most commonly needed for planning purposes includes information on land use and population, the characteristics of the transportation system, and information on travel demand. Inventories, surveys, and studies are made to determine traffic volumes, land uses, origins and destinations of travelers, population, employment, and economic activity. Collecting completely new data for a study may not be necessary. Before field collection of data begins, existing sources of data should be identified and the data examined to see if they are useful for the study in question. The information gathered is summarized for each traffic zone and for the existing highway and transit system.

Land Use and Population Data

Land use inventories provide basic data on land characteristics and activities including current land use and vacant land. In general, land use data are often collected to provide:

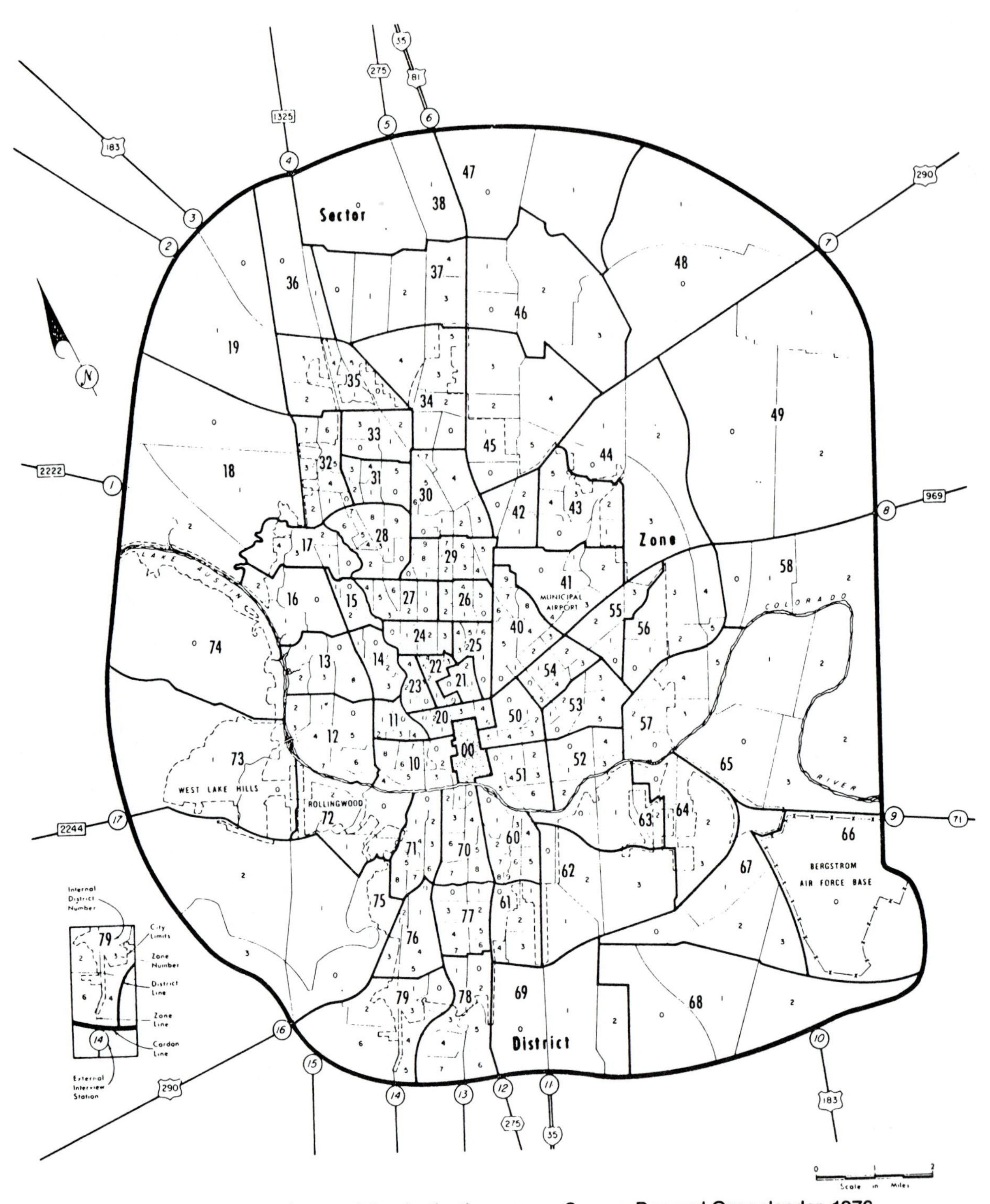

Figure 7-1 Zones for an origin–destination survey. *Source:* Box and Oppenlander, 1976.

- A basis to derive trip generation factors and trip generation forecasts
- Data for coordinating transportation facilities with other uses
- A ''universe'' of dwelling units from which a sample can be drawn for the home interview phase of the travel survey
- Data useful for the day-to-day planning activities of various agencies

A wide range of information on land is useful in the transportation planning process. This includes:

- Development trends over time
- Topography and physical constraints
- Current land use
- Vacant land by ownership
- Location of major travel generators
- Identification of community neighborhood boundaries
- Existing land use controls, including zoning, official maps, and subdivision regulations
- Identification of redevelopment areas

Many planning agencies maintain current inventories of land uses, vacant land, and economic activities. The development of geographic information systems has really revolutionized the archival and retrieval of land use data. The land use component of comprehensive plans typically contains information on zoning, activity systems, land values, environmental conditions, and aesthetic features of the local area. The UTPS package of the U.S. Census also contains information useful in establishing relationships between land use and trips.

Population distribution, density, average income, type of dwelling, and car ownership data are used for travel models. Historic patterns of distribution, migration, density, and growth trends combined with present conditions are used to prepare population forecasts. For simple studies or projects, data and relationship drawn from previous studies may be adopted (Homburger et al., 1985). Information on population projections may be available from the agencies responsible for planning and economic development. For larger projects, current, more detailed data may be required. Surveys may be conducted to check the accuracy and validity of available data or collect new data.

Collecting Land Use Data

Several types of land use surveys can be conducted to provide new data, or to update and supplement available information. The decision as to the type of survey must be based on its purpose or purposes; and the resources available. Land use data may be recorded in the field using several different techniques. One of the simplest methods is to record the data directly on maps or aerial photographs. This ''map-record'' type of land-use inventory may be appropriate in small towns or cities where the only data required are a simple use classification. Typical blueline prints at a scale of 1 inch = 400 feet or larger may be used. A second technique, *field listing,* is the most common method for transportation planning. Traditionally, this method has been carried out using a field form similar to that shown in Figure 7-2. A separate form is usually reserved for each block, with a line for each parcel or land use and dwelling unit. Observations may be made from the vehicle, or on foot if parking or traffic congestion presents a problem. Data from the forms are input directly into a computer database. Surveys of existing land use can be classified as ''inspection'' or ''interview'' surveys, depending on whether or not the dwellings and other places must be entered. Inspection surveys (often called ''windshield surveys'') are accomplished without entering the building. Combined inspection–interview studies are needed when exterior inspection does not yield enough information. Windshield surveys are frequently sufficient for transportation and traffic needs.

The field listing technique may be applied easily using a laptop computer and either database or spreadsheet software (depending upon the size and complexity of the application). Data collected in the field can, if necessary, be ported to another computer

Geographic location	S	Hse or building number	Non-residential use dimension	Room No.	Floor	% use of floor	Residence type	Description of Land Use Retail, services, wholesale, mfg, offices, other
x y Block			Frontage Width				0. Hse 2. Room 1. Apt 3. Trailer	
Geographic location	S	Hse or building number	Non-residential use dimension	Room No.	Floor	% use of floor	Residence type	Description of Land Use Retail, services, wholesale, mfg, offices, other
x y Block			Frontage Width				0. Hse 2. Room 1. Apt 3. Trailer	
Geographic location	S	Hse or building number	Non-residential use dimension	Room No.	Floor	% use of floor	Residence type	Description of Land Use Retail, services, wholesale, mfg, offices, other
x y Block			Frontage Width				0. Hse 2. Room 1. Apt 3. Trailer	
Geographic location	S	Hse or building number	Non-residential use dimension	Room No.	Floor	% use of floor	Residence type	Description of Land Use Retail, services, wholesale, mfg, offices, other
x y Block			Frontage Width				0. Hse 2. Room 1. Apt 3. Trailer	

Figure 7-2 Field list form.

system in the office. If the land use information is to be integrated with a number of other databases and will be retained and reused, a geographic information system (GIS) may be appropriate. GISs integrate the map (or geographically oriented system) with the database, providing a very powerful tool for tracking and analyzing spatially oriented information.

Land use data are traditionally presented on a map showing the land use by general category of use (residential, commercial, industrial, institutional, parking and recreation, transportation, utilities, agriculture, and water). Each category is given a difference color to provide visual differentiation. Statistical summaries may be prepared to show the total land area devoted to each category of use, and may be broken down into subareas, or traffic zones. Land use data collected for one study are valuable for later studies. Inconsistencies in categories of land use activities and data formats may make it difficult to compare results with subsequent studies. To minimize these problems in the United States, a standard system for identifying and coding land use activities was developed (Urban Renewal Administration and Bureau of Public Roads, 1965). The coding system provides four levels of detail on land use activity. Each level is subdivided into one-, two-, three- and four-digit categories.

Transportation Systems Data

Transportation facility inventories provide the basis for establishing the networks that will be studied to determine present and future traffic flows. Data can be organized around travel facilities, parking, and trip and travel data. In many cases, some data will be available from existing records of city, county, or state offices. Because data collection can be expensive and some data may be more essential than others, data needs must be evaluated carefully before any study is undertaken.

Travel Facilities Inventory

Travel facilities inventory consists of locating and describing each link of the transportation system. The description of each link includes a measure of the present capacity and the current use of the link, and a statement of its performance characteristics. This inventory includes:

- Classification of streets and highways
- Geometric characteristics such as link length, pavement width, right-of-way width, signals, green time, number of lanes, parking controls, speed limits, and adjacent land use
- Free-flow travel times over network links routes
- Existing capacity and levels of service

Classification of all streets and highways in the network encompassed by the study area should be completed in the early stages of the study. Jurisdictional classifications identify each participant and their respective responsibilities in regard to each segment of the overall transportation system. The functional classification of facilities is determined by the relative importance of the movement and access functions assigned to them. In the hierarchy of highway facilities shown in Figure 7-3, freeways, expressways, and major arterials constitute the major highway system, while collector and local streets comprise the local street system. The number of jurisdictions involved in a study area varies depending on the size and complexity of the area under study. Functional and jurisdictional classifications are done in conjunction with the geometric inventory of the streets.

Geometric characteristics needed include items such as link length, pavement width, right-of-way width, signals (and other traffic control), signal cycle length and phasing, number of lanes, speed limits, and adjacent land use. Much of this information may be part of existing traffic control device (TCD) and geometrics inventories. Speed

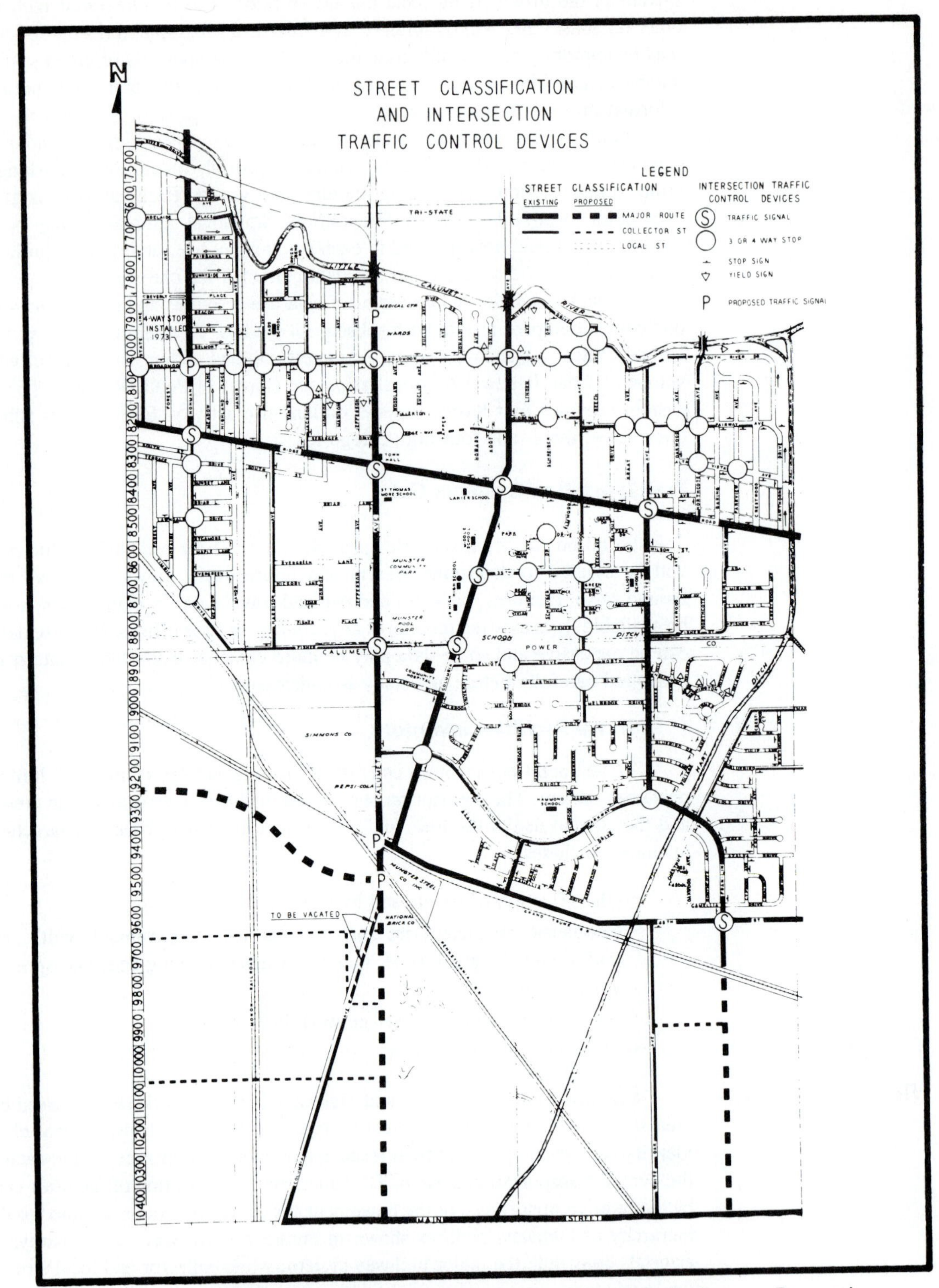

Figure 7-3 Functional classification of roadways. *Source:* Box and Oppenlander, 1976.

and travel time may also be needed for selected roadway and transit links. These data are collected as described in Chapters 3 and 4. Existing capacity and levels of service can be determined using the methods described in the *Highway Capacity Manual* (TRB, 1985). The data required for signalized intersections include basic geometrics, signal timing information, existing volumes, area type, and driver population (commuter or other).

Parking Data

Parking inventories consist primarily of information pertaining to the location, capacity, time limits, and other characteristics of existing parking spaces both along the curb and in offstreet areas, including those in alleys and between buildings. All legal parking spaces (public and private) should be included. As described in Chapter 10, the usual information collected includes:

- Number of parking spaces
- Time limits and hours of operation
- Ownership (e.g., public, private, restricted)
- Rates charged (if any) and method of collection
- Curb space regulations
- Type of facility (street, lot, or garage parking)

Trip and Travel Data

Trip and travel data are obtained from origin–destination (O-D) studies. These studies are designed to gather data on number and type of trips in an area, including movements of vehicles and passengers or cargo, from various zones of origin to various zones of destination. Information often collected includes the origin and destination of trips, the purpose, travel time, and length of trip, the mode of travel, and the land use at the points of origin and destination. The data are analyzed to define travel behavior and travel patterns within the area by time of day, mode of travel, and purpose of trip. Depending on study objectives, travel behavior may be studied for an average weekday or weekend, as well as for peak season travel (for resort areas). Travel demand for the base year is projected into the future to determine whether the current transportation infrastructure is adequate to meet future demands, to define the need for new facilities or improvements to the present system, and to evaluate different growth rates and patterns. O-D data are used to plan and program major street systems, street improvements, new street locations, freeway location and design, interchange location, public transit networks and coverage, and terminal facilities (bus, truck, and off-street parking).

ORIGIN–DESTINATION SURVEYS

Comprehensive O-D surveys are generally the basis for preparation of overall comprehensive transportation plans for an area. Because comprehensive plans are long range and slow in implementation, and because transportation facilities must be built for many years of usefulness, the O-D data must generally be projected to provide data on future transportation demands. Methods of projecting O-D data are beyond the scope of this chapter.

The scale of origin and destination studies varies widely. Large transportation planning studies may conduct home interviews to establish patterns for all trips made during a typical day throughout a large area. Large-scale planning studies may consist of a combination of several complementary surveys. Complete studies are very expensive and time consuming and are rarely conducted. More frequently, limited O-D studies may be conducted to supplement and update existing O-D data. Major O-D data are projected to a planning horizon or design year (usually, 15 to 25 years in the future) based on anticipated future economic and population growth, vehicle ownership and usage, transit availability and patronage, land use changes, and other factors. The scope of survey questions may be narrowed to a single trip or encompass all trips made during a 24-hour period. Small-scale studies may focus on a specific neighborhood, an individual trip generator (or group of generators), or limited sections of roadway. Studies may also be performed for limited sections of freeway to determine weaving and merging patterns, or to develop alternate routes (McShane and Roess, 1990).

Depending on the purpose of the O-D survey, studies focus on the basic "when" (time of day), "where" (origin and destination), "how" (modal choice), and "why" (trip purpose) people travel. Time-of-day data are used to establish potential peak periods and to estimate travel demand throughout the day. Trip purpose is used to establish trip patterns by trip purpose. Categories of data needed vary among studies; however, commonly selected categories include home-to-work, home-to-shopping, home-to-business, home-to-social/recreational, and other trip purposes. The trip origin is the beginning point of the trip unless that trip begins or ends at home. The origin of any home-based trip (e.g., home-based work trip) is classified as "home" regardless of the direction of the trip.

Trip types are also classified in their relationship to the study areas (e.g., whether the origin and/or destination are within the study area).

- *External–external* or *through trips* are those trips with neither origin nor destination in the study area. Travelers make no stops within the study area.
- *External trips,* further classified as *internal–external* or *external–internal* trips, have either their origin or destination outside the study area.
- *Internal–internal trips* have both their origin and destination within the study area.

EXTERNAL SURVEYS

Cordon surveys may be used to study external travel (e.g., trips that either pass through the zone or have one trip end outside the zone). The type of external survey to be conducted depends on the information to be collected as well as the size of the study area. Common types of studies include:

- Roadside interviews
- License plate surveys
- Postcard/mail-back surveys
- Vehicle intercept method
- Tag-on-vehicle method
- Headlight study

The study area may be a CBD or other major activity center. Data collected usually pertain only to auto and truck travel. Separate surveys of rail, bus, and air travel are made as desired to obtain additional travel information.

Establishing Cordons

Cordons should follow the boundaries of the study area. Adjustments are made to the line to minimize the number of roadways crossed and to cross roadways at midblock locations (to minimize the problem of vehicles at intersections). Stations are established at the cor-

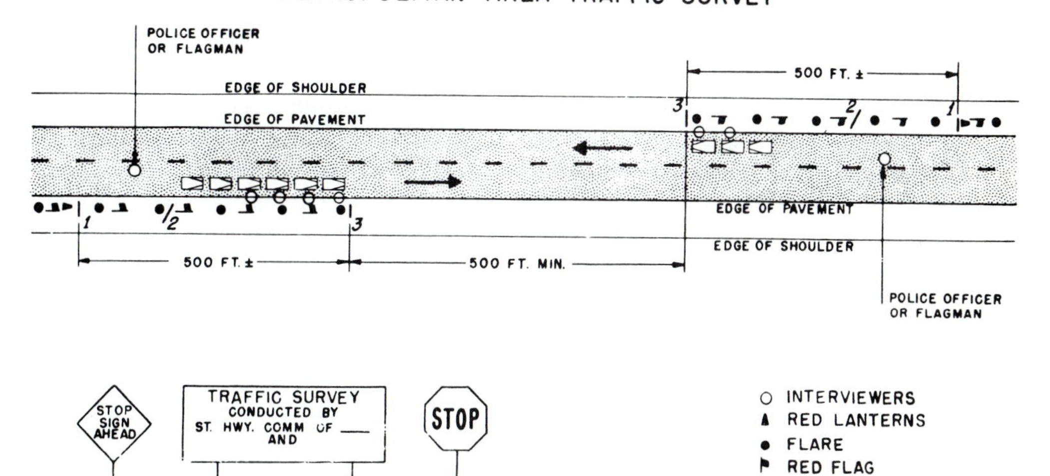

Figure 7-4 Layout of roadside interview station. *Source:* Box and Oppenlander, 1976.

don line on all intercepted streets; however, stations may not be placed on local streets known to carry negligible traffic volumes. Cordons must be large enough to define the area of interest, yet small enough to define an area useful for planning purposes. Cordons may define areas of similar land use.

Roadside Interviews

In comprehensive studies conducted for a large area or a metropolitan area, the roadside interview may be used for obtaining external travel information. The method is an integral part of the comprehensive O-D survey, with the interview stations being located along the external cordon boundary. This method has the advantage of permitting the observer to ask the motorist the purpose of the trip as well as the destination and origin. Interview stations are established at all major roads and most other roads crossing the cordon line encompassing the study area. Stations are located to intercept at least 95 percent of the crossing traffic and to minimize safety and congestion impacts. If the survey is concerned only with trip data on a single isolated route, driver interview taken at a single midpoint location might suffice. If data are desired on all traffic entering and leaving a small city, it is necessary to select interview locations on all routes radiating from the city.

Figure 7-4 illustrates a typical layout of a roadside interview station for a moderate-volume, two-lane, two-way road. The location should be fairly level and have a sight distance of over 800 feet. Where only part of the drivers can be interviewed, one or more bypass lanes are needed to avoid congestion. A paved shoulder of a rural route or a section of curb lane (cleared of parked cars) may be used for the interview, leaving the regular traffic lanes for use as a bypass.

Most stations are operated for 16 hours a day (6:00 A.M. to 10:00 P.M.). On major routes some stations are operated 24 hours. The 24-hour volume counts must be taken concurrently with roadside interviews. Data from the 24-hour count are used to expand survey data and to account for the sample of vehicles interviewed. Since freeways are normally too congested to permit stoppages for interviews, other techniques must be used to obtain travel information. Several alternatives are discussed below.

Stopping drivers usually requires the assistance of a police officer, slows traffic considerably, and may antagonize the public unless skilfully handled. Generally, a large portable sign explaining the project prepares the motorist for the delay and often enlists cooperation in answering the questions more readily. It may also be useful to involve the news media in explaining the need for the study. Every effort should be made to avoid congestion, not only for safety and maintenance of good public relations, but also because congestion may cause local drivers to detour around the interview station and thus distort the traffic flow patterns. Not every vehicle can be stopped on a high-volume route; usually, only a sample of drivers is interviewed. The required sample size is determined using the method described in Appendix B. Samples usually consist of 20 to 50% of the traffic. For a 50% sample it is satisfactory to interview drivers of three vehicles and pass the next three vehicles.

A typical crew at a location carrying 3000 to 5000 vehicles per day in both directions includes a party chief, two recorders, six interviewers, and one or two officers. The number of interviewers should equal the number of vehicles stopped in each group. Recorders count and record all traffic (by type and direction). A police officer may be used to hand each driver a card explaining the purpose of the survey as each car is stopped. Figure 7-5 shows a form for recording O-D information. Column 5 of this form may be used to record the purpose of the trip. The form should be modified to meet individual study needs. The sample data are then expanded to represent the total traffic volume and to obtain total estimated origins and destinations. This method has several advantages and disadvantages. The most complete and accurate information is usually obtained when personal contact is made between respondent and interviewer. The response rate is greater (relative to the voluntary return technique), thereby minimizing the survey bias, and samples can easily be chosen from a traffic stream to satisfy planned statistical standards. The roadside interview technique is more expensive than several other techniques, because more personnel are required. On high-volume facilities, there may be some traffic delays during the survey, especially during peak travel periods. This technique is often dangerous, especially on high-volume facilities, because survey personnel must operate on the highway and interfere with the regular flow of traffic.

A variation of the roadside interview is often used on comprehensive downtown parking studies to interview drivers who have just parked their cars. An interviewer is responsible for interviewing a sample of the parkers in a certain facility. Origin, destination, location parked, time of day, purpose of trip, and parking duration information are most often obtained. Details on this procedure are given in Chapter 10.

Postcard Studies

Where traffic is heavy and it is not possible to delay vehicles long enough to complete an interview, returnable postcards can be handed to drivers at the intercept stations. A survey may rely entirely on postcards; however, cards are often used in conjunction with interview studies. The postpaid, addressed cards are precoded with station identification and time. Drivers are asked to list the origin and destination of the trip and drop the card in any mailbox. The questions on the postcard should be simple. Normally, cards contain five to seven questions. Figure 7-6 shows a postcard questionnaire. As in roadside interview studies, 24-hour volume counts (by direction) should be taken concurrently. Data from the 24-hour count are used to expand survey data and to account for the sample of vehicles interviewed. A 25 to 35% return is common from this type of survey. A 30% return is considered excellent (ITE, 1992); a return of at least 20% is needed to maintain the accuracy of this type of study.

The postcard questionnaire technique is relatively inexpensive. Traffic delay is less than for direct interview, and untrained personnel can be used for handing out cards. A major disadvantage is possible sample bias due to better cooperation by some drivers. Care is required in location of distribution points to intercept a representative cross

Origin-Destination Study
Field Sheet

Location ____________ Station Number ____________

Time: Begin: ____________ Inbound ____________

End: ____________ Outbound ____________

1	2	3	4	5
Origin	Destination	Route Used	Parking	Other
Indicate by block, street, zone, or other city		Streets, zones, or highway	Location and type	

Date: ____________ Observer ____________

Figure 7-5 Origin–destination study field sheet. *Source:* Box and Oppenlander, 1976.

section of trips. Through trucks and passenger vehicles will not provide a high percentage return. It may be difficult to include all important vehicle movements, especially in large cities, and, like the interview method, it requires stopping traffic.

License Plate Studies

The basic concept is simple. As a vehicle passes each station, a portion (or all) of the license number is recorded, which permits vehicles to be tracked through the study area. Stations may be established on the boundaries of the study area and also at intermediate locations within the area. Plate numbers and the time are recorded for short periods of time (such as 1 minute). For the purpose of this study, the origin is the place where the vehicle is first observed, and the destination is where it was last observed.

DAN RYAN EXPRESSWAY ORIGIN AND DESTINATION STUDY

This Study is sponsored by the City of Chicago, County of Cook and State of Illinois

PLEASE ANSWER QUESTIONS AND DROP IN MAIL BOX - - NO STAMP REQUIRED

CARD NO. 3 5 9	STATION 8 4	DATE 5 2 3 3	HOUR	VEHICLE TYPE
Where Did This Trip Begin? Street Address, City	What Time Did You Leave Starting Point?	What Is The Purpose Of This Trip? To Work ☐ To Shop ☐ To Home ☐ To Others ☐	Please Check The Expressway Used (if any) On This Trip. If None, Please Check None. Lake Shore Drive ☐ Dan Ryan Expressway ☐ Congress Expressway ☐ Northwest Expressway ☐ Calumet Expressway ☐ Chicago Skyway ☐ None ☐	
Where Will This Trip End? Street Address, City	What Time Did You Arrive At Final Destination?	How Often Do You Make This Trip Each Week? 5 Times Or More ☐ 3 Times ☐ 1 Time ☐ Seldom ☐	What Streets Will You Use For The Longest Part Of Your Trip? East-West Street, North-South Street	
Where Will You Leave the Expressway on This Trip? Name of Street		Thank You Very Much For Your Cooperation.		

Figure 7-6 Postcard questionnaire. *Source:* Box and Oppenlander, 1976.

Observation points must be located carefully. If only information on through traffic is being collected, a minimum of two stations are needed: one representing the origin (for one direction) and the other the destination. To trace vehicle movements through an area, every reasonably heavy bypass point must be surveyed. In addition, intermediate stations may be used to establish routes or to give more detailed information on origins and destinations. The reduction of data requires considerable labor. A large amount of work is necessary to match the license plate numbers listed on the field sheets of each O-D station against nearby O-D stations to trace the route origin and destination of each vehicle. Simple computer programs and spreadsheets can be used in this process. Often, not more than 60% of the license numbers can be traced through this study. Trips may begin and end with the study zone and may never encounter a station. Errors in transcribing license numbers also reduce the number of matches made. The time between observations of a vehicle indicates fairly accurately whether stops are made in the business district, but knowledge of a time gap does not make it possible to ascertain the purpose of the stop.

License plate observations must be matched at a minimum of two locations. If 50% of all license plates were recorded at one location and 50% at another, it could generally be expected that only 25% of all license plates were observed at both locations. Sampling rates at various locations will differ widely depending on vantage points, volume, and skill of observers. Video cameras can be used to get a 100% sample at all locations.

McShane and Roess (1990) suggest a method for expanding license plate count to estimate total volume between points in this type of study. For this method, the total volume at all observation points must be recorded for the period of interest. Sample data are presented in Figure 7-7 and Table 7-1. The variables are defined as follows:

$$T_{ij}' = T_{ij}\frac{F_i + F_j}{2}$$

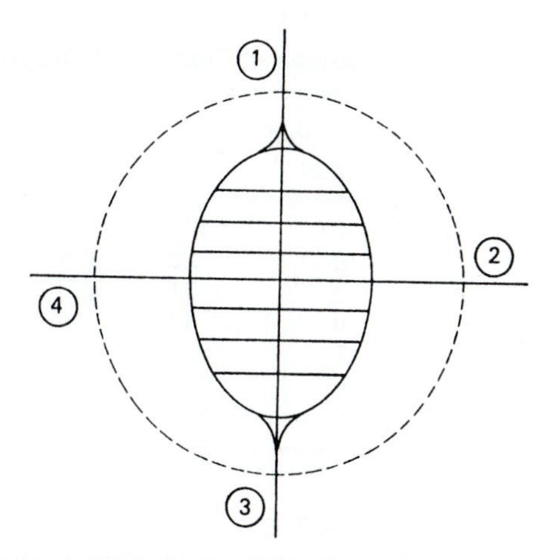

Destination Station	Origin Station				T_i	V_i
	1	2	3	4		
1	50	8	20	17	95	250
2	10	65	21	10	106	310
3	15	12	38	15	80	200
4	13	14	18	42	87	375
T_i	88	99	97	84	—	—
V_i	210	200	325	400	—	1135

Figure 7-7 Statistical expansion of license plate data. *Source:* McShane, William R. and Roess, Roger P., *Traffic Engineering*, © 1990, p. 107. Reprinted by permission of Prentice Hall, Englewood Cliffs, NJ.

$$F_1 = \frac{V_i}{T_i}$$

$$F_j = \frac{V_j}{T_j}$$

where

T'_{ij} = adjusted O-D volumes

T_{ij} = number of vehicles observed entering at i and leaving at j, obtained by matching license plate observations

V_i or V_j = volume counted at origin i or destination j

T_i = sum of observed vehicles T_{ij} at origin i for all destinations j

T_j = sum of observed vehicles T_{ij} at origin i for all destinations j

F_i, F_j = expansion factors

The O-D volumes of Figure 7-7 are expanded using these factors. Results are rounded to the nearest vehicle. The results are shown in Table 7-2. As an example, $T'(1,1)$ is calculated as

$$50\left(\frac{2.386 + 2.632}{2}\right) = 126 \text{ vehicles}$$

TABLE 7-1
Expansion of Illustrative Origin–Destination Data

Destination Station	Origin Station 1	2	3	4	T_i	V_j	F_j
1	126	19	60	64	269	250	0.93
2	28	158	66	39	291	310	1.07
3	37	28	106	55	226	200	0.88
4	44	44	70	191	340	375	1.07
T_i	235	249	302	349	1135	—	—
V_i	210	200	325	400	—	1135	—
F_i	0.89	0.80	1.08	1.15	—		—

Source: McShane and Roess, 1990, Table 6-13.

TABLE 7-2
First Iteration of Illustrative Origin-Destination Data

Destination Station	Origin Station 1	2	3	4	T_i	V_j	F_j
1	115	16	60	67	258	250	0.97
2	27	148	71	43	289	310	1.07
3	33	24	104	56	217	200	0.92
4	43	41	75	212	371	375	1.01
T_i	218	229	310	378	1135	—	—
V_i	210	200	325	400	—	1135	—
F_i	0.96	0.87	1.05	1.06	—		—

Source: McShane and Roess, 1990, Table 6-14.

The procedure should be iterated until the second-column O-D volumes are in less than 10% disagreement with observed volumes. This method does not work well in locations where a large number of trips begin and end between observation points.

Figure 7-8 shows a form that may be used to record license plates. In small surveys, only the last three digits of the license numbers need be recorded. Plate numbers may also be recorded on audiotape with a time queue. Data are later transcribed to paper forms or directly into a computer file. Video cameras may be used to record license plate data. Video has the advantage of providing a more permanent record of the data that may be played and replayed to obtain information. Most minicams also allow the time (and date) to be recorded directly on the video image (some to a tenth of a second). Cameras must be positioned carefully to record the plates of a wide variety of vehicles. Under certain lighting conditions, plates may be difficult to read. Reducing the data from video tape is time consuming and tedious. In a few years, new systems based on machine vision technology should allow license plate data to be reduced automatically from videotape. These systems, however, are not yet commercially available. The method is especially adaptable to locations where traffic is too heavy to be stopped for driver interview. License plate studies are also very useful in small-scale, limited O-D studies, or where the destination is known (CBDs, shopping centers, airports) (McShane and Roess, 1990). Tracing license plate numbers over a large number of exits and entrances is difficult. The method has the advantage of allowing the observers to obtain data without depending on the cooperation of individual drivers, as is required by interview methods. The likelihood of a biased sample because of poor driver cooperation is less with this method than with the methods described earlier. This method does not work well for large areas because of the extensive labor required, but is particularly adaptable to studies of single routes or facilities. Each study must be completed in 1 day, and it must be continuous. This method does not produce any information on the purpose of the trips, nor does it produce information on transit trips or vehicle parking.

License Plate Study
Field Sheet

Location ______________________ Direction of Traffic ______________________

Time: Begin: ______________________ Station Number ______________________

End: ______________________ Weather ______________________

License Number	Time	Truck or Bus	Out-of State ?	License Number	Time	Truck or Bus	Out-of State ?

Date: ______________________ Observer ______________________

Figure 7-8 License plate study field sheet. *Source:* Box and Oppenlander, 1976.

Vehicle Registrations

An alternate and far more detailed license plate technique involves recording the full license numbers, identifying vehicle ownership from registration records, and sending a mail-back questionnaire to each owner. This technique has the same advantages as the voluntary return postcard technique, of which it is a variation. It is less disruptive because traffic is not stopped. If video techniques are used to record license plate numbers, a smaller number of field personnel will be needed. Research indicates that the response rate to mailed-out questionnaires may be higher than that of the voluntary return-postcard technique. The disadvantages include the fact that personal contact is not made with respondents. Fewer questions can be asked. It is difficult to use economically and efficiently unless the agency responsible for motor vehicle registrations will provide addresses of drivers to whom questionnaires may be mailed.

Vehicle Intercept Method

The vehicle intercept method can be used for small study areas. Stations are established at all entrances and exits to the study area. Each entering vehicle is stopped and a precoded or colored card is handed to the driver with instructions to surrender the card as he exits the area. Exiting vehicles are stopped and the cards collected, or the notation that they had not received a card is made. Variations of this procedure are to use colored tape affixed to the bumper of the entering vehicle or to tape the colored card to the windshield. These variations eliminate the need for stopping vehicles at the exits from the area and permit the collection of data at intermediate locations (ITE, 1992).

Tag-on-Vehicle Method

The tag-on-vehicle method is a variation that does not depend on the complete cooperation of drivers. It may be used where traffic is too heavy for effective use of driver interview and where limited staff makes the use of the license plate recording methods impractical. A coded card is handed to the driver or fastened to the vehicle at it enters the route or area of study. The driver is informed of the nature of the survey and told that the card will be picked up when leaving the route or survey area.

When the vehicle leaves the route or area, the time, station, direction of travel and any other readily observed information are recorded on the precoded card. If traffic is too heavy, the cards may simply be bundled into groups according to the time intervals and the time written on the top card of each bundle. This method has many of the same advantages and disadvantages as the license plate method but requires some driver cooperation.

Lights-On Studies

In headlight or lights-on studies, individual vehicles are traced from one or two origin points to a maximum of two or three destination points, generally within ½ to 1 mile of each other. The study is conducted as follows:

1. Signs are posted asking drivers to turn on their headlights and leave them on until they pass an exit station.
2. The number of drivers complying is counted.
3. Lights-on vehicles and total vehicles are counted at destination points.
4. Signs asking drivers to turn off their lights and thanking them for their cooperation are posted at the exit to the study area.

A sample field setup and data analysis is shown in Figure 7-9. Flows from possible origins to destinations must be estimated and adjusted for drivers who do not comply. A method or estimating final O&D counts from the data is described in McShane and Roess (1990). This procedure works only during daylight hours and is effective for only relatively small study areas and under limited circumstances. Advance publicity is especially important to success because voluntary compliance by the driver is required. In a study of origins and destinations of vehicles on Shirley Highway approaching Washington, D.C., 85% of drivers complied with the "lights-on" request.

INTERNAL STUDIES

Internal studies provide information on trips made by residents of an area. These trips typically comprise the bulk of travel within an area. Variations of the home interview survey are a commonly used form of internal survey. Several other forms of internal sur-

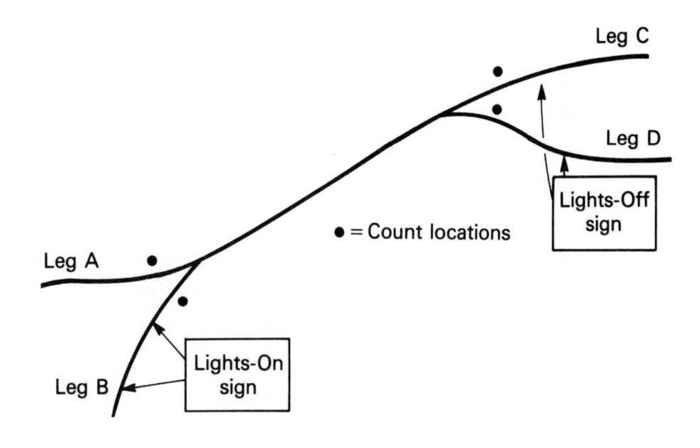

Time Period	*Counts*					
	A	*B*	*C(on)*	*C(off)*	*D(on)*	*D(off)*
5:00–5:15 A.M.	280	100	45	153	42	140
5:15–5:30	275	104	50	174	50	105
5:30–5:45	290	108	48	178	50	122
5:45–6:00	310	120	60	190	40	140
—	—	—	—	—	—	—
—	—	—	—	—	—	—
—	—	—	—	—	—	—
—	—	—	—	—	—	—
—	—	—	—	—	—	—
3:45–9:00	98	60	27	28	28	75

Figure 7-9 Application of lights-on study. *Source:* William R. McShane/ Roger P. Roess, *Traffic Engineering*, © 1990. Reprinted by permission of Prentice-Hall, Englewood Cliffs, New Jersey.

veys have been used, including controlled postcard surveys, multiple cordon surveys, and television as a replacement for the home interview (McGrath and Guinn, 1970). In addition, the license plate survey techniques discussed above can be adapted for internal studies. The types of studies discussed below include:

- Dwelling unit interviews
- Vehicle owner mail questionnaires
- Interviews at workplaces and special generators
- Transit route passenger questionnaires
- Truck and taxi surveys

This is a comprehensive type of survey to obtain information on all types of trips by residents in an area, including travel by public transit vehicles, trucks, taxis, and private vehicles. It is commonly made as part of a comprehensive metropolitan area O-D study. Manuals of detailed instructions and guidelines on this method have been prepared and should be consulted by anyone making this study. The basic concepts are described below.

Dwelling Unit Interviews

The study area is first divided into analysis zones that are as homogeneous as possible. Every dwelling unit in each zone is identified. Sample sizes are selected for each zone based on housing density, and the dwelling units to be surveyed are systematically selected within the zones. Each dwelling unit selected is initially contacted by mail to alert them to the studies and advise them that an interviewer will be contacting them in person. The mailing also indicates what information will be requested and approximately when the interview will be conducted. A trip diary should be included so that respondents can record information themselves. The interviewer can then confirm and supplement the self-recorded data.

Interviewers are hired, trained, and assigned to contact personally the residents of the selected dwelling units. They are instructed to interview only the selected units (i.e., they cannot substitute another dwelling unit if no one is at home at the selected unit or the selected unit is vacant at the time of the interview). A minimum of three repeat visits are made to residents who are not home during the initial contact. Interviews are usually conducted during the early evening hours (6:00 to 10:00 P.M.) when there is greater likelihood of people being at home.

The information collected includes social and economic data on all of the household residents as well as data on all trips made by each resident aged 5 or more during the preceding weekday. Dwelling unit data consist of the number of people living at the unit, the number of vehicles owned and/or operated by unit residents, and details on each person. Travel information is collected for each person living in the unit for trips made during the preceding weekday. Information includes time of trip, origin and destination, mode of travel, purpose of trip, and parking information for each trip by each person. Interviews are not generally conducted on Mondays because preceding weekday (Friday) trips are easily forgotten over the weekend.

Once the interviews have been completed, the data on travel trips are coded by origin and destination. Each trip end (origin or destination) is coded to the zone in which it occurred. The data are then expanded to the full sample using a zonal factor that is calculated by dividing the total number of dwelling units in the zone by the actual number of interviews completed successfully. Other factors are used to convert the data to an average weekday. The survey results are compared to the screen line counts and correction factors are developed as described later in the chapter.

Vehicle Owner Mail Questionnaires

This method involves mailing questionnaires or return-addressed, stamped postcards to owners of motor vehicles living within the survey area. This survey may be combined with driver interviews or postcard questionnaire surveys of traffic entering and leaving the study areas. The elements not obtained include taxi and transit patterns. The recipient is asked to record all trips made by motor vehicle on the day after the card is received, normally a weekday. A potential source of bias comes from the possibility of getting better response from certain vehicle owners. Different-colored cards may be used for private passenger vehicles and trucks.

Names and addresses of vehicle owners are obtained from state motor vehicle registration files. Cards are marked according to the address zone before mailing, and tabulations are made of the number of cards for each vehicle class sent to each zone. This enables proper estimation of the total trips based on the completed cards returned, as compared with the total vehicles listed for each zone. Truck fleet owners (three or more trucks) may be contacted personally, to assure better accuracy for this relatively small but important group. Some state highway departments have developed detailed manuals of procedure for this type of survey. An inexpensive variation of this method that has been used successfully is to send the questionnaire to the motorist along with the annual vehicle license certificate.

Interviews at Workplace and Special Generators

Questionnaires may be distributed to all employees of an activity center, such as a large industrial plant or a group of office buildings, or to people visiting an activity center such as an airport, shopping center, transit terminal, or other special generator. For employees, completed forms are picked up the same day that they are distributed. Information from large studies may be coded and punched into cards for electronic data processing. Data on where employees live, how they get to work, times of arrival and departure, parking information, and trip costs can all be obtained for auto drivers, auto passengers, and transit and taxi passengers. It is important to encourage a good response rate by keeping the questionnaire short. Preparation of questionnaires is reviewed in Appendix B.

This method is most effective when there are a few larger employers involved; however, it will work under other circumstances if the employers feel that the survey is important. After obtaining the cooperation of management, individual firms will often distribute and pick up the questionnaires within their own organizations. This type of study often yields information of direct value to the employer and thus helps to secure his cooperation. It is important to record the total number of forms distributed to each firm as well as the employment in each firm so that trip data for each company can be expanded properly. Similar surveys can be conducted at other special generators, such as airports and shopping centers. Depending on the circumstances, patrons can be interviewed or be given cards to complete and mail back.

Transit Route Passenger Questionnaires

This study is confined to the ultimate origins and destinations of passengers using a particular transit route. It is used primarily for planning improvements to the routes or schedules of transit units. One or two survey persons ride each transit vehicle and distribute a questionnaire card (and pencil) to each passenger who boards the vehicle. Passengers may be asked to complete the card and return it to field survey personnel when leaving the vehicle, or it may be mailed back. This study is best suited to lightly traveled lines where passengers are all seated. It is difficult for standing passengers to fill in the questionnaire cards. The results are expanded to represent 100% of the passengers, as based on the ratio of the total riders to the number of cards filled in.

When transit loads make it impossible to conduct this study properly on moving vehicles, return-addressed postcards may be used. The cards are handed to passengers as they board or alight, to be completed and returned at a later time. As with other postcard studies, the return rate may be low. Considerable care must be exercised in the analysis to make certain that the results are not biased. There may be a tendency for only the regular commuters or only those subject to most crowding, or other discomfort, to show interest in returning cards.

Truck and Taxi Surveys

Data on commercial trips (truck, taxi, and transit trips) are collected in separate surveys. Truck and taxi data are obtained by desk studies at the dispatch office of these vehicles because most of the data are available from office records. Studies of goods movement are discussed in Chapter 18.

PRESENTING O-D DATA RESULTS

A number of standard tabulations are recommended to summarize the basic trip data and to set forth the numbers and percentages of total trips made by car, transit, and taxi, as well as the numbers and percentages of trips made for various purposes. Other tables may be constructed to show the numbers of trips between zones by mode and purpose. Following the O-D survey, a set of trip tables is prepared that shows the number of trips between each zone in the study area. These tables can be subdivided by trip purpose, truck trips, and taxi trips. Tables are also prepared that list socioeconomic characteristics for each zone and the travel time between zones. Data presented in these tables may be graphed or plotted on maps for easy interpretation.

CHECKING SURVEY ACCURACY

There are several methods used to check the accuracy of all or portions of the comprehensive O-D survey. Some of these are listed below:

1. When routes of trips have been obtained in the interviews, two or three control points are selected. These should be important viaducts, bridges, or other well-known points of traffic flow constriction. Information obtained in the internal and external surveys, as expanded to represent 100% of traffic passing such points, is compared with actual field counts.
2. A natural barrier such as a river or a railroad track is selected as a screen line that divides the internal area into two parts. The estimated 100% count of traffic crossing the screen line as derived from the internal and external surveys is compared with actual field counts.
3. A cordon line comparison involves deriving the traffic at a cordon line station from the internal survey and comparing this with similar trips of area residents as derived from the external survey.
4. Public transit riding derived from the internal survey can be compared with total riding observed in the field or obtained from transit company records.
5. The accuracy of trip reporting may be estimated by selecting a zone in an area of high employment for which employment figures are available and comparing the number of persons employed in that zone with the number of work trips into the zones, as determined from the expanded interview data. In this comparison, work trips by all modes of travel must be included. Proper allowance must be made for those persons walking and bicycling to work, average absenteeism, and the likelihood of employees making midday shopping or business trips to and from the area.

If these comparisons reveal sizable discrepancies, adjustments should be made in the survey data to correct them. Figure 7-10 shows a comparison of screen line and O-D data.

ADDITIONAL SOURCES OF DATA

In addition to primary studies, where data are collected directly, there are a number of secondary data sources. The U.S. Census has proven the best secondary source of residential trip-production data (Dickey, 1983). More detailed information on household characteristics is available from the Urban Transportation Planning Package of the Bureau of Census. Traffic zones are often designed to follow Census tract boundaries. In those areas where traffic zones differ, the Census, for a small charge, will tabulate household characteristics using a block-to-zone aggregation table. Updates from the base year are sometimes necessary. Short of a survey, local records or city directories are the best source. Automobile registrations for a number of areas are commercially available. A few regions tabulate land use data regularly from local sources such as assessors' records, utility companies, and building or occupancy permits.

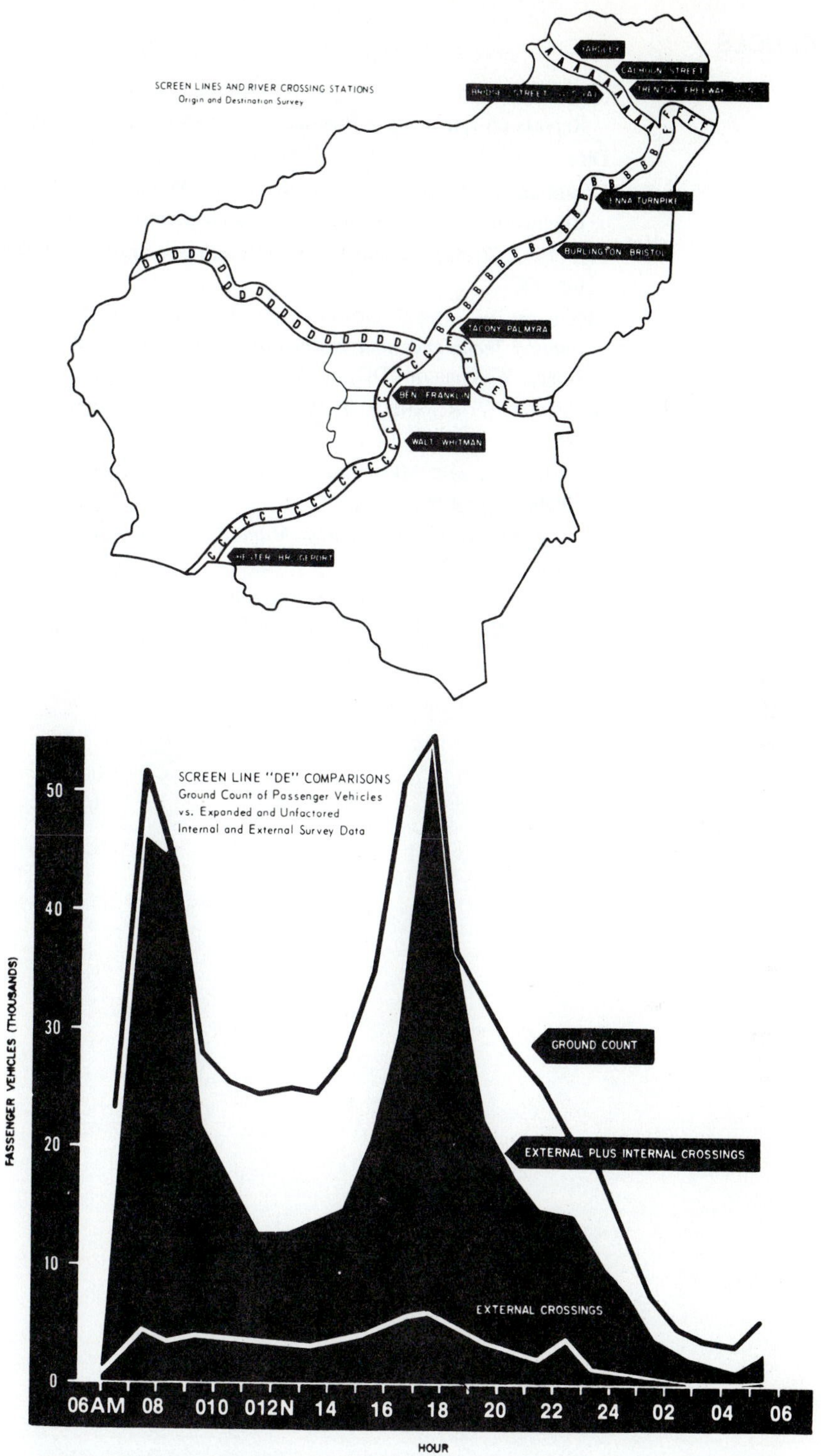

Figure 7-10 Comparison of screen line and O-D data. *Source:* Box and Oppenlander, 1976.

REFERENCES

ASCE (1986). *American Society of Civil Engineers Urban Planning Guide,* ASCE Manuals and Reports on Engineering Practice No. 49, American Society of Civil Engineers, New York.

DICKEY, J. W. (1983). *Metropolitan Transportation Planning.* 2nd ed., Hemisphere, New York.

HOMBURGER, W., L. KEEFER, AND W. R. MCGRATH, eds. (1982). *Transportation and Traffic Engineering Handbook,* Institute of Transportation Engineers, Washington, DC.

ITE (1992). *Transportation Planning Handbook,* Institute of Transportation Engineers, Washington, DC.

MCGRATH, W., AND C. GUINN (1970). *Simulated Home Interview by Television: Origins and Destination Techniques and Evaluation,* Transportation Research Record 41, Highway Research Board, Washington, DC, pp. 77–83.9.

MCSHANE, W. R., AND R. P. ROESS (1990). *Traffic Engineering,* Prentice Hall, Englewood Cliffs, NJ.

STOPHER, P., AND MEYBURG, A. (1975). *Urban Transportation Modeling and Planning,* D.C. Heath, Lexington, MA.

TRB (1985). *Highway Capacity Manual,* Special Report 209, Transportation Research Board, Washington, DC.

URBAN RENEWAL ADMINISTRATION AND BUREAU OF PUBLIC ROADS (1965). *Standard Land Use Coding Manual,* U.S. Government Printing Office, Washington, DC.

8

Environmental Impacts of Transportation Projects

Donna C. Nelson, Ph.D., P.E.

INTRODUCTION

Transportation systems affect the environment in many ways. The environmental impacts of transportation projects are coming under close scrutiny because of changing priorities across the United States and around the world. In the United States, governmental regulations require that an environmental impact statement (EIS) be prepared for any transportation project that will affect the human environment. For federal-aid transportation projects, the EIS must include studies of the social, economic, and environmental impacts of the proposed system (USDOT, 1975). A comprehensive EIS on a large project may include the following:

- Natural resources, including prime and unique farmlands, wetlands, threatened and endangered species, natural land forms, groundwater resources, and energy requirements
- Relocation of individuals and families, including the number of households displaced, neighborhood disruption, available housing, number of businesses displaced or affected, documentation of public participation, and any unusual circumstances
- Air quality studies, including microscale impacts, mesoscale impacts, analysis methodology, and a description of how consistent the project is with the state implementation plan (SIP)

- Noise impacts, including the identification of sensitive receptors (such as schools and hospitals), comparison of future noise levels with FHWA criteria and existing noise levels; noise abatement measures, and noise problems with no reasonable solution
- Wetlands and coastal zones studies to document the analyses, practical measures to minimize harm, and that there are no practical alternatives
- Social and economic impact studies of impacts on life-style, travel patterns, school districts, churches, recreation, businesses, minorities and ethnic groups, urban quality, and secondary impacts
- Water quality issues, such as erosion, sedimentation, use of deicing and weed control products, chemical spills, groundwater contamination, stream modifications, impoundment, as well as impacts on fish and wildlife
- Flood hazard studies: impacts on beneficial floodplain values, incompatible development, measures to minimize flood risks, and the evaluation of alternatives
- Construction effects: impacts of the construction process on air quality, noise, water, traffic detours, and the impact of spoil and borrow (the need to dispose of or find soil and other materials)

A detailed discussion of each of these studies is beyond the scope of this chapter. Many components of EIS studies fall beyond the usual practice of transportation and traffic engineers and should be conducted by a specialist in the specific area. The traffic impacts of new projects are discussed in Chapter 9. Many of the studies needed to supply basic data for the EIS (including volumes, speeds, traffic projections, and other basic studies) are discussed in other chapters in this book. The reader is referred to those chapters for detailed information on collection of that data.

NOISE IMPACT STUDIES

Highway noise studies are conducted to help determine the additional noise generated by the use of transportation systems in the community. To do this, the noise level generated by a new or improved facility is estimated and compared to measured ambient noise levels in the community in which the facility is proposed. The characteristics of environmental noise that are of particular concern are (Cohn and McVoy, 1982):

- Magnitude of the sound
- Frequency of the sound
- Temporal distribution of the sound
- Time variance of the sound

The *magnitude* of sound is perceived by the human ear as a short-duration fluctuation in atmospheric pressure. The level of sound pressure or the magnitude of a specific sound (or ambient sounds) is expressed in decibels (dB). The decibel scale is logarithmic; therefore, an increase of 1 dB reflects a tenfold increase in the sound pressure level. Sound pressure levels are generally adjusted to one of three scales: A, B, or C. The A-weighted sound-level scale is used to measure the magnitude of traffic noise because it most closely reflects the response of the human ear to transportation noise. The A scale (referred to as dBA) deemphasizes lower-frequency sounds.

The *frequency* (*f*) of a sound is determined by the number of times per second that the sound pressure fluctuates between positive and negative values on a sinusoidal wave. Noise from transportation sources are not usually pure tones but are broadbanded sounds with a wide frequency range.

Temporal distribution of noise is important because the time of day, day of week, and month or season affect the perception and effect of noise on the receiver. Noise levels are usually higher in the daytime than at night. However, nighttime noises may be perceived as more offensive. Noise levels acceptable at one time of day may not be acceptable at another time.

Time variance of sound refers to the fact that environmental noise is rarely stationary; that is, the magnitude often varies over very short periods of time. The passage of an automobile or airplane might cause the noise level in a quiet neighborhood to rise by 10 to 20 dBA for a short period of time. Several measures are currently in use to account for this variation.

- L50 is the sound level exceeded 50% of the time (i.e., the median sound level).
- The L10 level is the sound level exceeded 10% of the time.
- The L90 level is the sound level exceeded 90% of the time.

Noise Impacts

Although measurements of noise emissions from individual vehicles are commonly performed by traffic and environmental regulatory bodies, they do not generally fall within the sphere of interest of the traffic engineer. Two aspects of areawide noise levels are of concern: actual levels of traffic noise; and perceived annoyance of traffic noise. Interference with speech (including TV listening) and sleep are the most common complaints concerning transportation-related noise. There is no completely satisfactory measure of the subjective effects of noise. However, for the United States, the FHWA has developed maximum permissible noise-level criteria for all federally funded highways. These design goals require noise evaluations for new roadways and for existing roadways (when improvements affect noise levels). Evaluations must be performed and reported during the location planning and design phases. Table 8-1 presents a summary of the FHWA noise limits for various land use descriptions. Local areas may have additional noise criteria.

Predicting Noise Impacts

Prediction of traffic noise levels is difficult. *NCHRP Report 174* (TRB, 1976) includes a comprehensive design guide for highway noise computations. Noise prediction for highways is a three-step process using two different analysis techniques. A short nomograph–based approach is used to obtain a gross prediction of the expected noise levels. A computer-based or "complete" method is performed for a more detailed analysis of the proposed project. In the final step, noise mitigation alternatives are evaluated.

The short method is used to estimate potential problem areas when final horizontal and vertical roadway design parameters have not yet been determined, for example, during a location study where a number of alignments must be considered. The "complete" or computer-based method is used to predict highway noise levels at one or more locations in the study area. Noise contours are drawn (as shown in Figure 8-1) to identify locations that need special consideration, potential impact areas (where standards are exceeded), and roadway "noise" elements that contribute to excessive noise levels. Once the problem areas have been identified, the same method is used to evaluate noise control options. Possible options include changes in alignment and geometry and the use of roadside barriers or other shielding techniques in the right-of-way. A detailed discussion of the complete method is beyond the scope of this chapter.

The noise prediction worksheet for the short method is shown in Figure 8-2. Vehicle volumes (in vph) and average speeds are required for automobiles, medium trucks, and heavy trucks. A route map is used to identify all observer locations and the equiva-

TABLE 8-1
Noise Limits

Land Use Category	*Design Noise Level, L_{10} (dBA)*	*Description of Land Use Category*
A	60	(Exterior) Tracts of land in which serenity and quiet are of extraordinary significance and serve an important public need, and where the preservation of those qualities is essential if the area is to continue to serve its intended purpose. Such areas could include amphitheaters, particular parks or portions of parks, or open spaces that are dedicated or recognized by appropriate local official for activities requiring special qualities of serenity and quiet.
B	70	(Exterior) Residences, motels, hotels, public meeting rooms, schools, churches, libraries, hospitals, picnic areas, recreation areas, playgrounds, active sports areas, and parks.
C	75	(Exterior) Developed lands, properties, or activities not included in categories A and B.
D	55	For requirements on undeveloped lands, see FHPM 7-7-3(3).
E	55	Interior of residences, motels, hotels, public meeting rooms, schools, churches, libraries, hospitals, and auditoriums.

Source: FHPM, Vol. 7, Chap. 7, Sec. 3(3).

lent shortest distance in feet from the point where the noise level is to be calculated to the center of the traffic lane. Roadway geometry is used to determine shielding parameters, including line-of-sight distance, barrier position distance, the angle subtended, and breaks in the line of sight. In both methods, automobiles, light trucks, and heavy trucks are treated separately because their noise-generating characteristics are different. The noise ''source'' for automobiles and light trucks is considered to be at ground level because the primary radiated noise comes from the tires and pavement. The magnitude of the noise generated increases as vehicle speed increases. Of course, as the traffic volume rises, so does the noise level. Because light trucks are generally noisier than automobiles, light-truck volumes are multiplied by 10 before use in the calculations.

Measuring Ambient Noise Levels

Ambient noise is defined as the total noise composed of all natural and human-made noise sources that can be considered as part of the acoustical environment of the general area. Sources of ambient noise include aircraft and airports, railroad tracks, fire stations, emergency medical facilities (sirens), schoolyards and recreation areas, and areas where birds, crickets, or other noise-making wildlife congregate. The difference between predicted noise levels and existing ambient noise, together with information on the area itself, give an indication of the impact of the highway on the area. The methods and equipment described here are restricted to the type of field measurements necessary for highway noise studies.

Measurement Sites

Noise levels cannot be measured at every point in the study area. For projects covering large land areas, the study area is divided into representative sections enclosing

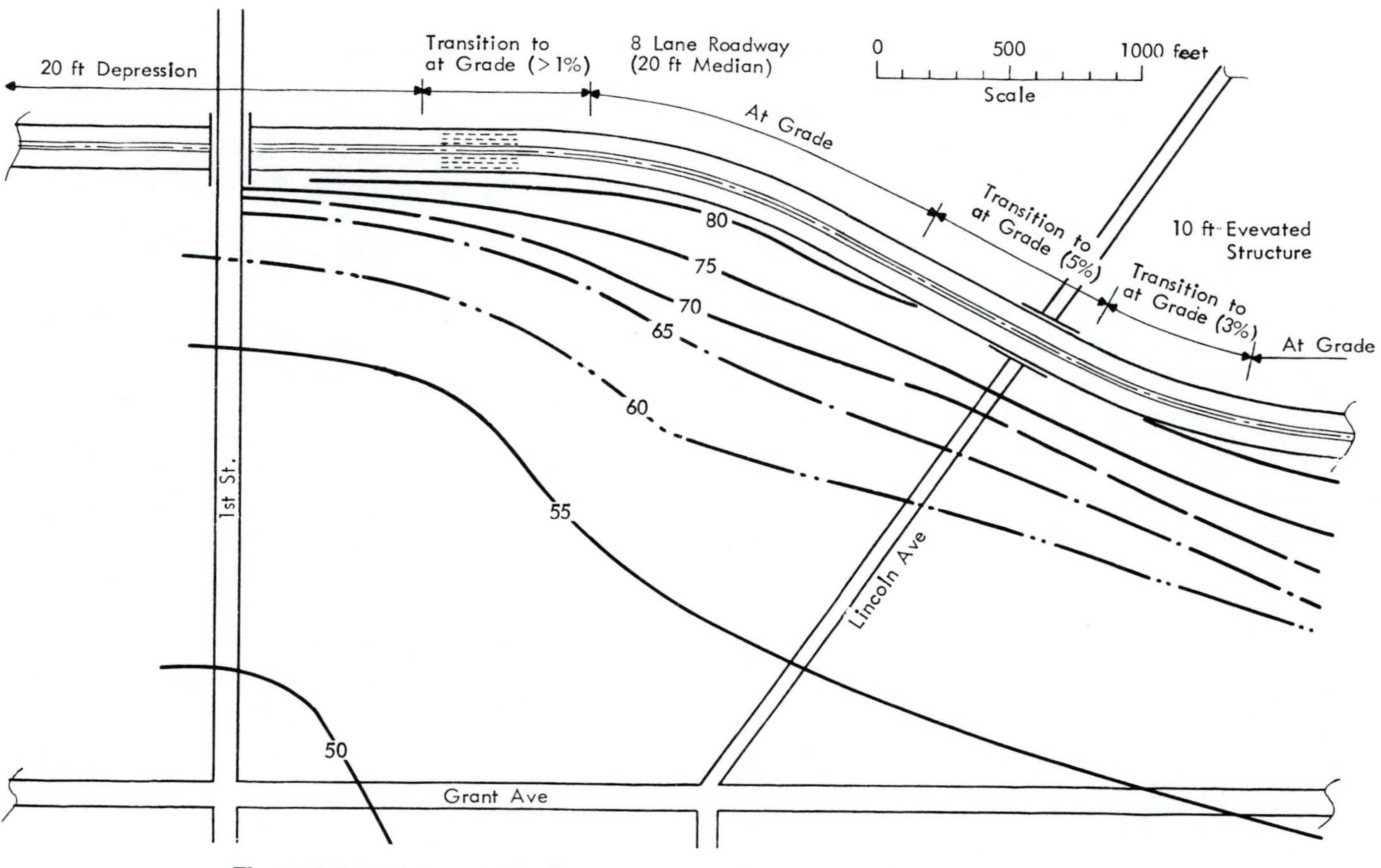

Figure 8-1 Noise contours. *Source:* National Cooperative Highway Research Program Report 174, *Highway Noise: A Design Guide for Prediction and Control,* Transportation Research Board, 1976.

NOISE PREDICTION WORK SHEET

Project Examples 1 and 2 Date January 1974 Engineer J. Smith

Step		Vehicle	Example 1A			Example 2A			Example 1B			Example 2B			Examples 2A and 2B			Complete Example 2		
			A	T_M	T_H	A	T_M	T_H	A	T_M	T_H	A	T_M	T_H	A	T_M	T_H	A	T_M	T_H
1	Traffic	Vehicle Volume, V(Vph)	3000			2000	100	100	-	-	-	-	-	-	2000	100	100	2000	100	100
2	Traffic	Vehicle Av. Speed, S (mph)	50			50	50	50	-	-	-	-	-	-	50	50	50	50	50	50
3	Traffic	Combined Veh. Vol.*, V_c (Vph)	-			3000			-			-			3000			3000		
4	Prop.	Observer-Roadway Dist., D_c (ft)	200			200			-			-			200			200		
5	Shielding	Line-of-Sight Dist., L/S (ft)	-			-			200			200			200			200		
6	Shielding	Bar. Position Dist., P (ft)	-			-			50			50			50			50		
7	Shielding	Break in L/S Dist., B (ft)	-			-			15		-	15		9	15		9	15		9
8	Shielding	Angle Subtended, θ (deg)							170°			170°			170°			170°		
9	Prediction**	Unshielded L_{10} Level (dBA)	66			66	-	68	-	-	-	-	-	-	66	-	68	66	-	68
10	Prediction**	Shielding Adjust. (dBA)	0			0		0	12.5		-	12.5		10	12.5		10	12.5		10
11	Prediction**	L_{10} at Observer (By Veh. Class)	66			66	-	68	-	-	-	-	-	-	53.5	-	58	53.5	-	58
12	Prediction**	L_{10} at Observer - TOTAL	66			70			-			-						59.2 or 59		

A = Automobiles, T_M = Medium Trucks, T_H = Heavy Trucks

* Applies only when automobile and medium truck average speeds are equal. $V_C = V_A + (10)V_{T_M}$

** If automobile-medium truck volume V_C is combined, use L_{10} Nomograph prediction only once for these two vehicle classes

Figure 8-2 Sample short method work sheet. *Source:* TRB, 1976.

similar noise environments. Particularly noise-sensitive locations are identified and studied separately. Four general categories of measurement site areas can be identified:

1. Sites near noise sources
2. Especially critical noise-sensitive sites
3. Residential areas
4. Remote areas use to establish the "noise floor"

Sites near noise sources are selected and studied to help calibrate and refine the preliminary noise level contours. Several sites should be selected with (as near as possible) a full view of any existing major roadways in the area. The measured values can be compared with the estimated noise levels for those positions. There should be moderately good agreement (approximately ±5 dB) between measured and calculated values. If measured levels and calculated levels do not agree reasonably well (and it is clear that ambient noise is made up largely of known traffic noise), this is an indication that either the calculated values do not properly represent the operational data or the measured levels do not correctly reflect the traffic and that some unusual effects should be sought out and explained.

Land use maps can be used to help identify the present noise sources in the area. Traffic noise prediction methods can be used to get a rough approximation of the existing noise environment due to traffic, and to estimate the noise levels expected from the new highway project. Preliminary noise contours for both the existing noise environment and the new project noise can be sketched on a land use map. From these data, noise-sensitive locations can be identified. Also, the area can be tentatively divided into smaller areas throughout which the existing noise environments are approximately uniform, and/or the anticipated noise impacts are approximately uniform.

Especially critical noise-sensitive sites such as schools, hospitals, and places of worship must be identified for ambient measurements. These areas must be quiet enough to allow clear speech indoors (and to some degree outdoors) and minimal disturbance of sleep. Other sites, such as school playgrounds and parks, music shells, and sports arenas require "acoustical privacy." Specific measurement sites should be located at the side of buildings or along the sides of outdoor areas that will face the proposed roadway. If future noise is a concern, additional sites may be selected at more remote locations. Upper-floor, outdoor noise readings should be taken if appropriate.

Residential areas include private residences, apartment buildings, hotels and motels, and nursing homes. Noise measurements should be conducted at representative sites in the affected area. If the residential area is too large to be described by a single ambient noise level or a small range of levels, it should be subdivided into appropriate smaller areas. For example, a large residential area near an existing traffic artery may be broken down into three groups of sites: one group located at the edge of the area adjacent to the existing highway, where the ambient noise is clearly due to the highway traffic; a second group located toward the interior of the residential area, where the arterial traffic is still a major factor in establishing the noise but where other noises of the community are beginning to make significant contributions; and a third group deep enough into the community that the only noise measured is from the community itself. For residential sites, ambient measurements should be made in the areas where human use typically occurs. Outdoor measurements are usually taken within 10 to 20 feet of the building.

Remote areas should be selected to determine the lowest ambient levels (the noise floor). These sites should represent the quietest regions in the area under consideration. It does not matter whether the ambient noise is due to natural or human-made sources. The sources of the sounds heard at these positions should be recorded on the data sheets.

Measurement Times

Ambient measurements should coincide with peak-hour traffic volumes for key sites. Depending on the purpose of the study, off-peak and nighttime measurements may also be taken. Under some circumstances a 24-hour study period may be desirable.

Noise Measurement Procedures

Noise can be measured directly using a sound level meter (SLM) or with a tape recorder and statistical analyzer. A simple SLM can be used to determine L_{10} values with reasonable accuracy.

Equipment

The sound-level meter is the simplest device available for highway noise measurements. The American National Standards Institute (ANSI) has set specifications in ANSI Standard S 1.4-1971 for four types of sound-level meters (SLM):

- *Type 1:* Precision
- *Type 2:* General purpose
- *Type 3:* Survey
- *Type S:* Special purpose

The type 1 SLM, the most accurate, exceeded the specifications generally required for highway noise measurements. In most cases, a type 2 sound meter is used for traffic measurement. The type 3 SLM, however, is too inaccurate for traffic study purposes. Operation of the sound meter also required a microphone (and cable), a preamplifier, a calibrator, a windscreen, a set of headphones, and a tripod. Other useful equipment includes a screwdriver to adjust the calibration level, a tape measure, a clipboard, a stopwatch, and data sheets.

A tape recorder can be used to record ambient noise in some cases (TRB, 1976). A high-quality full-range tape recorder is not essential for measuring highway noise. The overall frequency response may be flat (within ± 1 dB) and the frequency range between 50 and 5000 Hz. The tape speed should be 7.5 in./sec; however, a tape speed of 3.75 in./sec can be used if only sound-level measurements are needed. The procedure for using a tape recorder for noise measurements is described below.

Setup

The operations manual for the specific equipment should be read carefully. Generally, the microphone or sound-level meter is mounted on a tripod so that the person taking the readings has both hands free. In addition, hand-held microphones will bias the data. The microphone is usually positioned 4 feet above the ground for ground-floor ambient noise measurements. The microphone may be supported outside upper-level windows but as far as possible from the exterior wall of the building (at least 3 to 4 feet) for upper-level measurements.

The SLM should be calibrated in the field before and after each measurement session. It is a good precaution to calibrate sound meters periodically in the laboratory as well. Calibrators are standardized, stable sources that generate a predefined sound pressure level. When placed on the microphone, the meter deflection can be adjusted to a fixed value. The manufacturer's instructions must be followed carefully. A device that cannot be properly calibrated should not be used until it is repaired.

If a tape recorder is used, its frequency response should be measured in the laboratory periodically. The gain should be set permanently at a suitable value to avoid errors when the tape is played back. The tape recorder may also be calibrated using the

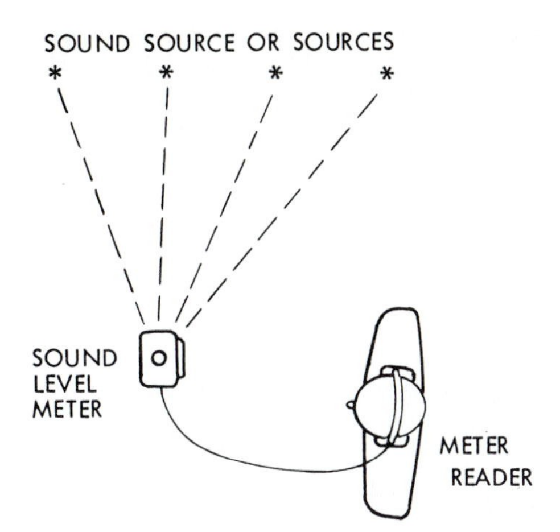

Figure 8-3 Observer position. *Source:* TRB, 1976.

techniques described above. The calibration tone should be recorded on the tape for approximately 30 seconds, and the position of the sound meter attenuator should be noted, preferably on the tape. The list of equipment used, including serial numbers, should also be recorded verbally on tape. The tape should be played back on the same machine that was used to record it.

Checklist

The following checklist provides an outline of the steps for the ambient noise measurement process (Dickey, 1990).

1. Gather and organize the required equipment.
2. Measure on the A-scale and on the ''fast'' mode.
3. Calibrate the sound meter before and after each set of measures.
4. When reading the meter, stand as far away as possible, standing so as to interfere the least with the sound field.
5. Avoid making measurement under extreme weather conditions. Make sure, by listening with the headphones, that no popping sound occurs. Repeat the measurements several days apart to determine if atmospheric conditions have affected the data.
6. When on the site, listen to the sounds of the neighborhood and make a list of those that may be considered exceptional or accidental. Include only the reasonable and representative in the ambient measurements.
7. Select ambient measurement sites representative of the four categories listed above: (a) noise-sensitive sites, (b) residential areas, (c) sites near noise sources, and (d) remote areas for noise floor.

Data Collection

The observer can be positioned as shown in Figure 8-3. The observer should be positioned to present a minimum body frontal area to the sound wave and the minimum possible interference with the sound field near the microphone. Data may be recorded on a form similar to that show in Figure 8-4. The A-levels are grouped in ''windows,'' each 2 dB wide. Work from left to right within each window. A watch (or stopwatch) is strapped to the top of a clipboard holding the data sheets. With the clipboard in one hand, the sweep second-hand or digital readout can be watched and the A-level recorded on the data sheet every 10 seconds. The earphones, which are connected to the sound level

AMBIENT NOISE SURVEY DATA SHEET

POSITION:
ENGINEER:
JOB NO. ________
DAY OF WEEK:________ DATE:________ TIME: BEGIN ________ FINISH:________
CAL: BEGIN ________ FINISH:________
NOTES AND SKETCH:
SKY: ________
WIND: ________
dBA L_{10}: ________
LIMITS, dBA: ________

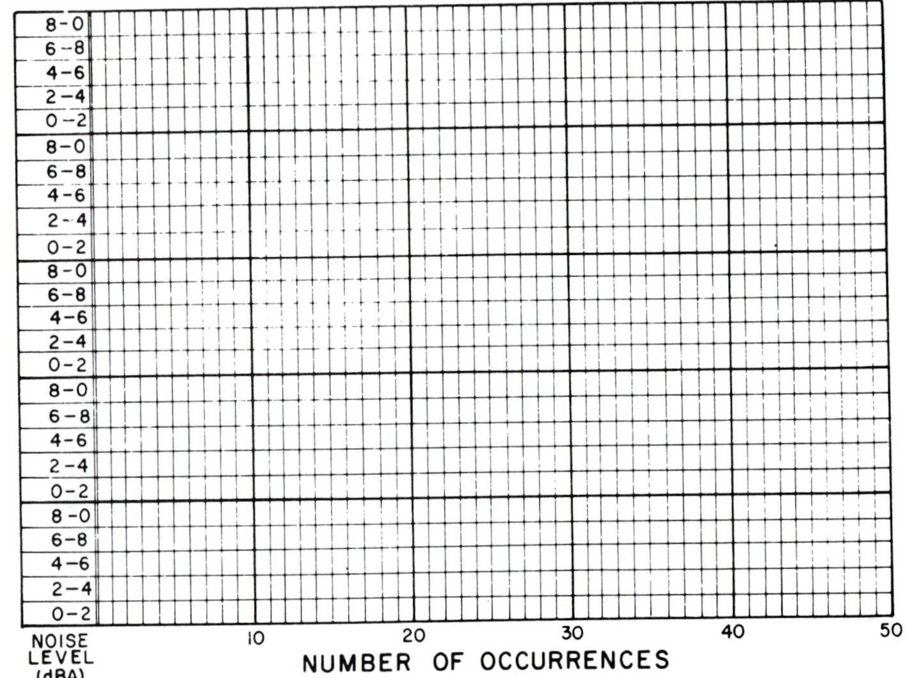

Figure 8-4 Sample field sheet for ambient noise measurement. *Source:* TRB, 1976.

meter output, should be worn at all times. The meter should generally be set on the A-scale and for fast response.

After 50 samples (8 minutes, 20 seconds), test the samples by the criteria listed below. If the samples meet the criteria, the measurement is complete. If not, another 50 samples must be taken and the test repeated.

1. Select the confidence level desired for the study. (The 95% interval is usually sufficient for highway noise evaluations.)
2. From Table 8-2, select the upper error limit, lower error limit, and L_{50} value.
3. Count down from the top of the data sheet (and from left to right within each window), and circle the samples found in step 2.

If these three test samples fall into one, two, or three adjacent windows, the measurement is complete, Otherwise, another 50 samples must be taken and the test made again. For a sample of 50 (with 95% confidence), the first, fifth, and tenth samples from the top would be selected. These three test samples constitute the L_{10}, flanked by its up-

TABLE 8-2
Statistical Criteria

Number of Samples	*Upper Error Limit*	L_{10}	*Lower Error Limit*
	95% Confidence Level		
50	1	5th sample	10
100	4	10th sample	16
150	7	15th sample	23
200	11	20th sample	29
250	15	25th sample	35
300	19	30th sample	41
350	24	35th sample	46
400	28	40th sample	52
	99% Confidence Level		
50	1	5th sample	11
100	2	10th sample	18
150	5	15th sample	25
200	9	20th sample	31
250	12	25th sample	38
300	16	30th sample	44
350	20	35th sample	50
400	24	40th sample	56
450	16	45th sample	62
500	17	50th sample	68
550	18	55th sample	74
600	19	60th sample	79
650	19	65th sample	85
700	20	70th sample	91
750	21	75th sample	97
800	21	80th sample	102

per and lower error limits. For the data sheet in Figure 8-4, the upper limit, the lower limit, and L_{50} fall within three contiguous rows; therefore, no more data must be taken at that observation point.

Results

After each group of 50 samples, L_{10} has been determined with 95% confidence to fall between the upper and lower error limit test samples. A-level values are assigned by selecting the highest A-level in the upper limit window. For uniformity, the L_{10} is chosen to be the center of the L_{10} window. For the example in Figure 8-4, the results would be stated as L_{10} = 49 dBA, within maximum limits of 46 and 50 dBA. The measurements are considered complete when this 95% error range is 6 dBA or less.

AIR QUALITY IMPACT STUDIES

Deteriorating air quality is a severe problem in many urban and suburban areas. The 1990 Air Quality Act will significantly update and refocus the role transportation agencies play in maintaining and improving air quality (Shrouds, 1991). Greater burdens are placed on "transportation" by increasing the contributions that transportation plans, programs, and projects must make toward air quality improvements in those regions that do not meet air quality standards (nonattainment areas) (Homburger et al., 1982). Through the act, Congress has directed that air quality planning must be carried out on a continuing basis and that it must become an integral part of regional development and transportation planning.

Air Quality Impacts

Air quality impacts are determined in respect to ambient concentration standards. The impacts of a proposed transportation project are usually analyzed by (Cohn and McVoy, 1982):

1. Projecting the amount of traffic expected to result from the project
2. Calculating the quantity of pollutants that will be emitted by the projected traffic
3. Estimating the resultant concentration of the pollutants of interest for a particular receptor site, using a dispersion model or some other analysis tool
4. Adding the traffic-generated pollutant concentration to an expected background concentration generated by other pollutant sources
5. Comparing the results to the ambient standard for various project alternatives

The emissions of interest in most transportation-related air quality studies include carbon monoxide (CO), hydrocarbons (HC), nitrous oxides NO_x), lead (where leaded fuel is still in use), and ozone.

Air quality impact analysis for a highway or an airport is commonly conducted to (Cohn and McVoy, 1982):

- Determine whether or not the proposed project is likely to cause a violation of the ambient air quality standards.
- Compare the relative impacts of the various project alternatives, including the null (or do nothing) alternative.
- Make an informal determination regarding consistency with the state implementation plan (SIP).
- Plan to defend the analysis against criticism from professionals with a vested interest in opposing (or supporting) the project.

There are two basic approaches used in assessing the air quality impacts of highways and airports: mesoscale and microscale analysis. *Mesoscale* or *emissions burden analysis* involves the calculation of total emissions generated by the various project alternatives and is most often used in connection with an assessment of photochemical oxidant impacts as required by many SIPs. For highway projects, the scope of the analysis usually includes emissions from the proposed project as well as from the surrounding highway network which might experience changes in traffic levels as a result of the proposed project. For a large highway project, a comprehensive mesoscale analysis might involve the calculation of emissions for hundreds of individual highway segments for each of several project alternatives for 3 or 4 years of analysis (e.g., the year of completion, 10 years after, and 20 years after). These calculations are generally performed with the aid of computer-based models.

Microscale analyses generally employ a dispersion model to predict concentrations at critical receptor sites. The object is to assess the air quality impact of a proposed action on a particular receptor (or receptors) and generally involves the use of a dispersion model. The pollutant of interest in microscale analysis is usually carbon monoxide, although lead emissions occasionally warrant a microscale investigation. The accepted procedures for conducting a microscale analysis vary somewhat from state to state and evolve with time. The state air pollution control agency and other reviewing agencies should be contacted before analysis is begun.

Emission Rates

The calculation of emission rates is an important step in the air quality impact analysis process. The way in which emission rates are developed for various pollutants depends on the analyses to be performed. Carbon monoxide (CO) standards are usually calculated on a peak-hour basis to determine a g/m·s (gram per meter per second) emission rate for use in a dispersion model. This will yield a concentration value for comparison with an ambient standard. CO emissions may also be calculated on a total daily or yearly basis for use in adjusting background concentrations, developing emission burdens for rollback calculations, or selecting a critical year for analysis.

Hydrocarbons (HC) emissions are usually calculated on a daily basis for a total-burden analysis to demonstrate conformity with state implementation plans. Total summertime HC emissions (from 6:00 to 9:00 A.M.) may also be calculated. The EPA believes that this time period is critical with respect to the formation of photochemical oxidants. Nitrogen oxide emissions are often calculated on a daily or yearly basis. Lead emissions from vehicles burning leaded fuel is a significant source of ambient lead in the environment. Average lead emissions are important in areas where leaded fuel is still used.

Air Pollutant Dispersion Models for Impact Analysis

If the amount of a pollutant is known or can be estimated, dispersion models can be used to predict concentrations at various locations. There are several models available that model the dispersion of pollutants. Some of the more common programs available for the highway environment include HIWAY 2, CAL3QHC and CALINE-3.

Input Data Requirements

Certain basic types of data are required to assess potential air quality impacts. Models most commonly require information on vehicular emissions, meteorology, traffic flow characteristics, ambient air quality, topography, land use, and geometric configuration of roadway. Vehicular emissions describe emission rates for each specific pollutant in terms of mass per unit distance per vehicle, generally in grams per vehicle mile. These emissions are a function of several parameters including vehicle type, speed, mechanical condition, and type of fuel. The most widely used method for determining vehicle emissions is that developed by the U.S. Environmental Protection Agency (for automobile, diesel and gasoline trucks, etc.). Ambient air temperature and the fraction of vehicles in warmed-up or cold operation condition are used to define an average composite emission factor for each specific situation. The procedures are updated periodically, and projections for future-year emission are based on current federal automotive emissions standards as prescribed by the Clean Air Act Amendments.

Construction Impacts

The air quality impacts from highway construction activities come from three basic sources: dust, traffic congestion due to construction, and emissions of construction equipment. When land is cleared of vegetation and the soil dries out, fine-grained soil material can be picked up by the wind. Some types of soils are more susceptible than others to this problem. The U.S. Department of Agriculture, Soil Conservation Service (SCS) has mapped much of the nation's soil and described the susceptibility of different soils to wind erosion. This information can be used to estimate the magnitude of the dust problem and to recommend mitigation techniques, including minimization of the amount of land cleared, replanting, use of dust suppression methods, and others. These types of controls are part of the standard specifications for most transportation projects.

Air quality impacts due to traffic congestion occur when access to existing highways must be restricted during construction. The impacts of these diversions can be assessed in much the same manner as any other highway project; however, due to their temporary nature they are usually determined to be insignificant. Impacts associated with construction also involve emissions from construction equipment. Emission rates for construction equipment are published by EPA and impacts can be analyzed using conventional modeling techniques. In most states, construction impacts are considered temporary and/or insignificant. Their discussion in air quality impact reports is usually a commitment to undertake all appropriate mitigation measures.

Air Quality Reports

Air quality reports should be written for professional reviewers. It should contain all the information necessary for the reviewer to assess the air quality impacts of the proposed project and should contain sufficient information for another professional to reconstruct the entire analysis. Every detail does not have to be included; however, the information regarding the model inputs, receptor locations, 1/8-hour conversion, and so on, should be made available simply in a convenient format. The report should also contain a summary suitable for inclusion in an EIS written in terms readily understood by the layperson (Cohn and McVoy, 1982).

Contents of Air Quality Reports

- Project description
- Map showing location
- Nature of project
- Alternatives to proposed activity
- Unusual topographic features
- Ambient air quality
- Monitoring? If so, when, where, how, and results
- Future background concentrations: 1-hour and 8-hour
- Microscale analysis
- Emissions calculations: specification of traffic volumes, vehicle mix, speed, operation modes, ambient temperature, and year of analysis
- Dispersion calculation for critical receptors, including model uses, receptor locations and how determined, wind direction, wind-speed stability, emission height, receptor height, 1-hour results, 8-hour results, and comparison with standards
- Lead analysis
- Mesoscale analysis (the regional emissions burden as appropriate)
- Emissions calculation assumptions
- CO, HC (6 to 9 A.M.), NO_x
- Definition of analysis area
- Analysis for year of completion
- SIP conformance
- Construction impacts
- Discussions of alternatives
- Conclusions

REFERENCES

COHN, L. F., AND G. R. MCVOY (1982). *Environmental Analysis of Transportation Systems,* Wiley, New York.

DICKEY, J. W. (1990). Models IV: Transportation system impacts, Chapter 8 in *Metropolitan Transportation Planning,* Hemisphere Publishing, Bristol, PA.

HOMBURGER, W., L. KEEFER, AND W. R. MCGRATH, eds. (1982). *Transportation and Traffic Engineering Handbook,* Institute of Transportation Engineers, Washington, DC.

SHROUDS, J. (1991). Course notes, *The 1990 Clean Air Act and the Transportation Engineering Profession,* Institute of Transportation Engineers, Washington DC.

TRB (1976). *Highway Noise: A Design Guide for Prediction and Control,* National Cooperative Highway Research Program Report 174, Transportation Research Board, Washington, DC.

USDOT (1975). *Environmental Assessment Notebook Series,* U.S. Department of Transportation, Washington, DC.

9

Traffic Access and Impact Studies

Donna C. Nelson, Ph.D., P.E.

INTRODUCTION

Typically, traffic access and impact studies are conducted to assess the transportation impacts of proposed developments and other land use changes. The proposed development could be a new office building, subdivision, factory, or shopping center. Proposed changes in land use might include the redevelopment of an existing area into an area that includes a mix of uses. Traffic impact studies project future transportation demands, assess the impact of changes in demand, and suggest ways for mitigating the adverse effects of land use changes in defined geographic areas. For these studies, *transportation demand* is defined as the need for movement of people and goods by all forms of transportation, including autos, car pools, transit, taxi, trucks, and bicycles, and movement of pedestrians in the vicinity of a proposed development.

The design and implementation of meaningful impact studies is a complex process. The results and recommendations of the studies are heavily dependent on the experience and knowledge of the persons conducting the study, as well as those reviewing the study. This chapter provides an outline of the analyses required with an emphasis on the collection and organization of materials and data for traffic access and impact analyses. A complete study draws upon data collection and analysis procedures described in a number of chapters in this book as well as several other references. These include *Traffic Access and Impact Studies for Site Development* (ITE, 1987a), *Trip Generation* (ITE, 1987b), and *Highway Capacity Manual* (TRB, 1985).

Computer-based methods are used for many of the analysis steps required for traffic access and impact analysis. For example, computer-based methods are typically used for level of service and capacity analysis, as well as for most steps in trip distribution, traffic assignment, and modal choice modeling. Available software packages will also estimate trip generation and parking generation for various land uses.

PURPOSE OF STUDIES

Traffic impact studies are conducted to evaluate the impacts of proposed land developments on an existing transportation network and to assist both public and private planners in making decisions with regard to the allowance or disallowance of major land use changes and new developments. These studies evaluate changes in traffic attributable to the proposed changes and translate them into transportation impacts in the vicinity of a development. Most studies also define the on-site and off-site transportation system improvements needed to accommodate the additional traffic generated by the development. The recommended road improvements and other transportation improvements are termed *mitigation measures.*

Often traffic impact studies are conducted within the context of a larger *environmental impact report* (EIR). State and federal law mandate these studies for all projects conducted under the auspices of a public jurisdiction that may have significant impacts on the natural and human environment within a reasonable area surrounding the project. The traffic impact study is incorporated into an *environmental impact statement* (EIS) or an EIR. An EIS/EIR has two primary purposes: determining and disclosing all significant environmental impacts of a proposed project; and identifying mitigation measures that could reduce or otherwise compensate for environmental disruptions. In many urban areas, traffic impacts and mitigations are routinely among the most visible and controversial aspects of environmental studies.

PREPARATION AND DESIGN OF STUDIES

Site traffic access and impact studies should be prepared under the supervision of a qualified and experienced person who has specific training in traffic and transportation engineering with several years of experience related to preparing traffic studies for existing or proposed developments. Also, traffic access and impact study reviews should be conducted by properly trained transportation engineers and/or transportation planners. All jurisdictions impacted by the proposed changes should also be offered an opportunity to review and comment on the study. Currently, many agencies require a registered professional engineer (P.E.) or traffic engineer (T.E.) to conduct or review these studies. The road authority and any agency affected by proposed changes should adopt an approval process and variance procedure to diminish future litigation concerns.

NEED FOR STUDIES

Warrants for traffic access and impact studies are addressed in a variety of ways. There is no general consensus on when a study should be done. However, data collected by the Institute of Transportation Engineers (ITE, 1987a) indicate that the need to conduct a traffic access/impact study is commonly determined by the following conditions:

- When a new development will generate (add) more than a specified number of peak-hour trips

- When a development will generate more than a specified number of daily trips
- When more than a specified amount of acreage is being rezoned
- When development contains more than a specified number of dwelling units or square footage
- At the judgment or discretion of public agency staff
- When development is in a sensitive area
- When changes are proposed in an area already suffering from congestion

The *Report on Traffic Access and Impact Studies and Impact for Site Development* (ITE, 1989) recommends that in lieu of other locally established thresholds, a traffic access/impact study should be conducted whenever a proposed development will generate 100 or more added (new) peak direction trips to or from the site during the adjacent roadway's peak hours or the development's peak hours. The rationale supporting this recommendation is that:

1. One hundred vehicles per hour are of a magnitude that can change the level of service of an existing intersection approach.
2. Left- or right-turn lanes may be needed to accommodate site traffic satisfactorily without adversely affecting through (nonsite) traffic.

Virtually any major traffic generator (which may include approved or anticipated developments) must be considered as a potential candidate for traffic impact analysis. Examples include high-density residential areas, offices, retail/commercial hotels, business parks, hospitals, schools, industrial facilities, and stadia.

STUDY TIMING

Transportation needs should be a major consideration for new or expanding developments throughout the planning stages, including site selection. Detailed formal studies, however, may only be required at specific development planning stages. Under normal circumstances there are several stages in the development process where traffic access/impact studies are potentially appropriate:

- Zoning and rezoning applications
- Redevelopment
- Land subdivision applications
- Environmental assessment
- Site plan approval
- Building permit application
- Formation of a special-purpose district
- Development agreements
- Amendments to comprehensive plans
- Permits for major driveways
- Annexations
- Signal warrants

Separate studies are not needed at each development stage. However, studies completed very early in the development process may need to be updated to include additional detail as the site plans become specific. In some cases, the planning process will result in a substantial reformulation of the development program and plan, resulting in a need for reanalysis and re-examination of study findings and conclusions. The initial study should be reviewed at each phase of the development to ensure consistency with the current development plan or to indicate the need for additional study because of substantial changes in impact over those predicted initially. For staged developments and projects, the original studies should be checked at each stage and updated as appropriate.

STUDY COMPONENTS

The major components of traffic access/impact studies include (1) definition of the scope and extent of the study, (2) collection of data on existing conditions, (3) site traffic forecasts, (4) nonsite traffic forecasts, (5) traffic assignment, (6) analysis, and (7) recommendations. The precise components and level of detail of an individual study will vary depending on the size, the type of land use and complexity of the multiple use development, the existing conditions of the local network, and the requirements of the approving agencies.

Scope of Study

The first step in the process is to identify the issues and needs of the particular study. It is critical to discuss the project with the reviewing agency's staff at an early stage in the planning process. The agency is a potential source of data for the study. The agency may be aware of projects approved for construction in the area, proposed changes in geometrics at key intersections, and other factors that will affect traffic patterns and the requirements for the study. Some reasonable agreement with the approving agency should be reached on the scope of study and appropriate assumptions for the analysis. Examples of issues that may be discussed with the reviewing agency include (ITE, 1987a):

- The components of a study needed to address issues associated with the particular site, the proposed development, and the existing transportation system
- The level of detail needed for the trip generation forecast (single or multiple use)
- If standard trip generation rates are appropriate or if a study must be conducted to determine generation rates
- If the analysis should include passby traffic and a modal split analysis
- If both internal circulation and external analysis should be done, and the level of detail required
- The extent to which approved projects adjacent to the development should be considered
- The use of areawide growth estimates and future traffic assignments
- If planned or phased transportation improvements should be considered, and how this should be done
- For a multiphase development, if the phases should be analyzed individually, and if so, what horizon years should be used
- Identification of area of influence and key intersections for analysis
- Selection of evaluation criteria for intersections based on delay or upon intersection capacity utilization (ICU)
- Estimation of air quality and noise impacts
- Consideration of pedestrian, bicycle, and transit requirements

Study Horizons

Suggested study horizons are shown in Table 9-1. The study horizons should be discussed and agreed to by the reviewing agency. Commonly, the target year of the impact study is at full build-out and occupancy of the project, or the horizon year of the planning studies

TABLE 9-1
Suggested Study Horizons

Development Characteristic	*Suggested Horizon*
Small development (< 500 peak-hour trips)	Anticipated opening year, assuming full build-out and occupancy
Moderate single-phase development (> 1000 peak-hour trips)	Anticipated opening year, assuming full development build-out and occupancy
	Five years after opening date
Large single-phase development (> 1000 peak-hour trips)	Anticipated opening year, assuming full build-out and occupancy
	Five years after full build-out and occupancy
	Adopted transportation plan horizon year if the development is significantly larger than that included in the adopted plan or travel forecasts for the area
Moderate or large multiple-phase development	Anticipated opening years of each major phase, assuming build-out and full occupancy of each phase
	Anticipated year of complete build-out and occupancy
	Adopted transportation plan horizon year if the development is significantly larger than that included in the adopted plan or travel forecasts for the area
	Five years after opening date if complete by then and there is no significant trip generation increase from adopted plan or area transportation forecasts (e.g., at least 15%)

Note: Peak-hour trips based on ITE *Trip Generation.*
Source: ITE, 1987a.

for a metropolitan area. The latter may provide a greater database to aid in the evaluation; however, some metropolitan plan horizon years have not been extended far enough into the future to encompass the major build-out of all components of the project.

Study Area Data

The study should incorporate all transportation and land development information that is considered current for the area. Suggested background data are shown in Table 9-2. As suggested by the ITE (1987a), specific data requirements vary according to the complexity and scope of the study. Studies will frequency include some or all of the following:

- Peak-period turning movements for site and street
- Adjustment factors to relate count data to design period
- Machine counts to verify peaking characteristics
- Primary traffic control devices
- Signal phasing and timing
- Roadway configurations, geometric features, and lane usage
- Parking regulations
- Posted speeds
- Driveways across from or adjacent to site
- Transit stops; passenger pickup and dropoff volumes
- Adjacent land use and zoning
- Pedestrian volumes on adjacent streets and crosswalks

The assembly and organization of available data should be accompanied by detailed reconnaissance of the project site, area roadways, and the surrounding vicinity. Inventory all relevant characteristics and information needed for the analysis, and observe existing

TABLE 9-2
Suggested Background Data for Review

Category	*Data*
Traffic volumes	Current and historic daily and hourly volume counts Recent intersection turning movement counts Seasonal variations Projected volumes from previous studies or regional plans Relationship of count day to both average and design days Pedestrian volumes
Land use	Current land use, densities, and occupancy in vicinity of site Approved development projects and planned completion dates, densities, and land use types Anticipated development on other undeveloped parcels Land use master plan Zoning in vicinity Absorption rates by type of development
Demographics	Current and future population and employment within the study area by census tract or traffic zone (as needed for use in site traffic distribution)
Transportation system	Current street system characteristics (including direction of flow, lanes, right-of-way, access control, and traffic control including signal timings) Roadway functional classification Route governmental jurisdiction Traffic signal locations, coordination, and timing Adopted local and regional plans Planned thoroughfares in the study area and local streets in vicinity of site, including improvements Transit service and usage Pedestrian and bicycle linkages and usages Available curb and off-site parking facilities Obstacles to the implementation of planned projects Implementation timing and certainty of funding for study area transportation improvements (whether or not it is funded in current capital improvement program)
Other transportation data	Origin–destination or trip distribution data Accident history (3 years if available) adjacent to site and at nearby major intersections if hazardous condition has been identified

Source: ITE, 1987a.

traffic conditions. Only those data needed to address issues to be studied must be collected. The data need to address both short- and long-term issues. Only current data should be used. In areas undergoing change, data should be less than 1 year old. In more stable areas, older data may be used provided that studies are conducted to verify that current conditions are reflected. Any adjustment factors applied to survey data in the report must be described and justified. The report should include data representing conditions appropriate for the analysis, such as average, design day, or seasonal peak traffic counts, and surveys factored to represent all members of a surveyed population. All procedures and factors should be summarized in the report. Methods of presenting these data are discussed below.

Study Area Definition

Define the study area to include all portions of the transportation network that may be affected by the proposed development. Focus the analysis on the segments of the surrounding transportation system where users are likely to perceive a change in the existing level of service. Roadway intersections are often a major area of impact. Include all known major intersections that will be affected by the proposed development. As an example, the key intersections and project location are clearly identified for a study as shown in Figure 9-1. The location and size of all approved projects in the vicinity of the project site should be presented as shown in Figure 9-2.

Existing Transportation System

The existing transportation system in the area of influence should be described as shown in Figure 9-3. This figure should show the existing roadway system serving the site, including all major streets, minor streets adjacent to the site, and site boundaries. The figure should show all transit, bicycle, and major pedestrian routes (if applicable), as well as right-of-way widths and signal locations.

Selection of Analysis Periods

In general, the critical time period for a given project is directly associated with the peaking characteristics of both project-related travel and areawide transportation system. The peaking characteristics of the adjacent street and highway system are determined by analyzing the traffic count data for the area and projecting peak demand periods for the development. Typical peak traffic flow hours for selected land uses are shown in Table 9-3. In general, observed peak periods occur during weekday morning (7:00 to 9:00 A.M.) and evening (4:00 to 6:00 P.M.) hours, although local area characteristics may result in other peaks. Some land uses, such as schools and entertainment facilities, often have schedules that do not conform to the "normal" 9 to 5 day; peak volumes for shopping centers may occur on weekends. If these land uses constitute a relatively large proportion of the traffic generated in the area, they may have a significant impact on the peaking characteristics. Additionally, peaking characteristics may change over time, especially in growing areas. Trip generation and general traffic levels also vary daily and seasonally. Land uses such as shopping centers, banks, and restaurants exhibit different daily patterns. For example, large shopping centers (over 400,000 square feet) should be analyzed for the period between Thanksgiving and Christmas, traditionally the busiest shopping season of the year.

To establish peaking characteristics in the vicinity of the site, hourly bidirectional counts must be obtained. The time period(s) that provides the highest cumulative directional traffic demands should be used to assess the impact of site traffic on the adjacent street system and to define the roadway configurations and traffic control measure changes needed in the study area. Intersection volume data are collected during this time period. For the count data plotted in Figure 9-4, traffic demand peaks between 5:00 and 7:00 P.M. for both directions. Turning movement counts should be conducted during this period.

Fifteen-minute turning movement counts should be obtained for intersections and driveways during the period. The peak hour can be identified more closely by selecting the four 15-minute periods that contain the highest total counts by direction. Existing directional volumes can be presented as shown in Figure 9-5. Other methods for presenting turning movement volumes include tables, graphic intersection summary diagrams, and intersection flow diagrams. Examples of these are given in Chapter 2. These data will be used for the level-of-service analysis for these locations.

SITE TRAFFIC FORECASTS

The potential traffic impacts of a planned development are forecast for the projected conditions in the horizon year(s) of the project. The steps in the process include trip generation, trip distribution, traffic assignment, and modal choice.

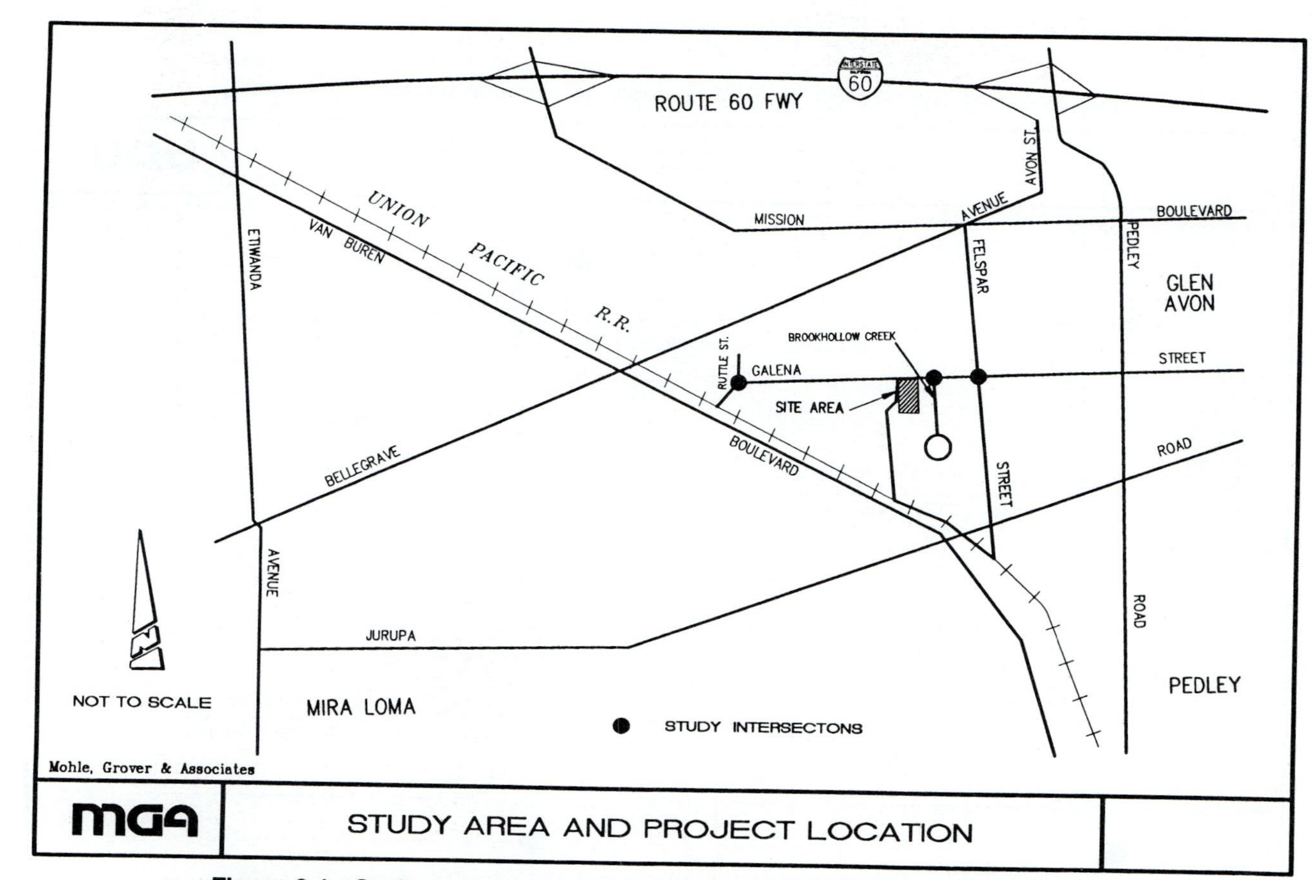

Figure 9-1 Study area and project location may be illustrated on a single map.
Source: Mohle, Grover and Associates, La Habra, CA.

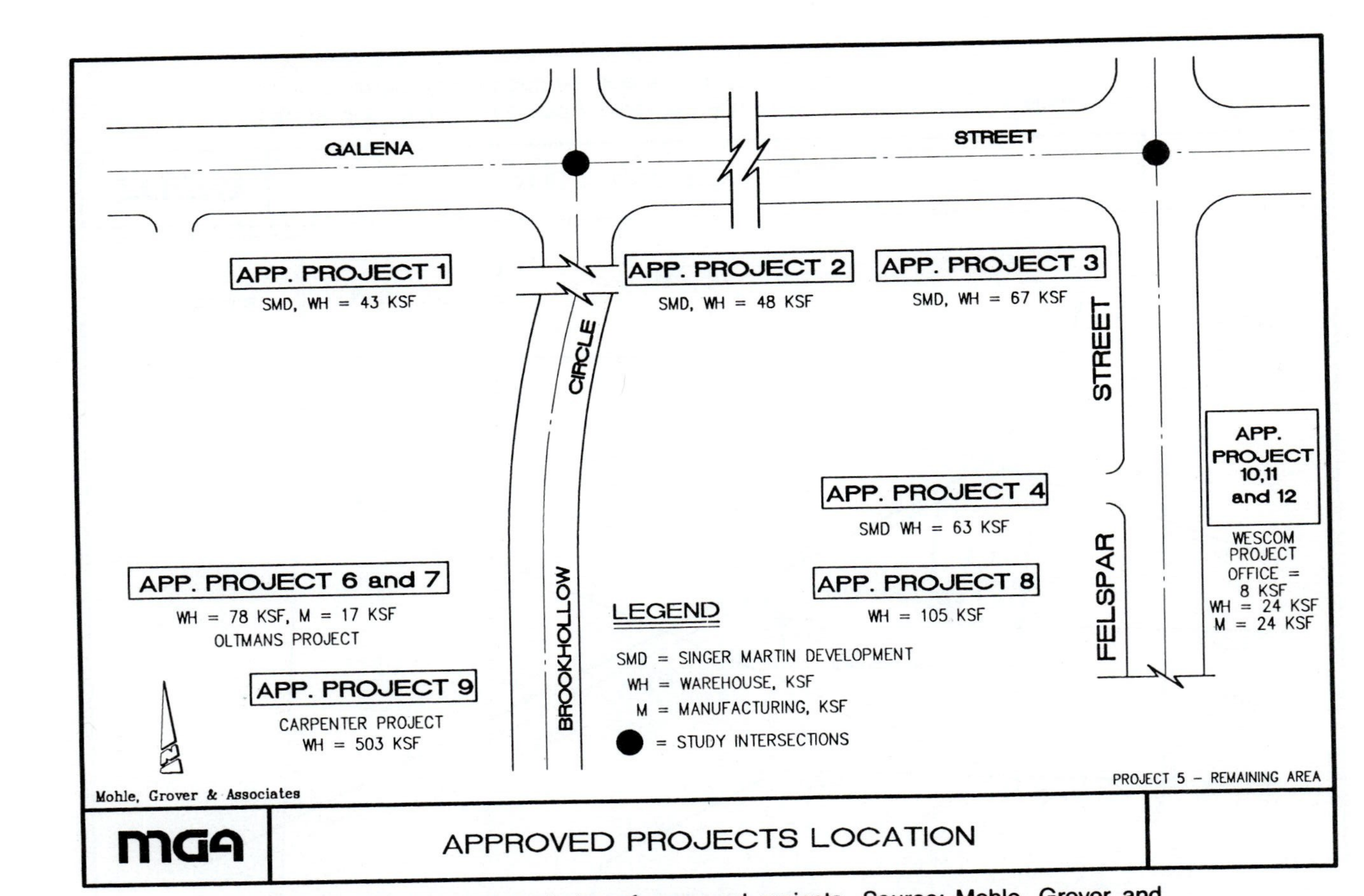

Figure 9-2 Location and size of approved projects. *Source:* Mohle, Grover and Associates.

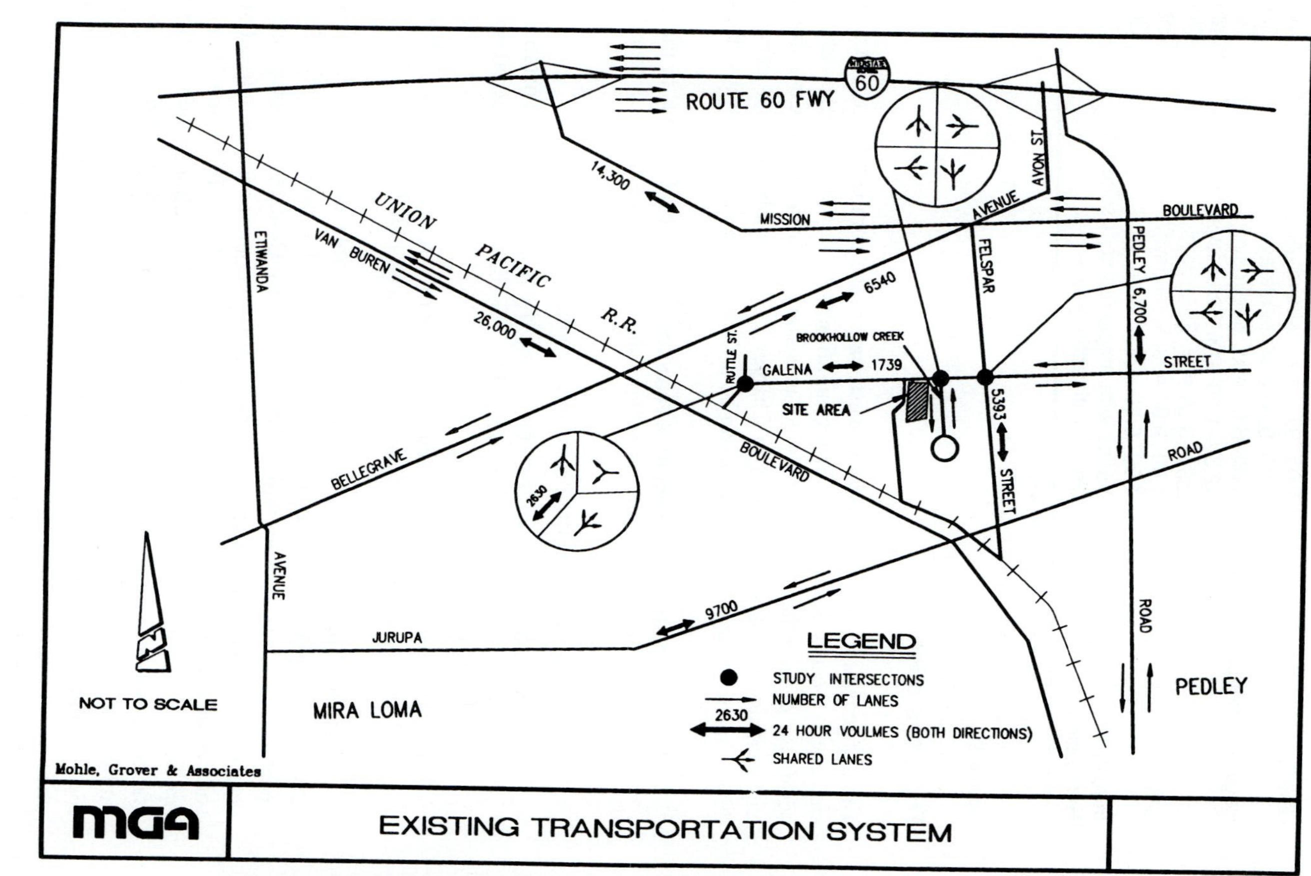

Figure 9-3 Existing transportation system for the influence area of the project.
Source: Mohle, Grover and Associates.

TABLE 9-3
Typical Peak Traffic Flow Hours for Selected Land Uses

Land Use	*Typical Peak Hours*[a]	*Peak Direction*
Residential	7:00–9:00 A.M. weekdays	Outbound
	4:00–6:00 P.M. weekdays	Inbound
Regional shopping center	5:00–6:00 P.M. weekdays	Total[b]
	12:30–1:30 P.M. Saturdays	Inbound
	2:30–3:30 P.M. Saturdays	Outbound
Office	7:00–9:00 A.M. weekdays	Inbound
	4:00–6:00 P.M. weekdays	Outbound
Industrial	Varies with employee shift schedule	
Recreational	Varies with activity type	

[a]Hours may vary based on local conditions.
[b]Period of maximum weekday traffic impact.
Source: ITE, 1987b.

Trip Generation

The trip generation process provides an estimate of the number of trips that will be generated due to the new development. Trip generation rates are applied to the various land uses within the development. It is usually not necessary to carry out the extensive data-based trip generation analysis described in Chapter 7. The number of trips generated by the development and approved projects can be estimated in a variety of ways. These include:

1. Determining trip generation rates for similar developments in the area for the same time of day and multiplying the rate per unit area on a prorata basis (see FHWA, 1985)
2. Using trip generation rates from a similar area
3. Obtaining trip generation rates from the Institute of Transportation Engineers' publication *Trip Generation* (ITE, 1987b) or other published sources (e.g., TRB, 1978)
4. Using the techniques available on microcomputer software such as the Quick Response System documented in Horowitz (1987)

Published national rates or local rates acceptable in the responsible jurisdiction may be used. Many state, regional, and local agencies have established their own trip generation database for sites within their boundaries. Contact the appropriate agencies to determine if such a database exists and if it is appropriate for the current study. Other sources of trip generation are *NCHRP Report 187* (TRB, 1978) and *Development and Application of Trip Generation Rates* (FHWA, 1985). National sources can be used as starting points in estimating the amount of traffic that may be generated by a specific building or land use. Whenever possible, these national rates should be adjusted to reflect local or forecasted conditions. These national sources should not be used without the application of sound judgment.

If existing local data samples are limited, collect additional local data to provide a credible sample size on which to base the trip generation estimate. Local trip generation data should be collected at sites that exhibit similar characteristics to the development being studied and that are self-contained, with adequate parking not shared by other activities. The following guidelines are suggested for collecting data on a similar site (ITE, 1987a).

- To obtain daily machine counts, select a generator where automatic counts can be made without double-counting turning vehicles and without counting through traffic. Directional counts should be in 15-minute periods.
- To check daily volume trends, take daily counts for at least a 3-day period.

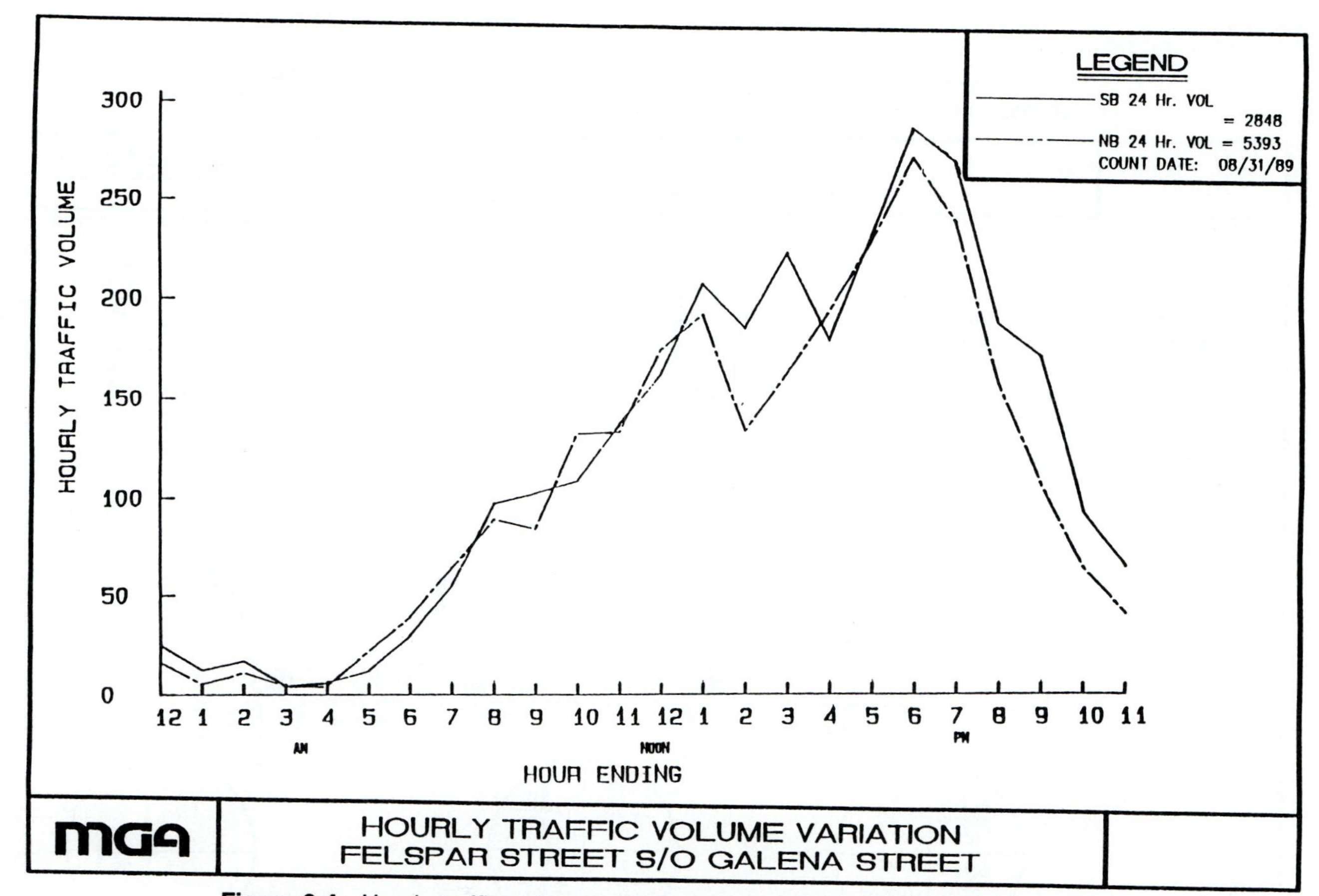

Figure 9-4 Hourly traffic volumes by direction are shown adjacent to the project.
Source: Mohle, Grover and Associates.

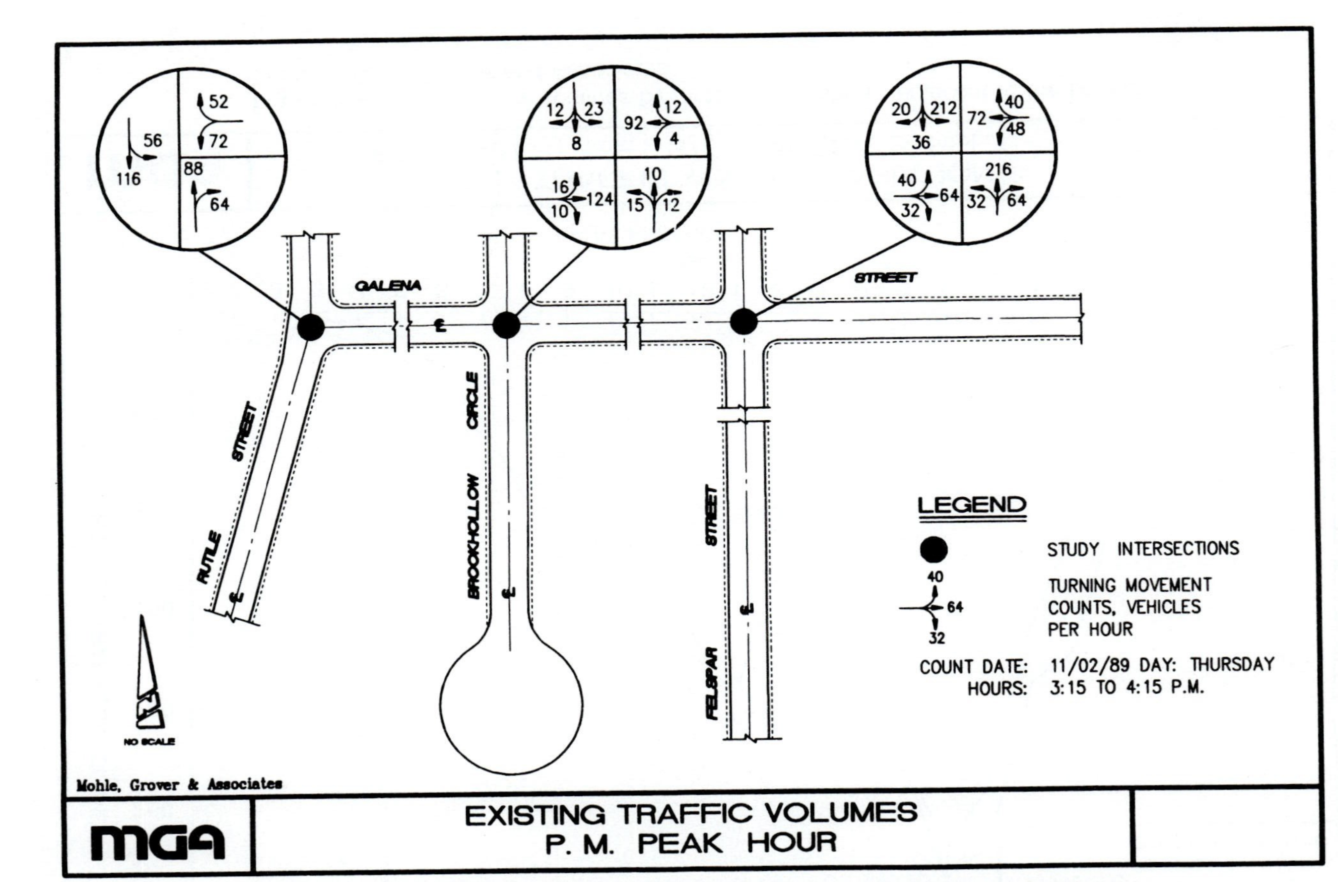

Figure 9-5 Existing traffic volumes for a peak hour can be shown by direction for intersections within the study area. *Source:* Mohle, Grover and Associates.

- If peak traffic hours of the development are unknown, conduct counts during a typical week of the year to provide data concerning the weekday and weekend peak hours.
- For uses that do not demonstrate substantial weekly or seasonal variations, select *average days* for the analysis.
- For developments that exhibit major seasonal variations, *design days* (approximating the 30th highest hour) should be selected.
- If only peak-hour data are needed, conduct manual counts for several hours on a typical weekday to record inbound and outbound vehicular traffic and compare these values with corresponding automatic counts at the same location to determine a counter factor for adjusting the raw automatic counts.
- Vehicle classifications and occupancies may also be counted if relevant to the analysis.
- Weekend counts may be needed to cover developments with peak activity on Saturdays or Sundays.
- Spot counts of pedestrians, 15 minutes for each intersection crosswalk.

If there is reason to believe that travel characteristics for the proposed development will be unique, annual counts or controlled interviews are needed to determine average weekday person trip ends by mode, the number of trips actually generated by the site, and the number of trips attracted to the site from traffic passing the site on the adjacent street. Information on site/development characteristics of the survey generator may be obtained through an interview with the site owner or manager, telephone conversations, mail-back questionnaires, and/or measurements as necessary. Obtain information on as many variables as possible to determine which is the most closely correlated to trip generation. The data used to apply ITE trip generation rates are listed in Table 9-4.

Trip Distribution and Assignment

Once the generated trips are estimated, they must be distributed to geographic origins or destinations and assigned to specific sections of the transportation network. With the most common method of predicting trip distribution and assignment, the forecast is based

TABLE 9-4
Dependent and Independent Variables Used for ITE Trip Generation Rates

	Data Required	
Location	Within CBD Urban (non-CBD) Suburban (non-CBD) Not given	Suburban CBD Rural Freeway Interchange area (rural)
Independent variables	Number of employees Number of persons Number of units Number of occupied units Building area (GSF) Percent occupied Net rentable area (ft^2) Gross leasable area (ft^2) Occupied gross leasable area (ft^2)	Acres Number of parking spaces Number of occupied beds Number of seats Shopping center percent outparcels/pads Other
Other data	Vehicle occupancy Percent by transit in 24 hours A.M., P.M. Percent by carpool/van in 24 hours A.M., P.M. Number of full-time employees by shift Parking cost on site, hourly, daily	

on existing trip patterns. Traffic generated at the project's driveways is distributed based on the directional split of current traffic on the roadways. At each intersection, project traffic is assigned in the proportion to existing turning movements. Examples of the trip distribution for a development and nearby approved projects are shown in Figures 9-6 and 9-7.

Areawide travel demand models are another source of trip distribution and assignment data. Use of travel demand models is most appropriate when substantial changes in land use, transportation facilities, or both are anticipated during the analysis period. Most modeling efforts are based in part on origin–destination surveys. The results of such surveys, stratified by trip purpose and translated into a matrix of trip exchanges among transportation analysis zones, can provide a geographic distribution of travel demand to and from the project zone. Many computer-based travel demand modeling packages permit the user to extract the traffic assignment for a given traffic zone.

Modal Split

Modal split refers to the distribution of all person trips generated by development among the various transportation modes available: automobile (drive alone or shared ride), transit, motorcycle, bicycle, and walking. In the 1960s and 1970s, *modal split* generally referred to automobiles and transit only. Now all forms of transportation are considered: automobile (drive alone and shared ride), carpools, transit (public and private), motorcycle, bicycle, and walking. In suburban areas of the United States and Canada the automobile is usually the dominant mode of access; however, other modes should not be ignored. The importance of transit increases in more intensely developed areas; in some large central business districts, transit is the dominant mode during peak periods. Most published trip generation data are based on surveys of vehicle trips. Such surveys are generally conducted in suburban areas where sites with separate parking lots make it relatively easy to track vehicles entering and leaving a given site. Thus if a traffic impact study is being performed in an urban area where the use of nonauto modes is significant, adjustments for modal split will be required. *Trip Generation* (ITE, 1987b) describes how to account for reductions in vehicular trip generation due to transit use.

One method of estimating modal split is by analyzing existing modal splits for similar developments in similarly located sites with similar transit service levels. Local transit agencies can often provide historical data on transit use in the project area. Assuming a constant level of transit service into the future, these modal split data can then be applied to the prospective trips generated by new development.

Areawide models are useful for projecting modal split for planned transit services. They can also project the effects of transit's modal share on major changes in the overall transportation and land use patterns of a community. Only limited data are available for estimating the percentage of trips made using modes other than auto and transit. Metropolitan planning organizations (MPOs) occasionally have survey data on prevailing use of walking and cycling for some trip purposes. A survey should be considered if it appears that such modes account for more than 5% of all trips generated. In the absence of information, mathematical models are used to forecast the probability that a tripmaker will use transit. These models are often described in terms of a disutility function that accounts for the generalized costs that are associated with modal choice. The QRS procedure (Horowitz, 1987) estimates modal split in this fashion.

NONSITE TRAFFIC FORECASTS

In addition to traffic generated by the site and approved projects, changes may occur outside the study area that will affect the transportation system in the horizon years. Estimates of nonsite traffic are required for a complete analysis of horizon-year conditions.

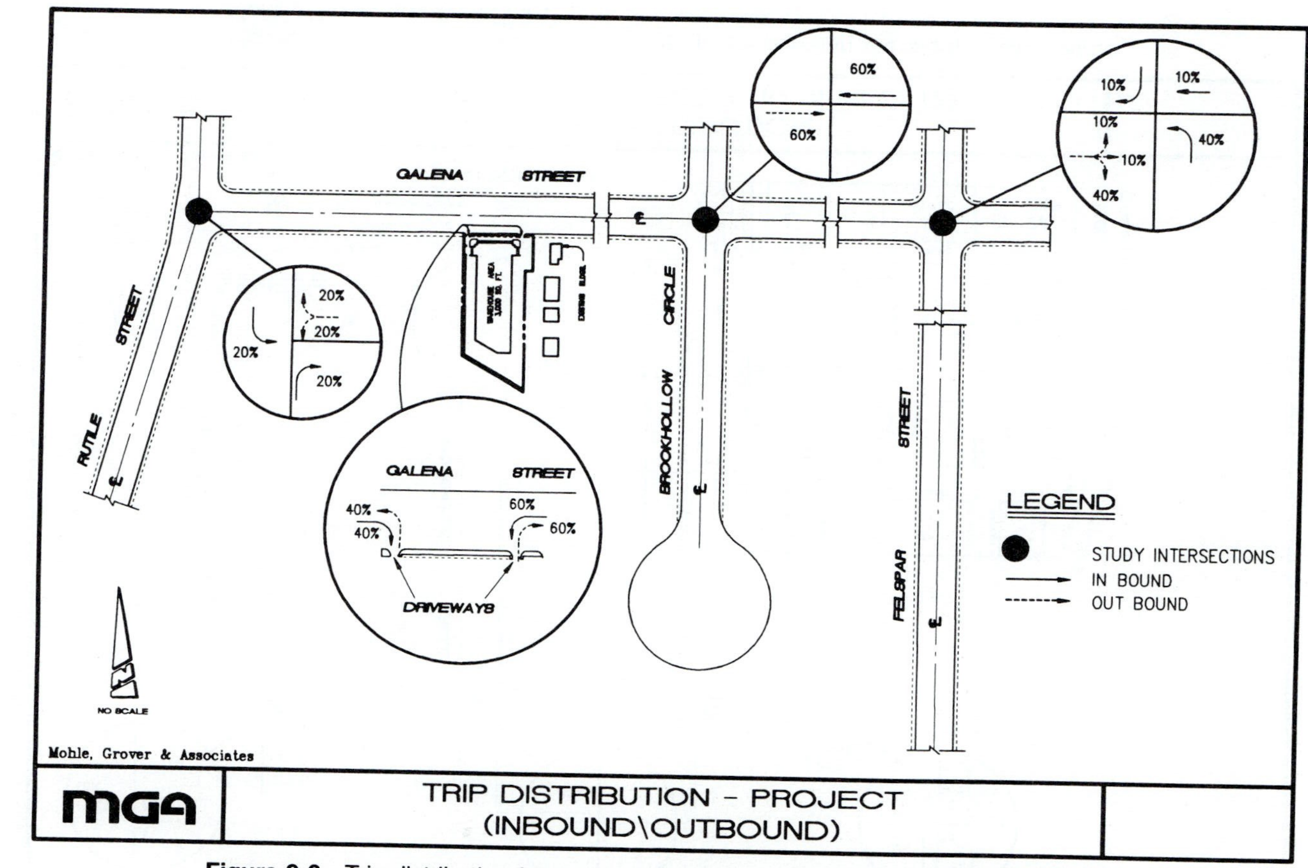

Figure 9-6 Trip distribution for the project. *Source:* Mohle, Grover and Associates.

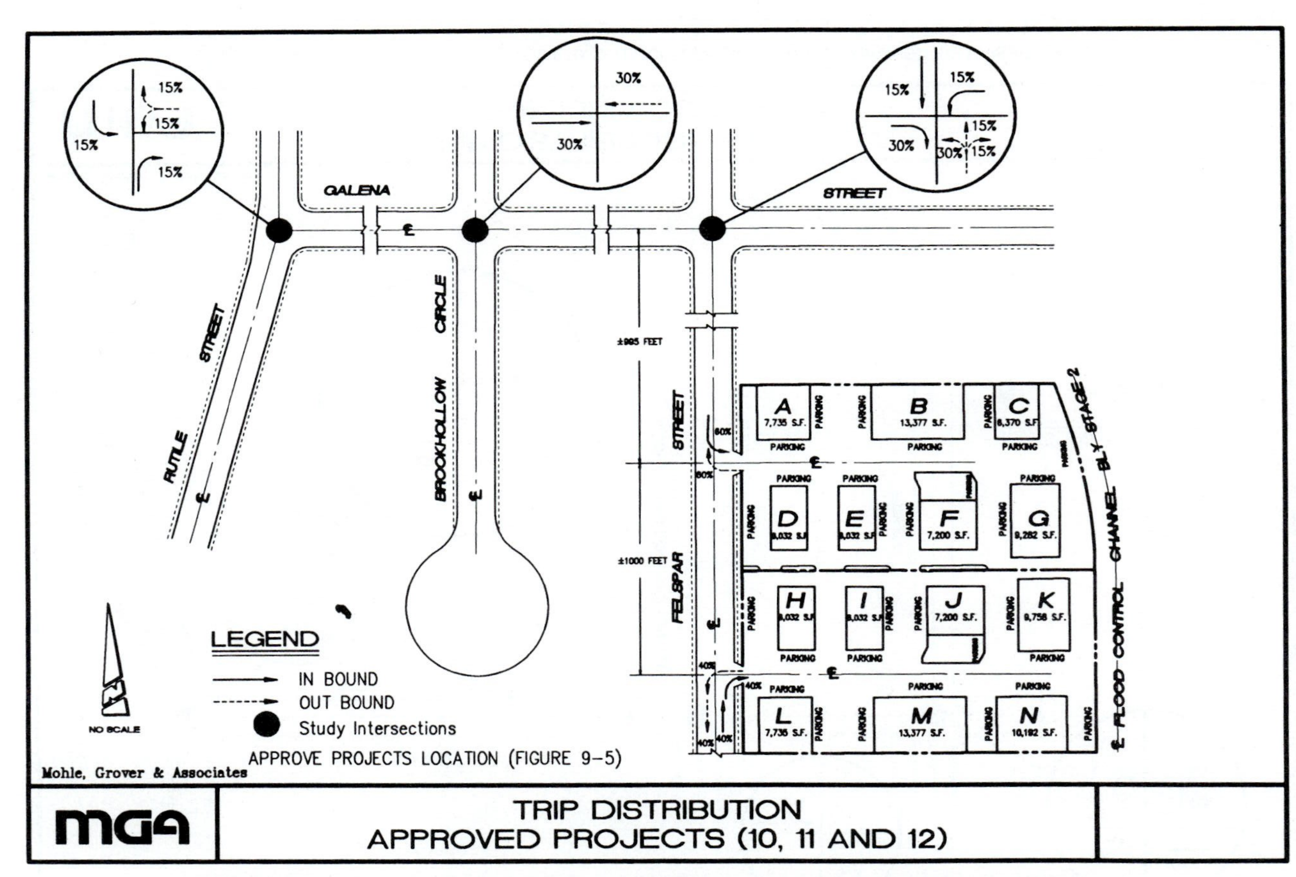

Figure 9-7 Trip distribution for approved projects within the study area. *Source:* Mohle, Grover and Associates.

These estimates represent the "base" conditions: that is, before the site has been developed (or redeveloped). Nonsite traffic consists of two components: (1) "through" traffic, consisting of all movements through the study area without an origin or destination in the study area; and (2) traffic generated by all other developments in the study area, with an origin or destination in the study area. Projects that have been proposed and approved (but not yet built) may be considered in the analysis. Future traffic demand estimates are developed by summing the contributions from the site, all approved (or potential) developments in the area, and from current traffic volumes adjusted for general growth in the area.

There are three principal methods of projecting off-site traffic: build-up, use of area transportation plan or modeled volumes, and trends or growth rates. Each method has its appropriate use and is based on the data that may be available or generated as part of the site traffic access/impact study. The advantages and concerns of each technique are summarized in Table 9-5.

Build-up Method

A popular technique for estimating the cumulative impacts of development projects and area growth is to estimate the trip generation of approved and potential projects as described above. These trips are added to base traffic volumes to produce the cumulative impact of known projects on traffic service levels. Trips already made in the area street system should be accounted for to avoid double counting. As part of the process, all transportation system changes that are programmed, committed, or highly likely during the forecast or study period according to local agencies and project probable changes in travel patterns are identified.

Areawide Travel Demand Model Forecasts

Most metropolitan areas maintain forecasts of future travel demand. These forecasts are based on computer-assisted models which reflect all officially anticipated land use and transportation network changes. These forecasts can be used in a cumulative analysis, with the following precautions:

1. The projection may be out of date and not reflect changes in local land use or highway plans subsequent to the traffic forecast.

TABLE 9-5
Techniques for Projecting On-Site Traffic

Build-up	Appropriate in areas of moderate growth Usually used when project has horizon of 10 years or less Often the best method when there is good local information on development approvals
Transportation plan	Often used with large, regional projects that will develop over a long period Often appropriate for areas of high growth Locally credible transportation plan data that are adaptable to the study year must be available
Growth rates	Typically used for small projects that will be built within a year or two Local record keeping of traffic counts must be good At least 5 years of data showing stable growth should be available Simple, straightforward approach Not appropriate for long-range horizons May result in over- or undercounting nonsite traffic growth

2. Regional traffic forecasts may only be available on a daily basis. If so, there is no straightforward way to derive the peak-hour volumes required for a detailed traffic impact analysis.
3. The land use and transportation network components of travel demand models are frequently too coarse in scale for the level of detail needed in traffic impact studies.

Trends or Growth Rate Method

Average growth rates can be used to project nonsite traffic. The growth rate method is one of the simplest methods to use; however, it often results in inaccurate projections, as it assumes that the patterns of growth rates in traffic volumes will continue through the study target year, or will change predictably. Growth in through traffic may be estimated by recent growth trends in traffic volumes in or near the study area or from traffic volume projections used in the area transportation plan. If area transportation plan volumes are used, all site volumes associated with the development that have already been projected must be subtracted. If the study year is not the same as the transportation plan projection year, straight-line or variable growth rates should be used to interpolate between current volumes and transportation plan forecast year volumes. The growth rate selected should be justified in the study report. The growth rates applied should be appropriate for the situations. For example, a 2 to 3% annual growth rate may be applied to the existing traffic in well-developed urban areas to reflect possible changes in land use. For a newly developing area a 5 to 7% annual growth rate may be applied to reflect the pace and type of development taking place.

This method requires that recent development trends and population growth rates will continue at approximately the same rate or at a rate that is predicable. If the study horizon is greater than 10 years, or growth rates are expected to change, another method should be used. This method is the least accurate because growth is never uniform throughout a community and its transportation network. Past traffic growth may not be indicative of future growth.

DETERMINING THE IMPACTS

The analyst can begin to assess the impacts of the proposed project once baseline data have been gathered and the key assumptions for trip generation, distribution, and assignment and modal split have been determined. In addition to the analyses of traffic facility capacity, a number of other factors should be considered (ITE, 1987a). These include:

- Safety
- Circulation patterns
- Traffic control needs
- Transit needs or impacts
- Transportation demand management
- Neighborhood impacts
- On-site and off-site parking adequacy
- Pedestrian and bicycle movements
- Service and delivery vehicle access
- Driveway location and operation
- Air quality and noise impact

The impact analysis and the development of improvement plans are conducted in an iterative process for each time horizon and key location. The intent is to show the relationship between operations and geometry and to assess deficiencies, as well as to identify alternatives for further consideration. Analyses should be conducted for predevelopment conditions and for conditions with the proposed project to gauge the incremental impacts of the project and the incremental needs it generates. As stated in the Introduction, a complete description of all the analyses required for a complete, comprehensive study is beyond the scope of this chapter. Important topics are discussed briefly below. Air quality and noise impacts are covered in Chapter 8.

Total Traffic Estimates

For each analysis period being studied, a projected total traffic volume must be established for each segment of the roadway system being analyzed. These projected total traffic volumes (consisting of site and nonsite traffic) are used to determine the ability of the transportation system to handle the increased demand. A capacity analysis should be performed for all key intersections and roadway sections. Projected cumulative traffic volumes can be shown as in Figure 9-8.

Capacity

Several methods, including the *Highway Capacity Manual* (TRB, 1985), intersection capacity utilization, and critical lane techniques, can be used to assess the adequacy of key intersections to handle addition demand generated by the new development (ITE, 1987b). Apply the same techniques for the existing situation and to the projected alternatives. For example, projected volumes at signalized intersections should be analyzed using the same capacity analysis techniques as those used to characterize existing conditions. All unsignalized intersections should be analyzed using the same unsignalized analysis technique, unless signalization is warranted and planned. Traffic signal warrants, such as those contained in the *Manual of Uniform Traffic Control Devices* (FHWA, 1988), may be used to determine whether signalization is desirable. If the analyses indicate that mitigation measures needed include improvements to the geometrics at the study intersections, improvements in levels of service can be represented as shown in Figure 9-9.

If signalization is proposed for access to a development along existing arterial roadways, it is important to assess the impact of the new traffic demand on operation of the through street. The HCM defines arterial roadway levels of service based on operating speeds and type of facility. In the simplest cases, the required analysis may consist of a time–space diagram to show that a proposed signal will, or will not, disrupt the platoon flows operating along the arterial.

Safety

The initial review of existing data within a study area should include recent (within 3 years) accident experience. This review is to identify locations where traffic safety should be given extra consideration. High accident locations on roadways serving the study site must be analyzed and measures to alleviate accident hazards must be considered. Accident rates vary, but any intersection with more than one accident per million vehicles entering is usually worthy of additional analysis (ITE, 1987a).The site plan should be reviewed to ensure that the internal circulation system and external access points are designed for pedestrian safety and to minimize vehicle/pedestrian conflicts. Locations for parking, and transit stops with their associated pedestrian flows to building access points, require thorough assessment to ensure safety.

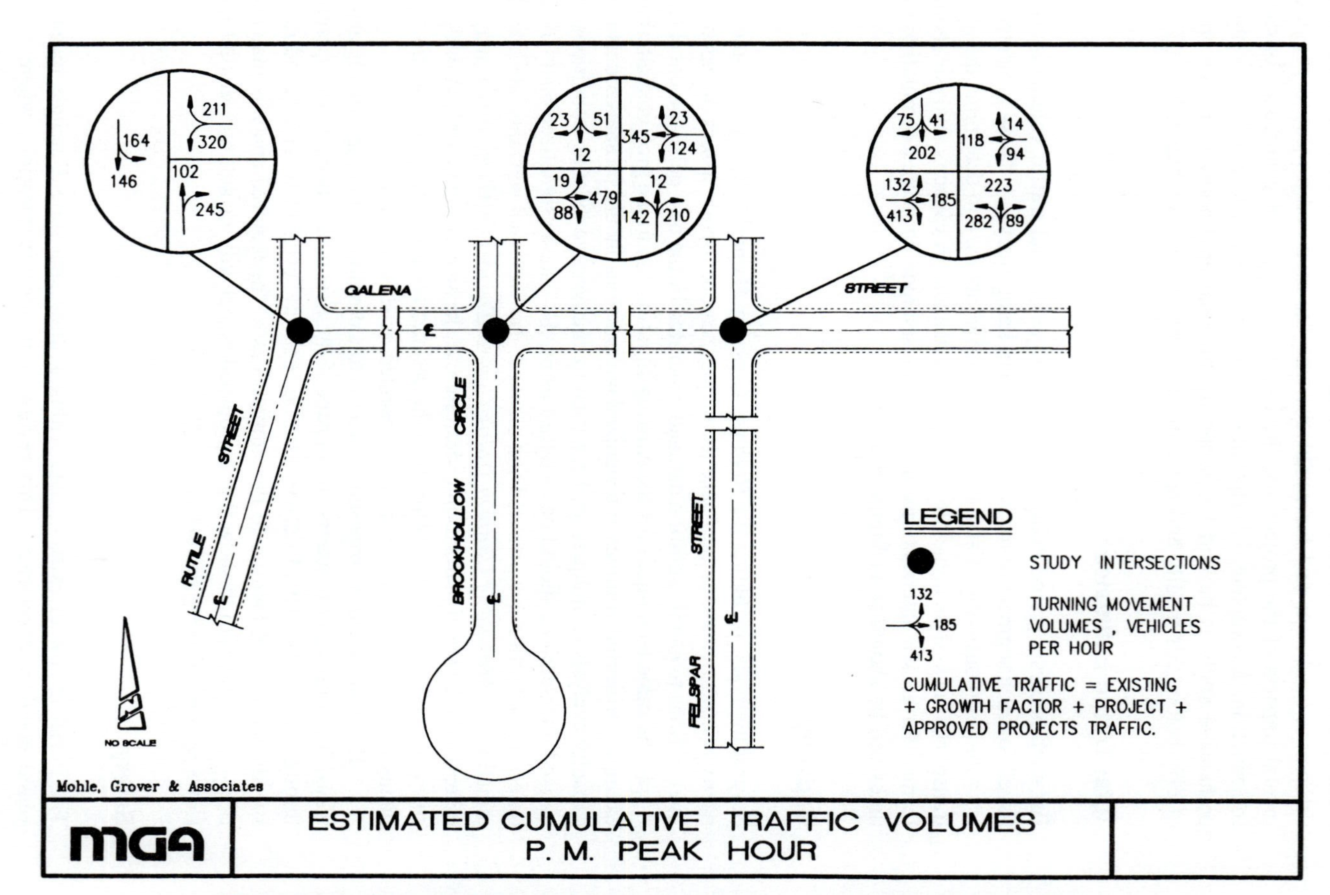

Figure 9-8 Estimated cumulative traffic volumes at study intersections. *Source:* Mohle, Grover and Associates.

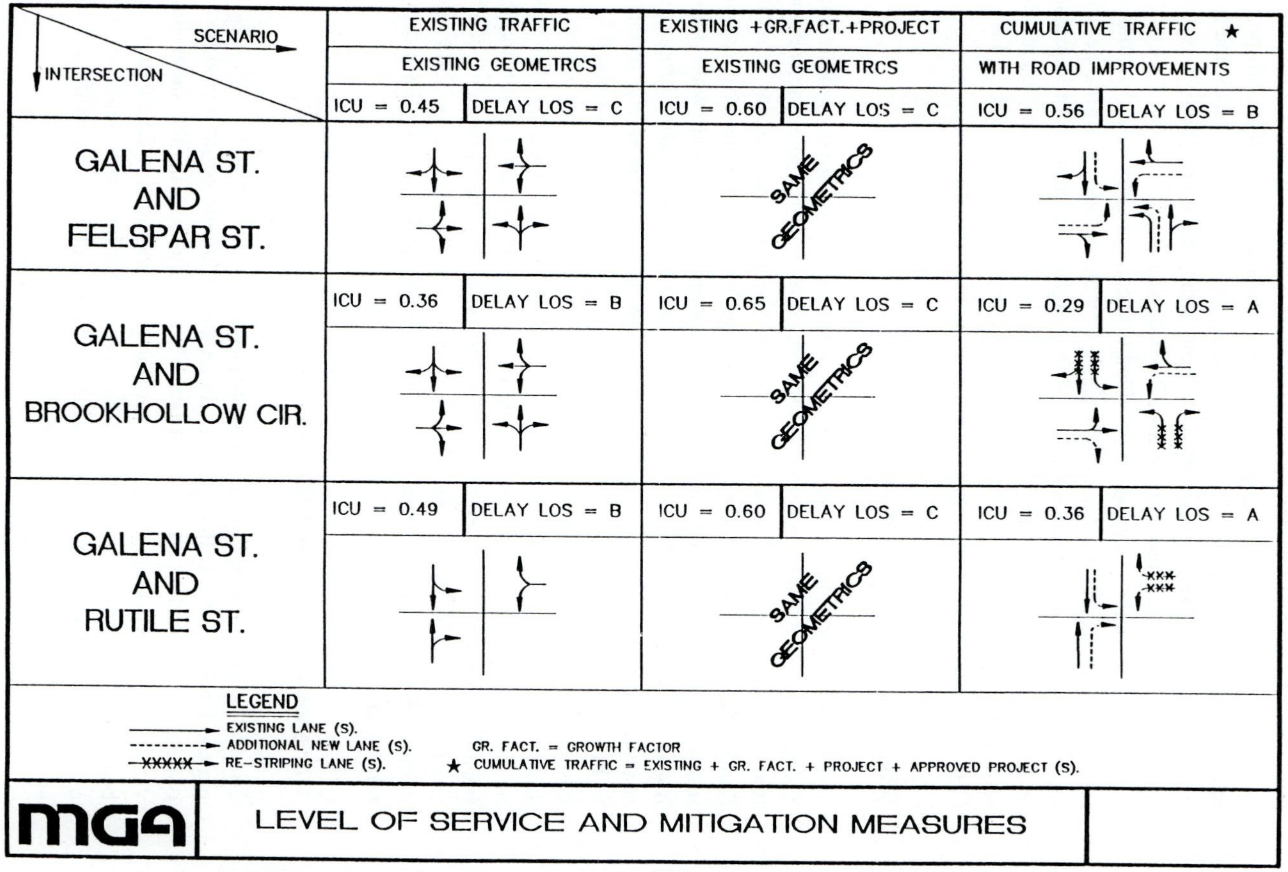

INTERSECTION / SCENARIO	EXISTING TRAFFIC		EXISTING +GR.FACT.+PROJECT		CUMULATIVE TRAFFIC ★	
	EXISTING GEOMETRCS		EXISTING GEOMETRCS		WITH ROAD IMPROVEMENTS	
GALENA ST. AND FELSPAR ST.	ICU = 0.45	DELAY LOS = C	ICU = 0.60	DELAY LOS = C	ICU = 0.56	DELAY LOS = B
			SAME GEOMETRICS			
GALENA ST. AND BROOKHOLLOW CIR.	ICU = 0.36	DELAY LOS = B	ICU = 0.65	DELAY LOS = C	ICU = 0.29	DELAY LOS = A
			SAME GEOMETRICS			
GALENA ST. AND RUTILE ST.	ICU = 0.49	DELAY LOS = B	ICU = 0.60	DELAY LOS = C	ICU = 0.36	DELAY LOS = A
			SAME GEOMETRICS			

LEGEND

EXISTING LANE (S).
ADDITIONAL NEW LANE (S).
RE-STRIPING LANE (S).

GR. FACT. = GROWTH FACTOR
★ CUMULATIVE TRAFFIC = EXISTING + GR. FACT. + PROJECT + APPROVED PROJECT (S).

MGA — LEVEL OF SERVICE AND MITIGATION MEASURES

Figure 9-9 Levels of service for existing geometrics and proposed improvements.
Source: Mohle, Grover and Associates.

On-Site Circulation

The assessment of internal circulation illustrates the relationship between external access points and building access locations, dropoff points, delivery points, and parking locations. Internal circulation should be efficient to minimize indirect travel while utilizing the least amount of land space and generating the fewest conflicts with pedestrians and other vehicles. Locations of internal roadways, number of travel lanes and turn lanes on internal roadways, maneuverability in parking areas, and accessibility to buildings are all factors that must be considered in a successful project.

Adequate parking should be provided to meet site-generated demands and be consistent with applicable local policies, which may be included in traffic demand management programs. Specific dimensions, parking angles, and parking ratio requirements are addressed in a variety of documents. Specific dimensions, parking angles, and parking ratio requirements are all issues addressed in detail in TRB (1971), ITE (1983), ULI (1982), Stover and Koepke (1987), and Wilbur Smith (1982).

Site Access and Off-Site Improvements

The 1984 design policy adopted by the American Association of State Highway and Transportation Officials (AASHTO) recognized that access points are intersections and that they should be designed with the same perspective as legs of any other intersection. Study recommendations and conclusions should provide safe and efficient movement of traffic to and from, within and past, the proposed development, while minimizing the impact to nonsite trips. Site access objectives are to serve abutting properties, preserve roadway capacity, maintain efficient traffic flow, and maintain safety. Among the factors to be considered are the following: site access for service vehicles should be examined based on size and operating characteristics; adequacy of site driveways and the internal site circulation scheme must be studied; and the design and location of driveways for the amount and type of traffic that will be using both the adjacent street and the driveway must be analyzed. Estimated driveway volumes, by turning movement, can be shown as in Figure 9-10 and used in the analysis of access points (driveways) and internal traffic circulation. Stover and Koepke (1987), ITE (1986), and USDOT (1982) deal with topics related to site access and off-site improvements.

Transportation Demand Management

Site traffic can often be reduced or spread over a longer period through a variety of techniques. Each of these techniques has the potential to reduce site traffic for certain types of development under certain conditions. These techniques include increasing transit service to areas of trip origins, carpool and van pool programs, reducing parking availability below normal demand or substantial increases in parking costs, significant transit subsidies; creation of a high-quality pedestrian environment on-site, modified work schedules, and a wide range of other techniques (ITE, 1987a).

Residential Neighborhoods

A variety of studies have found that the safety, neighborhood amenities, and overall livability of local or residential streets can be severely degraded by traffic volumes well before the physical capacities of such streets are reached. Consequently, if a project will add significant traffic to residential streets, the volumes should be quantified. *Residential Street Design and Traffic Control* (Deakin et al., 1989) and *Livable Streets* (Appleyard, 1981) suggest guidelines for measuring such impacts, and contain strategies and techniques for controlling the intrusion of nonlocal traffic into neighborhood streets.

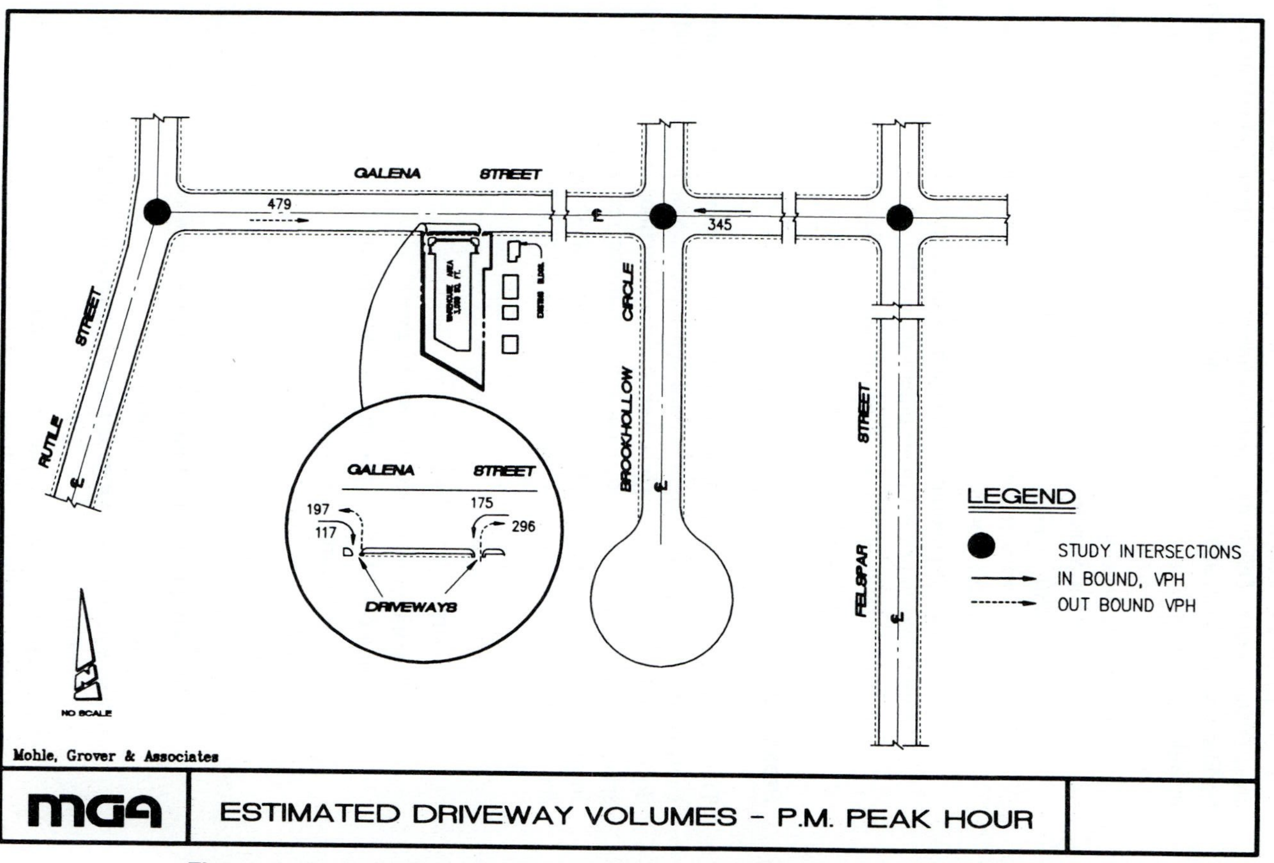

Figure 9-10 Estimated driveway volumes for proposed project. *Source:* Mohle, Grover and Associates.

STUDY REPORT

The purpose of a site traffic access and impact study is to assess the effects that a particular development will have on the surrounding transportation network, to determine what provisions are needed for safe and efficient site access and traffic flow, and to address related issues. The study report documents the purpose, procedures, assumptions, findings, conclusions, and recommendations of the study. Common uses for these reports include:

- To provide developers or designers with recommendations on site selection, site transportation planning, and traffic impacts
- To assist public agencies in reviewing the attributes of proposed developments in conjunction with requests for annexation, land subdivision, zoning changes, building permits, or other development review
- To establish or negotiate mitigation requirements where off-site impacts require improvements beyond those otherwise needed
- As a basis for levying impact fees or assessing developer contributions to roadway facility improvements

Presentation

The sample table of contents suggested (ITE 1987a) report provides a framework for organizing the report in a straightforward and logical sequence. Some studies are easily documented using this outline; however, additional sections may be warranted because of specific issues to be addressed, local study requirements, and the results of the study. Inapplicable sections may also be deleted from the report as needed. The documentation for a traffic access and impact study should include at a minimum:

- Executive summary
- Study purpose and objectives
- A description of the site and study area
- Existing conditions in the area of development
- Anticipated nearby development
- Trip generation
- Trip distribution
- Modal split
- Traffic assignment resulting from the development
- Projected future traffic volumes
- Assessment of the change in roadway operating conditions resulting from the development traffic
- Recommendations for site access and transportation improvements needed to maintain traffic flow to, from, within, and past the site at an acceptable and safe level of service

The report should lead the reader step by step through the various stages of the process and to the resulting conclusions and recommendations. The report may have several different audiences. Sufficient technical detail must be included to allow the staff of the reviewing agency to follow the path and methodology of the analysis. The report must also be understandable to nontechnical decision makers and interested citizens. The following suggestions are offered by the ITE (1987a).

1. Include a one- or two-page technical summary that concisely summarizes the study purpose, conclusions, and recommendations.
2. Clearly document assumptions. Assumptions based on published sources should be specifically referenced. If less-available sources are used, a more detailed explanation may be necessary.
3. Avoid technical jargon (or at a minimum, clearly define it).
4. Whenever possible, present data in tables, graphs, maps, and diagrams rather than narrative text. This enhances clarity and ease of review.
5. Do not attempt to include too much information in a single graphic. See Appendix D for some guidelines.
6. Discuss findings and recommendations with the reviewer prior to submittal of the final report.
7. Do not include political views or statements in the report; it should be an objective, technical analysis.
8. Inadequate reports should be returned to the preparer by the reviewing agency for completion or modification as needed.

REFERENCES

Appleyard, D. (1981). *Livable Streets*, University of California Press, Berkeley, CA.

Deakin, E., P. C. Bosselmann, D. T. Smith, Jr., W. S. Homburger, and B. Beukers (1989). *Residential Street Design and Traffic Control*, Institute of Transportation Engineers and Prentice Hall, Englewood Cliffs, NJ.

FHWA (1985). *Development and Application of Trip Generation Rates*, HHP-22, U.S. Department of Transportation, Federal Highway Administration, Washington, DC, January.

FHWA (1988). *Manual of Uniform Traffic Control Devices*, Federal Highway Administration, Washington DC.

Horowitz, A. J. (1987). *Quick Response System II Reference Manual*, U.S. Department of Transportation, Federal Highway Administration, Washington, DC, June 15.

ITE (1983). *Parking Generation*, 2nd ed., Institute of Transportation Engineers, Washington, DC.

ITE (1986). *Guidelines for Driveway Location and Design*, Institute of Transportation Engineers, Washington, DC.

ITE (1987a). *Traffic Access and Impact Studies for Site Development*, Institute of Transportation Engineers, Washington, DC.

ITE (1987b). *Trip Generation*, 4th ed., Institute of Transportation Engineers, Washington, DC.

ITE (1989). *Report on Traffic Access and Impact Studies for Site Development: A Recommended Practice*, Institute of Transportation Engineers, Washington, DC.

Stover, V., and F. Koepke (1987). *Transportation and Land Development*, Prentice Hall, Englewood Cliffs, NJ.

TRB (1971). *Parking Principles*, Special Report 125, Transportation Research Board, Washington, DC.

TRB (1978). *Quick Response Urban Travel Estimation Techniques*, National Co-operative Highway Research Program Report 187, Transportation Research Board, Washington, DC.

TRB (1985). *Highway Capacity Manual*, Special Report 209, Transportation Research Board, Washington, DC.

ULI (1983). *Dimensions of Parking*, 2nd ed., Urban Land Institute and National Parking Association, Washington DC.

USDOT (1982). *Access Management for Streets and Highways*, U.S. Department of Transportation, Washington, DC.

Wilbur Smith and Associates (1982). *Parking Requirements for Shopping Centers*, Urban Land Institute, Washington, DC.

10

Parking Studies

L. Ellis King, D. Eng., P.E.

INTRODUCTION

Parking is one of the three essential elements in urban transportation. With few exceptions, automobiles and trucks must be parked at least temporarily at each end of a vehicular trip. Even in areas served by public transit, the automobile is the preferred means of transportation and the demand for parking continues to increase. There are two general types of parking:

1. Owner-supplied facilities at homes, apartments, retail centers, industrial buildings, institutions, and offices, including public street curb where nonmetered parking is allowed
2. Commercial parking, including private lots or garages where fees are collected; and charge parking, such as meters at curbs or in public off-street lots or garages

Parking studies are typically conducted for two purposes:

1. To establish parking policy and parking regulations, such as zoning codes or specific developments
2. To check physical needs for revising or increasing the existing parking supply and location

The majority of parking studies are performed to determine the need to expand existing parking by comparing parking demands to parking space supply. Typical usage studies are summarized in Tables 10-1 to 10-3.

Many past parking studies have involved central business districts (CBDs), and much of the material in this chapter pertains to this type of comprehensive study, which may utilize all techniques described (BPR, 1957a). The CBD study typically results in collection of large amounts of material that can be handled readily by automated data processing equipment. Areas such as neighborhood business districts, industrial developments, office parks, densely populated apartment/condominium areas, and other specific generators may require a more limited specialized study (BPR, 1957b; HUFSM, 1970). Individual studies may also be required for hospitals, universities, sports arenas, cultural facilities, and special events (Weant and Levinson, 1990; Whitlock, 1982). Although these studies may be specialized and limited in scope, the study techniques and procedures are the same as those used for the more comprehensive CBD studies.

Similarly, major traffic routes may have need for a detailed study in relation to modification or removal of curbside parking, which creates a serious problem in most urban areas. Up to one-fifth of accidents in cities have been traced directly or indirectly to curb parking, and the detrimental effect of curb parking on traffic flow is well known (HRB, 1971). Of even greater concern is angle parking along the curb, which is still allowed in a few communities and has a direct or indirect influence on approximately 40 feet of roadway for each side of the street that has angle parking. This unsafe practice should be discouraged.

PARKING INVENTORY

The parking inventory assembles information about the location, number, and other pertinent characteristics of existing parking spaces at the curb and in off-street areas, including alleys and spaces behind buildings. Every available legal public and private parking space should be identified. The usual information needed is:

- Number of parking spaces.
- Time limits and hours of operation.
- Type of ownership; such as public, private, or restricted to employees or customers of a particular building.
- Rates (if any) and method of fee collection.
- Type of regulation at curb spaces, such as loading zone, passenger zone, handicapped zone, taxi zone, or bus zone.
- Type of facility: lot or garage.
- Probable degree of permanency; many informal, poorly maintained parking facilities are temporary and may be expected to be replaced with new construction in the foreseeable future.

In urban areas of over 50,000 population, basic parking or land use data may be available in files of the local transportation agency. The parking inventory is highly useful to the traffic engineer in day-to-day activities and is frequently used by zoning commissions, building departments, and other city agencies. The inventory is an essential prerequisite to any parking study and should be updated periodically, such as every 3 years.

Study Locations

If the inventory is for a CBD, the study must include the primary retail core, the secondary retail and office ring around the core, and the fringe area, where employees may be expected to park. For a neighborhood business district, parking may be expected up to

TABLE 10-1
Parking Spaces Available, Numbers of Vehicles Parking, and Daily Turnover by Type of Facility and Block

	Spaces Available				*Number of Vehicles Parking*					*Daily Turnover*[a]			
					Curb								
Block	*Curb*	*Lots*	*Garages*	*Total*	*Legal Space*	*Illegal Space*	*Lots*	*Garages*	*Total*	*Curb*	*Lots*	*Garages*	*Total*
021	17	0	0	17	117	12	0	0	129	6.9	0	0	6.9
022	28	106	36	170	264	12	292	47	615	9.4	2.8	1.3	3.5
023	72	63	0	135	373	20	127	0	520	5.2	2.0	0	3.7

[a]Includes legal curb, lots, and garages only, for period under study.

TABLE 10-2
Other Parking Usage Characteristics by Number of Vehicles Parked and Time Used in Different Facilities

	Legal Curb Space		*Illegal Curb Space*		*Total Curb Space*		*Lots*		*Garages*		*Total*	
Block	*Veh. Parked*	*Veh. Hours*	*Veh. Parked*	*Veh. Hours*	*Veh. Parked*	*Veh. Hours*	*Veh. Parked*	*Veh. Hours*	*Veh. Parked*	*Veh. Hours*	*Veh. Parked*	*Veh. Hours*
040	120	67	10	6	130	73	320	1500	52	520	502	2093
041	52	40	0	0	52	40	10	80	0	0	62	120

TABLE 10-3
Comparison of Observed Vehicle Hours of Parking, Space Hours Available, and Deficiency or Surplus in Existing Capacity

Block	Total Observed Vehicle-Hours of Parking	Overtime Vehicle-Hours of Parking	Vehicle-Hours of Parking with No Illegal Parking	Space-Hours Capacity of Available Legal Space	Space-Hours Capacity	
					Deficiency	Surplus
051	40	15	25	30	—	5
052	25	2	23	20	3	—
053	60	10	50	40	10	—

500 feet (150 meters) beyond the limits of the commercial zoning. A field inspection should be made to determine the actual extent of this parking, and the study boundaries adjusted accordingly. Similarly, field checks are needed to determine the study limits at a special generator, such as an industrial building, auditorium, hospital, stadium, and so on, which may extend for 1000 feet (300 meters) or more.

If the study involves a congested area, it is desirable to select natural boundaries, such as rivers, railroads, or major routes along which changes in land use occur. For a major route, it is generally necessary to include a distance of 300 to 500 feet (100 to 150 m) along each cross street since these areas are potential locations for curb parkers unable to park on the major route.

Personnel and Equipment

For a study of a small area, current aerial photographs at scales of about 1 inch = 50 feet (1:600) are most useful (Syrakis and Platt, 1969). A 1 inch = 100 feet scale (1:1200) may be used for larger areas, although considerable detail is lost if the area includes tall buildings. Land use maps may be available which facilitate the inventory. The field teams consist of one to two persons, who are usually on foot in the case of a business district, or in a vehicle for studies of small areas, and along major traffic routes. A roll-type measuring device is useful, although most measurements of curb and open parking facilities can be scaled from aerial photographs.

Method

Before beginning the inventory, a master coding system is established. Each block is given an identification number and extra-long blocks may be subdivided and given two numbers. If there is a prior transportation or origin–destination study for the area, which includes a separate number for each block of the parking study area, it may be convenient to use the same numbers.

Once a block number has been selected, numbers 1 through 4 are used for identification of curb faces in four-sided blocks, as shown in Figure 10-1. However, the maximum number of sides in the most oddly shaped block must be provided for. If a six-sided block is the worst condition, number 6 is reserved for the sixth curb face. Numbers 7 and upward are then left for identification of individual parking facilities within each block. Each block is numbered separately and should include informal areas off alleys at the rear of buildings, as well as the more obvious parking facilities.

At the beginning of the inventory, the exact number of off-street parking facilities within each block is usually unknown, and thus numbers must be assigned by inventory personnel who should be provided with the master map giving the block and curb face numbers. These numbers may readily be transferred to a map with a scale of approximately 1 inch = 200 feet (1:2400) for ease of use. As each off-street facility is inventoried, the field crew assigns it a number and the data are noted on a large-scale work

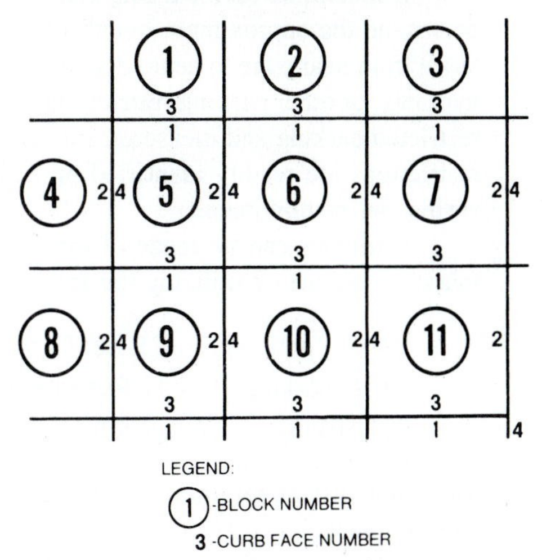

Figure 10-1 Block and curb face numbering system. *Source:* Box and Oppenlander, 1976.

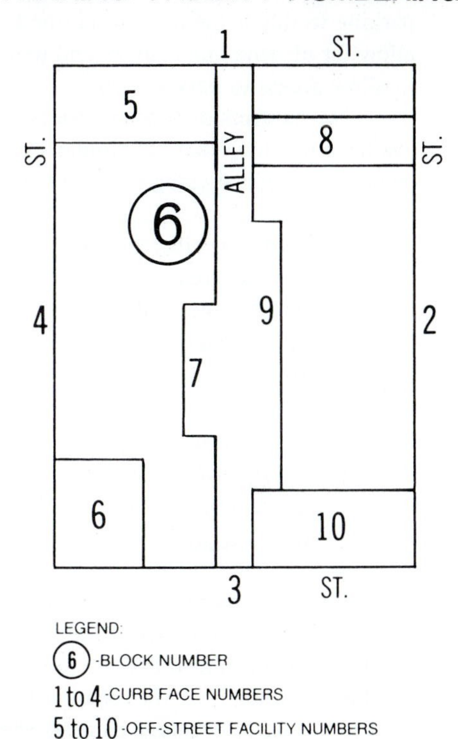

Figure 10-2 Parking facility numbering system. *Source:* Box and Oppenlander, 1976.

map or tabulated on a form. In either case the block number is followed by the facility number, such as 6-8, as shown in Figure 10-2.

The curb inventory should include special identification of angle parking, number of parking meters and their time limits, charges and enforced hours, and locations of no-parking zones and enforced hours. Locations of driveways should be indicated along with special notes made of any driveways which either temporarily or permanently no longer serve as access to a given tract due to building construction, fencing, or vacancy.

If individual curb parking stalls do not have pavement markings, it is necessary to determine the approximate length of each curb parking section. This length is the distance from near edge to near edge of crosswalk in a block that contains no driveways, fire hydrants, or other parking restrictions. However, the usual block contains several areas of restricted parking and measurements are needed within each separate section. If parking restrictions are readily identified on a scale aerial photograph, much of the field measuring can be eliminated.

Estimates can be made of the available number of parking spaces in a given unmarked distance by utilizing the following figures:

Parallel parking	23 feet (7 meters) per vehicle
Angle parking	12 feet (4 meters) per vehicle
90° parking	9.5 feet (3 meters) per vehicle

These dimensions for unmarked stalls reflect a lower operating capacity than if markings were in place.

Counts of individual stalls or linear measurements may be used to determine the number of parking spaces available in a lot or garage. If the facility is a private commercial type with attendant parking, determination of capacity is more difficult because vehicles may be stored in the aisles. In general, the operating capacity of an attendant-parking facility is the peak-condition loading when only sufficient space has been left to allow for off-street maneuvers and temporary parking of vehicles which must be moved to allow access to parked vehicles.

The numbers of parking spaces in each block of the survey area and the number of available parking spaces in similar facilities are tabulated and summarized for the entire study area. Typical categories for the overall summary are:

1. Curb and alley parking
   a. Metered
   b. Nonmetered
   c. Special zones
2. Parking lots
   a. Public
   b. Private
3. Parking garages
   a. Public
   b. Private

The block-by-block summaries are tabulated on an inventory form such as that shown in Figure 10-3. Under the "street and alley stalls" columns, individual types of facilities, such as those under a given time-limit restriction or a special regulation such as "loading zone," may be identified separately. In a parking report, maps are frequently used to illustrate inventory data, as shown in Figure 10-4.

Private lots or garages are those restricted to the use of employees, residents, or tenants of a building, taxicab or truck storage area, or other facilities not open to the general public. Lots and garages open to some public parking, as well as weekly or monthly rentals, are classed as public facilities. Where space is rented to individuals, it is classed as public space because the space may be rented in the future by other members of the general public. Employee spaces are not included since they are not usable by the general public.

When appropriate, the parking inventory may also gather data on assessed valuations of vacant lots or old buildings that might be replaced by parking facilities. Certain

PARKING INVENTORY

SUMMARY SHEET

AREA OF INVENTORY ____________________

DATE OF INVENTORY ____________________

BLOCK	FACILITY	STREET AND ALLEY STALLS						OFF-STREET PARKING		TOTAL STALLS
								PRIVATE	PUBLIC	

DATE ____________ COMPILED BY ____________

Figure 10-3 Parking inventory summary form. *Source:* Box and Oppenlander, 1976.

elements of traffic control, including one-way streets and alleys, and restrictions on turning movements at intersections, should also be included. Such controls affect routing of "windshield" surveys, as well as access to potential locations for development of new parking facilities. Inventories of small areas and along major traffic routes follow the general format of the CBD inventories except that aerial photographs and/or strip maps are used to a greater extent.

PARKING USAGE STUDIES

There are two general types of usage studies:

1. Accumulation or generation
2. License plate checks

These studies generally involve field checks that can be made without publicity or public knowledge of the work, while more specialized studies may require direct contact with the parkers.

Accumulation, or parking occupancy checks, plus turnover and duration studies, are useful in determining what (if any) curb parking improvements can be made to increase parking capacity. The duration analysis indicates where long-time parkers are using space inefficiently, and a relative measure of curb-use efficiency is provided by comparing hourly parking turnover rates with desirable rates to indicate efficient usage. Parking practices or regulations that contribute to uneconomical and inequitable use of street space are revealed in this analysis. For example:

1. The study may show that enforcement of existing time limits is needed to stop overtime parking.

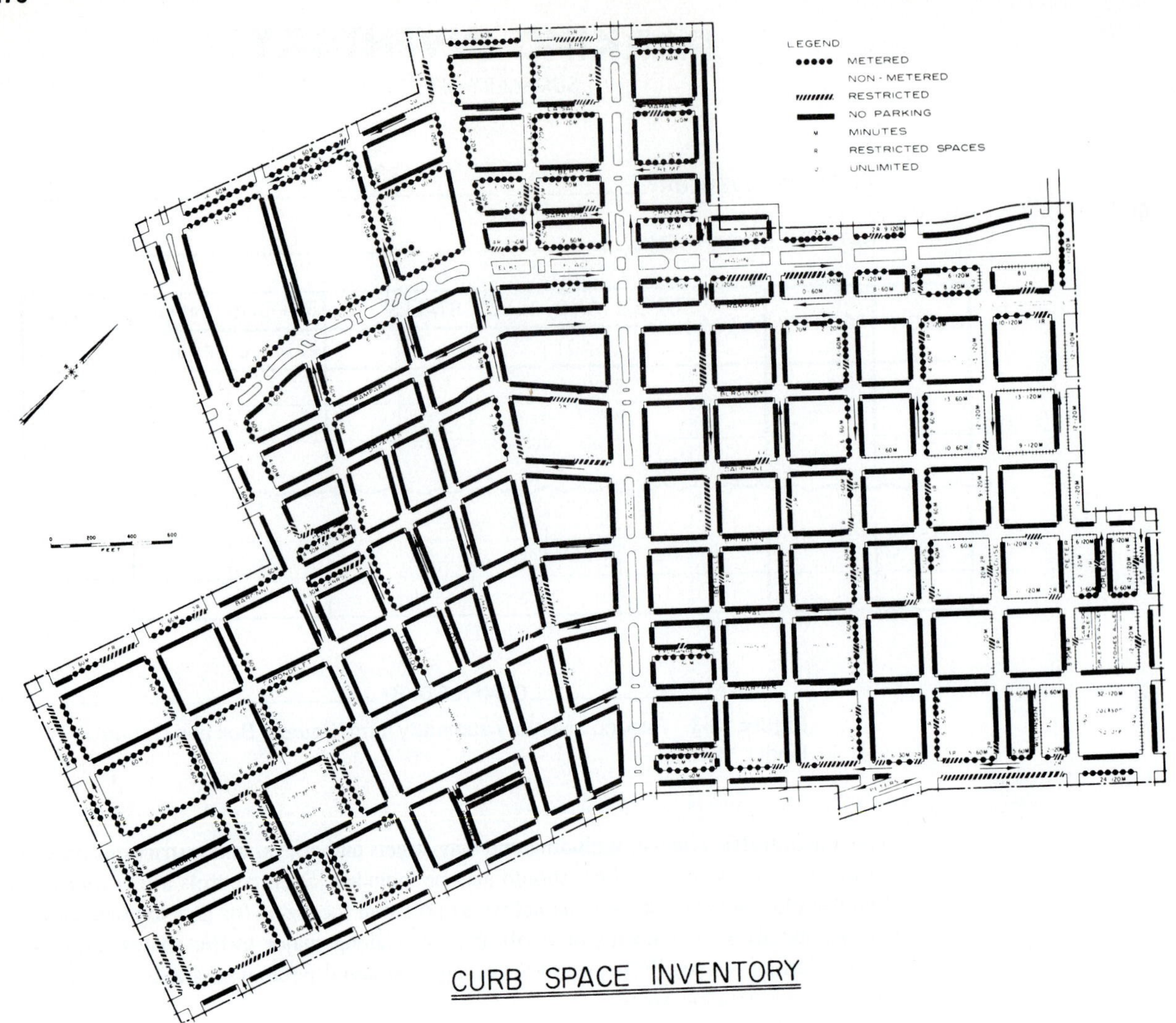

Figure 10-4 Curb space inventory map. *Source:* Box and Oppenlander, 1976.

2. It may indicate that existing time limits are too long or too short. Consistent 30-minute parking in a 15-minute zone might indicate a need to revise the regulation. Conversely, a preponderance of 30-minute parking in a 2-hour zone would indicate the desirability of reducing the limit to 60 minutes or less if an increased turnover is desired.
3. Hazardous or illegal parking may be revealed. The factual evidence of such practices may prove helpful in bringing about needed enforcement or voluntary public compliance with regulations.

Occupancy checks are useful in determining needed improvements for truck pickup and delivery service. Consistent double parking by commercial vehicles may indicate the need for additional loading zones or for better enforcement of regulations if available loading zone space is adequate. Police action may also be required if passenger vehicles are observed to be blocking the existing zones. Where it is proposed to prohibit parking, occupancy checks are helpful in determining the parking demand at the hours of the day during which the prohibition would be in effect. When considering changes from angle to parallel parking, installation of parking meters, or installation of curb passenger and freight loading zones, a study of the curb parking activity data is useful. These studies are also performed to determine parking efficiency at older industrial plants, and parking lots or garages in business areas.

ACCUMULATION AND GENERATION STUDIES

Accumulation and generation studies are made at relatively frequent intervals on different days of the week to determine hourly variations and peak parking demand. In a central business district, the studies are generally made on an hourly or 2-hourly basis between 6:00 A.M. and 8:00 P.M. In a business district, the retail shopping hours must be identified, and if stores are open only on certain evenings, the studies should include days of both early and late closing. In smaller communities, retail areas may produce peak activities on Saturday mornings or afternoons. When peak weekday parking demand hours have been determined, repeat checks should be taken at the peak for each weekday.

If the accumulation studies are being made at a particular generator, such as an office building, studies should begin prior to the morning opening of offices and immediately after closing hours. At industrial plants operating on more than one shift, parking checks are needed at time of shift overlap. It is typical for office and administrative personnel to have different hours from factory workers, and separate checks may be needed to determine these demands for parking.

Along major traffic routes, checks are made during the morning peak hours from 6:30 to 9:30 A.M., during several midday times, and across the evening rush period from 3:30 to 6:00 P.M. If residential properties abut the route, it is essential that checks be made during the late-night period since for residential parking, peak accumulation occurs between 1:00 to 5:00 A.M. Demand is relatively constant during this period and a single overnight check should be adequate. The traffic engineer is generally most interested in the peak parking demand times as related to current supply. In the case of a major traffic route study, where the intent may be to impose only morning or evening rush-hour restrictions, checks should be made during these time periods.

It is often desirable to prohibit all parking along a major route since this is the most effective way to provide for the total transportation of persons and goods, and to maximize public safety. Therefore, parking checks along major routes usually include hours other than just the rush periods. In all parking checks, it is important to avoid conditions of temporary parking restrictions, such as for street sweeping, street repairs, or snow plowing. Parking checks in retail areas are usually not taken during periods of abnormal demand such as the day following Thanksgiving. The effect of major sports activities should be considered in the area of such generators, and special checks are often made to determine peak parking needs.

If the accumulation studies are made to determine the peak parking demands of specific generator types in order to develop zoning code specifications, peaking characteristic information must first be secured from cooperative administrators or other reliable sources. The heaviest hours, days, and months must be determined and factors such as varying visitor hours at hospitals and nursing homes and shift overlap periods must be ascertained. In the generation-type study, it is also essential to know the occupancy or facility usage at the time of the checks. In hospitals and nursing homes, this includes the number of occupied beds as a percentage of total available beds, and in the case of subdivisions or apartment buildings, it involves the number of occupied dwellings. In offices, industrial plants, and retail centers, the amount of occupied leasable square footage of floor area as a percentage of the total leasable floor area is used.

A second precaution in the performance of generation-type studies is the need for identification of parked vehicles as related to specific establishments. When patron parking is comingled with other parking, it may be necessary to conduct interview-type studies or make specific observations of patron activities to secure reliable data. The parking needs of specific generators often vary among districts within an urban area, and the availability of public transit is a major factor. An office building in a CBD served by transit will have a lower parking demand than that of a similar building located in an outlying area. Parking-generation figures calculated for one district should not be applied arbitrarily to a dissimilar area. Representative trip generation rates for various land uses and building types are readily available (ITE, 1991). However, these average rates should be used with care when applied to a particular situation.

Parking-accumulation studies can be performed using maps that show the outline of the different parking facilities. In larger areas, a number of sectional maps may be required. The occupancy studies work well using a vehicle with driver and observer. Colored pencils are used to note the number of parked cars found in each facility, with a different color used for each time period. In larger facilities it may be easier to note the empty spaces rather than the occupied spaces.

Since part of the accumulation checks are made during periods of traffic congestion, it is important to plan the study route carefully. This includes preliminary checks of driving time required to cover the different areas and establishment of the most efficient travel pattern. In more complex CBD accumulation studies, where the route may be involved and congested with traffic, it is desirable to have the survey crews make several practice runs. In some of the more congested areas, it may be necessary to conduct the accumulation checks on foot. When it is impractical to count accumulation within garages by driving through them, the required data are gathered from garage operators or by walking through the facilities.

In some facilities, an accumulation analysis may be made by continuous counting of in and out vehicular movements. This requires knowledge of the number of vehicles within the facility at the beginning of the counts, plus a check of the number remaining at the end of the count, in order to verify accuracy of the counting. In and out data for facilities with automated gates or attendant-controlled gates may be recorded automatically and made available by the facility operator. This type of study may also include notation of license plate numbers as described in the next section.

Accumulation studies of residential parking are unable to determine whether vehicles are parked inside private garages. Checks are usually limited to vehicles observed in driveways and at curbs and it is important to alert data users to the incompleteness of the check. In larger areas, personnel limitations may make it impractical to perform all accumulation studies during a single day. As a general rule, it is desirable to conduct additional portions of the checks on the same day of succeeding weeks, provided that weather conditions and other influencing factors are the same.

Whenever possible, the complete area should be covered in the initial occupancy checks, even though this may mean a less frequent interval for each time check, to establish a control base. Supplemental studies can then be made at intermediate hours. The accumulation data from studies within a specific area are normally plotted on an hourly basis to show the changing demand. Figure 10-5 shows a typical accumulation curve. It is also customary to indicate the number of available parking spaces, including any hourly variations that may occur due to differing operating schedules and regulations. Accumulations may also be shown in tabular form for individual blocks and types of facilities.

While accumulation studies provide information on total numbers of vehicles by location, they yield no information on the length of time each vehicle is parked, nor on driver destinations. Information on length of time parked and turnover, or number of times each parking space is utilized during the day, is determined from a license plate check.

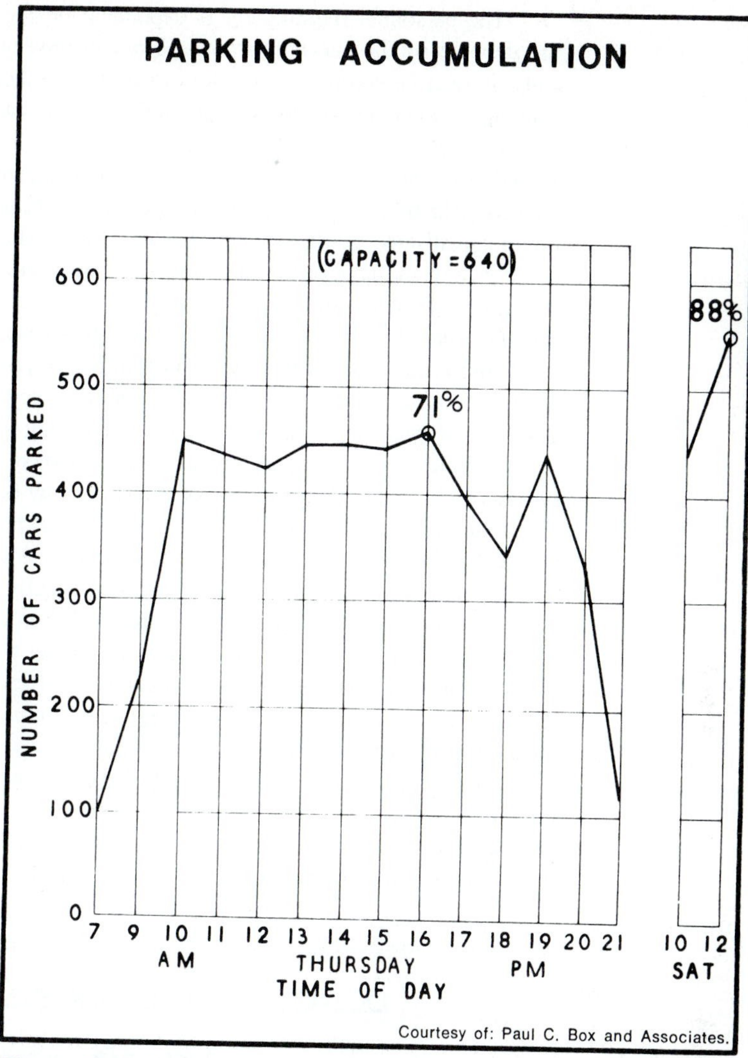

Figure 10-5 Parking accumulation curve. *Source:* Box and Oppenlander, 1976.

LICENSE PLATE CHECKS

License plate checks are used for detailed observation of curb parking. The primary purpose is to determine turnover, which is defined as the average number of cars parked per day during the study period in each space of a given block face. The equation for turnover is

$$T = \frac{\text{number of different cars parked}}{\text{number of parking spaces}} \tag{10-1}$$

Other reasons for license plate checks include securing data on length of stay, accumulation, illegal parking, and enforcement which requires special notation as to whether a ticket is found on an illegally parked car.

The study hours are selected to suit the purpose of the study and the license plate check is most often performed by a person on foot. Because the study is relatively expensive, a sampling technique is normally used. Several block faces are selected which typically represent the different time limits of parking found in the study area. Thus two or three block faces having 1-hour parking, two or three others with 2-hour parking, and so on, are checked.

The personnel requirement is dependent on the frequency of checking, which in turn depends on the parking turnover. Since turnover is unknown at the beginning of the study, it must be estimated from known or assumed characteristics of the situation. For example, if a given curb face is regulated by 1-hour parking, checks are needed at 20- or 30-minute intervals. If a curb face has 2-hour parking, checks would normally not be needed more often than every 30 minutes but no greater than 1 hour. Very short parking, such as a 15-minute limit, requires checks at 5-minute intervals. In the fringe areas, where unrestricted all-day parking occurs, hourly checks are generally sufficient.

Data from the accumulation checks on numbers of cars parked per block are useful in setting up routes and determining personnel requirements for license plate checks. Tape recorders have been used successfully in license plate checking, although there are severe limitations and large-scale opportunities for error in their use. The hand-held computer is also useful in conducting license plate surveys. Data recorded in the field can be downloaded into a parking database after returning to the office.

A field form for use at typical curbs is shown in Figure 10-6. The essential components include the identification of the facility by both code number and street names. The column headed "space and regulation" must have provision for illegal parking as well as legal spaces. This column can also be used as a curb inventory sheet. As indicated on the figure, a portion of the license number, usually the last three digits, of each parked vehicle is entered in the appropriate column. When the same vehicle is found at subsequent checks, the number is not repeated but rather a checkmark is used. This simplifies the calculation of turnover, since the number of different cars parked in the block is being determined. Office analysis is complicated if comparison for the same vehicle number must be made between two or more columns.

Figure 10-6 also shows other notations typically used in license plate checks. Expired meters may be indicated, and color notes may be used to identify vehicles that have no license plate. Trucks should be noted. Double-parked vehicles are identified by putting a diagonal line across the particular box in the column. The license plate of the curb-

LICENSE PLATE CHECK FIELD DATA SHEET

City ______ Date 10 MAY 1978 Recorded by JONES Side of Street W

Street WRIGHT between 5th and 6th

Codes: 000 last three digits of license number ✓ for repeat number from prior circuit; — for empty space

Space and Regulation	Time circuit begins											
	07	0730	08	0830	09	0930	10	1030	11	1130	1200	
5th												
X-WALK	—	—	—									
NPHC	—	—	—									
1 HR M	—	713	✓	✓/TK								
" M	631	✓	(✓)	971								
" M	512	342	✓	014								
DRIVEWAY	—	—	—	—								
"	—	—	—	613								
1 HR M	—	—	418	✓								
"	117	220	✓	989								
"	—	148	096	✓								
FIRE HYD	—	—	—	—								
1 HR M	042	—	216	✓								
N PHC	—	—	—	T14								
X-WALK	—	—	—	—								
6th												

Figure 10-6 Curb parking field form. *Source:* Box and Oppenlander, 1976.

parked vehicle is written in the upper part of the box, and the license number of the double-parked vehicle is written in the lower part of the box.

It is sometimes desirable to note full license plate numbers so that owners of vehicles may be identified. This information is used to check the extent to which store owners or employees are using curb space to park their private vehicles. It may also be desirable to designate commercial and out-of-state vehicles. The license plate numbers of trucks may be underlined in areas where the state plate does not give a separate identifying symbol such as a "T." Specialized notes may also be made on size of truck and type of deliveries being made. Knowledge of frequency and location of truck loading and unloading activities are needed in the planning of curb loading zones as well as for development of truck terminals, and the degree of truck activity during various hours is a factor when considering restrictions on such activity to increase street capacity.

One person can conduct license plate checks for approximately 60 parking spaces each 15 minutes. This permits coverage of two to four block faces on each trip after making the initial round, which generally requires additional time. The degree of parking turnover is a major factor influencing the actual time required and consideration must also be given to sections that the checker has to traverse where no parking is allowed. Other administrative elements include need for rest and food breaks for the checkers, personal security in some areas, and need for checkers to act in an unobtrusive fashion. While it is difficult to conceal the fact that a parking check is being made, the survey results may be biased if "meter feeders" are aware of the field study. When questions are asked, the checker should answer them courteously and give assurance that the study has no relation to law enforcement activity.

Where widespread license plate checks are being made, the usual method is to start routes at a common point to facilitate supervision. The field checker walks down one side of the street and back on the opposite side, covering all curb spaces along the route. In a rectangular street layout, routes may be started at one intersection and extend out around the block with the supervisor stationed at the starting point. Field checkers often work a 5-hour shift, but this length is dependent on the total length of the parking survey period.

The license plate check is most often used at the curb; however, it can also be used in lots or garages. In these cases, field observers go through each lot or garage at regular intervals, recording license numbers. Care must be taken to follow the same route within the facility to simplify comparison of license numbers during the summarizing process. If the off-street check is made as a part of the curb check, the intervals between trips are the same as in the curb check. If the off-street study is made separately, intervals can be increased by 20 to 30 minutes or longer since parking durations in off-street locations are commonly longer than at the curb. Any underlying schedules, such as nearby school class schedules or movie theater schedules, should be taken into consideration.

Where off-street facilities are too large to check every vehicle in the time allotted, or are operated by people who will not cooperate in allowing the observer to make the necessary observations, the "in-and-out" study is used. This study method requires observers at all entrances and exits of the facility and gives the maximum accuracy of turnover and duration data since every vehicle entering and exiting is observed and the license number recorded. In cases where the volumes of entering and/or leaving vehicles are unusually high, several observers or videotaping may be required.

The time limits of this study should coincide with those of the curb license plate check. At suitable intervals, the time of day is noted following the last recorded license plate number, on both the "in" and "out" sides of the form. When traffic volumes are low, the time of actual entry or exit may be recorded for each vehicle. During high-activity periods, time should be noted at 5- to 10-minute intervals. When possible, the number and license plates of the vehicles in a facility at the beginning and at the end of the study should be checked.

STREET PARKING

DURATION SUMMARY SHEET

LOCATION ____________________

DATE __________ TIME __________ WEATHER __________

DURATION OF PARKING-INDICATE RANGES BELOW	INDICATE CLASSIFICATION OF PARKING ZONE, BLOCK, OR OTHER CLASSIFICATION FOR EACH COLUMN BELOW															
	PASSENGER VEHICLES		TRUCKS		PASSENGER VEHICLES		TRUCKS		PASSENGER VEHICLES		TRUCKS		PASSENGER VEHICLES		TRUCKS	
	NO.	PER CENT	NO.	PER CENT	NO.	PER CENT	NO.	PER CENT	NO.	PER CENT	NO.	PER CENT	NO.	PER CENT	NO.	PER CENT
TOTAL VEHICLES																
TOTAL VEHICLE-HOURS																
AVERAGE DURATION																
PERCENT OVERTIME																

DATE __________ COMPILED BY __________

Figure 10-7 Parking duration summary form. *Source:* Box and Oppenlander, 1976.

Office summary and analysis of the license plate field check can provide information on parking accumulation. The peak accumulation for each block face is determined directly by a count of the parked vehicles. Figure 10-7 shows a typical duration summary sheet. The column headings may classify data by blocks, by type of parking zone (such as 1-hour curb parking, 2-hour, etc.), or by curb, lots, blocks, and grand total for the area. Parking durations may be estimated from the round-trip times of the parking activity checks. For example, if trips are made at 15-minute frequencies, a vehicle found on only one trip is assumed to be parked for 15 minutes. If it is seen on two successive trips, the vehicle is assigned to a 30-minute duration group, and so on. When the vehicle is seen on the first or last trip of the observer, its parking duration is unknown. All durations should be considered as starting within the time limits of the study.

Total vehicle hours are calculated by dividing the total number of vehicles seen on each trip by the number of checks made each hour. Thus if 10 vehicles were observed three times on 30-minute check intervals, the number of vehicle hours is $(3 \times 10/2)$ or 15 vehicle hours during the 1½ hours of the study. The average duration is found by dividing total vehicle hours by the total number of different vehicles observed $(15/10 = 1.5$ hours). This average is slightly higher than the true mean value, because all parked vehicles are not observed. The percentage of overtime parkers is found by adding the figures for all durations in excess of the legal parking time limit.

Figure 10-8 shows a typical turnover summary sheet that may be used to summarize the turnover for each block; for each classification of parking zone according to time limit; for lots, garages, and curb space; and for different areas. A summary is made for

STREET PARKING

TURNOVER SUMMARY SHEET

LOCATION ____________________

DATE __________ TURNOVER BETWEEN __________ AND __________

FACILITY (INDICATE BLOCK NO., CLASSIFICATIONS OF PARKING ZONE OR FACILITY)	NUMBER OF STALLS	OBSERVED TOTAL TURNOVER VEHICLES	OBSERVED HOURLY TURNOVER RATE-VEHICLES PER STALL PER HR.			

DATE __________ COMPILED BY __________

Figure 10-8 Parking duration summary form. *Source:* Box and Oppenlander, 1976.

individual blocks; for classification of time zones; for each lot or garage; for all curbs combined, all lots combined, and all garages combined; and finally, for a grand total. The left-hand column of Figure 10-8 is used to indicate classification. The next column contains the number of legal vehicle parking stalls that are being used all or part of the time on each normal weekday. The number of vehicles observed making use of the parking stalls is entered in the third column. Any vehicle making use of a space whether or not it entered or left during the period of the study is included in the turnover count. The fourth column is used to record the hourly turnover rate per stall. For instance, if 50 stalls had an observed total of 100 different vehicles in 8 hours, the rate would be 0.25 vehicles per stall per hour [100/(50 × 8)].

The three remaining blank columns may be used for various purposes. For example, it may be desired to show probable future number of spaces, allowing for new off-street facilities and prohibition of curb parking, and resulting possible turnover. Or it may be decided to show what turnover would be if the turnover rate were increased through enforcement of existing time limits. Column headings might be, in this case, "estimated turnover rate assuming compliance with time limits," "total turnover," and "increase in turnover compared to present." Daily or study period turnover is usually shown. Continuing the example, 100 cars parked in 50 stalls would yield a daily turnover of 2.0 for the 8-hour period.

PARKING INTERVIEWS

To determine trip origin, destination, purpose, and walking distance, it is necessary to make some type of direct contact with the parkers. This interview can be conducted by means of a return-type questionnaire, usually a postcard, or by personal interviewing

(FHWA, 1970). Such studies are used primarily in determining parking demand by location. Interview data are intended to reveal both the pattern and number of driver destinations reached on foot after parking, and to measure the demand for space under the assumption that all drivers would like to park near their ultimate destination.

Data on driver habits are collected, such as walking distances as related to trip purpose and parking fees charged. This information is essential in determining acceptable locations for new parking facilities. Duration of parking by various trip purposes indicates practical time restrictions which may be imposed on various classes of parkers. If revenue-producing facilities are to be developed, this information is needed to estimate income. Consideration of integrating public transit and parking facilities requires an evaluation of how many persons would use joint facilities, where they should be located, and what transit service would be necessary.

Postcard Interviews

The postcard interview is used to determine characteristics only at the peak parking hour and the handout or placement of cards under windshield wipers should be done in a short time span. The time required for card distribution should be determined experimentally before the study begins. Newspaper publicity is essential for any type of interview study, especially when postcards are being used, and every effort should be made to obtain maximum cooperation from the public. The personnel requirements for the postcard survey are much lower than those for the personal interview; however, the low rate of returns and the unknown degree to which data are biased are negative aspects of card usage. Figure 10-9 shows a typical parking-survey postcard.

Typically, 30 to 50% of the cards distributed are returned. To expand the sample, it is necessary to have a record of the number of cards placed, by location and time. For example, if five cards were returned from a given block face indicating that these parkers shopped at a particular store, and the return sample from this block face was 50%, it could be estimated that there were a total of 10 parkers at the peak time who actually

______ PARKING SURVEY

Dear Motorist
Please help improve parking in this community by answering these questions and mailing this post-paid card *today*
Please do *not mark* in gray boxes!

4 9 9 2

I parked my car at this location today because (check one)

1 - I am employed or have my business downtown
2 - I am on a shopping trip (primary reason to be here)
3 - I am on a personal business trip (bank, doctor, etc.)
4 - Other reason (Please Specify) ______

After parking my car here, I walked to (please give address or name of building visited) ______

I drove to this parking space from (place where trip started) (Nearest street intersection) ______
(City) ______

The length of time I parked here today was about
______ (hours) and ______ (minutes)

Comments ______

THANK YOU FOR YOUR COOPERATION

Figure 10-9 Parking survey postcard. *Source:* Box and Oppenlander, 1976.

went to the particular store. Because the sample of returns would be expected to vary from facility to facility, a separate calculation of sample size and expansion factor should be made for each parking facility.

One method of maximizing the return is to take special care in selecting questions so as not to ask for information that some drivers would be reluctant to reveal. The best illustration concerns the home address, which should never be asked since many persons do not want to give out this information and it is actually of no direct value in the parking survey. The needed location is some general part of the city or the name of the city for parkers from suburbs or other nearby communities. Appendix B should be consulted for additional suggestions.

Another method of contacting drivers is to record license plate numbers of parked vehicles, trace mailing addresses on the basis of these numbers, and mail return postcards or questionnaires to the owners. However, a time delay of several days is caused by lookup of numbers, mailing, and receipt of card by the vehicle owner. This results in fewer returns and less accuracy.

Personal Interviews

Personal interviews of parkers can be conducted at the curb or in a parking facility. They can also be conducted at the entrance or exit of a particular generator, such as department store, office building, or hospital. The exact questions vary on whether the interview is conducted at the origin or the destination of the parker, at the time of parking, or at the time of unparking. Figure 10-10 shows a typical form that is utilized when the interview is being conducted at either time of parking or unparking. Questions include trip purpose and destination of the parker. Interviews at an apartment garage would add "reside here" as a trip purpose. The parker may be asked to give an estimate of length of time parked. The form shown as Figure 10-10 does not include space for trip origin; however, this readily can be added.

Personal interviews can be used on a day-long basis, during several hours of the day, or only during the peak hour. The work is not only expensive but also personally demanding, and careful selection and training of interviewers are essential. Persons with previous experience in census taking are especially valuable for this activity. Each interviewer should wear a distinct armband, ribbon, or other identifying insignia to minimize confusion, apprehension, or antipathy on the part of parkers being approached for questioning. The words "Traffic Survey" or other suitable information should appear in large letters on the insignia.

The interviewer should be on location 15 minutes in advance of the starting time. One person can handle about 15 spaces, provided that turnover is not too rapid. Thus one or two persons are usually needed to handle one block face. In the case of a large off-street facility, it may be necessary to sample parkers by interviewing every second or third person. The study normally does not include vehicles stopping momentarily to discharge passengers, taxicabs parking in reserved curb zones, buses, and emergency vehicles. Truck drivers may be interviewed, depending on the information that is desired. Interviews are directed only to the driver of each vehicle.

Vehicles parked when the interviewer first arrives are recorded on the field form with an "in" time 5 minutes prior to beginning the official survey. The drivers are interviewed if they return to their vehicles during the survey period, and it is usually then possible to determine the approximate actual time of parking. The original entry is then deleted and replaced with the correct time.

Normal procedure calls for the interviewer to approach a vehicle quickly as it pulls into a parking space and record the "time parked" to the nearest minute. The interviewer then asks the driver the selected questions and notes the answers on the field sheet. When the driver returns and begins to unpark, the interviewer records the "time leaving" on the form if the survey is being made along a block face or at a point where the interviewer

PARKING INTERVIEW FORM

CITY ______________________ DATE __________ FACILITY NO. __________

RECORDED BY ______________________ TABULATED BY ______________________

Time Parked	Time Leaving	Purpose 1. Work 2. Shop 3. Business 4. Other	Destination (building name, street address, etc.)	Office Use Only		
				Duration		Walking Distance
				hr	min	

Figure 10-10 Personal interview form. *Source:* Box and Oppenlander, 1976.

can see the unparking activity. For such cases the interview form should also include a column for license plate numbers. For curb checks the interviewer records the time leaving only if the license number can be found quickly on the sheet. If not, only the license number and the time leaving column are filled out, and the cross-matching is done later in the office.

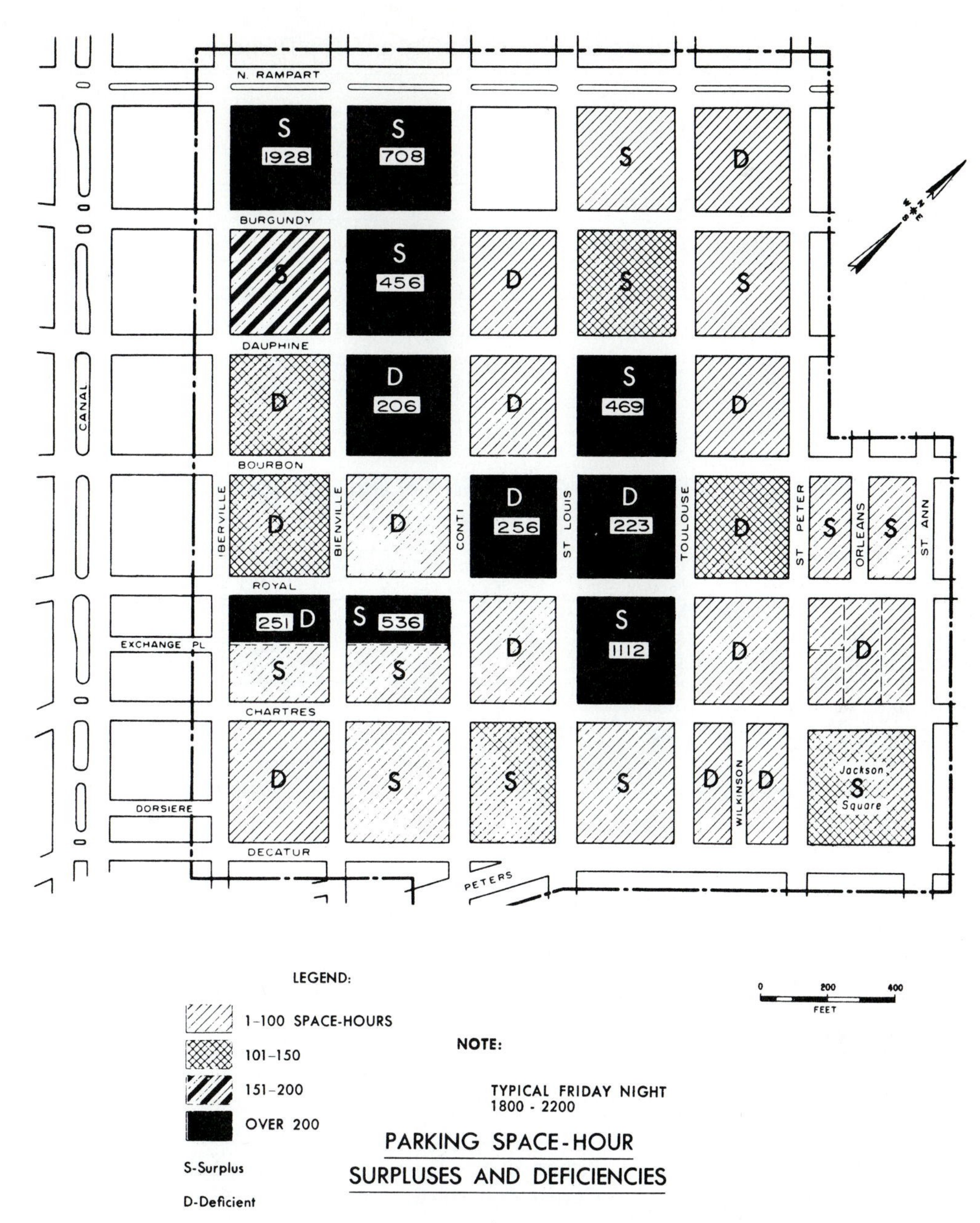

Figure 10-11 Parking space-hour surpluses and deficiencies. *Source:* Box and Oppenlander, 1976.

If the same driver parks several times in the same block during the study, she must be interviewed separately each time. If the driver refuses to cooperate or the interview is missed for other reasons, the interviewer writes "refused" or "missed" in the "purpose" column. Sometimes, trip destination and purpose can be determined by observing where the drivers walk and which building they enter.

Tabulation

Calculation of parking duration and walking distance is performed in the office. Walking distance is not airline, but rather, is determined along the pedestrian sidewalk system available to the nearest entry or exit from the building destination. In tabulating specific generator destination such as retail stores and office buildings, it will often be desirable

Parking by Type of Facility and Trip Purpose

	Purpose of Trip					
Type of Facility	*Shopping*	*Business*	*Work*	*Service or Sales*	*Other*	*Unknown*
Metered 1-hour zones						
2-hour zones						
Etc.						

Parking by Type of Facility and Distance Walked

	Distance Walked Feet								
Type of Facility	*Under 100*	*100-399*	*400-799*	*800-1199*	*1200-1599*	*1600-1999*	*Un-known*	*Total*	*Average Distance*
Metered 1-hour zones									
2-hour zones									
Etc.									

Parking by Distance Walked and Parking Duration

	Distance Walked Feet									
Duration	*Under 100*	*100-399*	*400-799*	*800-1199*	*1200-1599*	*1600-1999*	*Over 1999*	*Un-known*	*Total*	*Average Distance*
0-14 minutes										
15-29 minutes										
Etc.										

Figure 10-12 Typical parking data tabulation forms. *Source:* Box and Oppenlander, 1976.

to assign code numbers to these building. This may be done on a block-by-block basis by utilizing a number series beginning with 100 to distinguish clearly between parking facilities and pedestrian generators. In most cases all parking destinations within a given block will be consolidated for the purpose of determining parking-space hour surpluses and deficiencies. Such data may be illustrated on maps, as in Figure 10-11.

Other tabular information, such as trip purpose related to length of time parked, and distance walked related to both trip purpose and facility where parked, will also be needed. Depending on the application of the study, such as the development of a feasibility report for the construction of new revenue-producing parking facilities, it will be necessary to determine the number of parkers by both trip purpose and parking duration for each block. A variety of special forms may be used, as shown in Figure 10-12.

Parking by Trip Purpose and Walking Distance

Walking Distance-Feet	*Trip Purpose*						
	Shopping	*Business*	*Work*	*Service or Sales*	*Other*	*Unknown*	*Total*
Under 100							
100-399							
Etc.							

Parking by Trip Purpose and Parking Duration

Parking Duration	*Trip Purpose*						
	Shopping	*Business*	*Work*	*Service or Sales*	*Other*	*Unknown*	*Total*
0-14 minutes							
15-29 minutes							
Etc.							

Parking by Trip Purpose and Block of Destination

Block of Destination	*Trip Purpose*						
	Shopping	*Business*	*Work*	*Service or Sales*	*Other*	*Unknown*	*Total*

Figure 10-12 Continued.

REFERENCES

FHWA (1970). *Origin–Destination Surveys,* Federal Highway Administration, Washington, DC.

HUFSM (1970). *Planning for Parking Outside the CBD,* Highway Users Federation for Safety and Mobility, Washington, DC.

HRB (1971). *Parking Principles,* Special Report 125, Highway Research Board, Washington, DC, pp. 75–88.

ITE (1991). *Trip Generation,* Institute of Transportation Engineers, Washington, DC.

SYRAKIS, T. A., AND J. R. PLATT (1969). *Aerial Photographic Parking Study Techniques,* Highway Research Record 267, Highway Research Board, Washington, DC, pp. 15–28.

Parking by Duration and Block of Destination

Block of Destination	Parking Duration														
	Minutes			Hours											
	0-14	15-29	30-59	1-2	2-3	3-4	4-5	5-6	6-7	7-8	8-9	Over 9	Un-known	Total	Aver-age

General Summary of Parking Capacity and Usage

Block Number	Vehicle Stalls	Maximum Practical Total Turnover	Turnover Observed	Number of Driver Destinations

General Summary of Space Hour Capacity and Usage

Block Number	Supply		Usage			
	Available Spaces	Practical Space-Hours Capacity	Number of Vehicles Parked	Space-Hours Used	Driver Destina-tion	Space-Hours of these Drivers

Figure 10-12 Concluded.

U.S. Bureau of Public Roads (1957a). *Conducting a Comprehensive Parking Study,* Procedure Manual 3D, July; reprinted by Public Administration Service, Chicago.

U.S. Bureau of Public Roads (1957b). *Conducting a Limited Parking Study,* Procedure Manual 3C, July; reprinted by Public Administration Service, Chicago.

Weant, R. A., and H. S. Levinson (1990). *Parking,* The Eno Foundation for Transportation, Inc., Westport, CT.

Whitlock, E. M. (1982). *Parking,* The Eno Foundation for Transportation, Inc., Westport, CT.

11

Traffic Accident Studies

Joseph E. Hummer, Ph.D., P.E.

INTRODUCTION

Traffic accidents exact a terrible toll in the United States and other countries. The collection and analysis of data on traffic accidents are fundamental to the design of programs to reduce that toll. Analysts use accident data to help understand why accidents occur, to help identify accident-prone locations, to aid in deciding which safety programs or countermeasures should be implemented, and to assist evaluations of countermeasure effectiveness. In addition, certain programs of the U.S. federal government require that participating agencies have a program of accident data analysis.

In this chapter techniques for conducting studies of traffic accident data are discussed. First the accident report forms are described. Then we show how to prepare data for analysis. Next, the problems with and limitations on accident studies are discussed. Finally, specific types of analyses are described.

Many accident study techniques have improved dramatically in recent years, due to the widespread use of powerful and inexpensive computers and research into improved study methods. New techniques that appear practical for many jurisdictions and traffic engineers in the United States are discussed in this chapter. There are other new techniques that are not mentioned in this chapter because they require expertise, effort, and/or equipment not readily available to most jurisdictions and engineers. The premise of the chapter is that engineers wish to analyze traffic accident data that have been collected by others, principally the police. Thus we do not cover techniques for investigating individual accidents. However, engineers working with accident data should become familiar with accident investigation and reconstruction to provide greater understanding of the data.

Because of problems with and limitations on accident studies, engineers sometimes use other measures to stand in for accidents, called *surrogates.* One of the most widely used surrogates is the traffic conflict, which is recorded when a vehicle has to brake or swerve to avoid colliding with a second vehicle. Traffic conflict studies, which are discussed in Chapter 12, can be an effective supplement to other information in judging the safety of some types of locations.

ACCIDENT REPORTS

Accident data used by traffic engineers are recorded primarily by the police on report forms soon after an accident. One police report form is filled out per accident. Most states have a standard accident form used by all police forces within the state. Figure 11-1 shows the accident report form from the state of Michigan. The form requests information on the drivers and passengers, the vehicles, the roadway, and the conditions at the time of the accident. Most forms require a sketch of the accident showing vehicle paths and objects struck and a narrative describing the accident.

Accident type is one of the most important fields on the report form for traffic engineers. Various jurisdictions code accident type in various ways. Some states, including Michigan (Figure 11-1), code the accident type when the report is being computerized in the central office, based on the diagram and narrative on the police accident report form. Other states rely on the police officer to code the accident type. The accident codes used by police in North Carolina are given in Table 11-1. Police in North Carolina record accident type in terms of the first harmful event and the most harmful event for each of the involved vehicles. The person filling out the report has 23 choices, including several choices of "other." They could code the accident as a run-off road, noncollision, collision with an object other than another motor vehicle in the traffic stream, or a collision between motor vehicles in the traffic stream. For motor vehicles that collided in the traffic stream, North Carolina employs a typical coding system for accident type, including right turn, left turn, rear-end, head-on, sideswipe, angle, and backing accident types. Some accident report forms would include separate codes for sideswipe accidents between vehicles moving in the same direction and vehicles moving in opposite directions. However, some accident report forms do not have codes for different categories of left- and right-turn accidents as North Carolina's does.

The driver and passenger injury codes are very important for many studies. The most common coding scheme for the extent of injuries incurred by participants in an accident includes five categories:

1. F (fatality): The person died within 30 days of the accident as a direct result of injuries received during the accident.
2. A: The person experienced serious, incapacitating, nonfatal injuries during the accident. Broken bones, massive losses of blood, or more serious injuries are rated A.
3. B: The person experienced a visible but not serious or incapacitating injury during the accident.
4. C: The person complained of pain or momentary loss of consciousness due to an injury during the accident, but no visible sign of injury was evident to the investigator.
5. No injury.

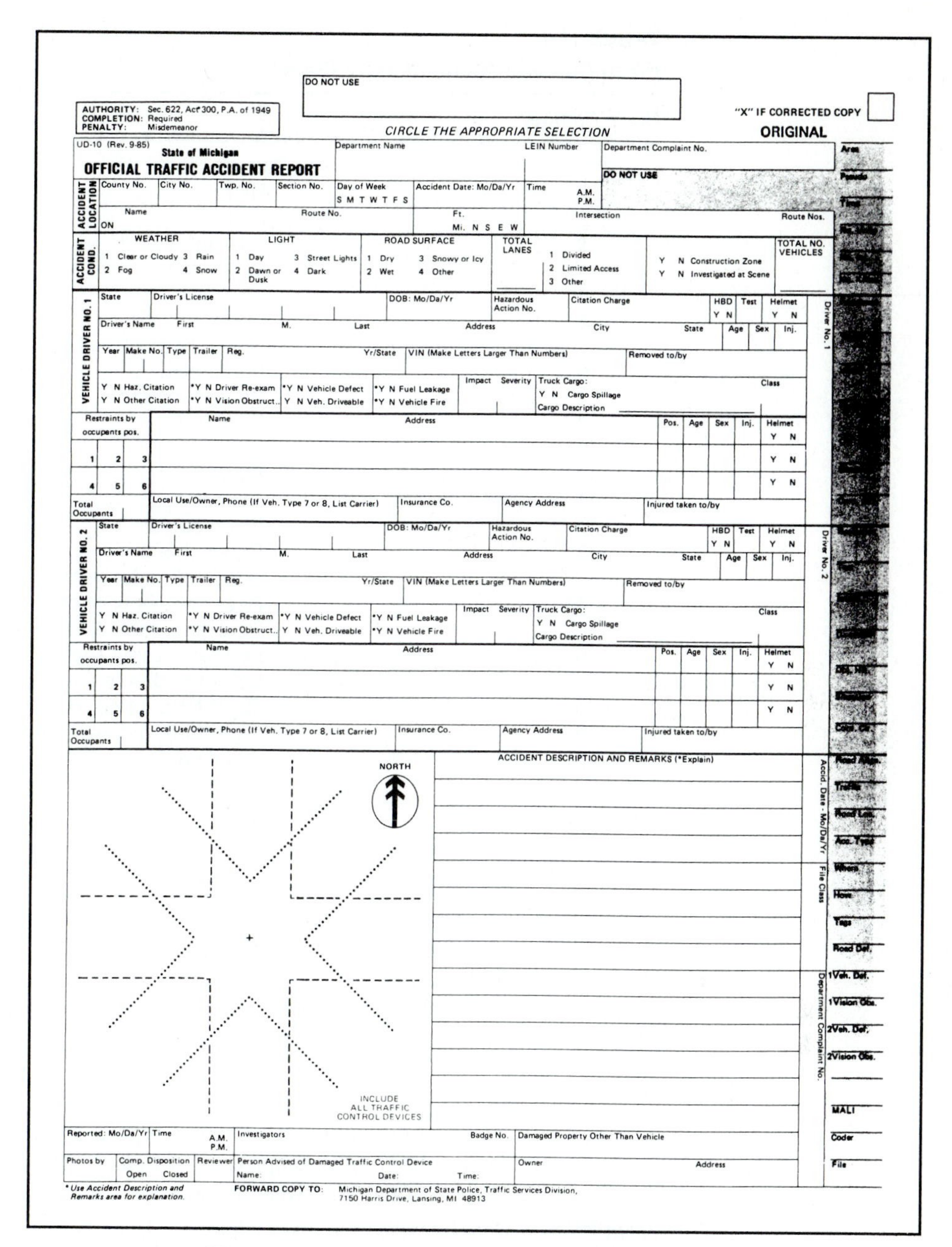

DO NOT USE

AUTHORITY: Sec. 622, Act 300, P.A. of 1949
COMPLETION: Required
PENALTY: Misdemeanor

"X" IF CORRECTED COPY

CIRCLE THE APPROPRIATE SELECTION

ORIGINAL

UD-10 (Rev. 9-85)

State of Michigan

OFFICIAL TRAFFIC ACCIDENT REPORT

Department Name | LEIN Number | Department Complaint No.

DO NOT USE

ACCIDENT LOCATION: County No. | City No. | Twp. No. | Section No. | Day of Week S M T W T F S | Accident Date: Mo/Da/Yr | Time A.M. P.M.

Name ON | Route No. | Ft. Mi. N S E W | Intersection | Route Nos.

ACCIDENT COND.: WEATHER 1 Clear or Cloudy 3 Rain 2 Fog 4 Snow | LIGHT 1 Day 3 Street Lights 2 Dawn or Dusk 4 Dark | ROAD SURFACE 1 Dry 3 Snowy or Icy 2 Wet 4 Other | TOTAL LANES 1 Divided 2 Limited Access 3 Other | Y N Construction Zone Y N Investigated at Scene | TOTAL NO. VEHICLES

VEHICLE DRIVER NO. 1

State | Driver's License | DOB: Mo/Da/Yr | Hazardous Action No. | Citation Charge | HBD Y N | Test | Helmet Y N

Driver's Name First M. Last | Address | City | State | Age | Sex | Inj.

Year | Make No. | Type | Trailer | Reg. | Yr/State | VIN (Make Letters Larger Than Numbers) | Removed to/by

Y N Haz. Citation | *Y N Driver Re-exam | *Y N Vehicle Defect | *Y N Fuel Leakage | Impact | Severity | Truck Cargo: Y N Cargo Spillage Cargo Description | Class
Y N Other Citation | *Y N Vision Obstruct. | Y N Veh. Driveable | *Y N Vehicle Fire

Restraints by occupants pos. | Name | Address | Pos. | Age | Sex | Inj. | Helmet Y N
1 2 3 | Y N
4 5 6 | Y N

Total Occupants | Local Use/Owner, Phone (If Veh. Type 7 or 8, List Carrier) | Insurance Co. | Agency Address | Injured taken to/by

VEHICLE DRIVER NO. 2

State | Driver's License | DOB: Mo/Da/Yr | Hazardous Action No. | Citation Charge | HBD Y N | Test | Helmet Y N

Driver's Name First M. Last | Address | City | State | Age | Sex | Inj.

Year | Make No. | Type | Trailer | Reg. | Yr/State | VIN (Make Letters Larger Than Numbers) | Removed to/by

Y N Haz. Citation | *Y N Driver Re-exam | *Y N Vehicle Defect | *Y N Fuel Leakage | Impact | Severity | Truck Cargo: Y N Cargo Spillage Cargo Description | Class
Y N Other Citation | *Y N Vision Obstruct. | Y N Veh. Driveable | *Y N Vehicle Fire

Restraints by occupants pos. | Name | Address | Pos. | Age | Sex | Inj. | Helmet Y N
1 2 3 | Y N
4 5 6 | Y N

Total Occupants | Local Use/Owner, Phone (If Veh. Type 7 or 8, List Carrier) | Insurance Co. | Agency Address | Injured taken to/by

NORTH

INCLUDE ALL TRAFFIC CONTROL DEVICES

ACCIDENT DESCRIPTION AND REMARKS (*Explain)

Reported: Mo/Da/Yr | Time A.M. P.M. | Investigators | Badge No. | Damaged Property Other Than Vehicle

Photos by | Comp. Disposition Open Closed | Reviewer | Person Advised of Damaged Traffic Control Device Name: Date: Time: | Owner | Address

* Use Accident Description and Remarks area for explanation.

FORWARD COPY TO: Michigan Department of State Police, Traffic Services Division, 7150 Harris Drive, Lansing, MI 48913

Driver No. 1 | Driver No. 2 | Accid. Date - Mo/Da/Yr | File Class | Department Complaint No.

Area | Pseudo | Time | Dir. Hit | Road Alignment | Traffic | Road Loc. | Acc. Type | Where | How | Type | Road Def. | 1Veh. Def. | 1Vision Obs. | 2Veh. Def. | 2Vision Obs. | MALI | Coder | File

Figure 11-1 State of Michigan accident report form.

Researchers have developed other scales of injury with finer gradations than the scale shown above, but these have gained limited acceptance on standard accident report forms. An example of another injury scale is the Maximum Abbreviated Injury Scale (MAIS), discussed by Miller et al. (1985), which rates injuries from "1 = minor injury" through "5 = critical injury" and "6 = untreatable, immediately fatal injury."

Another interesting and useful code on most accident report forms is the vehicle identification number (VIN). The VIN is a 17-digit number unique to each vehicle that is assigned to most automobiles and other motor vehicles. Coded within the VIN is information such as vehicle weight, wheelbase, and engine size. The VIN provides information about the vehicles involved in accidents that is not otherwise recorded on the report forms. Computer programs are available to decode the VIN for further analysis.

TABLE 11-1
North Carolina Accident Type Codes

Code Number	Accident Type
	Ran off road
1	Right
2	Left
3	Straight ahead
	Noncollision
4	Overturn
5	Other
	Collision of motor vehicle with:
6	Pedestrian
7	Parked vehicle
8	Train
9	Bicycle
10	Moped
11	Animal
12	Fixed object
13	Other object
	Collision of motor vehicle with another motor vehicle
14	Rear end, slow or stop
15	Rear end, turn
16	Left turn, same roadway
17	Left turn, different roadways
18	Right turn, same roadway
19	Right turn, different roadways
20	Head on
21	Sideswipe
22	Angle
23	Backing

For many accidents, other data are available in addition to, or instead of, a police accident report. Many jurisdictions accept accident reports by drivers or witnesses. A driver or witness form is usually shorter than a police accident report form. For some accidents, police or other investigators collect supplemental information. Photographs, sketches, calculations, notes, and other information may be on file, especially for more severe accidents.

Police officers usually complete accident reports during the shift when they investigated the accident. Once an officer submits a report, a supervisor or other personnel may check it. They will return incomplete or obviously inaccurate reports to the officer for correction. Reports filed by drivers or witnesses are rarely checked or corrected for completeness or accuracy.

PREPARING ACCIDENT DATA FOR ANALYSIS

Locating the Data File

Hard-Copy Filing Systems

In some studies, the analyst must retrieve accident data from hard copies of the reports. Accident reports are usually on file in several locations. The local station of the investigating police officer or the personnel receiving the report will keep a copy. A central office in the municipality, such as the police agency headquarters, will maintain another copy. The police agency will submit a third copy to a central state office with responsibility for collecting and storing accident data. The responsible agency varies from state to state, and could be the state police, the department of motor vehicles, or the department of transportation. Agencies usually keep forms on file for 2 to 5 years before destroying them. Some agencies will microfilm reports before destroying the hard copy.

Agencies file hard-copy forms several different ways. Most police agencies file accident reports by serial number. This method is convenient for police, attorneys, insurance investigators, and others interested in particular accidents. Filing reports by date or driver name is also a common practice that is convenient for those interested in particular accidents. Accident reports that are filed by the location of the accident are the easiest for engineers to retrieve. Separate files for accidents that happened at intersections and accidents that happened at nonintersection (midblock) locations are convenient. Figure 11-2 describes how a convenient accident report file by location in an urban jurisdiction would be organized. Intricate hard-copy filing systems by location exclusively for traffic engineering studies are unnecessary in most jurisdictions. This is true for jurisdictions that are so small that accident reports are easy to retrieve in any filing system or for jurisdictions that have computerized files that can be easily sorted by location.

Computer Filing Systems

All states, most large jurisdictions, and many small jurisdictions keep accident reports in computerized files. If a relevant computer file is available, analysts should use it instead of a manual file. All but the smallest of agencies should use computer files because they can save labor for a relatively low cost. Inexpensive microcomputers can store and quickly process the accident data from a small city for several years, and many programs are available for typical applications. The advantages of computer filing systems are likely to grow in the future as more agencies apply recent technological advances in data entry and storage such as optical scanners and compact disks.

For most computer files, operators key in data from accident report forms manually. In the United States, a state agency that is responsible for accident data collection and storage usually performs data entry. The agencies routinely check for missing data or erroneous keystrokes. The costs of entering accident data into the computer are significant. A recent survey showed that the mean value per accident record was $5.60 in 1987 dollars (Batz, 1989). That survey also indicated that the states computerize between 45 and 250 data items per accident, with a mean of 110, and that most states do not include the narrative description of the accident in the computer file. A computerized version of the sketch of the collision was still beyond the means of the responsible state agencies.

Most states can provide analysts with copies of the accident data they have processed on magnetic tape. States will erase data items identifying individuals to protect privacy when providing data to outside agencies, a private company, or individual. Complete files of accidents reported in a calendar year are usually available by June of the following year. Many states destroy computerized accident data after 4 or 5 years.

There are various ways in which accident data in a computer file relate locational information. In most states, police and motorist report forms ask for the name of the highway where the accident occurred and the distance and direction of that point from some known reference point such as an intersection or milepost. The form may also request the address of the accident location. The information on the form is usually translated into the standard location system of the state at the time the data are entered into the computer. Instead of a highway name, the state agency may enter a code number identifying the highway. Instead of the distance from the intersection nearest the accident, the state agency may record the distance from the beginning of the highway or the distance from an established reference point such as a major intersection.

The U.S. Department of Transportation also maintains computerized accident databases that are useful for some studies. The National Highway Traffic Safety Administration (NHTSA) maintains the Fatal Accident Reporting System (FARS). FARS is an almost complete census of all accidents in the United States that result in one or more fatalities. The General Estimates System, which is also maintained by NHTSA, contains

A. Basic elements
 1. Legal-size, one-third tab file folders are desirable.
 2. At high-accident frequency locations, accordion vertical file pockets are useful.
 3. Many jurisdictions have streets with identical names, but differentiated by suffixes. Therefore, abbreviated suffixes (St., Ave., Ter., Pl., Rd., Hwy., etc) should be used to assure proper identification.

B. Intersection file
 1. A file folder is made for each intersection in the jurisdiction.
 2. Names of intersecting streets may be tabulated in alphabetical order, and in numerical order, with folder tabs made in sequence:
  Archer St. and 1st Ave.
  Archer St. and 2nd Ave., etc.
  Archer St. and Baker Ave.
  Archer St. and Bakersfield St., etc
 3. If a significant group of streets are numbered, the intersection sequence may use numbers *first,* and then alphabetical sequence.
  1st and Archer St.
  1st and Brown St., etc.
  2nd and Archer St.
  2nd and Brown St., etc.
  George St. and Green St., etc.
 4. *All* intersections involving more than two different street names must have empty cross-reference folders.
  Archer St./Collins Ave./George St. (basic file).
  Archer St./Collins Ave.—SEE Archer/Collins/George.
  Collins Ave./George St.—SEE Archer/Collins/George.
 5. For loop streets, intersecting another street at two different places, set up N and S, or E and W file folders:
  1st Ave. and Baker Ave., N
  1st Ave. and Baker Ave., S
 6. Accident reports are filed in date sequence, with latest date in front.
 7. At very high-accident intersections, a separate folder may be useful for each year.

C. Midblock file
 1. A file folder is prepared for each different name street in the jurisdiction.
 2. Names are tabulated in numerical or alphabetical sequence.
 3. If address numbers progress in opposite directions from a base street, set up separate folders for N and S or E and W as suffixes.
 4. If part of the address numbers along a given street are, for example, running northward, and the street changes direction so that numbers pick up westward sequence, then separate folders should be set up:
  Archer St., N
  Archer St., W
 5. A very active driveway (such as a signalized access to a regional shopping center or major industrial plant), may warrant assignment of a specific address and a separate file folder:
  #1010 Archer St. (driveway)
  #4050 Green Rd. (driveway)
 6. For high-accident frequency streets, the midblock files may ultimately require subdivision into two or more folders, utilizing address breakpoints:
  Archer St. (0 to 3000 N)
  Archer St. (3001 N to 6000 N)
 7. Midblock accident reports are filed in *address sequence* within each folder, not by date.

Figure 11-2 Convenient location-type accident file system for a city or urban county. *Source:* Box and Oppenlander, 1976.

records from detailed independent investigations of a small, statistically selected sample of accidents across the United States. The other notable federal database is the Highway Safety Information System (HSIS), which the Federal Highway Administration developed. HSIS consists of five relatively high-quality databases from geographically scattered states. The police report accident data have been processed to correct some errors and to allow easy merges with traffic, roadway, and other files also maintained within HSIS.

Reducing Accident Data

Choosing a Time Frame

Once a database has been chosen, the analyst must separate the data needed for the particular study from the remainder of the file. One of the first steps in this reduction process is to choose an analysis time frame and discard the data from outside the time frame. In some studies the choice of a time frame is self-evident (e.g., the study is to see whether the accident rate in 1990 differed from the rate in 1991). When analysts must choose a time frame, 3 years is the most common choice. Three years represents a compromise between the desire for larger samples and the desire for time frames within which conditions were unlikely to have changed a great deal. Also, a 3-year time frame eliminates many problems with discarded data. Time frames of up to 4 years are common, but for periods longer than 4 years, analysts must use special care to ensure that changes in background conditions can be tolerated within the scope of the study. Time frames of less than 2 years may be necessary, but the small sample size will constrain the study.

It is good practice to check that accident data for particular locations have not been biased by construction or other major but temporary traffic events during the time frame selected. Analysts can consult highway agency records to check for construction, but such records are often incomplete or inaccurate. Supervisory and/or experienced area highway personnel can estimate whether major construction projects or other traffic events have taken place. However, those estimates should not inspire great confidence without corresponding documentation.

Analyzing Certain Locations

Many studies focus on one location or a limited set of locations on the highway network. One of the analyst's important tasks in these studies is to reduce the database to accidents that occurred at locations of interest. Analysts usually summarize accidents into those that occurred at spots and those that occurred in sections. Spots are short segments of highways that help identify problem "point" locations such as intersections, curves, and short bridges. The highway cross section and features at a spot should be noticeably different from surrounding spots. Sections are longer, relatively homogeneous segments of highway convenient for studying cross sections, pavement surfaces, and other longitudinal features. Spot lengths of 0.2 to 0.3 mile and section lengths of 1 to 2 miles are recommended for most agencies (Zegeer, 1982).

Spots and sections that float are usually more desirable than those that begin at fixed increments. For example, suppose that many accidents occurred near a highway feature at the 0.6-mile mark. It is likely that the police would code some of these accidents as occurring at the 0.5-mile mark and some as occurring at the 0.7-mile mark. If the agency used a fixed spot length of 0.3 mile, a spot would end at the 0.6-mile mark, so some of these accidents would fall in one spot and some would fall in the next. By contrast, a floating spot definition might have 0.3-mile-long spots beginning every 0.1 mile, overlapping each other. Thus there would be a spot including all accidents coded as occurring at the 0.5-, 0.6-, and 0.7-mile marks, and the agency would obtain a more accurate estimate of the true number of accidents at the feature (Zegeer, 1982).

Analyzing accidents at intersections requires a precise definition of the limits of the intersection. Some agencies include accidents with an intersection if they were coded as occurring on the approaches up to a fixed distance, usually 100 or 200 feet, from the intersection. A more time-consuming but technically defensible definition, especially in areas with chronic congestion, is to examine each accident on an approach to determine whether the intersection was an influencing factor or not. With the latter definition, the intersection size varies and basically extends to the back of the queue on the approaches.

Reducing Accident Data Based on Other Variables

The accident types or vehicle intent patterns analyzed during a particular study should always be those which are most closely related to the subject under study. The subjects of some studies obviously relate directly to certain accident types. For example, a study of accidents occurring on an approach with a new left-turn signal treatment should examine left-turn accidents or accidents involving left-turning vehicles, among other accident types. For some studies it is not obvious which accident types should be examined, so previous studies and engineering judgment must play a role in the selection.

Analysts may wish to reduce their databases based on several variables. They may have an interest in certain times of day, months of the year, weather, types of vehicles, driver conditions (i.e., only drivers coded as ''had not been drinking''), and so on. While reducing the database, analysts should be cautious not to delete data that may become useful later. It is generally less expensive to sort a larger database several times than to trim down to a small database and discover that needed information has been lost.

Merging Accident and Other Data

Some state and local agencies maintain files on roadway features, traffic volumes, intersection characteristics, construction and maintenance histories, traffic control devices, or other traits. The ability to merge these files with accident data files is essential for some analyses. Indeed, the ability to merge large files quickly is one of the main advantages of computerized accident data files. An example of a common type of merge is to combine accident and traffic flow data to produce accident rates (accidents per 100 million vehicle miles for highway sections and networks, accidents per million entering vehicles for spots).

To execute a merge, analysts need a data element in common between the files to be merged. In manual merges, this common element will most often be the name(s) of the highway(s). In computer merges, analysts most often use the codes and mileages identifying the spots or segments. It is best if these codes are common to many different files. It is also desirable that distances are measured in a uniform way (from the same points with the same units) in different data files. Some state departments of transportation have built these common codes and measurements into all their inventory and accident data files.

In the merging process, analysts must change the format of one of the files. Accident files are usually arranged in a format of one record (line of data) per accident. Traffic, highway characteristic, and other data files, however, are usually arranged in a format of one record per segment or feature. Typically, it is the accident data format that must be changed. Analysts might create a file summarizing accidents that occurred in particular segments before merging the data with a file providing annual average daily traffic by segment. Collapsing accident data in this way loses some information. When summarizing accident data for a merge, analysts should not purge too much of the file. It is less expensive to have unused information in a computer file than to discover that needed information has been lost during the merge and must be reconstructed from the original file.

RECOGNIZING AND DEALING WITH PROBLEMS IN ACCIDENT DATA

The greatest limitation to recognize when analyzing accident data is that the number of accidents occurring at a location or to a particular driver is not the same thing as the "safety" of the location or driver. Accidents are rare and somewhat random events—high-accident locations or drivers may be unlucky rather than unsafe. At best, accident data provide only an estimate of the safety of the location or driver (i.e., the number of accidents that would occur in an infinite time span) if conditions are the same in the future as they were in the past. However, because of various other problems and limitations on the use of accident data, even this estimate is difficult to obtain. The following paragraphs outline some of these other problems and limitations that could affect accident studies.

Unreported Accidents

Motorists do not report all traffic accidents to the police. The major reason that accidents are not reported is that they were not serious enough. Most states have thresholds of property damage below which police decline to investigate an accident. For many states, this threshold is around $500 (1990 dollars). Some states do not prepare reports on any accidents that do not involve an injury. Some states have established dual thresholds—a higher threshold for police reports and a lower threshold for reports prepared by motorists.

Accidents are unreported for reasons besides a reporting threshold. Sometimes motorists do not report accidents for fear of higher insurance rates. Police probably will not file a report on an accident that occurred on private property. Police may also be too busy with other activities to investigate accidents they are informed of. This is more common in larger cities. There are two key points for analysts regarding unreported accidents. First, analysts relying on reported accidents will underestimate the total number of accidents at a location or set of locations. Recent estimates (Hauer and Hakkert, 1988) are that:

- The number of fatalities is generally known within ±5% of the true number.
- The number of injuries requiring hospitalization is underreported by about 20%.
- Only about half of all injuries occurring in accidents are reported.
- Motorists report fewer than half of all property damage only (PDO) accidents.

Analysts should include an awareness of this underestimation in reports by writing in terms of "reported accidents" rather than merely "accidents."

The second key point regarding unreported accidents is that the extent of underreporting varies by accident type, driver age, location, time, and many other variables. Analysts making comparisons across such variables must assume that the underreporting is constant, assume that the underreporting is negligible within the scope of the particular study, or correct for the varying underreporting. Analysts sometimes correct for the most obvious reasons for underreporting, such as changing reporting thresholds, but rarely for less obvious reasons. As often as practical, agencies should determine the proportion of accidents reported for important variables and avoid naively assuming uniform reporting.

Erroneous Data

Accident reports often contain errors, and the number of errors grows with every step of further processing. In fact, researchers have found an average of between 1 and 2.2 discrepancies per accident between the accident report and the coded file (as related by Hauer and Hakkert, 1988). The errors are not evenly distributed throughout the report, so it is very difficult to generalize about erroneous data.

Range and logic checks will reveal some errors in accident files. Analysts conduct range checks with programs that compare the values of a variable to acceptable limits for that variable. The program will flag a particular report if its value is outside the limits. Logic checks compare the values of two or more variables within an accident record, looking for internal consistency. For example, left-turn accident types are impossible on freeways except where vehicles can make U-turns through median openings. When range and logic checks turn up erroneous data, analysts could delete the offending accident report or delete the offending value. If the error occurs in an important variable or if a limited sample of accidents is available, the analyst may need to consult a hard copy of the accident report. Consider deleting the entire accident report only if the error is so gross that the accident is most likely irrelevant to the analysis (i.e., the accident occurred in another city) or if the program flags many errors in the same record.

Some errors are important for many analyses and should be checked for routinely. Errors in accident location are common—agencies are typically unable to locate 10 to 40% of reported accidents, especially in urban areas. Distance estimates are a prime culprit, but misspelled street names, confusion involving multiple street names and numbers for a given facility, and unusual geometric features such as freeway interchanges also cause location errors. Manual double-checking of accident locations is sometimes necessary. The accident-type variable is very useful but often erroneous. At intersections, vehicle intent codes may be more accurate than accident-type codes because they are less ambiguous (Hauer et al., 1988). Using vehicle intent codes at intersections provides another advantage as well, which will be discussed in a later section of this chapter. Variables on the accident report form where police officers, drivers, or witnesses describe highway features are also notoriously erroneous. If analysts need more than very simple information on highway features for a study, they should look for it from a source other than the accident record, such as a visit to the accident site.

Accident Cause

Engineers often think of the events leading to or causing a traffic accident as a chain, where the removal of a link in the chain would have prevented the accident. Links in accident cause chains might include the decision to make a trip, the choice of a route, vehicle defects such as worn tires, slick pavements due to rain or snow, objects blocking driver vision, distractions in or outside the vehicle, or the presence of fixed objects on the roadside. Vehicle and highway factors are usually present on cause chains, but driver factors are the most common links.

Most accident reports contain only the last and most obvious links of the accident cause chain. Police who are investigating accidents do not have the time or, in many cases, the training to seek links farther up the chain. Police often use favorite standbys such as "driver inattention" when other causes are obscure. Most analysts believe driver and witness reports are virtually worthless for providing credible information on accident causes. Indeed, it is very difficult for even highly trained investigators with sufficient time to construct an accurate chain. Thus, in the case of a single-vehicle, run-off-road accident, the object that was struck on the roadside will be prominent in the report. Analysts may tend to seek treatments for this "cause" while ignoring driver, vehicle, and highway factors that preceded the vehicle leaving the road. That tendency can be resisted by searching the accident report for other clues to links on the chain that are not prominent. Another way to attack the accident cause problem is to analyze smaller samples of more detailed accident data when designing ways to break the chain.

Limitations of the Report Form

Accident report forms have resulted from balancing competing interests. One of the major interests is the police, whose time is required to complete the report. The report form therefore is likely to be missing elements that traffic engineers would like to use. Usually, traffic engineers consider information on the highway to be insufficient on accident report forms. Information on the vehicle is also lacking for some analyses, especially analyses involving trucks and other large vehicles and their contents. Driver and passenger information is usually complete for traffic engineering uses. Analysts should check the standard form and the coded accident files to ensure that the information needed has been collected before launching a study.

ANALYZING ACCIDENT DATA

The previous sections of this chapter dealt with how to select and prepare accident data for an analysis. After those tasks, the analyst is ready to make some type of inference about the "safety" of the highways or drivers of interest. In the following section four different types of accident analyses are described: summarizing numbers of accidents and trends, identifying accident-prone locations, selecting accident countermeasures, and evaluating existing countermeasures. Those four types of studies encompass most of the current uses of accident data by practicing traffic engineers. Other types of studies are possible and useful but typically fall within the researcher's domain.

The focus in the following paragraphs is on identifying and correcting accident-prone locations, because that is the primary task of most transportation engineers in state and local agencies who conduct accident studies. Those interested in identifying and correcting accident-prone drivers or vehicles can, in most cases, use the same methods while substituting the words "driver" or "vehicle" for "location."

Numbers of Accidents and Trends

Engineers, policymakers, the news media, and others often want to know the number of accidents of some type that occurred at a location or set of locations during some time period. Such summaries are useful for comparing highway safety with other competitors for funds, for noting trends with time (i.e., annual summaries), or for grasping the magnitude of the problem. The numbers are usually very easy to obtain, either by manual counts or by simple computer programs, once the file has been prepared. Appendices D, E, and F provide information on how to prepare graphics, text, and presentations.

Analysts preparing summaries of the numbers of accidents or trends should provide the audience with more than just a number. The main tasks facing an analyst presenting numbers or trends are to put them in perspective and alert the audience to any suspicions regarding accuracy. The audience might need to know that the number of accidents is a random variable that fluctuates through time and that many accidents are unreported, for example. In displaying a trend of accidents through time, the analyst should note for the audience any major factors that may have influenced the trend. For example, the reporting threshold may have changed, inflation may be significant (it takes less damage to require $500 of repair than it used to), or travel habits may have changed.

Summaries of accident data must emphasize for the audience whether statistics presented are "injury accidents" and "fatal accidents" or the "number of injuries" and the "number of fatalities." Either type of measure could be important to the aims of a particular study. Analysts rate the severity of a particular accident based on the most severe injury experienced by anyone in the accident. Thus if in an accident at least one person experienced a B injury but no F or A injuries occurred, an analyst would label the accident a B-injury accident. There could have been several B injuries and several C injuries recorded for the B-injury accident. The severity rating of an accident thus says little about the numbers of persons injured or killed. If none of the participants in an accident reported injuries, analysts refer to the accident as a PDO (property damage only) accident.

Involvements account for the number of individual vehicles in an accident. A three-car accident includes three involvements; a single-vehicle accident includes one involvement. Analysts comparing different classes of vehicles or drivers must not confuse accidents with involvements when looking at multiple-vehicle accidents. For example, consider a location where older drivers make up 30% of the driver population and accident data show that at least one older driver is involved in 50% of the two-vehicle accidents. Many analysts would conclude that older drivers are overrepresented and start designing countermeasures for them. However, a close look at involvements rather than accidents shows that older drivers are not overrepresented in this case. If older drivers are no more likely to be involved in accidents than other drivers, the following statements are true:

- The probability that an older driver has collided with another older driver is $0.3 \times 0.3 = 0.09$.
- The probability that an older driver has collided with a younger driver is $0.3 \times 0.7 = 0.21$.
- The probability that a younger driver has collided with an older driver is $0.7 \times 0.3 = 0.21$.
- The probability that a younger driver has collided with another younger driver is $0.7 \times 0.7 = 0.49$.

Therefore, the total probability that an accident involves one or more older drivers is $0.09 + 0.21 + 0.21 = 0.51$. Since 0.51 is so close to the actual proportion of older drivers involved in accidents (0.50), older drivers are most likely not overrepresented. Analysts comparing driver gender, vehicle types, and other similar factors must separate involvements and accidents.

Identifying Accident-Prone Locations

Due to limited budgets, it is essential that agencies making highway safety improvements direct resources to real problem locations. Good litigation risk management also demands that agencies identify accident-prone locations through a logical process. Thus engineers have developed procedures to identify accident-prone locations using accident data, and this section presents the most useful of those procedures. A discussion of which procedure a particular study should use follows the presentation of all the procedures.

Spot Maps

A spot map is a simple and effective way to determine accident-prone locations qualitatively. Engineers create spot maps by marking the location of each relevant accident on a map. For a small area or a limited number of accidents, a map hung on a wall with pins to mark accidents is useful. Computer graphics easily allow spot maps of large areas or with large numbers of accidents to be updated and displayed on a monitor. Analysts can use different colors and sizes of pins or graphical symbols to represent different types or severities of accidents or to represent "multipliers" of accidents (i.e., one large pin equals ten accidents). Spot maps are very useful for specialized situations such as pedestrian accidents or parked-car accidents.

Accident Frequency

Some agencies identify accident-prone locations through lists of locations (spots, sections, intersections, etc.) ranked by the number of reported accidents. The primary virtues of this approach are that it is simple and that it makes intuitive sense. If the agency goal is to minimize total accidents, attacking the locations with the most accidents seems logical.

Accident Rates

Agencies have also identified accident-prone locations through lists of locations ranked by accident rate. Agencies usually compute highway section accident rates in terms of accidents per 100 million vehicle miles using the formula

$$\text{RSEC} = \frac{100{,}000{,}000\,A}{365\,T \times V \times L} \tag{11-1}$$

where

RSEC = accident rate for the section
A = number of reported accidents
T = time frame of the analysis, years
V = AADT
L = length of the section, miles

For spots, agencies usually calculate the accident rate in terms of accidents per million entering vehicles from the equation

$$\text{RSP} = \frac{1{,}000{,}000\,A}{365\,T \times V} \tag{11-2}$$

where RSP is the accident rate for the spot, V is the AADT or, for intersections, the sum of the average daily approach volumes, and A and T are as defined in equation (11-1).

Obviously, ranking locations by accident rate requires traffic volume data. The time period of the volume data should match the time period of the accident data being analyzed. Analysts may use volume data somewhat outside the accident data time period if they adjust the volumes for temporal variation (growth). Analysts should use volume data that are 5 years or more removed from the time period of the accident data with extreme caution, if at all.

Accident rates account for exposure, which is the chance that an accident will happen to a particular driver, vehicle, or highway segment. Many studies have used AADTs or entering volumes as a general measure of exposure. There are methods to estimate exposure that are not based on traffic volumes but these methods are more appropriate for research studies. Engineers have developed computer programs to search accident files and rank locations automatically by accident rate or frequency. For example, the FHWA developed HISAM to manage medium-sized and small accident databases and to perform such tasks as ranking locations. HISAM runs on a microcomputer, is somewhat user friendly, and requires a minimum of field data to install and maintain. Harkey and Ruiz (1989) describe an application of HISAM. Another program, called WINDOW, will flag high-frequency and rate locations using the floating segment technique. Mak et al. (1986) relate details of an application of WINDOW. The McTrans Catalog lists several other programs for microcomputers that aid in the analysis of accident data (Center for Microcomputers in Transportation, 1992).

Accounting for Severity

Analysts can adjust accident frequencies or accident rates to reflect the greater costs of injury and fatal accidents. One common method of taking severity into account before ranking locations is to compute the number of equivalent property damage only (EPDO) accidents. The method uses the number of PDO accidents that mean the same to society as a fatal accident or an injury accident. Different agencies use different equivalency factors. Kentucky used a factor of 9.5 PDO accidents per fatal or A-injury accident, and 3.5 PDO accidents per B-injury or C-injury accident (from Zegeer, 1982). There is no methodological reason that prevents agencies from applying frequency, rate, or other methods to only injury and fatal or only fatal accidents. However, small sample sizes of injury and especially fatal accidents sometimes makes analysis difficult.

Classic Statistical Method

After ranking locations of interest by frequency or rate (adjusting for severity if necessary), analysts could choose the top n locations for further detailed analysis. They could choose n arbitrarily, where there appears to be a real "break" in the list, by tradition, according to the labor available to perform the detailed analyses, or for some other reason. However, a more scientific means of selecting locations is to assume that the number of accidents at locations of interest follows a standard normal probability distribution. Then analysts can select those locations that appear to be significantly higher than the mean frequency or rate. Using this procedure, analysts will select a location if it satisfies the inequality

$$\mathrm{OB}_i > XA + K \times S \tag{11-3}$$

where

OB_i = accident frequency or rate at location i
XA = mean frequency or rate for all locations under consideration
K = constant corresponding to a level of confidence in the finding
S = sample standard deviation for all locations (see Appendix C for the formula to calculate the sample standard deviation)

Agencies commonly use 90, 95, and 99% levels of confidence, which correspond to K values of 1.282, 1.645, and 2.327, respectively. The forces listed above that guide the nonscientific choice of n locations also guide the choice of a particular level of confidence. The greater the number of sites chosen, by whatever method, the greater the probability that a truly hazardous location is flagged and the greater the probability that a truly nonhazardous location is also flagged.

EXAMPLE 11-1: Application of the Classic Statistical Method

Roadway Section 33 has an accident rate of 210 accidents per 100 million vehicle miles (mvm). The mean accident rate for all sections in the jurisdiction is 89 per 100 mvm, and the standard deviation corresponding to this mean rate is 64 per 100 mvm. Should an analyst flag Section 33 as hazardous with 95% confidence?

Solution: For these data, the inequality above holds true as follows:

$$\begin{aligned} \mathrm{OB}_i &> XA + K \times S \\ 210 &> 89 + (1.645 \times 64) \\ &> 190 \end{aligned}$$

Therefore, one should flag Section 33 as hazardous with 95% confidence using the classic statistical method.

Rate Quality Control Method

A variation on the classic statistical method described above is the rate quality control method. A number of agencies have used the rate quality control method for several years. It differs from the classic statistical method in several important ways, including:

- It applies only to accident rates, not frequencies.
- It assumes that the number of accidents at a set of locations follows a Poisson distribution.
- It compares the rate of a particular location to the mean rate at similar locations rather than at all locations.

The rate quality control method can apply to spots or sections. For spots, analysts use accidents per million vehicles. For sections, analysts use accidents per mvm or per 100 mvm.

The rate quality control method flags a location as hazardous if it satisfies the following inequality:

$$\text{OBR}_i > XS + K\left(\frac{XS}{V_i}\right)^{0.5} + \frac{1}{2V_i} \qquad (11\text{-}4)$$

where

OBR_i = accident rate observed at location i

XS = mean accident rate for locations with characteristics similar to those of location i

V_i = volume of traffic at location i, in the same units as the accident rates are given

and where K is as defined for equation (11-3).

The question of which locations are similar enough to include in the computation of XS is difficult. Generally, agencies have used relatively broad definitions of similarity to compute XS. For example, one agency used statewide average rates based only on intersection type (i.e., arterial meeting collector in an urban area) and traffic volume to provide XS (Zegeer, 1982).

EXAMPLE 11-2: Application of the Rate Quality Control Method

Roadway Section 33 had 40 reported accidents in 3 years, and the agency responsible for the section estimated that travel on the section was 19 mvm during that time. The mean accident rate for all sections in the jurisdiction similar to Section 33 is 140 per 100 mvm. Should an analyst flag Section 33 as hazardous with 95% confidence?

Solution: The rate quality control method requires the same units for each variable in the inequality, so the analyst should convert XS to 1.4 accidents per mvm to be consistent with the units given for V_i. OBR_i is 40 accidents/19 mvm = 2.1 accidents per mvm. For these data, the rate quality control inequality above holds true as follows:

$$\text{OBR}_i > XS + K\left(\frac{XS}{V_i}\right)^{0.5} + \frac{1}{2V_i}$$

$$2.1 > 1.4 + 1.645\left(\frac{1.4}{19}\right)^{0.5} + \frac{1}{2 \times 19}$$

$$> 1.9$$

Therefore, the agency should consider Section 33 hazardous with 95% confidence using the rate quality control method. Notice that the mean rate used for comparison is much higher in this example than in Example 11-1, yet the rate quality control method still identifies the section as hazardous.

Bayesian Methods

Researchers have identified problems with the classic statistical method and the rate quality control method. They point to the arbitrary nature of the choice of a confidence level. They also cite the fact that the methods are unable to combine information from previous studies or information about the location characteristics with current accident information during an analysis. Finally, they worry that those methods do not flag truly hazardous sites often enough. In response to those concerns, researchers have developed methods of identifying accident-prone locations based on Bayesian statistics. Bayesian statistics use the concept of a conditional probability, which is the probability that something is true given the knowledge that something else has occurred. In accident studies analysts want to know the probability that a location is truly hazardous given the accident history, traffic volumes, and physical characteristics of the location. Analysts interested in Bayesian methods should consult Hauer and Persaud (1984, 1987) for example applications identifying hazardous freeway ramps and railroad grade crossings.

Equations for Different Accident Patterns at Intersections

Agencies typically use equation (11-2) to compute accident rates for intersections. However, Hauer et al. (1988) showed that using the total entering volume for an intersection to compute an overall intersection accident rate in equation (11-2) is not logical. To understand why, consider two identical intersections with identical total entering volumes. Intersection 1 has a high percentage of through traffic, while intersection 2 has a high percentage of left-turning traffic. Because left-turning traffic is involved in accidents more frequently than through traffic, intersection 2 will probably rank above intersection 1 using equation (11-2) and will probably receive more attention and safety improvements. However, intersection 2 does not necessarily have a more correctable safety problem than intersection 1. In fact, intersection 2 may function very well compared to other intersections with high proportions of left-turn volume, while intersection 1 may compare very poorly to other intersections with high proportions of through volume. Thus there is a need for a procedure that ranks intersections while taking into account the types of traffic volumes rather than the overall entering volume.

Hauer et al. (1988) developed a procedure that can rank intersections by hazard and accounts for different types of volume. Their procedure uses the vehicle intent variable on the accident form instead of the accident-type variable, because the former is considered to be more reliable and less ambiguous. For instance, many accidents involving left-turning vehicles will have accident type coded as ''angle'' because the vehicles collided at right angles. Analysts must have turning movement counts on hand for each intersection they want to examine with the procedure. The major steps of the procedure include:

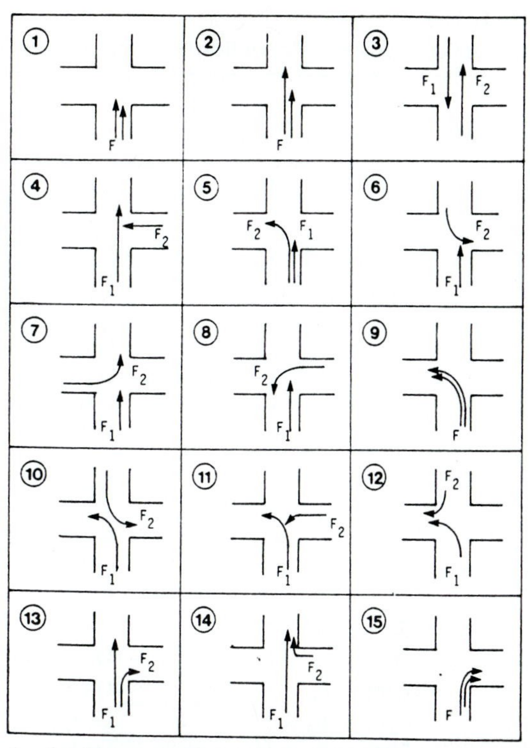

Figure 11-3 Accident patterns at intersections by vehicle movement. *Source:* Hauer, E., J. C. N. Ng and J. Lovell, "Estimation of Safety at Signalized Intersections," *Transportation Research Record 1185,* National Research Council, Washington, DC, 1988.

1. There are 15 typical multivehicle accident patterns on each approach to a four-legged intersection as shown on Figure 11-3. For each intersection to be examined with the procedure, calculate the expected mean number of accidents for each pattern on each approach using the equations given in Table 11-2.
2. Sum all the estimated means from step 1 for each intersection.
3. Adjust the sums from step 3 to get the estimated mean number of accidents for the entire time period of interest. The equations in Table 11-2 are for an A.M. peak of 2 hours, a P.M. peak of 2 hours, an off-peak of 5 hours, and a day of 9 hours. One can assume 260 weekdays per year.
4. For each intersection, compute the difference between the recorded number of accidents (that conform to one of the 15 types on Figure 11-3) and the adjusted sum from step 3.
5. Compare the differences computed in step 4 between all intersections. The intersection with the greatest difference has the greatest hazard that is unrelated to the magnitude of the traffic volume.

The equations in Table 11-2 apply only to weekday, daylight, two-vehicle accidents at signalized intersections. The equations for patterns 1, 2, 4, and 6 are based on larger samples and are therefore more reliable than the equations for the other patterns. The traffic volumes to use for the 15 patterns are self-explanatory except for pattern 4, where F_1 is the larger of the two flows. The comparison in step 5 above is simple but crude; Hauer et al. describe a more complicated and precise way to determine whether the intersection has experienced more than its share of accidents.

TABLE 11-2
Intersection Accident Prediction Equations from Toronto, Ontario

Pattern	Model Form[a]	Time	Coefficient Estimates b_0	b_1	b_2
1	$\hat{E}\{m\} = b_0 \times F$	A.M.	0.1655×10^{-6}		
		P.M.	0.2178×10^{-6}		
		Off	0.2164×10^{-6}		
		Daily	0.2052×10^{-6}		
2	$\hat{E}\{m\} = b_0 \times F$	A.M.	0.0987×10^{-6}		
		P.M.	0.0933×10^{-6}		
		Off	0.1080×10^{-6}		
		Daily	0.1014×10^{-6}		
4	$\hat{E}\{m\} = b_0 \times F_2^{b_2}$	A.M.	19.020×10^{-6}		0.1536
		P.M.	1.4127×10^{-6}		0.6044
		Off	9.7329×10^{-6}		0.3860
		Daily	8.1296×10^{-6}		0.3662
6	$\hat{E}\{m\} = b_0 \times F_1 \times F_2^{b_2}$	A.M.	0.2383×10^{-6}		0.5163
		P.M.	0.0940×10^{-6}		0.3091
		Off	0.0718×10^{-6}		0.4127
		Daily	0.0418×10^{-6}		0.4634
3	$\hat{E}\{m\} = b_0 \times F_2^{b_2}$	Daily	8.6129×10^{-9}		1.0682
5	$\hat{E}\{m\} = b_0 \times F_1^{b_1} \times F_2^{b_2}$	Daily	0.3449×10^{-6}	0.1363	0.6013
7	$\hat{E}\{m\} = b_0 \times F_1^{b_1} \times F_2^{b_2}$	Daily	0.2113×10^{-6}	0.3468	0.4051
8	$\hat{E}\{m\} = b_0 \times F_2^{b_2}$	Daily	2.6792×10^{-6}		0.2476
9	$\hat{E}\{m\} = b_0 \times F^{b_1}$	Daily	6.9815×10^{-9}	1.4892	
10	$\hat{E}\{m\} = b_0 \times F_2^{b_2}$	Daily	5.590×10^{-12}		2.7862
11	$\hat{E}\{m\} = b_0 \times F_1^{b_1} \times F_2^{b_2}$	Daily	1.3012×10^{-9}	1.1432	0.4353
12	$\hat{E}\{m\} = b_0 \times F_1^{b_1} \times F_2^{b_2}$	Daily	0.0106×10^{-6}	0.6135	0.7858
13	$\hat{E}\{m\} = b_0 \times F_1^{b_1} \times F_2^{b_2}$	Daily	0.4846×10^{-6}	0.2769	0.4479
14	$\hat{E}\{m\} = b_0 \times F_1^{b_1} \times F_2^{b_2}$	Daily	1.7741×10^{-9}	1.1121	0.5467
15	$\hat{E}\{m\} = b_0 \times F^{b_1}$	Daily	0.5355×10^{-6}	0.4610	

[a] $\hat{E}\{m\}$ is the expected mean number of accidents per hour; F, F_1, and F_2 are average hourly traffic volumes for flows shown in Figure 11-3.

Source: Hauer et al., 1988.

EXAMPLE 11-3: Using the Equations for Different Accident Patterns at Intersections

The northbound approach to intersection 19 had 11 weekday accidents of patterns 1, 2, 4, and 6 during 9 daylight hours (two in A.M. peak, two in P.M. peak, five in off-peak) in 3 years. The responsible agency estimated average hourly traffic movements for those same 9 hours in the middle of the 3-year period as 500 northbound through vehicles, 410 westbound through vehicles, and 200 southbound left-turning vehicles. If agency policy is to flag as hazardous approaches that have more than double the expected number of accidents over 3 years, should it flag this approach?

Solution: Table 11-2 is used to estimate the expected number of accidents. The expected number of pattern 1 accidents per hour is

$$\begin{aligned}\hat{E}\{m\} &= b_0 \times F \\ &= 0.2052 \times 10^{-6} \times 500 \\ &= 0.000103\end{aligned}$$

Similarly, the expected daily number of pattern 2, 4, and 6 accidents is 0.000051, 0.000074, and 0.000243, respectively. The expected number of accidents of patterns 1, 2, 4, and 6 is therefore 0.000103 + 0.000051 + 0.000074 +

0.000243 = 0.000471 accident per hour. Over 3 years the expected number of accidents of interest is 0.000471 accident per hour × 9 hours per weekday × 260 weekdays per year × 3 years = 3.3. Since the number of accidents recorded was more than twice the number of expected accidents, the agency should flag the approach as hazardous.

One hundred and forty-five signalized intersections in Toronto, Ontario provided the data used to calibrate the equations in Table 11-2. Analysts who want to calibrate equations such as those in Table 11-2 for their own areas should consult Hauer et al. (1988). Those analysts will need statistical expertise and certain statistical programs.

Choosing a Method

No single method to identify accident-prone locations is universally superior. The best approach for an analyst is to select a particular method for a particular analysis, or to use several methods for large studies when adequate resources are available. Smaller agencies and studies with limited resources will tend to choose frequency and rate methods. Both methods have serious flaws. Using frequencies results in identification of too many high-volume urban locations, since a primary factor related to accident occurrence is traffic volume. These high-volume locations also may be places where the search for realistic and effective countermeasures is especially difficult. Using rates results in identification of too many low-volume rural and local street sites, because a chance occurrence of an accident or two divided by a low volume results in a high rate. Using frequencies and rates together helps to mitigate these biases somewhat. Many agencies therefore rank by rate those locations that have experienced some minimum frequency of accidents.

Methods to account for severity can supplement other methods and reveal locations that experience extreme numbers of severe accidents. However, severity methods such as the EPDO method introduce yet another arbitrary judgment and volatile source of variation into the analysis. Also, since underreporting levels vary by severity, the choice of logical EPDO values is difficult. Severity methods should not serve as the only means of identifying locations for further review.

The rate quality control method is generally superior to simple frequency and rate and classic statistical methods in that it identifies truly hazardous locations correctly more often. However, the rate quality control method requires many more resources than other simpler methods because agencies need average rates for different classes of locations. If an agency has a reliable source for average rates or sufficient resources to collect such data, it should use the rate quality control method.

The Bayesian techniques of identifying accident-prone locations hold great promise for the future but need further refinements before they gain widespread use. Bayesian methods are computationally intense, and user-friendly programs are not yet available for applications. Research shows that the rate quality control method is almost as good at identifying accident-prone locations and is much easier to apply (Higle and Hecht, 1989). Analysts should consider using the procedure based on 15 typical multivehicle accident patterns at intersections instead of accident rates based on total entering volume. The procedure is computation intensive and data intensive, requiring turning movement counts at each included intersection. Analysts must also be able to trust the accuracy of vehicle intent code on the accident report form. However, the procedure is more accurate than the method based on total entering volume and the procedure provides a "head start" for generating countermeasures since it looks at individual accident patterns.

To restate points made earlier, accident data analysis is only one part of the process of identifying hazardous locations. Accidents are random occurrences, and an accident history is not equivalent to the "safety" of a location. Accident data are also flawed and limited in other important ways. Analysts must coordinate an accident study with a direct look at the characteristics of the locations of interest to have the best chance of correctly identifying hazardous locations.

Selecting Countermeasures

Collision Diagrams

Once an agency has identified a location as "accident-prone" or otherwise worthy of improvement, a search for affordable and effective countermeasures begins. The first stage in selecting countermeasures is to look for overrepresented clusters of particular kinds of accidents. No mathematical procedure to find overrepresented clusters has gained widespread use. The procedure described above using 15 accident patterns at intersections developed by Hauer et al. (1988) can help but is generally too computation intensive and data intensive for most agencies. Collision diagrams are the main tools used by agencies to identify these clusters.

A collision diagram is a schematic, not-to-scale, graphical representation of the accident pattern at a particular location. Collision diagrams can be useful for many types of locations but are most often used at spots such as intersections. Collision diagrams can quickly show analysts where concentrations of accidents are located, the types of accidents that predominate, and other useful information. Each accident is usually plotted separately, on the approach and near the place where the first harmful event is said to have occurred. An arrow in the direction of travel represents each vehicle involved in an accident. Symbols used with the arrows, as shown with the sample collision diagram in Figure 11-4, describe vehicle types, intended vehicle movements, accident severity, and accident type. Again, the intended vehicle movement can be more valuable during the analysis than the accident type. Some police reports describe vehicles that helped cause but were not damaged in an accident. Arrows with a different type of line than that used for involved vehicles can represent these noninvolved vehicles. The type of fixed objects struck may also be useful and may be shown.

If the diagram becomes too crowded when each accident is shown separately, analysts can use symbols to represent sets of accidents of a particular kind. A bold arrow representing 10 accidents of a particular kind is common, for example. The diagram loses interesting details by summarizing the data in this way, however. Twenty to thirty accidents fit comfortably on an 8.5- by 11-inch sheet, and analysts can use oversize sheets for spots with more accidents or longer periods. Labels on the arrows could indicate the date of the accident, the day of the week, the time of the accident, the road surface condition at the time of the accident, and the lighting condition at the time of the accident. Collision diagrams should show military time for clarity and brevity. Abbreviations such as "CD" and "WN" can represent "clear, day" and "wet, night," for example, to show important information concisely.

Light condition and road condition are two of the most important variables used to identify overrepresented clusters of accidents. A cluster of night accidents, for example, might indicate the need for countermeasures such as reflectors, delineators, and street lighting. Light condition is a reasonably accurate data item that analysts can check against the time of the accident to find errors. Analysts often check the night-to-day accident ratio to gauge whether a location or cluster of accidents is overrepresented at night or during the day. Analysts must be cautious, however, because the night-to-day ratio will fluctuate a great deal with low numbers of accidents. Most accident report forms include codes for "dawn" or "dusk" light condition at the time of the accident. These codes present a dilemma since the samples of dawn or dusk accidents are often too small to analyze separately. It is customary in many agencies to eliminate dawn and dusk accidents when computing night-to-day ratios and when necessary during other analyses. Finally, since the amount of daylight per day varies throughout the year, light condition

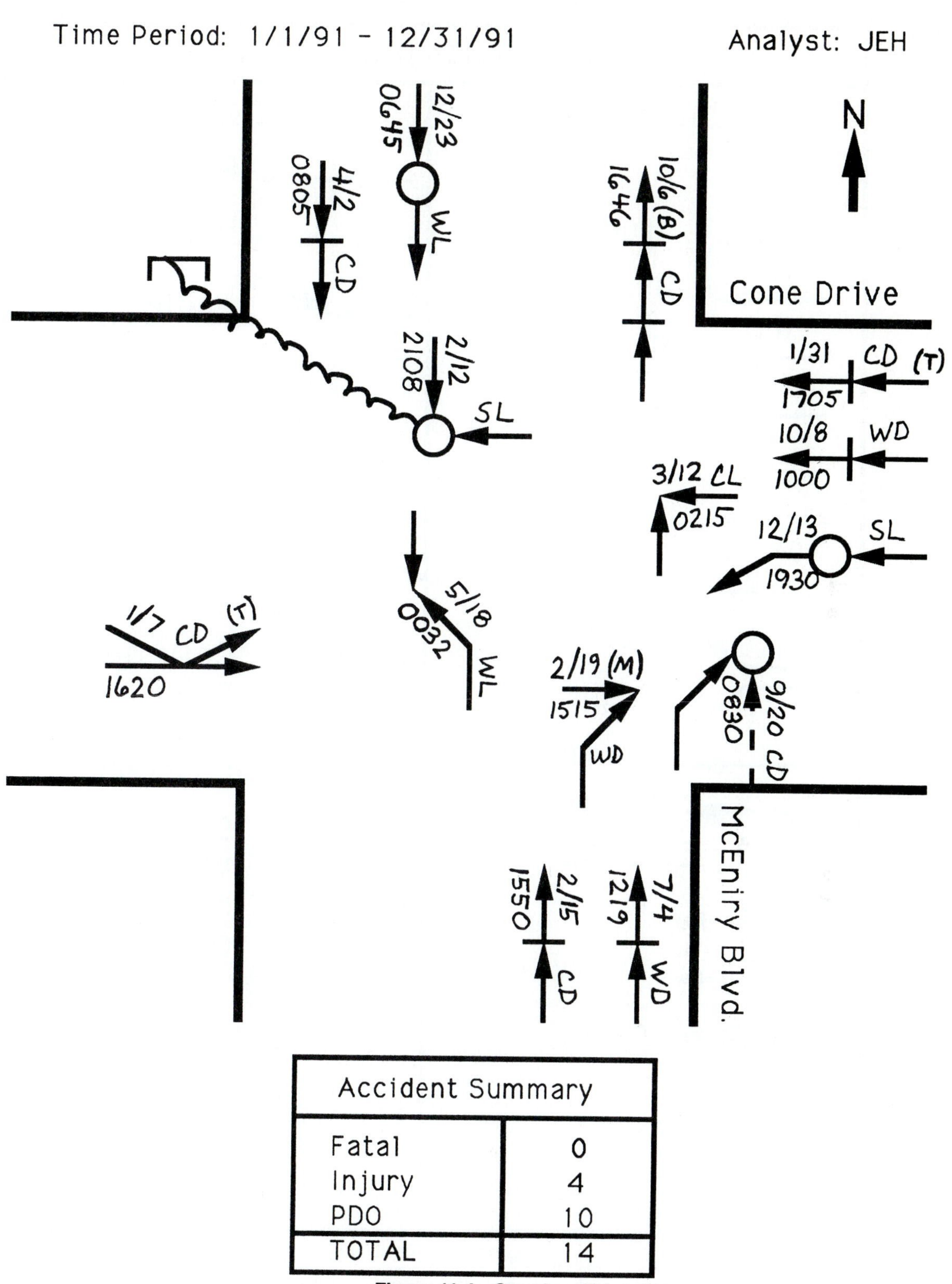

Accident Summary	
Fatal	0
Injury	4
PDO	10
TOTAL	14

Figure 11-4 Sample collision diagram.

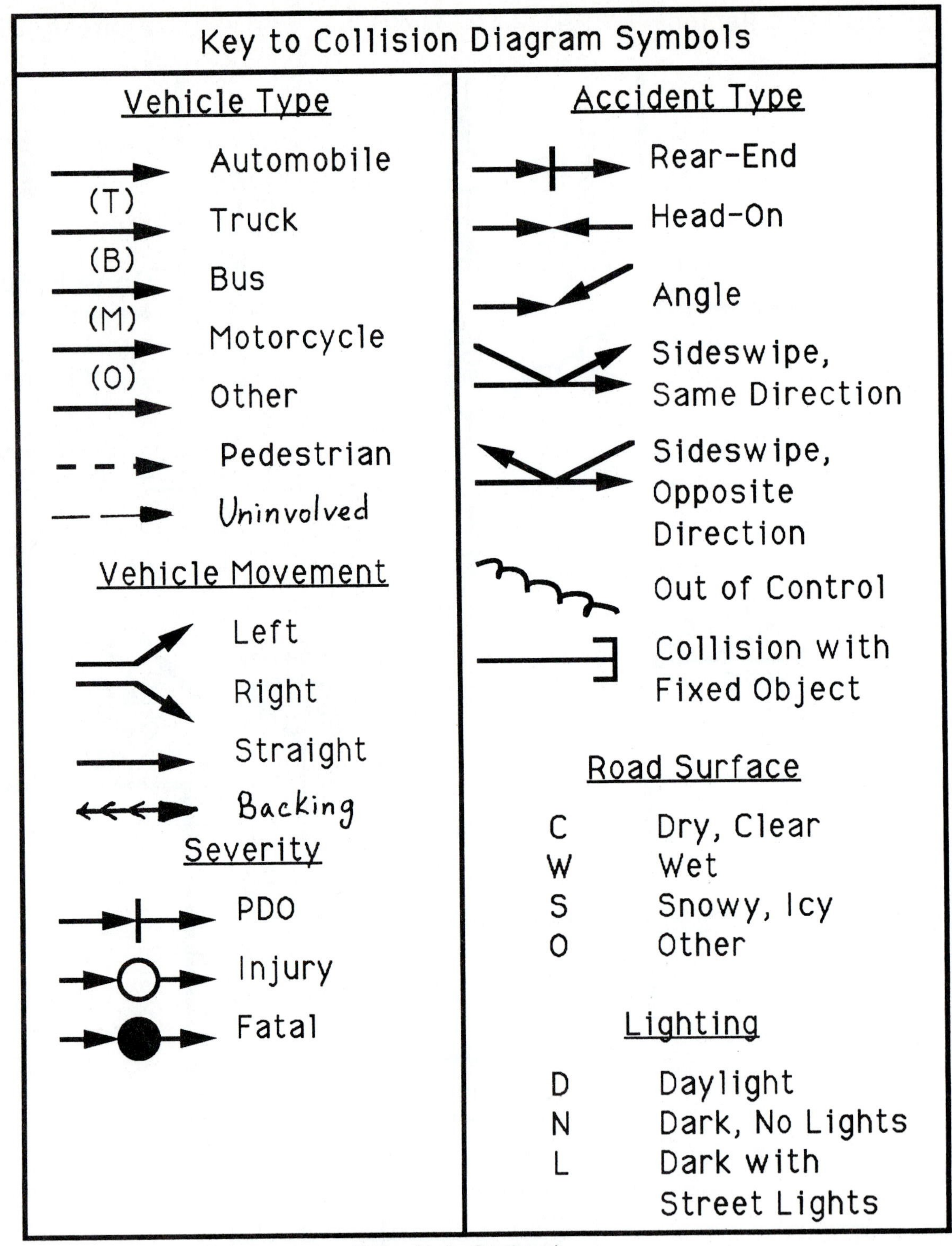

Figure 11-4 Continued

measures such as night-to-day ratio will also vary throughout the year even if other factors are constant. Analysts making comparisons using light condition should therefore use the same months of the year for all entities being compared.

Collision diagrams contain very few details about the particular location. Street names, outlines of the edges of the pavement, and a direction orientation are all that is necessary besides the symbols for accidents and fixed objects. Sometimes, collision diagrams contain a table summarizing the accidents. Computer programs are available to help prepare collision diagrams. Some of these programs work directly from an agency's computerized accident files. The programs are often flexible and allow the user to specify which accidents to include and which variables to show on the drawing. The flexibility of computer collision diagrams is a major advantage. For example, analysts can quickly obtain separate diagrams of night accidents, injury accidents, or any subset of the total reported accident picture. The output from computer-generated collision diagrams can be plotted or shown on the monitor and computer plots are usually easier to read than manual drawings. Use caution that the information on computer-generated collision diagrams is complete and accurate—some collision details may have to be added manually.

Analysts often use condition diagrams with collision diagrams to generate countermeasure ideas. Condition diagrams are scale drawings of the location of interest that show the layout, lane and roadway widths, grades, view obstructions, traffic control devices, crosswalks, parking practices, light standards, major roadside fixed objects, and other potentially important features. Condition diagrams must reflect the same time period as the collision diagrams they accompany. Chapter 6 shows a typical condition diagram.

Generating a List of Countermeasure Possibilities

After an analyst has identified the predominant clusters of accidents at a location, the next stage of countermeasure selection, is to generate a list of possible countermeasures. Analysts identify possible countermeasures through:

- Detailed investigations of accidents (looking especially at accident cause chains)
- Reviews of site plans and condition diagrams
- Site visits
- Other transportation engineering studies, such as spot speed studies and intersection sight distance studies
- The practices and previous experiences of the agency

Another major source for possible countermeasures is the technical literature. Many references are available that suggest certain countermeasures for certain highway situations. Table 11-3, for example, provides lists of countermeasures for several typical accident types. Highway locations vary greatly, the state of the art in accident countermeasures changes rapidly, and there are limits on the quantity and quality of accident data at many locations. Therefore, generating lists of possible countermeasures for particular sites is, in some ways, more art than science. Ideas for countermeasures come from experience and a thorough knowledge of the technical literature more than other factors.

Analyzing Countermeasure Alternatives

The final stage of countermeasure selection is to narrow the range of possibilities to one or more measures to be implemented. The analyst will use many of the same strategies during this stage as outlined above for generating a list of possibilities. At this third stage, however, available budget and countermeasure cost-effectiveness become important. Data such as in Table 11-4 (see page 216) are available to suggest the cost-effectiveness of different countermeasures. Analysts must use such data cautiously because countermeasure applications vary greatly and published data usually apply to only "average" conditions. One popular way to analyze countermeasure alternatives is through an economic analysis. In an economic analysis, every cost and saving for the agency and for the public due to each alternative is quantified in monetary terms. Quantifying every effect of each countermeasure can be difficult, and it is especially difficult to quantify the effects of accidents. Many large agencies have specified their own

TABLE 11-3
General Accident Countermeasure Ideas

Accident Pattern	*Probable Cause*[a]	*Possible Countermeasures*[b]
Left turn, head-on	A	1–11
	B	3, 6, 12–15
	C	16, 17
	D	3
	E	15
Rear-end at unsignalized intersection	A	4, 13, 18
	E	15
	F	14
	G	15, 19–22
	H	23
	I	10, 24
	J	25
Rear-end at signalized intersection	A	3, 4, 13, 18
	G	15, 19–22
	H	23
	J	25, 26
	K	12, 14, 15, 27–32
	L	16, 17, 33
	M	34
Right angle at signalized intersection	B	6, 12, 14, 15, 35, 36
	E	15, 16, 37
	H	23
	K	14, 27–32, 38
	L	11, 16, 17, 33, 39, 40
	N	14
	O	2, 11
Right angle at unsignalized intersection	B	6, 10, 12, 14, 15, 24, 35, 36, 41, 42
	E	15, 16, 37
	H	23
	N	14
	O	10, 43
	P	44, 45
Pedestrian–vehicle	B	12, 25, 35, 46
	E	14, 15, 45, 47
	H	23
	I	10, 25, 26
	L	11
	P	26
	Q	47, 48
	R	49
	S	14, 15, 47, 50
	T	51–53
Wet pavement	G	15, 19–22, 62
	T	53
Ran off roadway	E	15
	G	15, 19–22
	H	23
	K	54
	U	55–58
	V	14, 53, 59
	W	60
	X	6
	Y	61
Fixed object	E	15
	G	20, 22, 55, 62
	H	23
	T	53
	U	14, 63
	Z	58, 64–67
	AA	68
Parked or parking vehicle	E	15
	T	69
	BB	35
	CC	70
	DD	45, 50, 71
	EE	1, 43
Sideswipe or head-on	E	15, 72, 73
	T	53
	U	1, 55
	W	60
	X	6, 13, 74
	Y	61
	FF	38, 75
Driveway-related	A	13, 18, 35, 55, 72, 76
	B	12, 15, 23, 35
	E	15
	H	23
	GG	77–81
	HH	43, 79, 82
	II	6, 10, 74
Train–vehicle	B	12, 14, 24, 83–85
	E	15
	G	62
	K	23, 54
	T	36, 42, 53
	JJ	11
	KK	86
	LL	87
	MM	88
Night	K	14, 23, 59
	V	14, 59, 89
	X	14, 53, 59, 89
	FF	44, 90

TABLE 11-3
General Accident Countermeasure Ideas (continued)

[a]Key to probable causes:

A	Large turn volume
B	Restricted sight distance
C	Amber phase too short
D	Absence of left-turn phase
E	Excessive speed
F	Driver unaware of intersection
G	Slippery surface
H	Inadequate roadway lighting
I	Lack of adequate gaps
J	Crossing pedestrians
K	Poor traffic control device (TCD) visibility
L	Inadequate signal timing
M	Unwarranted signal
N	Inadequate advance intersection warning signs
O	Large total intersection volume
P	Inadequate TCDs
Q	Inadequate pedestrian protection
R	School crossing area
S	Drivers have inadequate warning of frequent midblock crossings
T	Inadequate or improper pavement markings
U	Inadequate roadway design for traffic conditions
V	Inadequate delineation
W	Inadequate shoulder
X	Inadequate channelization
Y	Inadequate pavement maintenance
Z	Fixed object in or too close to roadway
AA	Inadequate TCDs and guardrail
BB	Inadequate parking clearance at driveway
CC	Angle parking
DD	Illegal parking
EE	Large parking turnover
FF	Inadequate signing
GG	Improperly located driveway
HH	Large through traffic volume
II	Large driveway traffic volume
JJ	Improper traffic signal preemption timing
KK	Improper signal or gate warning time
LL	Rough crossing surface
MM	Sharp crossing angle

[b]Key to possible countermeasures:

1	Create one-way street
2	Add lane
3	Provide left-turn signal phase
4	Prohibit turn
5	Reroute left-turn traffic
6	Provide adequate channelization
7	Install stop sign
8	Revise signal phase sequence
9	Provide turning guidelines for multiple left-turn lanes
10	Provide traffic signal
11	Retime signal
12	Remove sight obstruction
13	Provide turn lane
14	Install or improve warning sign
15	Reduce speed limit
16	Adjust amber phase
17	Provide all-red phase
18	Increase curb radii
19	Overlay pavement
20	Provide adequate drainage
21	Groove pavement
22	Provide "slippery when wet" sign
23	Improve roadway lighting
24	Provide stop sign
25	Install or improve pedestrian crosswalk TCDs
26	Provide pedestrian signal
27	Install overhead signal
28	Install 12-inch signal lenses
29	Install signal visors
30	Install signal back plates
31	Relocate signal
32	Add signal heads
33	Provide progression through a set of signalized intersections
34	Remove signal
35	Restrict parking near corner/crosswalk/driveway
36	Provide markings to supplement signs
37	Install rumble strips
38	Install illuminated street name sign
39	Install multidial signal controller
40	Install signal actuation
41	Install yield sign
42	Install limit lines
43	Reroute through traffic
44	Upgrade TCDs
45	Increase enforcement
46	Reroute pedestrian path
47	Install pedestrian barrier
48	Install pedestrian refuge island
49	Use crossing guard at school crossing area
50	Prohibit parking
51	Install thermoplastic markings
52	Provide signs to supplement markings
53	Improve or install pavement markings
54	Increase sign size
55	Widen lane
56	Relocate island
57	Close curb lane
58	Install guardrail
59	Improve or install delineation
60	Upgrade roadway shoulder
61	Repair road surface
62	Improve skid resistance
63	Provide proper superelevation
64	Remove fixed object
65	Install barrier curb
66	Install breakaway posts
67	Install crash cushioning device
68	Paint or install reflectors on obstruction
69	Mark parking stall limits
70	Convert angle to parallel parking
71	Create off-street parking
72	Install median barrier
73	Remove constriction such as parked vehicle
74	Install acceleration or deceleration lane
75	Install advance guide sign
76	Increase driveway width
77	Regulate minimum driveway spacing
78	Regulate minimum corner clearance
79	Move driveway to side street
80	Install curb to define driveway location
81	Consolidate adjacent driveways
82	Construct a local service road
83	Reduce grade
84	Install train-actuated signal
85	Install automatic flashers or flashers with gates
86	Retime automatic flashers or flashers with gates
87	Improve crossing surface
88	Rebuild crossing with proper angle
89	Provide raised markings
90	Provide illuminated sign

Source: FHWA 1981.

TABLE 11-4
Average Cost-Effectiveness of Various Safety Improvements

Type of Improvement to Construction Classification	*Indexed Cost of Evaluated Improvements (millions)*	*Percent Reduction in Accident Rates after Improvements*			*Cost per Accident Reduced (thousands)*		*Benefit/ Cost Ratio*
		Fatal	*Injury*	*Fatal + Injury*	*Fatal*	*Fatal + Injury*	
Intersection and traffic control	562.3	37	15	15	344.5	18.7	4.0
Channelization turning lanes	297.7	48	23	24	510.9	19.7	2.8
Sight distance improvements	7.8	44*	31	32	371.4	22.8	3.6
Traffic signs	19.6	34	3	4	59.3	15.7	20.9
Pavement markings and/or delineators	33.7	15	(1)*	(1)*	751.8	—	1.6
Illumination	13.2	45	8	9	122.6	16.8	10.3
Traffic signals upgraded	63.1	40	22	22	412.1	8.6	4.0
Traffic signals, new	127.3	49	21	21	344.5	10.5	5.1
Structures	307.7	50	28	29	752.2	92.1	1.7
Bridge widened or modified	79.8	49	22	23	1,077.9	103.8	1.2
Bridge replacement	156.9	72	47	49	1,201.6	159.3	1.1
New bridge construction	26.2	77*	40	43	1,637.8	223.2	0.8
Minor structure replacement/improved	39.0	36	20	21	277.1	39.8	4.5
Upgraded bridge rail	5.6	72*	41	45	189.5	39.4	6.5
Roadway and roadside	1,971.3	31	13	13	722.2	54.0	1.8
Widened travel way	511.0	9*	7	7	4,041.3	174.2	0.4
Lanes added	212.9	(2)	13	13	—	66.8	0.1
Median strip to separate roadway	56.8	73	17	19	382.4	72.8	3.2
Shoulder widening or improvement	88.3	28	11	12	497.9	37.9	2.6
Roadway realignment	329.9	61	32	34	1,193.2	111.1	1.1
Skid resistant overlay	468.4	26	18	19	837.0	30.3	1.8
Pavement grooving	12.6	34*	15	15	377.3	14.5	3.8
Upgraded guardrail	149.7	42	8	9	151.7	31.5	8.1
Upgraded median barrier	7.4	45*	28	29	192.8	10.8	7.0
New median barrier	58.9	62	0*	3*	224.7	213.6	5.4
Impact attenuators	10.7	31*	36	36	390.0	8.0	4.0
Flatten side slopes/regrading	36.5	(25)*	9	8*	—	102.2	—
Bridge approach guardrail transition	4.8	61*	44	45	200.9	25.8	6.3
Obstacle removal	15.0	49	22	23	202.5	16.0	6.4
Railroad–highway crossings	331.2	89	63	67	570.0	114.2	2.2
New flashing lights	55.0	91	74	77	551.4	103.5	2.2
New flashing lights and gates	138.1	92	84	86	591.8	115.3	2.1
New gates only	63.2	90	78	80	432.8	96.7	2.8

Note: Numbers in parentheses indicate increased accident rates.

*No significant change at the 95% confidence level.

Source: USDOT, 1989.

TABLE 11-5
Two Independent Estimates of Accident costs

Source	Event	Cost per Person
National Safety Council, 1991 (1990 dollars)	Fatality	$410,000
	A-Injury	38,200
	B-Injury	8,900
	C-Injury	2,900
	PDO	3,500 (per accident)
NHTSA, National Center for Statistics and Analysis (1986 dollars)	Fatality	360,000
	5-Injury (MAIS scale)	280,000
	4-Injury	65,000
	3-Injury	15,000
	2-Injury	6,500
	1-Injury	3,100
	PDO	580 (per vehicle)

Sources: National Safety Council, "Estimated Costs of Traffic Accidents, 1990," Chicago, Ill., 1991; NHTSA, National Center for Statistics and Analysis.

accident cost statistics. Table 11-5 shows two independent estimates of the costs of traffic accidents by accident severity. Analysts must check the assumptions underlying accident cost estimates very carefully, because they can differ greatly between agencies. Once all effects are quantified, analysts use formal procedures that account for the time value of money, taxes, inflation, and other factors to arrive at the suggested alternative(s). The techniques of economic analysis are beyond the scope of this manual; refer to AASHTO (1977) or one of the many texts on those techniques for assistance.

Countermeasure Evaluation

Agencies frequently use accident data to evaluate highway improvements, whether or not the improvements were installed with enhanced safety as a goal. Indeed, a program of regular countermeasure evaluation is essential for intelligent future countermeasure selection. Analysts evaluate countermeasures using one of the experimental techniques described in Appendix A with accidents as the measure of effectiveness. The most important point regarding countermeasure evaluation using accident data is mentioned here and in Appendix A: that "before and after" studies are often misused. Analysts conduct before and after studies of improvements at a set of locations by comparing the accident history of the locations before improvement to the accident history of the locations after improvement. As currently conducted by many practitioners, before and after studies using accident data suffer from two serious flaws and provide incorrect and misleading results. The flaws are (1) the failure to control for the effects of changing conditions during the lengthy time periods required to amass before and after accident statistics, and (2) the failure to correct for regression to the mean. Regression to the mean occurs when locations with high accident counts during one time period experience more normal counts during the next time period even if no causative factor changes. The most rigorous way to overcome these two flaws is to use randomly selected control sites where the agency measures accidents but does not install the improvements. Appendix A describes how to conduct a before and after experiment with control sites and how to conduct other common countermeasure evaluation studies.

The Federal Highway Administration has developed several aids for countermeasure evaluation using accident data. Among the most helpful aids is the *Accident Research Manual,* which is a detailed guide on the analysis of accident data (Council et al., 1980). The computer program HISAFE (Datta et al., 1987) is also helpful, providing a convenient and quick evaluation tool.

REFERENCES

AASHTO (1977). *A Manual on User Benefit Analysis of Highway and Bus-Transit Improvements,* American Association of State Highway Transportation Officials, Washington, DC.

BATZ, T. M. (1989). *An Overview of New Jersey's Accident Processing Costs Based on a National Survey,* Transportation Research Record 1238, Transportation Research Board, Washington, DC, pp. 44–52.

BOX, P. C., AND J. C. OPPENLANDER (1976). *Manual of Traffic Engineering Studies,* 4th ed., Institute of Transportation Engineers, Washington, DC, p. 49.

CENTER FOR MICROCOMPUTERS IN TRANSPORTATION (1992). *McTrans Catalog,* University of Florida Transportation Research Center, Gainsville, FL, June.

COUNCIL, F. M., ET AL. (1980). *Accident Research Manual,* FHWA/RD-80/016, U.S. Department of Transportation, Federal Highway Administration, Washington, DC, February.

DATTA, T., ET AL. (1987). *Computer Programs for Safety Analysis: HISAFE User's Manual and Operator's Guide,* FHWA/RD-87/074, U.S. Department of Transportation, Federal Highway Administration, Washington, DC, June.

FHWA (1981). *Highway Safety Engineering Studies Procedural Guide,* USDOT, Washington, DC, June.

HARKEY, D. L., AND R. RUIZ (1989). *HISAM: An Accident Data Base Manager,* Transportation Research Record 1238, Transportation Research Board, Washington, DC, pp. 37–44.

HAUER, E., AND A. S. HAKKERT (1988). *Extent and Some Implications of Incomplete Accident Reporting,* Transportation Research Record 1185, Transportation Research Board, Washington, DC, pp. 1–10.

HAUER, E., AND B. N. PERSAUD (1984). *Problems Identifying Hazardous Locations Using Accident Data,* Transportation Research Record 975, Transportation Research Board, Washington, DC, pp. 36–43.

HAUER, E., AND B. N. PERSAUD (1987). *How to Estimate the Safety of Rail–Highway Grade Crossings and the Safety Effects of Warning Devices,* Transportation Research Record 1114, Transportation Research Board, Washington, DC, pp. 131–140.

HAUER, E., ET AL. (1988). *Estimation of Safety at Signalized Intersections,* Transportation Research Record 1185, Transportation Research Board, Washington, DC, pp. 48–58.

HIGLE, J. L., AND M. B. HECHT (1989). *A Comparison of Techniques for the Identification of Hazardous Locations,* Transportation Research Record 1238, Transportation Research Board, Washington, DC, pp. 10–19.

MAK, K. K. ET AL. (1986). *Automated Analysis of High-Accident Locations,* Transportation Research Record 1068, Transportation Research Board, Washington, DC, pp. 59–64.

MILLER, T., ET AL. (1985). *Sensitivity of Resource Allocation Models to Discount Rate and Unreported Accidents,* FHWA/RD-85/092, Federal Highway Administration, McLean, VA.

NATIONAL SAFETY COUNCIL (1991). *Estimated Costs of Traffic Accidents, 1990,* NSC, Chicago.

USDOT (1989). *The 1989 Annual Report on Highway Safety Improvement Programs,* U.S. Department of Transportation, Washington, DC.

ZEGEER, C. V. (1982). *Highway Accident Analysis Systems,* National Cooperative Highway Research Program Synthesis of Highway Practice 91, Transportation Research Board, Washington, DC.

12
Traffic Conflict Studies

Joseph E. Hummer, Ph.D., P.E.

INTRODUCTION

Traffic conflicts are interactions between two or more vehicles or road users when one or more vehicles or road users take evasive action, such as braking or weaving, to avoid a collision (Parker and Zegeer, 1988). Engineers use traffic conflicts as a supplement to traffic accident studies in estimating the traffic accident potential at an intersection or other location. Traffic conflicts are useful because the study results are often available much sooner than the results of traffic accident studies (when several years' data may be needed). Traffic conflict studies can also provide much more detailed information than traffic accident studies. However, conducting traffic conflict studies is not simple. When performed improperly, they may provide misleading information.

Traffic conflict studies require a relatively small investment of time and other resources and require no special equipment. Trained observers watch traffic and note on a form when a conflict occurs. Observers usually require a week or less of training. On a single intersection approach, one or two persons are usually needed for $\frac{1}{2}$ to 3 days. Besides training the observers, engineers establish study guidelines and analyze results. Research sponsored by the Federal Highway Administration during the 1980s improved the state of the art of traffic conflict studies. Migletz et al. (1985) demonstrated that traffic conflicts predict future traffic accidents about as well as accident records. Manuals by Parker and Zegeer (1988, 1989) provide excellent detail on how to conduct traffic conflict studies.

PURPOSES OF STUDIES

Traffic conflict studies supplement traffic accident studies in several ways. The magnitude of the traffic safety problem at a particular location can be estimated from traffic conflicts. One possible result of a traffic conflict study at an intersection is a mean rate of traffic conflicts of a particular type per day. This rate may then be compared to a standard or certain percentile rate from a sample of similar intersections. Treatments may be needed at the location if the observed mean rate is higher than the comparison rate.

The use of traffic conflicts for estimating the magnitude of a safety problem is restricted due to the lack of good comparison conflict rates. It is time consuming to collect a database of traffic conflict rates for comparison purposes. Glauz and Migletz (1980) and Migletz et al. (1985) provided comparison rates for some common types of intersections. However, if the intersection types for the published conflict rates do not apply to the location under study, traffic conflicts may not be helpful in estimating the magnitude of the safety problem.

Traffic conflict studies are very useful in determining the types of safety problems that exist at a location. Once the type of problem is known, possible countermeasures can be identified. There are, for instance, 14 basic types of traffic conflicts at intersections. The relative overabundance of one of these 14 types at an intersection reveals a particular problem. Again, a database of typical rates must be available against which to compare a location's rates. Lists of countermeasures that may reduce the occurrence of a type of conflict are available (Parker and Zegeer, 1988). At many highway locations, it is impossible to obtain enough relevant accident data to make such a detailed diagnosis.

Traffic conflict data are often collected and analyzed to evaluate the effectiveness of a safety-related countermeasure. Countermeasure evaluation using traffic conflicts is attractive because traffic conflict data are available to the analyst before traffic accident data. In fact, a before-and-after study with traffic conflicts may not need control sites to overcome the history and maturation threats to experiment validity (see Appendix A for details on experiment design). To evaluate countermeasures, the conflict types being studied should be closely related to the countermeasures that have been implemented. If not, the true effectiveness of the countermeasures will remain unknown (Parker and Zegeer, 1988).

TYPES OF TRAFFIC CONFLICTS

As stated previously, traffic conflicts are interactions between two or more vehicles or road users when one or more vehicles or road users take evasive action, such as braking or weaving, to avoid a collision. Thus, actions taken by vehicles or road users in response to traffic control devices, highway geometry, or weather are not traffic conflicts. A driver braking to join a queue at a red signal is not involved in a traffic conflict. Another driver braking to avoid a rear-end collision with a slow-moving vehicle during a green signal phase is involved in a traffic conflict. Observers use brake lights, squealing tires, or vehicle front ends that dip or dive as indications that braking occurred and a conflict was possible. A collision or near miss during which no evasive actions were observed also counts as a traffic conflict. Traffic conflicts can involve motor vehicles, pedestrians, bicycles, and other road users. Rates of pedestrian and motor vehicle conflicts can be very high at intersections with appreciable pedestrian volumes.

Researchers have identified 14 basic types of conflicts at intersections, as shown in Figures 12-1 to 12-14. These conflict types apply at signalized intersections, unsignalized intersections, and driveway openings. However, not all the 14 types apply at every intersection. In most conflict studies, observers record only the conflict types that are related to the study purpose rather than all 14 types. Traffic conflict types are not well defined for nonintersection locations such as weaving sections, diverges, or merges. Pre-

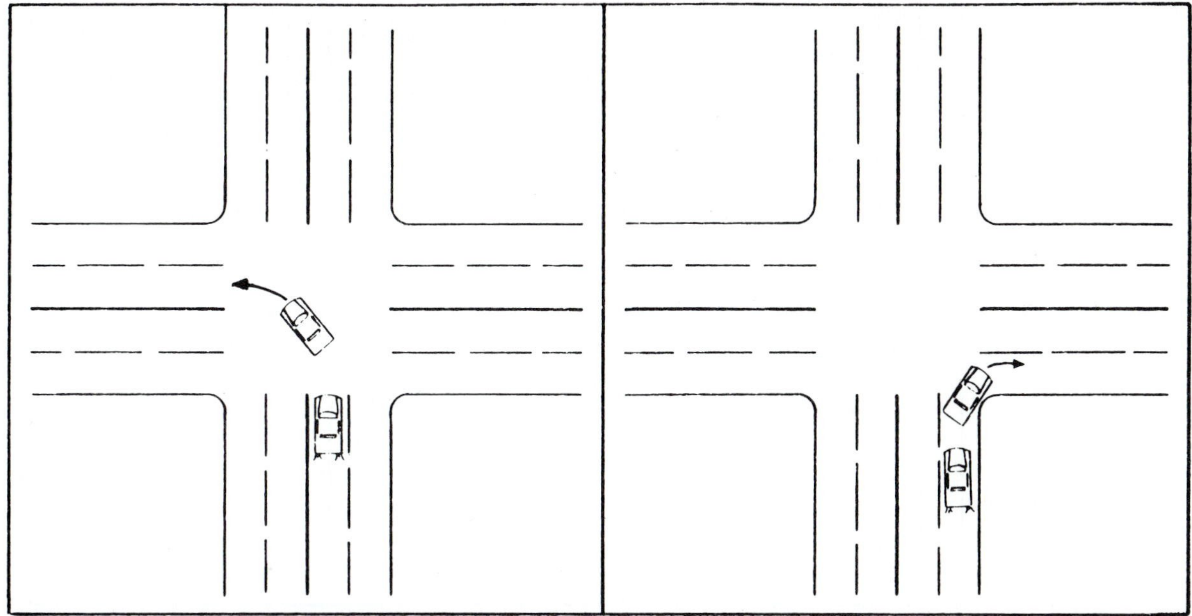

Figure 12-1 Left-turn same-direction conflict. *Source:* Glauz, W. D., and D. J. Migletz, "Application of Traffic Conflict Analysis at Intersections," *National Cooperative Highway Research Program Report 219,* Transportation Research Board, Washington, DC.

Figure 12-2 Right-turn same-direction conflict. *Source:* Glauz and Migletz, 1980.

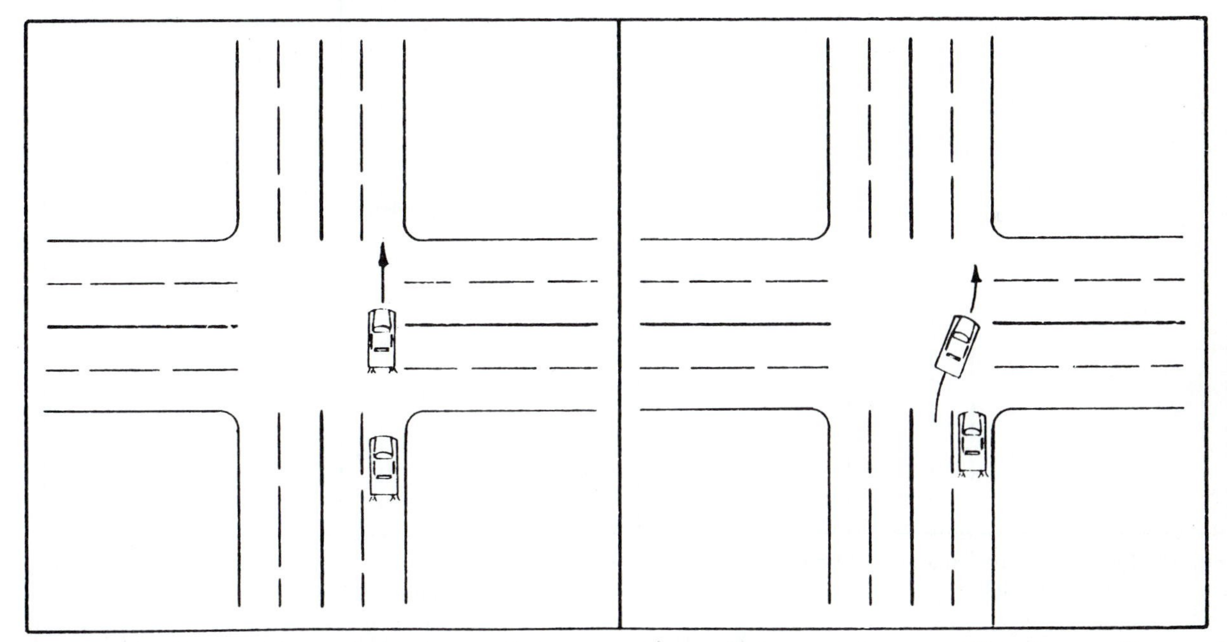

Figure 12-3 Slow vehicle same-direction conflict. *Source:* Glauz and Migletz, 1980.

Figure 12-4 Lane-change conflict. *Source:* Glauz and Migletz, 1980.

liminary observations or pilot tests are necessary to inform observers which conflicts to record at nonintersection locations.

Secondary conflicts occur because of the prior occurrence of another traffic conflict (Parker and Zegeer, 1988). Secondary conflicts consist primarily of "slow-vehicle, same-direction" (Figure 12-3) or "lane-change" (Figure 12-4) conflicts involving a third vehicle in response to a conflict between two other vehicles. Usually, a maximum of one

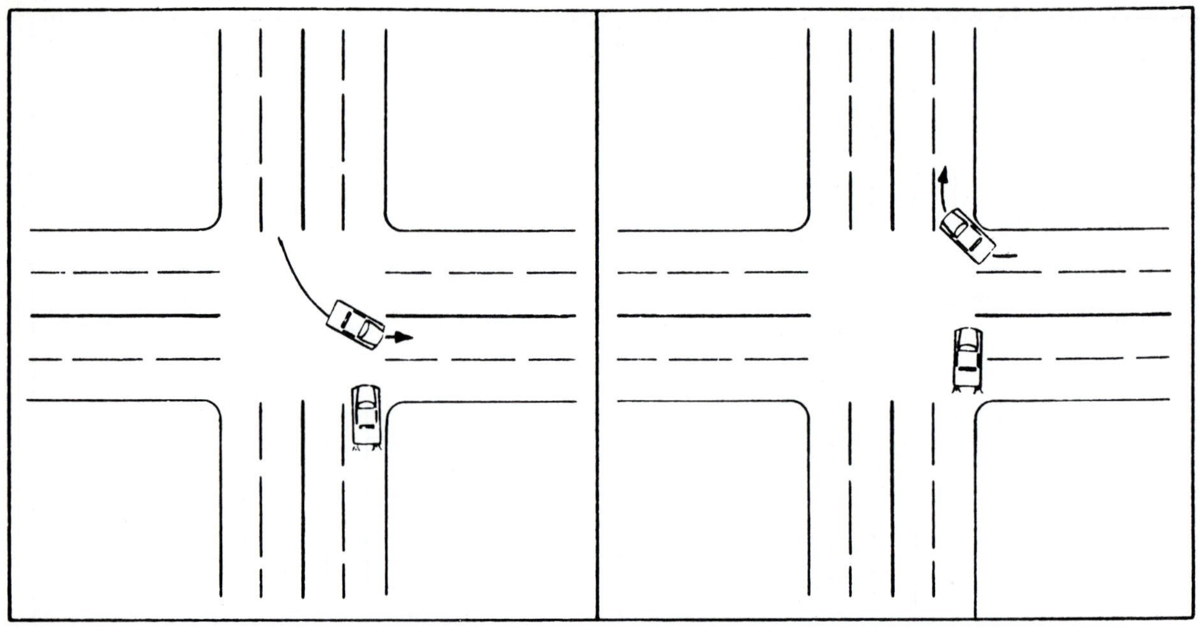

Figure 12-5 Opposing left-turn conflict. *Source:* Glauz and Migletz, 1980.

Figure 12-6 Right-turn cross-traffic-from-right conflict. *Source:* Glauz and Migletz, 1980.

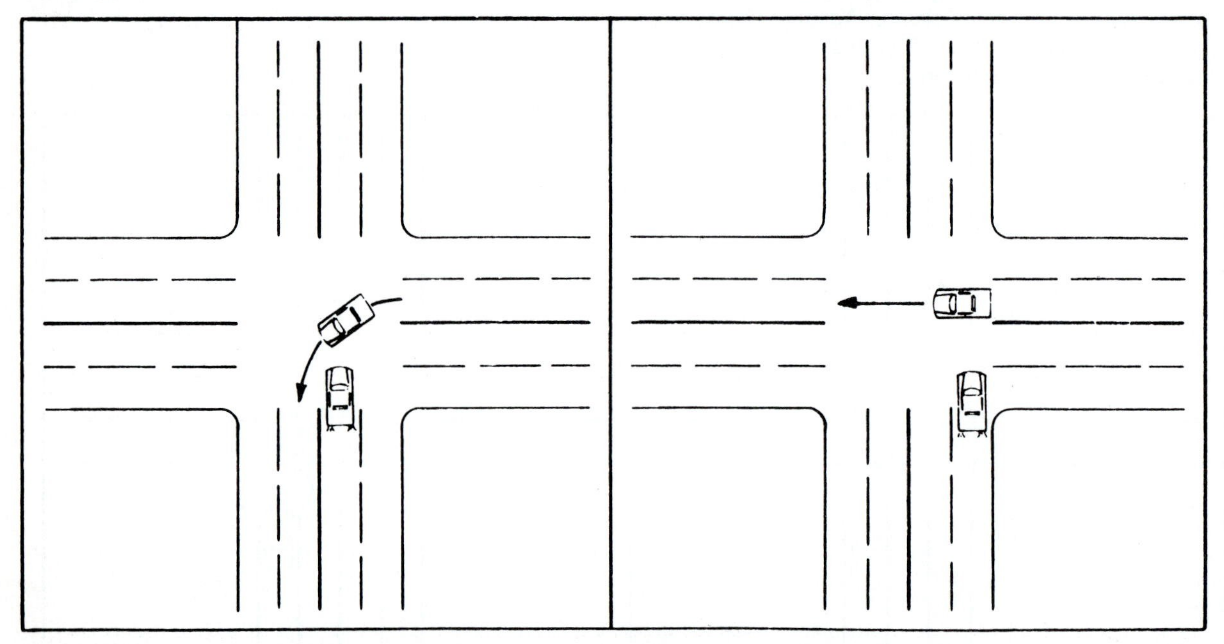

Figure 12-7 Left-turn cross-traffic-from-right conflict. *Source:* Glauz and Migletz, 1980.

Figure 12-8 Through cross-traffic-from-right conflict. *Source:* Glauz and Migletz, 1980.

secondary conflict is recorded for each main conflict even if more than one secondary conflict occurs. For example, if an entire platoon of vehicles braked in response to a lead vehicle executing a right turn, only two conflicts should be recorded: a ''right-turn, same-direction'' conflict and a secondary conflict.

Traffic Events

Traffic events are unusual, dangerous, or illegal nonconflict maneuvers. Typical traffic events include vehicles running red signals, executing right turns on red without a full

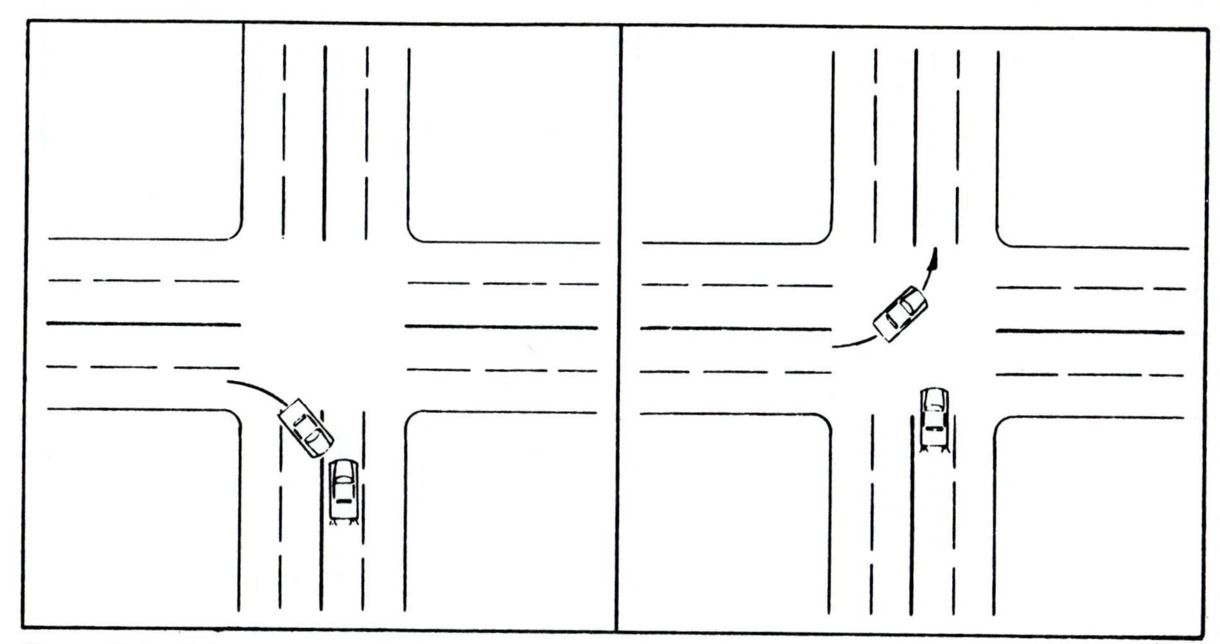

Figure 12-9 Right-turn cross-traffic-from-right conflict. *Source:* Glauz and Migletz, 1980.

Figure 12-10 Left-turn cross-traffic-from-left conflict. *Source:* Glauz and Migletz, 1980.

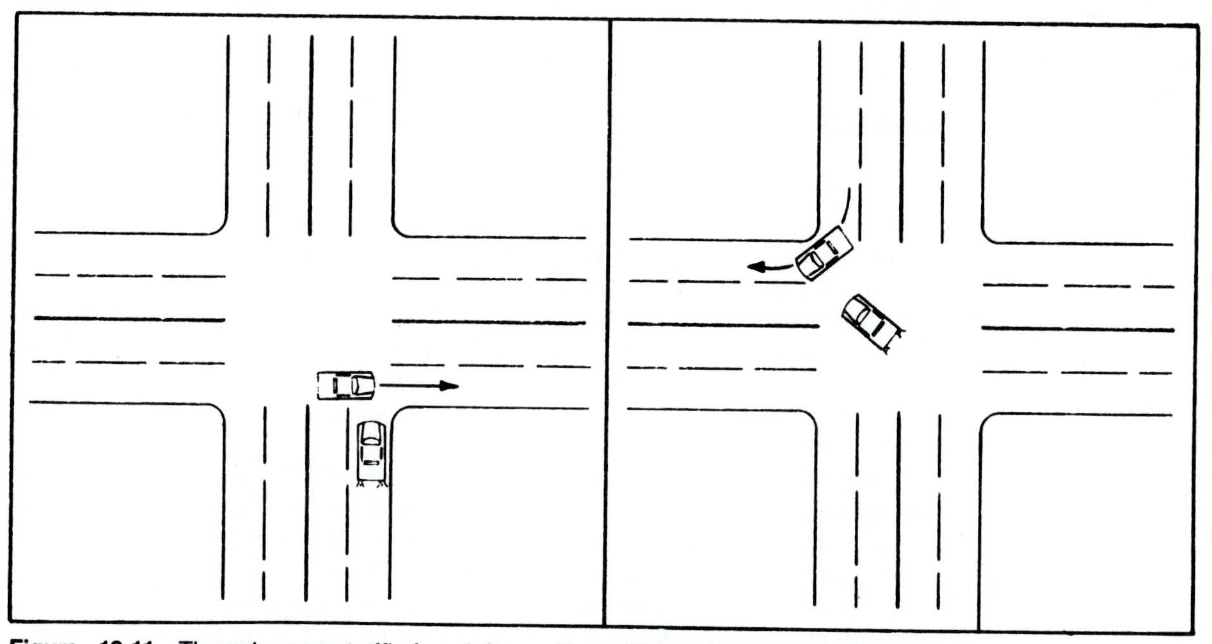

Figure 12-11 Through cross-traffic-from-left conflict. *Source:* Glauz and Migletz, 1980.

Figure 12-12 Opposing right-turn-on-red conflict. *Source:* Glauz and Migletz, 1980.

stop, weaving across painted gore areas, and slowing considerably in travel lanes. Traffic events are defined very loosely, and some types of traffic events have not been researched thoroughly. Engineers should be certain that the traffic events under scrutiny are useful estimators of traffic accidents. Engineers should also be sure that an observer has a clear idea of the actions that constitute a traffic event. For example, an engineer could define a "running red signal" traffic event as when "the front tires of a motor vehicle that proceeds through the intersection without stopping cross the stop bar when the signal is red." Pilot testing is necessary to ensure that there are no gaps in the definition (i.e., actions that cannot be appropriately coded).

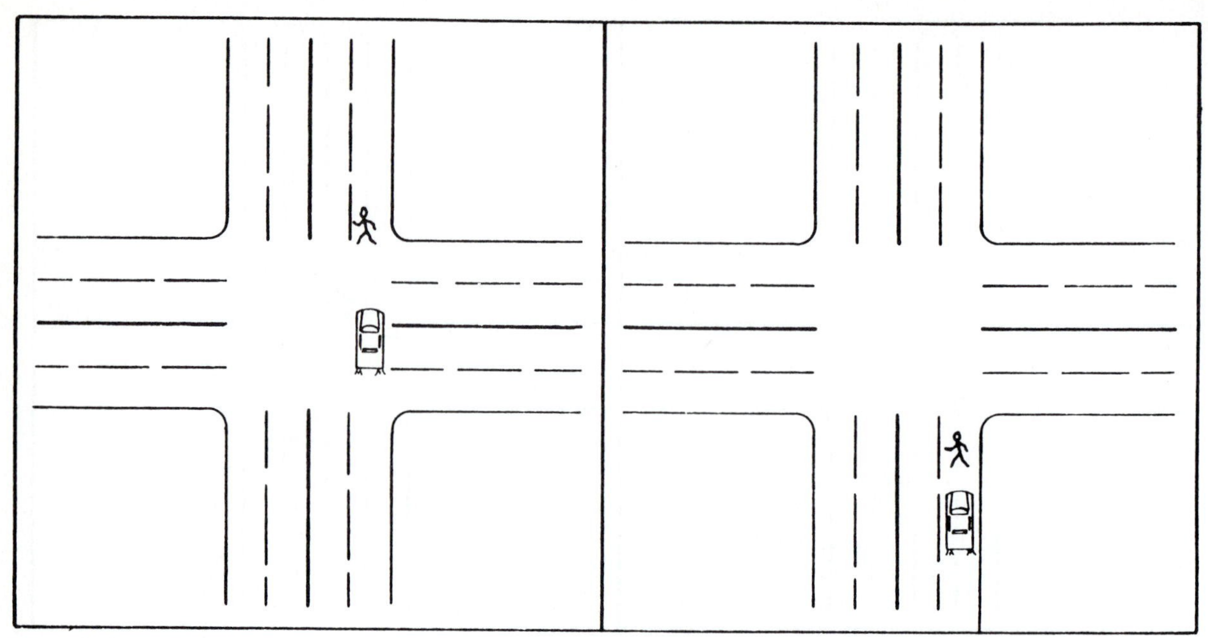

Figure 12-13 Pedestrian far-side conflict. *Source:* Glauz and Migletz, 1980.

Figure 12-14 Pedestrian near-side conflict. *Source:* Glauz and Migletz, 1980.

STUDY METHOD

Personnel and Training

Personnel requirements for traffic conflict studies at intersections depend on whether turning movement counts are needed. Turning movement counts are necessary if the traffic conflict rate to be estimated is per vehicle rather than per unit of time (discussed below under the heading "Sample Size"). Turning movement counts are also necessary if strict experiment controls are used, such as in a before-and-after experiment when the traffic volume is suspected to have changed between the data collection periods. Turning movement counts are not essential (but may still be useful) for rates per unit of time with some study purposes such as countermeasure generation and estimation of the magnitude of the safety problem at a location.

One or two persons per intersection approach are sufficient to conduct traffic conflict studies. One person can record traffic conflicts on one intersection approach if turning movement counts are not necessary. One person should also be able to observe traffic conflicts and count turning movements where three or fewer movements with low or moderate volumes are made. If there are more than three movements or volumes are heavy, additional observers will be needed. Observers should not have to look away from the location where conflicts are being watched to record turning movements. If turning movements are to be counted at an intersection too far downstream for an observer to see both the approach conflict area and the intersection, a second observer counting only turning movements is needed.

A traffic conflict observer should watch only one intersection approach or one end of a weaving area at a time. Consequently, when it is desirable to study traffic conflicts at an entire intersection or weaving area either larger crews of observers must be used or particular observers must stay longer at the location. For example, if traffic conflicts per vehicle are desired on all four legs of a busy intersection and a full day of observation is needed on each approach to gather the appropriate sample, staffing options include:

1. A crew of six (four conflict observers and two turning movement counters) is scheduled for 1 day.
2. A crew of three (two conflict observers and one turning movement counter) is scheduled for 2 days.
3. A crew of two (one conflict observer and one turning movement counter) is scheduled for 4 days.

A traffic conflict study requires very little equipment. Observers will need forms, clipboards, pens, watches, and a place to sit (a vehicle or a folding chair). Electronic or manual turning movement recorders may easily be modified for traffic conflict studies with a template or by labeling keys so that each key is associated with a conflict type. However, traffic conflict totals tend to be smaller than turning movement counts so that electronic boards are usually not required. Observers have used audiotape recorders to record data during traffic conflict studies. The extra time necessary to listen to the tape and code the data later makes this alternative less desirable. Finally, videotape can be used during traffic conflict studies. Videotape creates a permanent record so that close calls can be reevaluated. The disadvantages of video, including the extra labor to record and view the tape and the technical problems associated with lighting and fields of view, usually outweigh the advantages. A well-designed data form remains the best choice for most traffic conflict studies.

Training Observers

The goal of observer training is to create observers who are consistent with themselves over time, with each other, and with the established definitions of conflicts and events. Without the confidence that what is called a conflict now will be called a conflict later, traffic conflict studies degenerate into simply observing traffic. Conclusions drawn from an inconsistent traffic conflict study are misleading because comparisons to standard rates or rates from other locations must be made.

The amount of training needed by personnel learning to observe traffic conflicts is controversial. Parker and Zegeer (1988) recommend a 1-week training period for experienced traffic technicians and a 2-week training period for others. However, many professionals view those training periods as economically infeasible. In practice, the amount of training provided varies widely depending on the experience, professional status, and motivation of the trainee and the quality of the instruction. Observers should train until they achieve consistent performance.

Table 12-1 shows the activities that make up the 2-week training period recommended by Parker and Zegeer (1988). Appropriate training activities include lectures, reading, discussion, videotape viewing, supervised practice, and unsupervised practice. The *Observer's Manual* by Parker and Zegeer (1989) is available for training preprofessional personnel. "What if" examples are useful to provoke discussions during training. The instructor poses a hypothetical traffic situation verbally, on a chalkboard, on a handout, or on videotape, and the trainees state whether the situation should be classified as a conflict, and, if so, what type. An experienced traffic technician or engineer can learn to observe traffic conflicts alone by reading and unsupervised practice. However, it is preferable for two or more people to learn the technique together so they can discuss key points and compare practice data.

Observer consistency can be estimated by having two or more observers record conflicts independently at the same location and time. If an observer at a particular type of location produces results that are consistent with other observers, especially with experienced traffic conflict observers, his or her training may be considered complete for that type of location. Engineers setting up such a consistency test must make sure that the observers see the same portion of roadway and are not influencing each other. An

TABLE 12-1
Suggested 10-Day Training Schedule for Conflict Observers

Period	*Topic*
Day 1	Introductory remarks
	Orientation to the training program
	General background on traffic safety
	History of traffic conflicts
	Overview of a traffic conflict survey
	What the survey is
	How the results are used
	How the survey is conducted
	Contents of the *Observer's Manual*
	Traffic counting
	Turning movements
	Use of mechanical counting boards
	Introductory field work
Day 2	Presentation of traffic conflict definitions
	General definition
	Same-direction and opposing left-turn conflicts
	Group field observations at a signalized intersection
	Discussion
Day 3	Definitions of cross-traffic conflicts
	Group field observation at an unsignalized intersection
	Discussion
	Use of videotape to illustrate conflict situations
Day 4	Small-group field practice
	Question-and-answer session
	Special conflict types
Day 5	Simulated limited conflict counts
	Discussion
	Intersections with unusual geometrics
Day 6	Use of other data forms
	Field collection of other data
	Discussion
Day 7	Simulated full conflict counts (8-hour day)
	Discussion
Day 8	Review of the concepts and procedures
	Analysis of day 7 data
	Discussion of problem areas
Day 9	More field practice
Day 10	Analysis of day 9 data
	More field practice

Source: Parker and Zegeer, 1988.

informal and usually sufficient method for judging consistency is simply to look over the completed data forms from different observers. A variety of statistical tests are available for formally estimating consistency. A correlation coefficient, a Z-test for proportions, or a paired comparison t-test can be used to estimate the difference between two observers. A group standard deviation, a chi-square test, or an analysis of variance can be used to estimate consistency among a group of three or more observers. An F-test can be used to compare the variances between two groups of observers or between two trials with the same group of observers. Appendix A contains a discussion of experiment design (a formal comparison between observers is an experiment) and Appendix C contains a discussion of common statistical tests.

There are few standards for observer consistency to indicate whether more training is needed. Based on total conflicts at an intersection approach in 15-minute periods, Parker and Zegeer (1988) state that a correlation coefficient of 0.95 between two observers is desirable. They also recommend that observers whose conflict counts consistently fall one or more standard deviations above or below the group mean should be singled out for more training. However, if the group standard deviations are small, the group is ready for unsupervised data collection and singling out particular trainees for further training would waste time.

Sample Sizes Required

The sample size requirements for a traffic conflict study depend on the type of conflict rate to be analyzed. Engineers use two basic types of rates: rates per unit of time, and rates per vehicle observed. Often, an agency customarily uses only one type of rate. If there is no customary type of conflict rate, the choice should be based on the relative advantages of each type. Rates per unit of time are advantageous because tables of such rates are available for intersections that are convenient for comparisons. Table 12-2 shows some typical rates per unit of time; Glauz and Migletz (1980) and Migletz et al. (1985) have provided others. Since existing data are necessary to judge the magnitude of the safety problem at a location or to generate lists of countermeasures, those studies generally use rates per unit of time. Also, rates per unit of time are advantageous because turning movement data may not be needed. The major advantage of rates per vehicle is that the amount of time spent collecting data to achieve a given level of precision is usually lower, especially when conflicts are relatively rare. This is because rates per vehicle are treated as proportions (each vehicle observed either does or does not instigate a conflict), whereas rates per unit of time are treated as frequencies. Another important advantage of a conflict rate per vehicle is that no previous knowledge of the variance of the mean conflict rate is necessary to estimate the needed sample size.

Once the type of rate is chosen, the necessary sample size can be calculated. The equation for determining the sample size needed to estimate a mean conflict rate per unit of time, if previous estimates of the variance of the mean and the mean are available, is

$$\mathrm{NT} = \left[\left(100 \times \frac{t}{\mathrm{PC}}\right)^2\right] \times \frac{\mathrm{var}}{\mathrm{mean}^2} \qquad (12\text{-}1)$$

where

NT = number of units of time that must be observed

t = a constant corresponding to the desired level of confidence, from Table 12-3

PC = permitted error in the estimate of the mean conflict rate, percent (if the mean hourly conflict rate is 6 and PC = 50, the precision of the estimate is 6 ± 50% of 6 or 3 to 9 conflicts per hour)

var = expected variance of the conflict rate, from previous studies or Table 12-2

mean = expected mean of the conflict rate, from previous studies or Table 12-2

Equation (12-1) can be used for conflict rates per any unit of time if prior estimates of the mean and variance of the rate are available. If a previous estimate of the variance of the mean is available but not the mean itself, the equation becomes

$$\mathrm{NT} = \left(\frac{t}{\mathrm{PQ}}\right)^2 \times \mathrm{var} \qquad (12\text{-}2)$$

where NT, t, and var are as defined for equation (12-1) and PQ is the permitted error in the estimate of the mean conflict rate, in conflicts per unit of time (if the level of confidence corresponding to t is 90% and PQ is 7 conflicts per hour, the actual conflict rate will be within 7 conflicts per hour of the estimated rate 90% of the time).

TABLE 12-2
Typical Conflict Rate Statistics for Intersections with Four Approaches

	Conflicts/Hour		*Conflicts/Day*			
					Percentile	
Conflict Type	*Mean*	*Variance*	*Mean*	*Variance*	*90th*	*95th*
	Signalized with Entry Volumes Greater Than 25,000 Vehicles/Day					
Left-turn same direction	7.6	22	83	12,000	270	360
Slow vehicle	61	34	670	24,000	870	940
Lane change	1.7	[b]	18	160	35	43
Right-turn same direction	20	11	220	7,600	470	510
Opposing left turn	2.0	1.2	22	380	48	60
All same direction	90	74	990	67,000	1300	1500
	Signalized with Entry Volumes 10,000 to 25,000 Vehicles/Day					
Left-turn same direction	12	22	130	10,000	270	340
Slow vehicle	34	11	380	4,900	470	500
Lane change	0.7	[b]	8	53	17	22
Right-turn same direction	11	12	120	2,400	190	220
Opposing left turn	2.6	1.2	29	210	49	56
All same direction	59	95	640	25,000	860	930
	Unsignalized with Entry Volumes 10,000 to 25,000 Vehicles/Day					
Left-turn same direction	12	21	130	12,000	270	350
Slow vehicle	14	5.2	150	5,900	260	290
Right-turn same direction	5.6	11	62	1,200	100	120
Opposing left turn	0.8	1.2	9	40	17	21
Right turn from right	0.8	1.1	9	99	21	29
All same direction	29	77	320	29,000	540	640
Through cross traffic	0.6	[b]	7	16	12	14
	Unsignalized with Entry Volumes 2500 to 10,000 Vehicles/Day					
Left-turn same direction	6.4	22	71	1,000	110	130
Slow vehicle	9.3	5.5	100	9,600	220	300
Right-turn same direction	5.3	11	58	2,200	120	150
Opposing left turn	0.3	[b]	4	8	8	9
Right turn from right	0.5	1.1	6	12	10	12
All same direction	21	77	230	18,000	410	490
Through cross traffic	1.1	[b]	12	75	24	29

Note: Basic intersection conflict types not shown had mean hourly rates less than 0.5. Statistics are based on sample counts conducted in the Kansas City metropolitan area on all four approaches of signalized intersections and on the approaches with the right-of-way at unsignalized intersections. Counts were taken during the daylight, in dry weather, and do not include secondary conflicts.

[a]"All same direction" includes left turn same direction, slow vehicle, lane change, and right-turn same direction conflict types. "Through cross traffic" includes cross traffic from left and cross traffic from right conflict types.

[b]Not available.

Source: Glauz and Migletz (1980); Migletz et al. (1985).

EXAMPLE 12-1

For an approach to a signalized intersection with a total entry volume of 35,000 vehicles per day, an engineer wants an estimate of the mean number of same-direction conflicts in an hour. The engineer would like to be 90% sure that the estimate is within 10% of the actual rate. Using equation (12-1) and Tables 12-2 and 12-3,

$$\text{NT} = \left(100 \times \frac{1.64}{10}\right)^2 \times \frac{74}{90^2}$$

$$= 2.5 \text{ hours of observation}$$

TABLE 12-3
Constant Corresponding to Level of Confidence

Constant, t	Confidence Level (%)
1.28	80.0
1.50	86.6
1.64	90.0
1.96	95.0
2.00	95.5
2.50	98.8
2.58	99.0

If conflict rates per vehicle are desired, the sample size necessary to achieve a certain precision in the estimate of a mean conflict rate is

$$NV = \frac{p \times q \times t^2}{PP^2} \qquad (12\text{-}3)$$

where

t is as defined for equation (12-1), and where

NV = number of vehicles that must be observed

p = expected proportion of vehicles observed that are involved in conflicts

q = expected proportion of vehicles observed that are not involved in conflicts

PP = permitted error of the estimate of the proportion of vehicles involved in conflicts, in a proportion between zero and 1.

If the level of confidence corresponding to t is 95% and PP is 0.01, the actual conflict rate per vehicle will be within 0.01 of the estimated rate 95% of the time. The sum of p and q must be 1.0 in equation (12-3).

A conservative estimate of the sample size (i.e., a larger sample will be gathered than probably is necessary to achieve a given precision) can be provided from equation (12-3) without prior knowledge of p and q. If p and q are unknown, they are assumed as 0.5 and equation (12-3) reduces to

$$NV = 0.25 \times \frac{t^2}{PP^2} \qquad (12\text{-}4)$$

Since conflict rates per vehicle are usually much closer to zero than to 0.5, use an estimate of p and equation (12-3) rather than equation (12-4) to reduce the sample size estimate dramatically.

EXAMPLE 12-2

An engineer wants an estimate of the mean rate of right-turn same-direction conflicts per approach vehicle. The engineer would like to be 95% sure that the estimate is within 0.01 of the actual rate. Using Table 12-3 and the conservative equation (12-4),

TABLE 12-4
Coefficient of Variation Corresponding to Number of Conflicts for Inverse Sampling

Coefficient of Variation Guaranteed (%)	*Number of Conflicts Observed*
50	6
33	11
25	18
20	27
15	46
10	102
5	401
3	1100

$$NV = 0.25 \times \frac{1.96^2}{0.01^2}$$
$$= 9600$$

so 9600 approach vehicles would have to be observed to achieve the desired precision. If the engineer is certain that the conflict rate is less than 5%, equation (12-3) can be used and the needed sample size is much smaller:

$$NV = 0.05 \times 0.95 \times \frac{1.96^2}{0.01^2}$$
$$= 1800 \text{ vehicles}$$

Inverse sampling is another sample size formulation for proportions that may be useful for traffic conflict studies using rates per vehicle. Inverse sampling applies when the proportion to be estimated is known to be small (i.e., less than 0.10), but no reliable estimate of the proportion is available. The sample size formulation given in equations (12-3) and (12-4) may require a larger sample of vehicles than is necessary for a typical estimate. Inverse sampling depends on the coefficient of variation, which is the ratio of the standard error of the proportion to the proportion being estimated. The lower the coefficient of variation, the more precise the estimate. Cochran (1977) has shown that

$$CV < \frac{\sqrt{m}}{m - 1} \tag{12-5}$$

where CV is the coefficient of variation and m is the number of occurrences of the event to be observed (i.e., the number of conflicts). Table 12-4 is based on inequality (12-5). To achieve a certain coefficient of variation, one keeps observing until m conflicts are counted. For instance, observing an approach until 27 conflicts of a particular type are recorded guarantees a coefficient of variation of less than 20%. Table 12-4 illustrates how difficult it is to get a precise estimate of a very low proportion.

Conducting the Study

Conflict observers sit upstream of the feature of interest. Observers record each conflict that happens between their position and the feature of interest and ignore conflicts observed in other places. The distance between the observer position and the intersection or other feature depends primarily on the type of feature, the purpose of the study, the visibility afforded by different positions, and the speed of vehicles being observed. Observers are typically positioned 100 to 300 feet from intersections in urban areas with cluttered roadsides and relatively low approach speeds. Observers are 300 feet or more from intersections during studies in suburban areas with uncluttered roadsides and relatively high approach speeds. Observer position is constant during repeated visits to the same site. The distance between observers and features of interest should also be as consistent as possible during studies comparing different sites. Observers should try to be concealed from approaching traffic while keeping visible the area to be observed. Usually, an observer sitting in a vehicle, parked legally on or off the roadway, is sufficiently concealed. If no legal parking space is available, positioning observers on folding chairs behind utility poles, trees, or any fixed roadside object is adequate.

Conflict studies are conducted in daylight with dry weather and pavement unless the study is specifically oriented to other conditions. Weekdays between 7:00 A.M. and 6:00 P.M. are the usual hours for conflict studies. In fact, daily statistics, such as given in Table 12-2, are based on this 11-hour period. During a study, similar time periods should be used at each site. Conflict studies are scheduled to avoid periods of recurrent congestion, since conflict data collected under stop-and-go conditions are invalid. Observers should also avoid unusual traffic conditions such as construction or maintenance. If unusual traffic conditions suddenly occur during an observation period (a signal malfunctions, an accident occurs, a maintenance crew arrives, etc.), observers should note the time and nature of the condition. Observers should stop temporarily if it appears that typical traffic conditions will be quickly restored, or observers should quit for the day if it appears that the unusual condition will last a long time.

Observers must maintain a high level of concentration during a traffic conflict study. Frequent breaks allow observers to regain concentration and allow tasks such as recording data, clearing counters, and changing forms to be performed without distracting from observing. Parker and Zegeer (1988) recommend a 20- or 25-minute observation period followed by a 10- or 5-minute break during each half hour of a conflict study. Before conflicts are observed, it is wise to record data on the physical features of the site. A condition diagram (see Chapters 6 and 11) of the site may be helpful in diagnosing a safety problem from conflict data. Photos of the site may also be useful.

Figures 12-15 and 12-16 contain forms for recording traffic conflicts (the forms are reproduced in Appendix G). The form in Figure 12-15 includes 12 of the 14 basic types of conflicts expected on intersection approaches. Only the two types of basic pedestrian conflicts are omitted from the form. Observers enter one line on the form in Figure 12-15 after each observation period (i.e., each 20- or 25-minute period between breaks). A mechanical or electronic counter may be helpful in keeping a running total of conflicts by type during an observation period. Inappropriate columns of the form in Figure 12-15 should be voided in advance to prevent the mistaken entry of data.

Figure 12-16 shows a form that requires that one line be entered for each conflict. For each conflict, observers record the time, the position of each involved vehicle (actor), the movement of each involved vehicle (action), and comments that help describe the conflict or event. Analysts should develop codes for actors and actions to meet the individual needs of each study. Figure 12-17 shows a sample form with actor and action codes that was developed for a conflict study concentrating on left turns. With a form such as that given in Figure 12-16, no counter is needed, observers have much more flexibility in coding unusual occurrences, and analysts have very detailed data to work with. Also, recording the times of conflicts makes it easier to check observer consistency during training. However, the form requires more data reduction and may distract observers (while writing, they may miss some conflicts) at sites with high numbers of conflicts.

Several key points apply equally to traffic conflict data forms with one line per observation period (Figure 12-15) and one line per conflict (Figure 12-16). First, extensive testing of and practice with a form is essential. Second, observers must complete all "header" information or the data on a form are useless. A critical piece of header

INTERSECTION TRAFFIC CONFLICTS SUMMARY

Location____________________ Leg Number(s)____________________

Day__________ Date__________ Observer(s)__________ Length of Recording Period__________

C = Conflict SC = Secondary Conflict

Count Start Time (Military)	Time Period	Approach Volume	Left-Turn Same Direction		Right-Turn Same Direction		Slow Vehicle		Lane Change		Opposing Left-Turn		Right-Turn From-Right		Left-Turn From-Right		Through From-Right		Right-Turn From-Left		Left-Turn From-Left		Through From-Left		Right-Turn On-Red		All Same Direction		All Through Cross-Traffic		Other
			C	SC	C	SC	C	SC	C	SC	C	SC	C	SC	C	SC	C	SC	C	SC	C	SC	C	SC	C	SC	C	SC	C	SC	
Total																															
C + SC																															
Daily Count																															
Rate Per 1,000 Veh																															

Figure 12-15 Traffic conflict form with one line per time period. *Source:* Parker and Zegeer, 1988.

ACTOR CODES	ACTION CODES	Name: Date: Time Period: Intersection: Direction (leg with actor 1): Weather:

Time	Actor 1	Action	Actor 2	Action	Comments

Figure 12-16 Conflict form with one line per conflict. *Source:* Hummer et al., 1989.

information that is often forgotten is the direction of the approach leg being studied. Third, every form should have a place for comments, which are important during data analysis.

Data Reduction and Analysis

Before data are reduced to an analyzable format, the data forms must be checked for comments or descriptions of unusual events. If an event is described that probably biased a certain section of data, analysts should omit that section. A check with the observers may be necessary to clear discrepancies in the data.

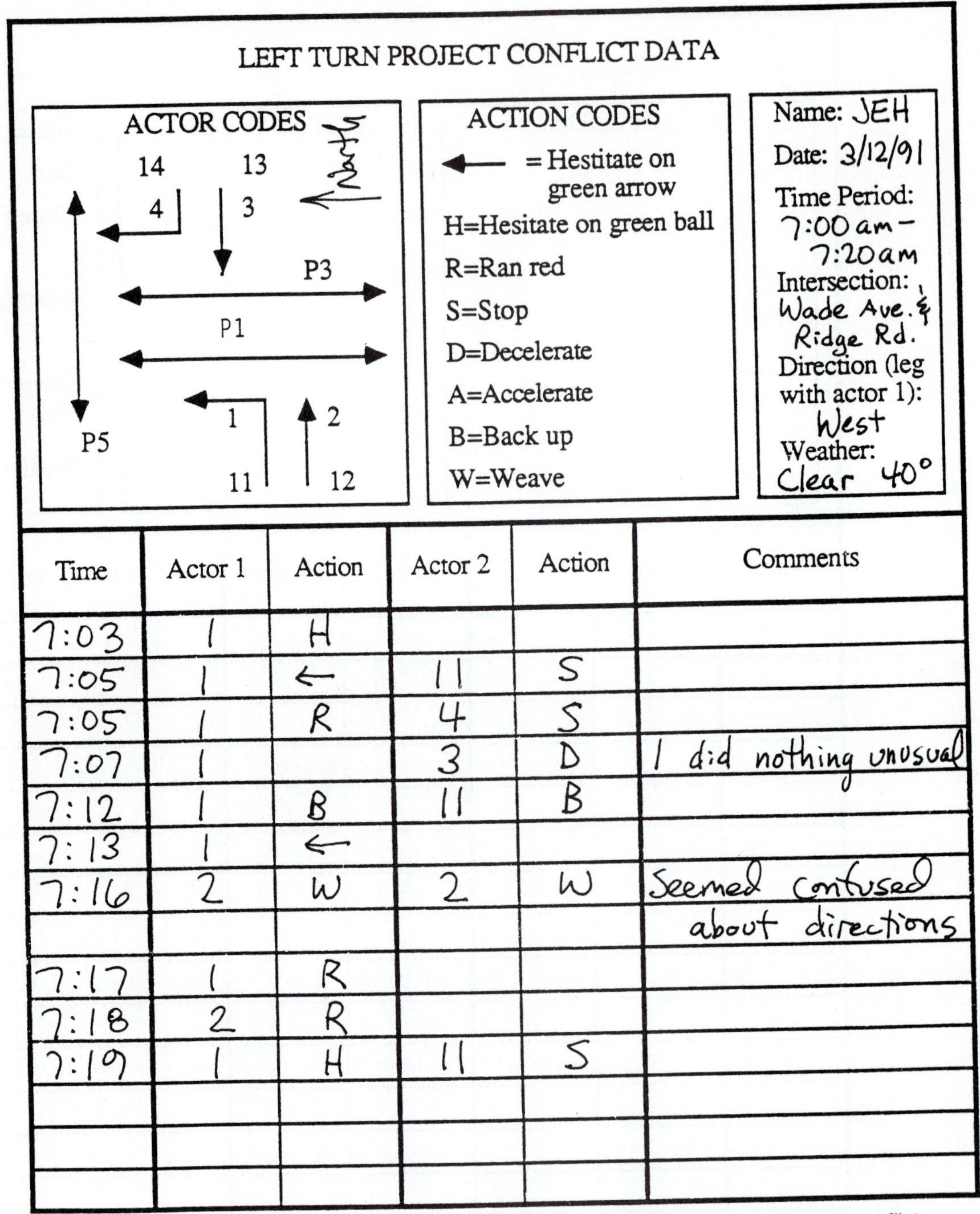

LEFT TURN PROJECT CONFLICT DATA

ACTOR CODES

ACTION CODES

← = Hestitate on green arrow
H=Hesitate on green ball
R=Ran red
S=Stop
D=Decelerate
A=Accelerate
B=Back up
W=Weave

Name: JEH
Date: 3/12/91
Time Period: 7:00 am – 7:20 am
Intersection: Wade Ave. & Ridge Rd.
Direction (leg with actor 1): West
Weather: Clear 40°

Time	Actor 1	Action	Actor 2	Action	Comments
7:03	1	H			
7:05	1	←	11	S	
7:05	1	R	4	S	
7:07	1		3	D	1 did nothing unusual
7:12	1	B	11	B	
7:13	1	←			
7:16	2	W	2	W	Seemed confused
					about directions
7:17	1	R			
7:18	2	R			
7:19	1	H	11	S	

Figure 12-17 Example of a completed conflict form with one line per conflict.

Traffic conflict data are reduced by summing the totals of each type of traffic conflict. If the data collection form resembles that shown in Figure 12-15, reduction consists of summing each column. If the form resembled that shown in Figure 12-16, the number of lines corresponding to particular conflict types must be determined. Unless the study is extremely large, this is best accomplished by manually scanning the completed data forms instead of by a computer program. A manual scan allows flexible scoring of conflicts by type and allows the analyst to obtain a good feel for the data. During data reduction it may also be appropriate to sum counts from individual approaches into a total for an intersection. Engineers should examine all conflicts totals for reasonableness. If the totals are not reasonable, engineers should then ask, "Why not?" and "What can be done to assure that the data are worth analyzing?"

If conflicts per unit of time are of interest, analysts need to adjust for unobserved time periods. For example, a particular study calls for an hourly conflict rate. The observers used 20-minute data collection blocks with 10-minute breaks between blocks so 40 minutes of data were collected per hour. Therefore, multiplying the number of conflicts in two adjacent blocks by 60/40 provides the needed rate per hour. Analysts should adjust for unobserved time periods with data from similar time periods that were observed, not by assuming constant conflict rates throughout a long period. A total of 30 conflicts recorded from noon to 3:00 P.M. should not be expanded to a standard 11-hour day (7:00 A.M. to 6:00 P.M.) total of 110, because the 3 hours observed did not cover either the A.M. or P.M. peak hours. It is usually permissible to use a 3:00 to 4:00 P.M. count of x conflicts and a 5:00 to 6:00 P.M. count of y conflicts to estimate that the 4:00 to 5:00 P.M. count would have been $(x + y)/2$ conflicts.

Conflict rates per vehicle observed are produced by combining the conflict sum and the appropriate turning movement count. One should only combine conflict and turning movement data if they are from the same time and the same approach.

Once the data are reduced, the purpose of the study should guide analysts. For many studies, mean rates of particular types of conflicts and the accompanying standard deviations are necessary. Appendix C provides descriptions of and formulas for common statistical procedures. If the study purpose was to determine the magnitude of the safety problem or to generate countermeasures, comparison conflict rates are needed. For agencies without their own comparison rates, Table 12-2 provides typical rates for some types of conflicts at intersections.

REFERENCES

COCHRAN, W. G. (1977). *Sampling Techniques,* 3rd ed., Wiley, New York, pp. 76–77.

GLAUZ, W. D., AND D. J. MIGLETZ (1980). *Application of Traffic Conflict Analysis at Intersections,* National Cooperative Highway Research Program Report 219, Transportation Research Board, Washington, DC.

HUMMER, J. E., R. E. MONTGOMERY, AND K. C. SINHA (1989). *An Evaluation of Leading versus Lagging Left Turn Signal Phasing, Final Report,* FHWA/IN/JHRP-89/17, Joint Highway Research Project, Purdue University, West Lafayette, IN.

MIGLETZ, D. J., W. D. GLAUZ, AND K. M. BAUER (1985). *Relationships between Traffic Conflicts and Accidents,* FHWA/RD-84/042, Federal Highway Administration, Washington, DC.

PARKER, M. R., AND C. V. ZEGEER (1988). *Traffic Conflict Techniques for Safety and Operations: Engineer's Guide,* FHWA-IP-88-026, Federal Highway Administration, McLean, VA.

PARKER, M. R., AND C. V. ZEGEER (1989). *Traffic Conflict Techniques for Safety and Operations: Observer's Manual,* FHWA-IP-88-027, Federal Highway Administration, McLean, VA.

13

Pedestrian Studies

H. Douglas Robertson, Ph.D., P.E.

INTRODUCTION

There are several types of pedestrian studies designed to capture some aspect of pedestrian behavior or performance. Engineers use the results of these studies to determine if traffic signals are warranted, develop exposure data for calculating pedestrian accident rates, locate and design sidewalks and crosswalks, design and implement pedestrian safety improvements, and analyze roadway crossings to determine appropriate controls and control operations (e.g., school crossing protection or signal timing).

The behavior or performance of the pedestrian is generally measured by one or more of the following:

- Volume
- Walking speed
- Gaps in traffic
- Conflicts with vehicles
- Understanding of and compliance with traffic control devices
- Exhibited behaviors (e.g., running, stopping, retreating, looking)

This chapter focuses on the common methods for capturing these measures. Discussed are issues of study design, sampling, equipment, personnel, field procedures, and applications. Examples of field data collection and summary forms are shown throughout the chapter. Appendix G provides additional sample forms suitable for copying.

VOLUME STUDIES

Pedestrian volumes are obtained by recording the number of pedestrians passing a point, entering an intersection, or using a particular facility such as a crosswalk or sidewalk. Counts are usually samples of actual volumes, although agencies may conduct continuous counts for certain situations or circumstances. Sampling periods usually range from a few minutes to several hours. The length of the sampling period is a function of the type of count being taken and the eventual uses of the pedestrian volume data. Agencies usually count pedestrians in good weather, unless the purpose of the study involves certain environmental conditions. The two basic methods of measuring pedestrian volume are manual observation and mechanical or automatic recording.

Manual Observation

Purpose and Application

Most types of pedestrian counts are taken manually through direct observation. Several types of counts require classifications and are more easily and accurately obtained with trained observers. Examples include pedestrian counts by age group, sex, physical handicap, and type of behavior. Other reasons for conducting manual counts are time and resources. A number of studies that use pedestrian volumes often require less than 10 hours of data at any given location. Thus the effort and expense to set up and take down automated equipment, plus the time required to manually reduce the data, are usually not justified. Count expansion techniques, such as the one described later in this chapter, offer a way to obtain reliable estimates from manual short counts for less cost than continuous sampling.

Equipment

Tally Sheets. The simplest means of conducting manual counts is to record each observed pedestrian with a tick mark on a prepared field form. An example of a field sheet for a crosswalk pedestrian count is shown in Figure 13-1. The form allows for any desired classification. A watch or stopwatch is required to cue the observer to the desired count intervals. Observers tally their raw counts and summarize or key them into a computer upon return to the office.

Mechanical Count Boards. Mechanical count boards consist of various combinations of accumulating hand counters mounted on a board to facilitate the type of count being made. Typical hand counters are accumulating push button devices with three to five registers, as described in Chapter 2. To conduct a four-crosswalk intersection count, a counter is positioned on each side of a board to represent each crosswalk at the intersection. Pedestrian counts may be in conjunction with vehicle turning movement counts or for various classifications of pedestrian data only. A register is added to the board for each classification and each crosswalk (approach). Many configurations of registers and counters are possible. Figure 13-2 shows examples of mechanical count boards.

Each button represents a different category of vehicle or pedestrian being counted. The observer pushes the correct button each time a vehicle or pedestrian passes. A watch or stopwatch is required to cue the observer to the desired count intervals. When observers reach the end of an interval, they read the counter, record the data on a field form, and reset the counter to zero. Observers summarize the data from the field form or key them into a computer upon return to the office.

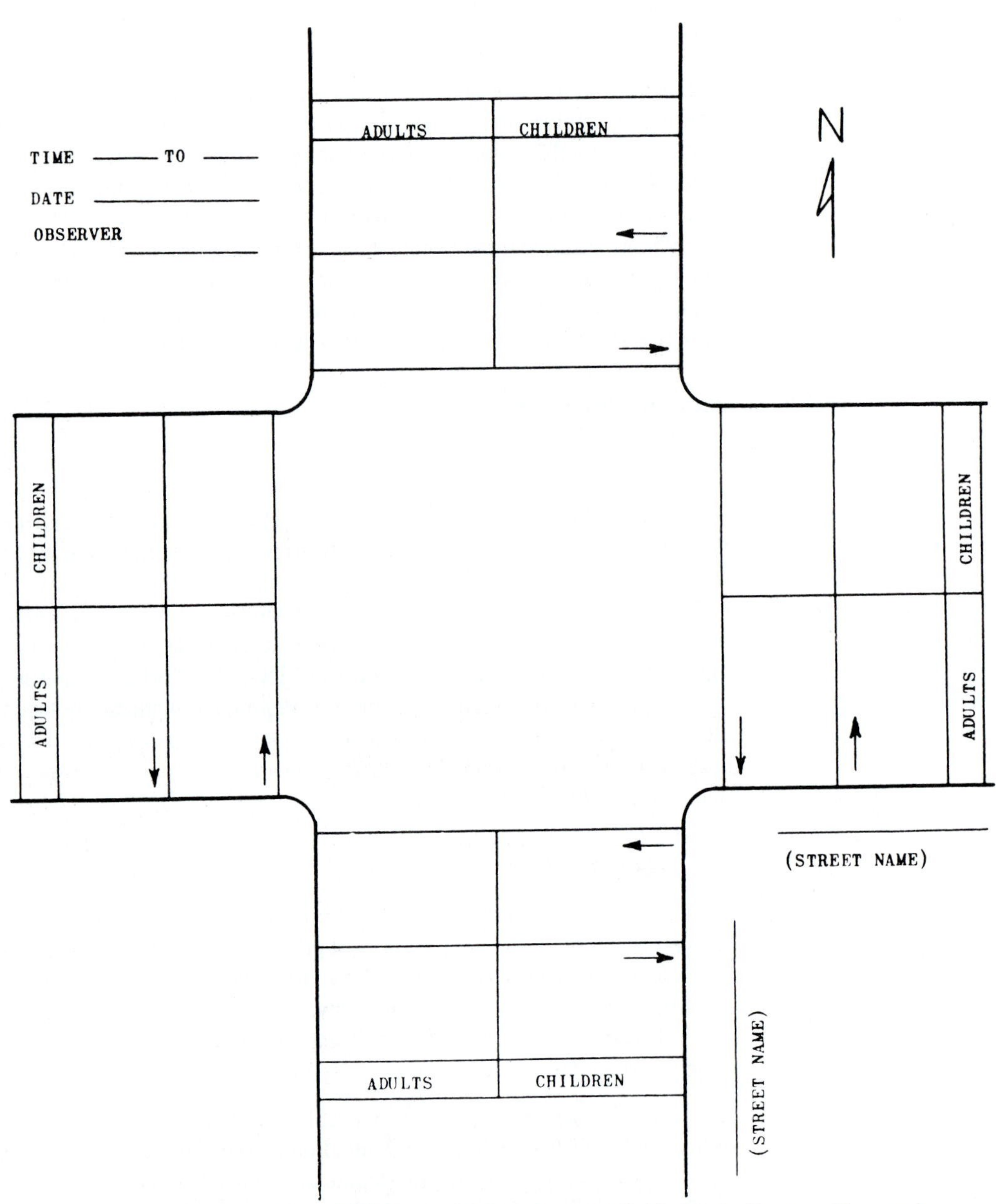

Figure 13-1 Example pedestrian count field sheet. *Source:* Box and Oppenlander, 1976, p. 24.

Electronic Count Boards. Battery-operated electronic count boards or hand-held computers are the latest devices to aid in the collection of pedestrian count data. They operate in a similar fashion to mechanical count boards with a few important differences. They are usually lighter in weight, more compact, and easier to handle. They contain an internal clock that separates the data by whatever interval is chosen. Most important, they preclude the need for manual data reduction and summary. Data may be transferred directly from the field to a computer in the office via a modem or dumped into a computer upon return from the field. Regardless of the transfer means, the data are summarized, the data are analyzed, and the results are displayed in a selected presentation format by

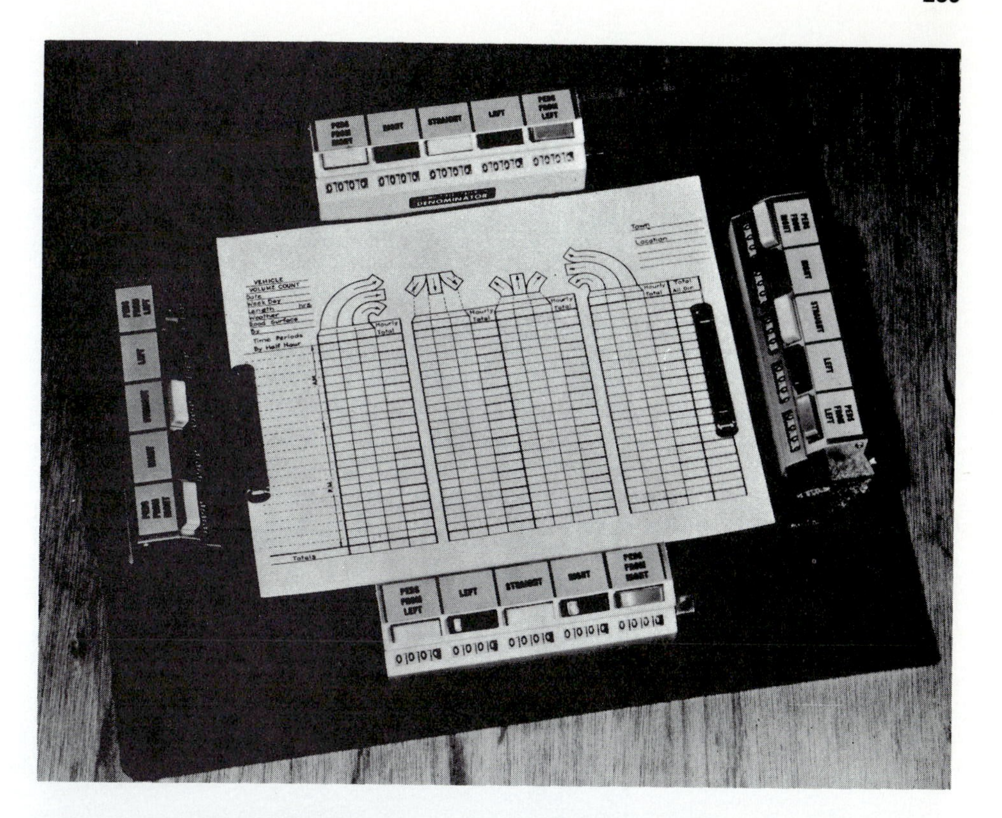

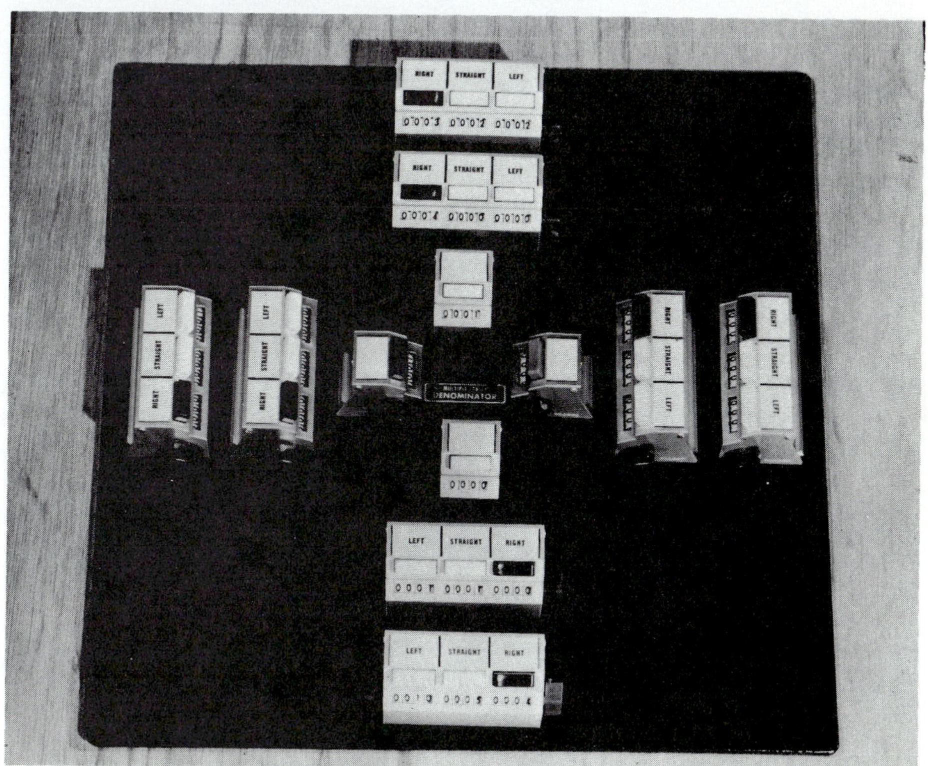

Figure 13-2 Example of a mechanical count board. *Source:* Box and Oppenlander, 1976, p. 27.

Figure 13-3 Electronic count board used for pedestrian and traffic counting. *Source:* "PC-Travel Sample Reports," JAMAR Sales Company, Inc., Ivyland, PA, 1990.

means of computer software. The use of software eliminates the manual data reduction required with tally sheets and mechanical count boards.

Many electronic count boards are capable of handling several types of traffic studies involving pedestrian count data. Examples include crosswalk, classification, gap, gap acceptance, and pedestrian behavior studies. For agencies requiring more than occasional manual pedestrian and traffic counts, the electronic count board or hand-held computer is a cost-effective, labor-saving tool. Figure 13-3 illustrates a typical electronic count board.

Personnel Required

Trained observers are required to perform accurate manual pedestrian counting. They must be relieved periodically to avoid fatigue and degraded performance. Breaks of 5 minutes every half hour or 10 minutes every hour are typical. If the data collection period is more than 8 hours, breaks of 30 to 45 minutes should be allowed every 4 hours.

The size of the data collection team depends on the length of the counting period, the type of count being performed, the number of crosswalks being observed, and the volume level of pedestrians. One observer can easily handle a four-way signalized intersection with single approach lanes and low volumes as long as special classifications and/or directional counts are not required. As any or all of the foregoing parameters increase, the complexity of the counting task increases and additional observers will be needed. The exact number needed can be determined by conducting pilot studies at the locations of interest.

Duties may be divided among observers different ways. At a signalized intersection, one observer may record the north and west crosswalks while a second observer watches the south and east crosswalks. In that way, only one crosswalk is active for each observer at any given time. Another way to divide duties is for one observer to record certain classes of pedestrians, while the other observer counts total pedestrian volumes. At complex sites, individual crosswalks or classifications may be assigned to individual observers. Also at complex sites, one observer may have the sole job of relieving the other observers on a rotating schedule basis.

Field Procedures

Preparation. An accurate and reliable pedestrian count begins in the office. A locally developed checklist is a valuable aid even to experienced teams to ensure that all preparations for the field study have been completed before the team arrives at the site to be counted. Figure 1-1 is an overall checklist that analysts can modify or add to for local conditions. Preparations should start with a review of the purpose and type of count to be performed, the count period and time intervals required, any information known about the site (i.e., geometric layout, vehicle or pedestrian volume levels by time of day, signal timing, etc.). This information will help determine the type of equipment to be used, the field procedures to follow, and the number of observers required. If the purpose of the study requires ''good weather'' conditions, analysts must prepare criteria for canceling the count or procedures for dealing with inclement weather.

The selection of equipment will dictate the type of data forms needed, if any. Header information should be filled in to the extent possible in the office and the forms arranged in the order they will be used by each observer in the field. The preparation checklist should include equipment items such as pencils, batteries, stopwatches, and blank videotapes. Returning to the office to retrieve forgotten items may delay the start of the study or cause it to be postponed. An inadequate number of forms to complete the study could also invalidate the study resulting in wasted resources. An office review of the study procedures and a check of the proper operation of all equipment completes the preparation stage of the study.

Observer Location. Observers must be positioned where they can most clearly view the pedestrians they are responsible for counting. Observers should be located well away from the edge of the travel way, both as a personal safety precaution and to avoid distracting drivers. A position above the level of the street and clear of obstructions usually affords the best vantage point. If several observers are counting at the same site, they must maintain visual contact with one another and be able to communicate so as to coordinate their activities.

Protection from the elements is an important consideration for the observer. Proper clothing to suit prevailing weather conditions is paramount. Safety vests should be worn if the observer is near traffic at any time. Observers may count from inside vehicles as long as their view is unobstructed. Outside, observers may use chairs to prevent fatigue and umbrellas for protection from the sun, as long as these devices are not distracting to drivers and pedestrians. A sign indicating that a traffic count is under way usually satisfies driver curiosity.

Data Recording. Keeping the data organized and correctly labeled is the key to successful pedestrian counts. The counts may produce a large number of data forms. Each must be clearly labeled with information such as the count location, the observer's name, the time of study, and the conditions under which the counts were made. The form itself should clearly indicate the movements, classifications, and time intervals.

The observer must concentrate his or her attention on accurately recording each count in the proper place on the form or with the proper button. Special care must be taken with electronic count boards or hand-held computers to ensure that they are properly oriented to the geographic and geometric layout of the intersection. Time intervals must be accurately maintained and coordinated when two or more observers are working together. When mechanical count boards are used, the observers must have time to record the accumulated counts and reset the counters at the end of each interval. Two procedures may be used to accomplish this: the short-break and alternating count procedures (McShane and Roess, 1990). These procedures are described in Chapter 2.

Automatic Counts

Purpose and Application

There are some applications of pedestrian volume data that do not require complex classifications, or on the other extreme, are so complex that they must be recorded for slow or still-motion analysis. The simple counts may be needed for extended periods of time (i.e., days, weeks, or even months) in remote locations on paths and trails. The use of observers for such purposes would be cost prohibitive. When complex behavior classifications are required, the actions (e.g., head movements) may be too quick for observers to see and record. Automatic video recording provides a means of gathering these pedestrian data at a reasonable expenditure of time and resources.

Equipment

There are several types and models of automatic volume data collection equipment. This equipment generally includes two basic components: sensors to detect the presence of pedestrians and a data recorder. Sensors may employ infrared light transmission and detection, photocells, pressure pads, or heat-sensing devices. To date, none of these methods have proven to be very practical or otherwise acceptable for common use in other than remote areas (Cameron, 1977; Mudaly, 1980; Vozzolo and Attanucci, 1982).

Photographic data collection techniques have been used by researchers for some time. Seldom has the purpose been to count only pedestrians. Usually, other types of behavioral data are also being sought. In recent years videotaping has largely replaced filming as the principal visual data collection medium. Taping or filming provides an accurate and reliable means of recording volumes, as well as other data, but requires time-consuming data reduction in the office (Berger and Knoblauch, 1975; Older and Grayson, 1972).

Personnel Required

The only personnel required for making automatic counts are those needed to install and recover the equipment. Crew sizes of one to two are usually sufficient to deploy and recover most pedestrian counting equipment.

Field Procedures

Preparation. As mentioned previously, field work should never be undertaken without proper preparation in the office. A locally prepared checklist is an invaluable aid even for the most routine task. The purpose of the count will drive the type of equipment to be used and the deployment procedures to follow. All equipment should be checked to see that it is functioning properly. An ample supply of accessory items (i.e., nails, clamps, tapes, adhesive, chains, locks, and batteries) and all necessary tools should be provided. Analysts also cannot forget weatherproofing for cameras.

Selecting the Count Location. The street or highway on which the count will be made and the general location—midblock or intersection—where the counters or cameras will be placed is decided in the office and is a function of the type of study being performed. The exact locations of the cameras, count recorders, and sensors are usually determined in the field. In the case of cameras, the most important factor is the field of view. The location of the camera should take into account adverse weather and reduced visibility from shadows.

Installation and Retrieval. The primary concern during installation and retrieval operations is the safety of the field crew. Most of the automatic equipment used for pedestrian data recording is placed on sidewalks or paths or away from the roadway traveled by vehicles. The crew's vehicle should be clearly visible to traffic and parked away from the traveled way. All crew members should wear reflective clothing at all times. Deployments and recoveries should be accomplished during periods of low traffic volume and good visibility. Installation techniques vary and in many cases are product specific. Information of this nature is generally available from the product manufacturer.

The second concern is the security of the data recording equipment. Most automatic counters can be physically secured to trees, lamp posts, or sign posts. In most cases, cameras must be manned. In addition to providing security, the operator also changes tapes or film and ensures that the camera is properly aimed and focused during data collection.

Data Reduction and Analysis

Following collection, raw data must be placed in a form suitable for analysis. This reduction of the data usually consists of converting tally marks to numbers, summarizing the data by calculating subtotals and totals, and arranging the data in a format for performing analyses. Electronic count boards or hand-held computers preclude the data reduction steps and automatically produce the data in a summary format. The analysis may range from a simple extraction of descriptive information to a sophisticated statistical treatment of the data, depending on the type of study being conducted.

Count Periods

The counting period selected for a given location depends on the planned use of the data and the methods available for collecting the data. The count period should avoid special events unless the purpose is to study an event. Count periods may range from an hour to several days. Manual and video count periods are usually for less than 1 day. Typical count periods for warrant analysis and behavioral studies include:

- 2 hours; peak period
- 4 hours; morning and afternoon peak periods
- 6 hours; morning, midday, and afternoon peak periods
- 12 hours; daytime (e.g., 7:00 A.M.–7:00 P.M.)

Count intervals may be 1, 5, 10, 15, 30, or 60 minutes long. Typical intervals for pedestrian counts are 15 and 60 minutes. Pedestrian counts at signalized intersections may be summarized by signal cycle.

Sample Counts and Count Expansions

All counts are samples. Count periods are also a sample of the overall long-term pedestrian flow. Time and resources do not permit the continuous counting of every path and crossing on all existing streets and highways. Consequently, sample counts are taken over shorter time periods at specific locations. These counts are then adjusted and/or expanded to produce estimates of the expected pedestrian flow at the count location or

Step 1. Select the time period for which pedestrian volume will be estimated.
Step 2. Select the count interval.
Step 3. Develop the data collection plan.
Step 4. Collect the data.
Step 5. Select the expansion model coefficient and exponent.
Step 6. Compute estimated volumes.
Step 7. Determine estimated volume ranges.

Figure 13-4 Procedure for expanding pedestrian counts.

similar locations. Recent research has suggested that sample counts may be taken and estimates of pedestrian volumes calculated based on very short-term counts (Davis, 1988).

Expansion of Short Counts. An expansion model technique developed for the Federal Highway Administration (FHWA) uses short counts in conjunction with empirically derived models for a variety of time periods and counting intervals to expand the short counts into pedestrian volume estimates for the time period of interest. Ranges for these estimates, based on the standard error about the mean, may also be calculated. The accuracy of the technique is sufficient for screening purposes or to confirm volume levels based on qualitative evaluations. The technique is briefly summarized below. A user manual containing details and examples is available from FHWA (Mingo, 1988a).

Expanded Short Count Procedure. The seven-step procedure is summarized in Figure 13-4. The first step is to select the time period for which pedestrian volumes will be estimated. To estimate hourly volumes for use in signal warrant analysis, 1-hour time periods are selected. For uses involving average daily volumes, choose 2-, 3-, or 4-hour periods that represent variations in pedestrian flow throughout the day.

The second step is to select the count interval. The choices are 5, 10, 15, or 30 minutes of counting in the middle of each time period selected in step 1. The trade-off in the choice of count interval is between economy and accuracy. Obviously, the accuracy of the estimate increases with the length of the count interval, which in turn increases the cost. If, for example, a likely outcome is being verified, a shorter count interval may be satisfactory.

In step 3, develop the data collection plan (i.e., choose the order of counting and the specific time periods for each location). In step 4, collect the data. One observer is usually sufficient depending on the volume level. At an intersection, the observer simply conducts short counts of each crosswalk in turn according to the schedule. Check the user manual for specifics on dealing with signalized locations.

Step 5 is to select the expansion model coefficient and exponent. From Table 13-1, choose the proper values, based on time period and count interval, of a and b for the expansion model:

$$\text{volume} = a \times \text{count}^b \tag{13-1}$$

where

volume = estimate of pedestrian volume for the 1-, 2-, 3-, or 4-hour period of interest
Count = number of pedestrians counted during the count interval
a, b = derived parameters (from Table 13-1)

Step 6 is to compute estimated volumes. For example, if 25 pedestrians were counted during a 10-minute interval in the middle of the time period from 7:00 and 9:00 A.M., the estimated volume for the 2-hour period would be

TABLE 13-1
Expansion Model Parameters

Time Period (hr)	Counting Interval (min) 5 a	5 b	10 a	10 b	15 a	15 b	30 a	30 b
1	19.9	0.786	9.8	0.847	5.8	0.900	2.4	0.963
2	43.0	0.769	20.9	0.823	14.7	0.824	6.1	0.892
3	60.2	0.785	32.2	0.818	17.4	0.884	9.5	0.890
4	62.4	0.811	44.9	0.762	27.1	0.809	15.6	0.813

Source: Mingo et al., 1988a, Table 5.

TABLE 13-2
Range Factors for 1-hour Predictions (percent)

Pedestrian Volume Level	Count Interval (min) 5	10	15	30
0–100	34	35	27	16
101–200	35	26	19	13
>200	27	22	15	9

Source: Mingo et al., 1988a, Table 1.

$$\text{volume} = 20.9 \times 25^{0.823} = 296 \text{ pedestrians}$$

Since the numbers calculated from the expansion models are estimates, the range of values within which the actual volumes are likely to fall may be established with step 7, determine estimated volume ranges. Tables 13-2 through 13-5 contain range factors (in percent) by pedestrian volume level and count interval for each of the four time periods, respectively. For the example above, the factor would be (from Table 13-3) ± 32%. Thus the actual volume for the 2-hour time period would likely lie between 201 and 391 pedestrians. This range may seem large; however, there are many situations where this level of accuracy is sufficient. Simply knowing whether a crossing location has a low, moderate, or high pedestrian volume may be adequate to select the proper pedestrian control or accommodation.

WALKING SPEED STUDIES

Walking speed is a parameter used in a number of pedestrian studies. Examples include gap acceptance, school crossing, and signal timing studies. Walking speeds are affected by a number of factors, including:

- Volume of pedestrians
- Age of pedestrians
- Sex of pedestrians
- Level of physical fitness of pedestrians
- Density of oncoming pedestrians
- Steepness of grade
- Width of crossing
- Width of path
- Distance of oncoming vehicle
- Speed of oncoming vehicle
- Weather condition

TABLE 13-3
Range Factors for 2-hour Predictions (percent)

Pedestrian Volume Level	Count Interval (min)			
	5	*10*	*15*	*30*
0–500	42	32	24	22
>500	24	25	23	19

Source: Mingo et al., 1988a, Table 2.

TABLE 13-4
Range Factors for 3-hour Predictions (percent)

Pedestrian Volume Level	Count Interval (min)			
	5	*10*	*15*	*30*
0–500	35	37	34	26
>500	32	27	24	22

Source: Mingo et al., 1988a, Table 3.

Walking speeds typically range from 2.5 to 6.5 feet per second. The *Manual on Uniform Traffic Control Devices* (MUTCD) assumes a normal walking speed to be 4.0 feet per second, but also suggests adjustments in warrant criteria when the "predominant" walking speed is less than 3.5 feet per second (USDOT, 1988). Analysts should conduct walking speed studies where a significant number of pedestrians walk at a slower or faster pace than 4.0 feet per second.

The study should be performed at the location of interest under the conditions of interest. One or more observers may be used based on how much the conditions vary over time and the number of classes of data desired. The observers should be positioned where they have a clear field of view and do not distract passing pedestrians. Observers mark a measured distance along the path traveled by the pedestrians and then simply time individual pedestrians through the speed trap. A sample of 100 observations is generally adequate. Analyze the data by first calculating each individual average walking speed by dividing the trap distance by the observed time, then classifying the observed speeds, and finally plotting the cumulative percentage of observations by class. This will produce a cumulative speed curve from which values of various speed percentiles may be derived. The 15th percentile speed is a generally accepted value to use in timing signals for pedestrians (Kell, 1991).

GAP STUDIES

Purpose and Application

Gap studies refer to the determination of the number of available gaps in traffic passing a point that are of adequate length to permit pedestrians to cross. In this context a *gap* is defined as the time that elapses from when the rear of a vehicle passes a point on a roadway until the front of the next arriving vehicle (from either direction) passes the same point. Gaps are normally expressed in units of seconds.

Gap studies consist of measuring the predominant pedestrian group size, determining the length of a minimum adequate gap, measuring the gap sizes in the traffic stream, and determining the sufficiency of adequate gaps. The principal application of the study results is in analyzing roadway crossings by pedestrians to determine appropriate traffic controls and safety improvements. The results of gap studies are used in traffic signal warrant analyses and school crossing studies. In addition to the techniques described be-

TABLE 13-5
Range Factors for 4-hour Predictions (percent)

Pedestrian Volume Level	*Count Interval (min)*			
	5	*10*	*15*	*30*
0–750	34	30	29	26
>750	33	27	26	21

Source: Mingo et al., 1988a, Table 4.

low, the procedures for determining gap acceptance characteristics for drivers of vehicles entering or crossing roadways described in Chapter 5 may be applied to pedestrian gap acceptance studies.

Measuring Predominant Group Size

Pedestrians waiting to cross a roadway will generally arrange themselves in rows one behind the other. Group size is comprised of row width and number of rows. When the group starts to cross, they enter the roadway (step off the curb) with approximately 2 seconds of headway between rows. Since the factor of interest is the amount of time it takes the entire group to enter the crossing, it is only necessary to determine the predominant number of rows waiting to cross at the time crossings begin. Thus group size is represented by the number of rows entering the crossing. The width of the rows and the total number of pedestrians in the group are inconsequential.

Distinguishing distinct rows may be somewhat difficult at first. With some training and experience, however, observers manage easily. A sample of 30 to 50 groups is usually sufficient to establish the group size (i.e., number of rows per group). This measurement should be made during the time and under the conditions of interest for the gap study. One observer is required for each crossing point to be sampled. The crossing point for the heaviest direction of pedestrian flow should be the one sampled. The observer should be unobtrusively positioned perpendicular to the crossing and parallel to the roadway with a clear view of the crossing point.

The top portion of the field sheet shown in Figure 13-5 may be used to record the sampling of number of rows. Observe each group as they enter the crossing. Place a tick mark in the tally column corresponding to the number of rows in the group. Stragglers are not included. Groups will form naturally when gaps are inadequate to accommodate random arrivals. When the group sampling period is complete, count the tally marks and record the frequency of each corresponding group size in the total column. The sum of the total column will be the total number of groups sampled. Enter the cumulative totals for each group size in the cumulative column as shown in Figure 13-5. The 85th percentile group size is generally used to define the predominant number of rows. Therefore, in the computation column of the field sheet, multiply the number of groups by 0.85 to calculate the 85th percentile group size (i.e., number of rows). If desired, another percentile may be used. Enter the result in the space for the predominant number of rows, N. This value is used in the calculation of the minimum adequate gap.

In the example shown in Figure 13-5, a total of 38 groups of pedestrians were observed. The 85th percentile of this sample is 32.3, which corresponds to groups containing five rows. Appendix G contains a blank form like that shown in Figure 13-5 suitable for copying.

Determining the Minimum Adequate Gap

A minimum *adequate gap* is defined as the time (in seconds) for one or a group of pedestrians to perceive and react to the traffic situation and cross the roadway from a point of safety on one side to a point of safety on the other (Pline, 1992). This minimum safe

GAP STUDY FIELD SHEET

Group Size Survey   Location EAST BLVD AT DILWORTH
Date 7/23   Time: From 8:15 am To 9:00 am
Crossing Distance 48 ft   Walking Speed 3.5 ft/sec

No. of Rows	Tally	Total	Cumulative
1					3	3
2	~~||||~~ ||||	9	12
3	~~||||~~ ~~||||~~		11	23
4	~~||||~~ |||	8	31
5						4	35
6					3	38
7			

N = 0.85 x 38 = 32.3 or 5 rows   H = 2 sec   R = 3 sec

Minimum Acceptable Gap, G = W/S + (N-1)H + R = 25 sec

Available Gap Survey   Date 7/24
Time: From 8:15 am To 9:00 am   Duration 45 min

Gap (sec)	Tally	Total
10		
11		
12		
13		
14		
15		
16		
17		
18		
19		
20	Discard gaps less than 25 sec	
21		
22		
23		
24		
25						4
26	~~||||~~ ~~||||~~ ||	12
27	~~				~~		6
28	~~||||~~ ||||	9
29						4
30	~~||||~~	5
31		0
32				2
33		0
34						4
35			1
36					3
37		0
38			1
39		
40		

Total Adequate Gaps 51

Figure 13-5 Sample field sheet for a gap study.

crossing time (gap in traffic) is a function of crossing distance, walking speed, predominant number of rows in the group, time headway between rows, and the group startup time. This relationship is shown in the following equation:

$$G = \frac{W}{S} + (N - 1)H + R \tag{13-2}$$

where

G = minimum safe gap in traffic, seconds
W = crossing distance or width of roadway, feet
S = walking speed, ft/sec
N = predominant number of rows (group size)
H = time headway between rows, seconds
R = pedestrian startup time, seconds

Commonly assumed values for some of these variables include

S = 4.0 or 3.5 ft/sec
H = 2 sec
R = 3 sec

From the data in Figure 13-5:

$$G = \frac{48}{3.5} + (5 - 1)2 + 3 = 24.7 \text{ sec}$$

The result is normally rounded to the nearest second. In this case, the minimum gap in traffic that would allow the safe crossing of the 85th percentile group size is 25 seconds. If the roadway to be crossed is divided such that the median provides a safe haven for the pedestrians crossing, a minimum adequate gap would be determined for each half of the crossing.

Measuring Gap Sizes

The next part of the field study is to measure the time lengths of the gaps in traffic. This may be done simply with a stopwatch and the bottom portion of the field sheet shown in Figure 13-5. Only the gaps that exceed the minimum adequate gap are of interest; therefore, it is not necessary to record every gap. The observer can develop a "feel" for gaps that are close to or exceed the minimum adequate gap by observing the distance between and speed of vehicles while measuring the gap time. With some experience, the observer will be able to capture the majority of adequate gaps.

Measured gaps are rounded to the nearest second. A tick mark is placed in the tally column corresponding to the measured gap size that equals or exceeds the minimum adequate gap. The tally marks are then totaled for each gap size. The sum of these totals is the number of gaps of sufficient length to accommodate the safe crossing of 85% of the pedestrian groups using the crossing at a day and time and under the conditions similar to those of the study. The example shown in Figure 13-5 indicates that a total of 51 adequate gaps were recorded.

Gaps may also be measured using electronic count boards or hand-held computers in place of the stopwatch and tally sheet. The observation procedure is essentially the same as described above. Internal clocks in the computer record the times. Observers push the appropriate buttons to record gaps in the traffic. The primary advantage with this technique is that computer software reduces the data, thus saving time.

To evaluate the study results, analysts compare the number of gaps equal to or exceeding the minimum adequate gap to the number of minutes the gap measurement study is conducted. The appropriate criteria are then applied to the result. The length of the study depends on the type of application for which the gap study results are being used. For example, warrant 3 in the *Manual on Uniform Traffic Control Devices* (MUTCD) for traffic signals requires that in addition to the stated minimum pedestrian volumes, there shall be fewer than 60 gaps per hour in the traffic stream of adequate length for pedestrians to cross during the same period when the pedestrian volume criterion is satisfied. Another MUTCD criterion states that a traffic signal may be warranted when the number of adequate gaps in the traffic stream when school children are crossing is less than the number of minutes in the same period (USDOT, 1988). If the analyst applied this criterion to the data shown in Figure 13-5, the signal would not be warranted since the number of adequate gaps (51) exceeded the number of minutes in the study (45).

PEDESTRIAN BEHAVIOR STUDIES

Research and development of pedestrian safety countermeasures and design considerations for pedestrian accommodations have prompted studies of observed pedestrian behavior. These studies provide an understanding of the needs of pedestrians and identify the human factors relationships that are critical to the mobility and safety of pedestrians. The studies may be grouped into three general categories: pedestrian/vehicle conflicts, understanding of and compliance with traffic control devices, and exhibited behavior studies. Each type is described briefly below, followed by a description of the generic procedure for conducting these types of studies.

Types of Pedestrian Behavioral Studies

Pedestrian/Vehicle Conflicts

A pedestrian/vehicle conflict occurs when a driver and/or pedestrian has to take some action, such as a change in direction, speed, or both, in order to avoid a collision. Researchers have met with difficulty in establishing a relationship between pedestrian accidents and pedestrian/vehicle conflicts. The existence of a surrogate measure for accidents would allow potentially hazardous situations to be dealt with before accidents occurred. While some evidence of such a relationship has been uncovered, the complexity and relatively rare occurrence of pedestrian accidents has to date prevented a clear result. A recent study devised a method for developing pedestrian/vehicle accident prediction models (Davis et al., 1989). Details for applying this methodology are contained in a user manual available from the Federal Highway Administration (Mingo et al., 1988b). Despite the difficulty in predicting accidents with them, pedestrian/vehicle conflicts are a useful measure of relative differences among pedestrian safety alternatives. A number of studies have used conflicts as a measure of effectiveness for identifying pedestrian safety problems, evaluating traffic control devices, and comparing pedestrian accomodation designs (Robertson, 1977a; Robertson et al., 1977b; Robertson, 1983; Zegeer et al., 1984).

Understanding and Compliance

One simple way to determine pedestrians' understanding of traffic controls devices is to ask them. Engineers have often used surveys to identify pedestrian problems and evaluate alternative control devices (Robertson, 1977b). Appendix B provides guidelines for survey design. Another way to measure pedestrian understanding is to observe pedestrian compliance with traffic control devices. While some pedestrians may understand a device and choose to ignore it, compliance is often an indicator of the degree of understanding by the pedestrian, particularly when coupled with other measures. Compliance is usually measured by observing and recording violations, such as crossing with a red signal. Research studies have used compliance to good effect as a measure of effectiveness (Robertson, 1977a, c; Robertson 1983; Zegeer et al., 1982).

Exhibited Behavior

In addition to pedestrian/vehicle conflicts and compliance with traffic control devices, other pedestrian behaviors have proven reliable to varying degrees in identifying problems and evaluating safety countermeasures. Examples of these behaviors include failure to look left and right before and while crossing, hesitating in the roadway, running, and returning to the curb after starting to cross. These types of behaviors represent undesirable actions that reflect some degree of threat to the pedestrian. Pedestrian accommodations and/or traffic control devices that reduce these behaviors are generally regarded to be safer. Use of these measures is documented in a number of research studies (Berger and Robertson, 1976; Snyder and Knoblauch, 1971; Robertson, 1983; Knoblauch, 1984).

Pedestrian Behavioral Study Procedures

Behavioral studies are similar to each other in the sense that they are all guided by the same fundamental components of experimental design. However, behavioral studies often differ from each other in the situation of interest and the objectives, so that each study must be tailor-made in some ways.

Required Characteristics of Behavioral Measures

For a behavior to be useful to the experimenter, it must possess certain characteristics. First, the behavior must be definable in terms of objective, observable events so that coding is reliable. Second, it must occur with sufficient frequency to permit an efficient data collection schedule. Third, the behavior must have an association with pedestrian safety or flow, either theoretical, empirical, or assumed. The behaviors must also be sensitive, which implies the ability of the measures to discriminate reliably between certain variables of interest (to discern a difference where one exists). The behaviors used in the study must be meaningful and believable to the users of the study results.

Choose the Measures of Effectiveness

Selection of measures of effectiveness (MOEs) depends on the purpose and objectives of the study, the situation and conditions at the sites to be studied, and the resources (time and money) available. It is not necessary to include all of the types of behavioral measures. The following outline typical behavioral MOEs.

Conflicts. Pedestrians crossing a roadway may encounter through vehicles, right-turning vehicles, and left-turning vehicles. These basic conflicts occur when the projected paths of the pedestrian and the vehicle intersect and either the pedestrian or the vehicle must change direction and/or speed to avoid a collision. Variations and refinements to these basic conflicts have been used. In addition, the severity of the conflict, as determined by the strength of the deceleration or acceleration, the speed differential, and how closely spaced the involved parties are, has been used effectively in further defining conflict MOEs.

Violations. Both pedestrians and vehicles may not comply with traffic control devices and the rules of the road. Vehicles may run a red signal or stop sign, and pedestrians may anticipate the walk or green signal or start to cross during a clearance or prohibited signal indication.

Other Behaviors. Pedestrians may look at potentially conflicting traffic, run, or hesitate while crossing a roadway. Carefully defined, these and behaviors like them can serve as MOEs for a variety of countermeasures.

Design the Experiment

The experimental design for behavioral studies is much the same as for other studies. The issues revolve around sampling, site selection, scheduling data collection, and developing the analysis plan. Appendix A contains further guidance.

Prepare for Data Collection

Data for behavioral studies are collected through manual observation or by film or video. Manual observation is the most common method used because of the expense of reducing video data. If the behaviors chosen are difficult to observe, video may be the only feasible method. Engineers usually must custom design data collection forms for each specific study, since the MOEs vary. The training of observers is perhaps the most critical aspect of performing behavioral studies. This is true for both manual and video data collection. The behaviors are coded by observing the actions of the pedestrians and vehicles and recording the MOEs of interest. It is critical that each observer code the same behavior (MOE) the same way. This is referred to as *interrater reliability.* Agencies can check interrater reliability by having two or more data collectors observe the same events, independently code what they see, and compare their results. Differences are resolved by a trained observer. The data collectors must practice until they reach an agreement level of 95% or higher on every MOE. This training is best done using a videotape of a pedestrian crossing similar to that to be studied. Electronic count boards or hand-held computers may be used in place of tally sheets. Observers must mark each button for the MOE it represents. Computer software provides for quick and accurate data reduction.

Collect and Analyze the Data

Engineers must carefully supervise data collection to ensure that the data collection plan and schedule are followed closely. Observers must be relieved frequently to avoid fatigue and subsequent errors in judgment. Observers must carefully note conditions at the data collection sites to prevent an atypical event or situation from confounding the study. It is important that changes in pedestrian and driver behaviors be explained by exactly what caused them. Analysis of the data should follow a preconceived plan. Most behavioral studies employ a "before/after with control" type of design. The assistance of a statistician in designing an analysis plan is generally wise. Appendix C contains guidance on data analysis.

REFERENCES

BERGER, W. G. AND R. L. KNOBLAUCH (1975). *Urban Pedestrian Accident Countermeasures Experimental Evaluation,* Vols. I and II, DOT HS-801 346/347, U.S. Department of Transportation, Washington, DC, February.

BERGER, W. G. AND H. D. ROBERTSON (1976). *Measures of Pedestrian Behavior at Intersections,* Transportation Research Record 615, Transportation Research Board, Washington, DC.

Cameron, R. M. (1977). Pedestrian Volume Characteristics, *Traffic Engineering,* Institute of Transportation Engineers, Washington, DC, January.

Davis, S. E., L. E. King, and H. D Robertson (1988). *Predicting Pedestrian Crosswalk Volumes,* Transportation Research Record 1168, Transportation Research Board, Washington, DC.

Davis, S. E., H. D. Robertson, and L. E. King (1989). *Pedestrian/Vehicle Conflicts: An Accident Prediction Model,* Transportation Research Record 1210, Transportation Research Board, Washington, DC.

Kell, J. H. (1991). Transportation planning studies, Chapter 2 in *Transportation Planning Handbook,* Prentice Hall, Englewood Cliffs, NJ.

Knoblauch, R. L. (1984). *Pedestrian Characteristics and Exposure Measures,* Transportation Research Record 959, Transportation Research Board, Washington, DC, pp. 35–41.

McShane, W. R. and R. P. Roess (1990). *Traffic Engineering,* Prentice Hall, Englewood Cliffs, NJ, pp. 87–88.

Mingo, R., H. D. Robertson, and S. E. Davis (1988a). *Measuring Pedestrian Volumes: A User's Manual,* Vol. III, FHWA/IP-88-030, Federal Highway Administration, Washington, DC.

Mingo, R., S. E. Davis, L. E. King, and H. D. Robertson (1988b). *Pedestrian/Vehicle Accident Prediction Model: A User's Manual,* Vol. IV, FHWA/IP-88-031, Federal Highway Administration, Washington, DC, March.

Mudaly, S. S. (1980). Improved microprocessor controlled pedestrian data acquisition system, *Electronics Letter,* Vol. 16, No. 1, January.

Older, S. J., and G. B. Grayson (1972). *Perception and Decision in the Pedestrian Task,* Transportation and Road Research Laboratory, Crowthorne, England.

Pline, J. L. (1992). Traffic studies, Chapter 2 in *Traffic Engineering Handbook,* Prentice Hall, Englewood Cliffs, NJ.

Robertson, H. D. (1977a) *Pedestrian Signal Displays: An Evaluation of Word Message and Operation,* Transportation Research Record 629, Transportation Research Board, Washington, DC.

Robertson, H. D. (1977b). Pedestrian preferences for symbolic signal displays, *Transportation Engineering,* Institute of Transportation Engineers, Washington, DC, June.

Robertson, H. D. (1977c). *Urban Intersection Improvements for Pedestrian Safety,* Vol. IV, *Pedestrian Signal Displays and Operation,* FHWA-RD-77-145, Federal Highway Administration, Washington, DC, December.

Robertson, H. D. (1983) *Signalized Intersection Controls for Pedestrians,* University Microfilms International, Ann Arbor, MI.

Robertson, H. D., W. G. Berger, and R. F. Pain (1977). *Urban Intersection Improvements for Pedestrian Safety,* Vol. II, *Identification of Safety and Operational Problems at Intersections,* FHWA-RD-77-143, Federal Highway Administration, Washington, DC, December.

Snyder M. and R. L. Knoblauch (1971). *Pedestrian Safety: The Identification of Precipitating Factors and Possible Countermeasures,* Operations Research, Inc., Silver Spring, MD, January.

USDOT (1988). *Manual on Uniform Traffic Control Devices for Streets and Highways,* Federal Highway Administration, Washington, DC, pp. 4C-4 to 4C-5.

Vozzolo, D. and J. Attanucci (1982). *An Assessment of Automatic Passenger Counters: Interim Report,* Multisystems, Cambridge, MA, September.

Zegeer, C. V., K. S. Opiela, and M. J. Cynecki (1982). *Effect of Pedestrian Signals and Signal Timing on Pedestrian Accidents,* Transportation Research Record 847, Transportation Research Board, Washington, DC, pp. 62–72.

Zegeer, C. V., M. J. Cynecki, and K. S. Opiela (1984). *Evaluation of Innovative Pedestrian Signal Alternatives,* Transportation Research Record 959, Transportation Research Board, Washington, DC, pp. 7–18.

14

Traffic Control Device Studies

Donna C. Nelson, Ph.D., P.E.

Introduction

The *Manual of Uniform Traffic Control Devices* (MUTCD) defines *traffic control devices* (TCDs) as "all signs, signals, markings, and devices placed on, over, or adjacent to a street or highway by authority of a public body or official having jurisdiction to regulate, warn or guide traffic" (FHWA, 1988). The general purpose of traffic control devices is to provide visual information to the road user. TCDs are used to help ensure the safe, orderly, and efficient movement of all types of traffic.

Devices are classified into groups that regulate, guide, or warn traffic. *Regulatory devices* inform the road user of regulations that are in force, instruct the road user to take some action, prohibit or permit the road user from making certain maneuvres, and assign the right-of-way. *Warning devices* provide notice of unexpected conditions. They draw attention to the presence of geometric features with potential hazards, major changes in roadway character, obstructions or other physical hazards in or near the roadway, and areas where hazards may exist under certain conditions. They inform the motorist of regulatory controls ahead and advise drivers of appropriate actions. *Guide devices* typically are used to identify routes, provide traveler directions, delineate the roadway, and provide information on facilities, services, points of interest, and political boundaries.

In the United States, the *Manual of Uniform Traffic Control Devices* (MUTCD) defines the basic principles that govern the design and usage of traffic control devices. The MUTCD presents traffic control device standards for streets and highways open to public travel, regardless of the type, class, or governmental agency having jurisdiction.

While the MUTCD is not a statute, it carries the power of a statute in defining national standards. Many jurisdictions adopt the MUTCD without revision; others modify or eliminate specific designs, applications, or requirements by state legislative action. Frequently, modifications reflect more stringent requirements than the minimum expressed in the MUTCD. Equivalent state and local manuals that meet or exceed the MUTCD's minimum requirements also carry the power of a statute (FHWA, 1990).

The MUTCD sets out general requirements for the design, placement, operation, and maintenance of effective TCDs (Doughty, 1982).

- *Design:* TCDs must be designed with the combination of physical features (size, color, and shape) needed to command attention and convey the correct message.
- *Placement:* Devices are placed to fall within the road user's cones of vision so that they are able to command attention and allow time for driver response.
- *Operation:* Devices must be employed in a way that meets traffic requirements in a uniform and consistent manner, fulfills a need, commands respect, and allows time for response.
- *Maintenance:* Devices must be maintained to retain legibility and visibility. Devices that are obsolete or are no longer needed should be removed.

Application and operation of TCDs should be uniform. Similar devices should be used for similar situations and in similar locations to minimize road user confusion and gain their confidence. The inappropriate or overuse of TCDs can lead to a number of problems including driver disregard, increased delay, excess fuel consumption, increased vehicle emissions, and increased accidents. Control devices should supplement each other by providing a meaningful message to motorists and should be designed and placed so that they stand out from the environment.

TCD STUDIES

TCDs are studied for a wide range of reasons. Typically studies are conducted to:

- Support warrants for the installation or removal of TCDs.
- Determine the effectiveness of existing TCDs.
- Assess the condition of TCDs.
- Assess ongoing maintenance and improvement programs.

TCD studies may be conducted at any location where excessive delays, excessive speed, or other traffic problems have been observed; where there have been citizen comments or complaints; or where analysis has indicated that traffic control will be needed to accommodate future demand. Most state agencies have developed their own procedures and guidelines for conducting engineering studies based on state policies. In addition, both federal and state funding programs carefully define the type of project studies required to be eligible to receive financial assistance in implementing traffic control and safety improvements (FHWA, 1990).

Types of Studies

A wide variety of studies may be conducted to collected information regarding TCDs. The most common studies include roadway conditions, accident studies, volume studies, spot speed studies, delay studies, gap distribution, and TCD inventories. The planning and implementation of many of these studies are described in earlier chapters. The information presented below focuses on the application of these studies to the study of TCDs.

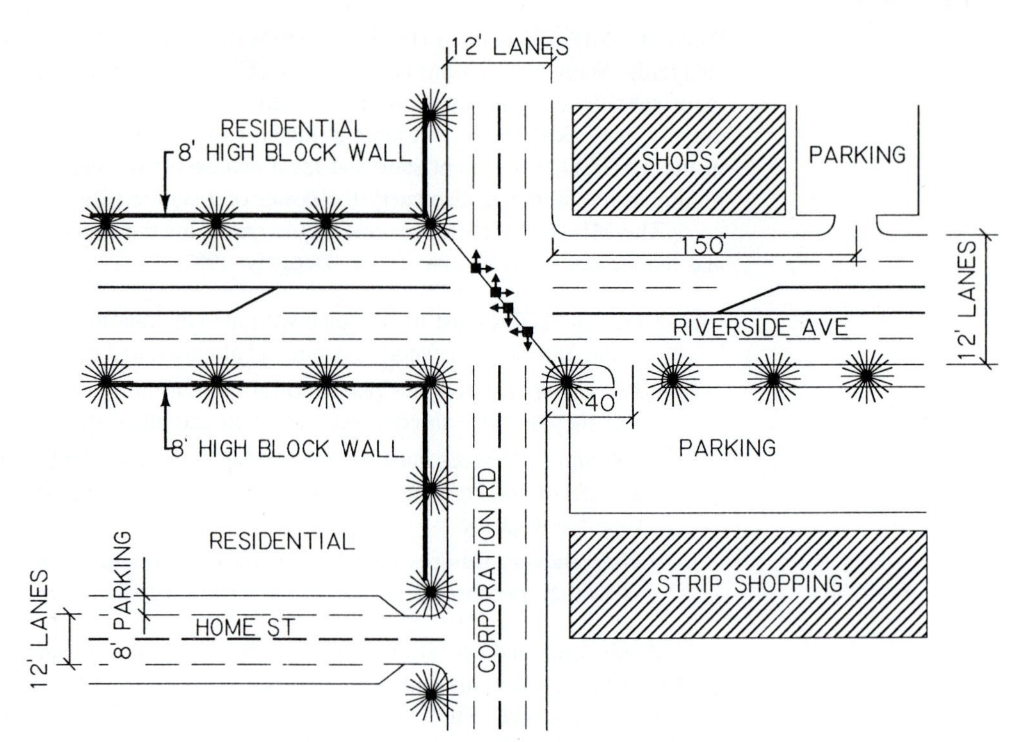

Figure 14-1 Roadway conditions diagram.

Roadway Conditions

Most traffic studies require a thorough description of the study site. The extent and detail of the information needed depends on the analysis to be performed. A conditions diagram and location plan shows the details of the physical layout, including such features as intersection geometrics, channelization, grades, sight-distance restrictions, bus stops and routes, parking conditions, pavement markings, signs, signals, street lighting, driveways, location of nearby railroad crossings, distance to nearest signals, utility poles and fixtures, and adjacent land use. An example is shown in Figure 14-1.

Accident Studies

Accident information is used to identify and analyze high accident locations, conduct before and after studies, evaluate requests for additional traffic control, evaluate requests for additional traffic control, evaluate roadway features, identify and rank improvement projects, establish and maintain traffic regulatory devices, and identify need for police surveillance and enforcement (FHWA, 1990). The information typically required includes collision diagrams, spot accident location maps, and accident rates. Collision diagrams summarize accident experience of a specific location by accident type, location, direction of vehicle movement, accident severity, time of day, date, and day of week. A single collision diagram should summarize data for at least one year (preferably 3 to 5 years). The construction of collision diagrams similar to the one shown in Figure 14-2 is described in detail in Chapter 11.

Spot accident location maps are used to identify high accident locations. A spot map is a quick visual method of identifying locations and accidents that may warrant detailed analysis. Accident locations are marked on a map. This is an excellent application for geographic information systems (GIS) technology. The type, severity, and time of accident may be coded using colors and or symbols.

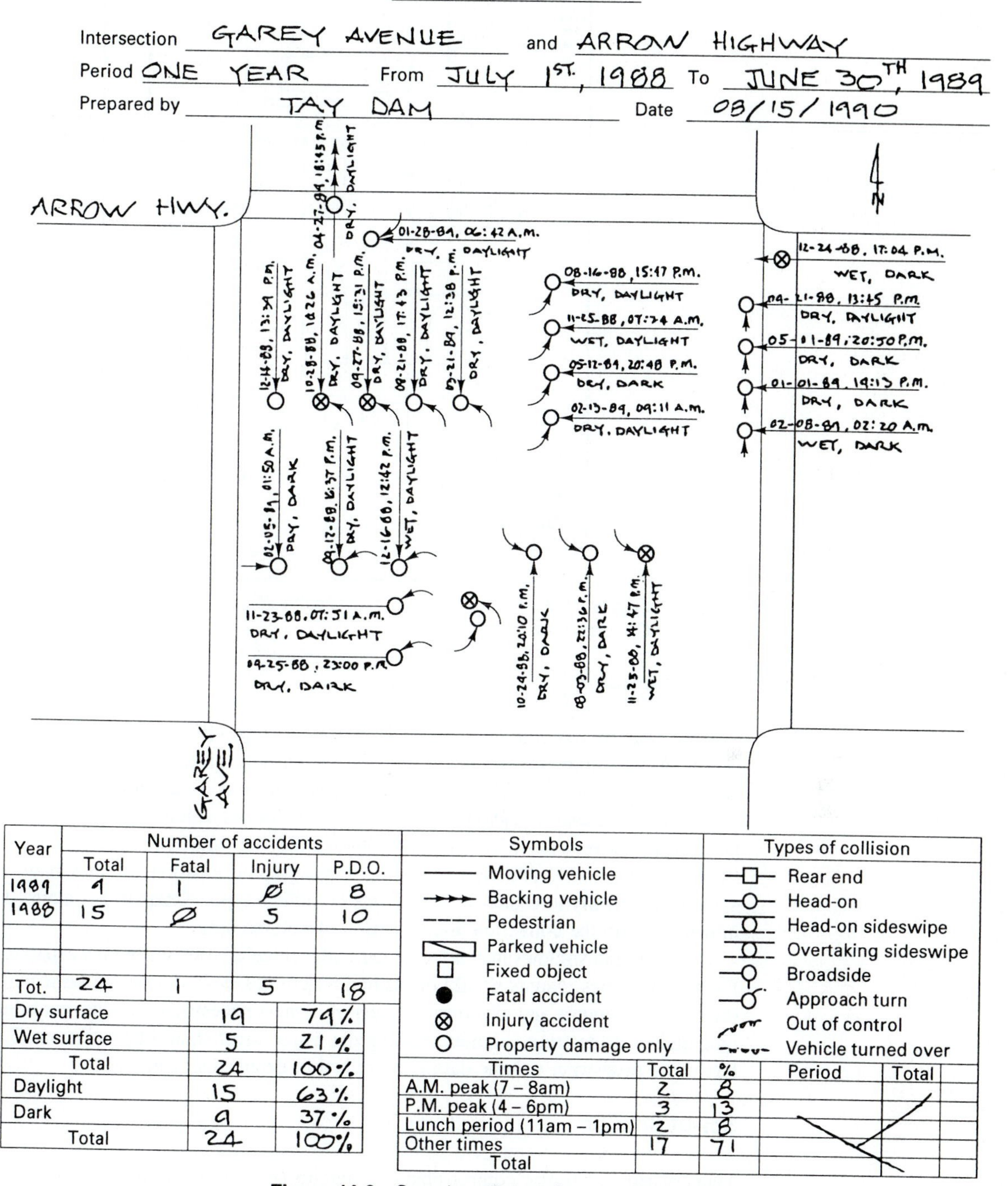

City of Pomona,
Public Works Department
Traffic Engineering Division

COLLISION DIAGRAM

Intersection GAREY AVENUE and ARROW HIGHWAY

Period ONE YEAR From JULY 1ST, 1988 To JUNE 30TH, 1989

Prepared by TAY DAM Date 08/15/1990

Year	Number of accidents			
	Total	Fatal	Injury	P.D.O.
1989	9	1	Ø	8
1988	15	Ø	5	10
Tot.	24	1	5	18

Dry surface	19	79%
Wet surface	5	21%
Total	24	100%
Daylight	15	63%
Dark	9	37%
Total	24	100%

Symbols	
———	Moving vehicle
→→→	Backing vehicle
- - - -	Pedestrian
▭	Parked vehicle
□	Fixed object
●	Fatal accident
⊗	Injury accident
○	Property damage only

Types of collision	
—□—	Rear end
—○—	Head-on
	Head-on sideswipe
	Overtaking sideswipe
	Broadside
	Approach turn
	Out of control
	Vehicle turned over

Times	Total	%	Period	Total
A.M. peak (7 – 8am)	2	8		
P.M. peak (4 – 6pm)	3	13		
Lunch period (11am – 1pm)	2	8		
Other times	17	71		
Total				

Figure 14-2 Sample collision diagram. *Source*: City of Pomona, CA.

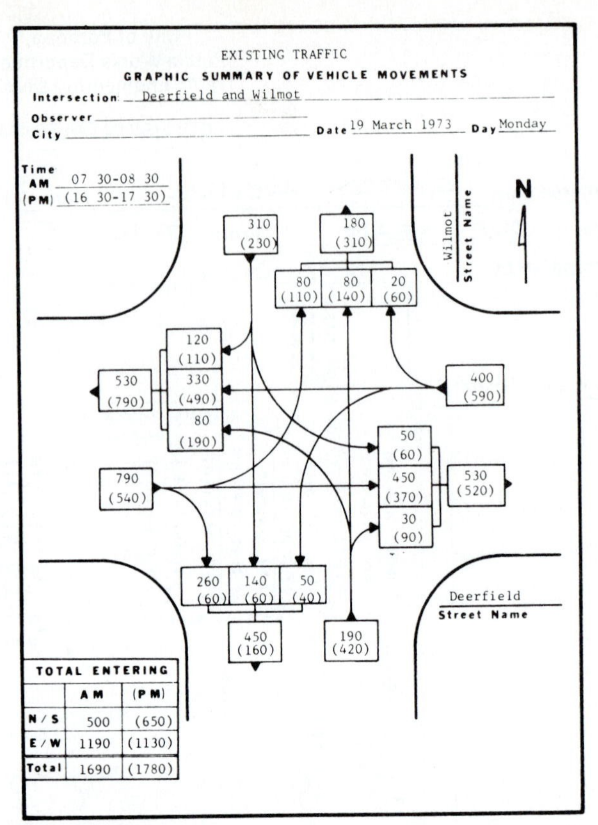

Figure 14-3 Presentation of intersection volumes. *Source:* Box and Oppenlander, 1976.

Volume Studies

Volume data are required for most TCD studies. The specific volume data required and details of data collection are determined by the purpose of the study. Traffic counts are an important factor in evaluating improvements and recommendations. Vehicular volumes can be presented as shown in Figure 14-3. Vehicular volumes may be collected and grouped by movement or by approach. Studies may be limited to specific time periods (such as peak hour), direction of travel, or geographic location. Studies may be conducted specifically to establish axle counts or vehicle types. Chapter 2 contains detailed information on the collection and reduction of volume data.

Pedestrian volumes may be recorded with vehicular volumes or on a separate sheet. For signal studies, counts should be taken on each crosswalk during the same periods as the vehicular counts and also during hours of highest pedestrian volume. For other warrant studies, pedestrian volumes crossing the major street should be sufficient. The collection of pedestrian volume data is described in detail in Chapter 13.

Speed Studies

Many aspects of traffic control planning require speed distribution information. Speed distributions are commonly used to establish maximum and minimum speed limits; to determine the need for posting safe speeds at curves; to determine the proper location of regulatory, warning, and guide signs; to establish the boundaries of no passing zones; and for the analysis of special operational situations (e.g., work zones and school areas).

Spot speed studies are made by measuring the individual speeds of a sample of vehicles passing a given point (spot) on a street or highway. These individual speeds are used to estimate the speed distribution of the entire traffic stream at the location under

43	53	54	61	63
58	41	58	43	38
46	63	34	31	37
47	52	53	46	44
45	42	47	48	49
45	32	61	37	36
57	36	54	62	43
48	39	62	37	59
51	55	48	42	47
44	34	54	39	37
46	43	47	51	47
53	50	37	32	32
52	47	50	47	54
36	37	42	61	64
57	58	48	59	53

Mean 47.6 Samples (n)= 75
Sum 3567

Category (1)	Actual Samples (2)	Upper limit of Range	Frequency Distribution	Cumulative Frequency Distribution
<35 kph	6	35	8%	8%
35.1-40	12	40	16%	24%
40.1-45	12	40	16%	40%
45.1-50	16	50	21%	61%
50.1-55	13	55	17%	79%
55.1-60	0	60	11%	89%
>60.1	8	60+	11%	100%
Total	75			

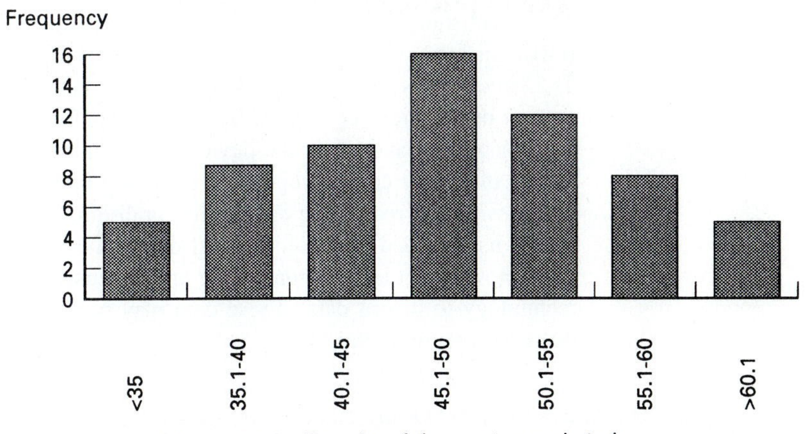

Figure 14-4 Results of the spot speed study.

the conditions prevailing at the time of the study. The results of a spot-speed study are shown in Figure 14-4. Spot speed studies are usually conducted during off-peak average hours; however, the period during which speeds are measured depends on the purpose of the study. When spot speeds are needed to determine the 85th percentile speed of specific roadway sections and/or intersection approaches, measurements for low-speed intersection approaches (15 to 25 mph) should be made 150 to 200 feet before the intersection. On high-speed approaches (50 to 65 mph), the checkpoint should be located 800 to 1200 feet before the intersection (Kell and Fullerton, 1982). Spot speed studies are discussed in Chapter 3.

Delay Studies

TCD studies may also require information on the amount of delay encountered by vehicles at signalized and unsignalized intersections. Two methods, stopped-time delay and the travel-time method, are commonly used to measure the delay at intersections. The stopped-time delay method consists of determining the amount of time that vehicles are actually stopped at the intersection. The amount of stopped time can be determined by visual observation, the use of delay meters, or from a timed series of photographs taken from a suitable vantage point. Various methods are used to determine the travel time through an intersection. This can be accomplished by observation and stopwatch timing, by using a test vehicle through the area, by using a 20-pen recorder with road tubes at critical points, and by a timed series of photographs taken from a suitable vantage point. The study of delay at intersections is discussed in Chapter 5.

Gap Distributions

The ability of vehicles to enter a major street from a side street or driveway, and often, the ability of pedestrians to cross at an unsignalized location depends on the distribution of gaps in the traffic stream. If gaps of adequate length are infrequent, it may result in unacceptable delay for vehicles attempting to enter the traffic stream. Some newer road counters can record the gap distribution as well as the axle count, by time of day. Gap distribution can also be observed in the field by a person with a portable or "notebook" computer. Gap acceptance and gap studies are discussed in Chapter 5.

TCD Inventories

The establishment and maintenance of TCD inventories is a necessary part of day-to-day traffic operations. Chapter 6 focuses on the planning and implementation of inventories. TCD inventories for signs, markings, and signals should contain, as a minimum, information on location, condition, TCD type, MUTCD or state code, message (for signs), mounting type, and when the device was installed and last inspected. A traffic control device inventory can be conducted manually using a trained person or crew to inspect each device and record the data on its condition, location, if it meets standards, and if it serves its intended purpose. Manual data collection may work well for small jurisdictions; however, it is time consuming and expensive. Larger jurisdictions are using video technology to record data. These data may be stored in video libraries or on computer disks for easy retrieval.

ESTABLISHING THE NEED FOR TRAFFIC CONTROL DEVICES

A warrant is a set of criteria that is used to define the relative need for a particular device and is intended to assure motorist and pedestrian safety and convenience. Warrants are often expressed as numerical requirements, such as the volume of vehicular or pedestrian traffic. Some are, however, discussed as general policy statements rather than as absolute warrants. A warrant normally includes a means of assigning priorities among several alternative choices. In the MUTCD (1988), two principles guide the development and implementation of warrants:

1. That the most effective traffic control device is that which is the least restrictive while still accomplishing the intended purpose.
2. That driver response to the influences of a TCD has previously been identified by observation, field experience, and laboratory test under a variety of traffic and driver conditions.

The MUTCD presents warrants as a series of guidelines that should be used to help evaluate the situation at hand, not as absolute values. The satisfaction of a warrant does not guarantee that a TCD is needed. Similarly, failure to fully satisfy a specific a warrant is not positive proof that the device could not serve a useful purpose. The application of warrants is effective only when combined with knowledgeable engineering judgment. The MUTCD describes warrants, design, and placement criteria for a wide range of TCDs. Warrants for signs, markings, and signals are summarized below. In addition, state and local jurisdictions may have developed their own warrants for applications not included in the MUTCD, including loading zones and speed bumps.

Signals

When justified and properly designed, a traffic signal installation may:

- Reduce the frequency of certain types of accidents.
- Effect orderly traffic movement.
- Provide for the continuous flow of a platoon of traffic through proper coordination at a definite speed along a given route.
- Allow other vehicles and pedestrians to cross a heavy traffic stream.
- Control traffic more economically than by manual methods.

Traffic signals do not always have a positive effect on roadway operations. An unjustified, poorly designed, improperly operated, or poorly maintained traffic signal may result in increased accident frequencies, excessive delay, motorist disregard, decreased capacity, and circuitous travel by alternative routes. Experience has indicated that although the installation of signals may result in a decrease in the number of right-angle collisions, there may be an increase in the number of rear-end collisions. Consequently, a thorough study of traffic and roadway conditions should precede the installation and the selection of equipment.

Warrants

The MUTCD describes 11 warrants for the installation of signals. The requirements of these warrants are summarized below. The MUTCD or the equivalent local manual contains the complete warrants and should be consulted in evaluating any warrants for signal installation. Traffic signals should not be installed unless one or more of the signal warrants are met.

Warrant 1: Minimum Vehicular Volume

Warrant 1 is satisfied when for each of any 8 hours of an average day, the traffic volumes given in Table 14-1 exist on the major street and on the higher-volume minor-street approach to the intersection. These major- and minor-street volumes are for the

TABLE 14-1
Minimum vehicular volumes for warrant 1

Number of lanes for moving traffic on each approach		*Vehicles per hour on major street total both approaches*	*Vehicles per hour on higher volume minor street approach (one direction only)*
Major street	*Minor street*		
1	1	500	150
2 or more	1	600	150
2 or more	2 or more	600	200
1	2 or more	500	200

Source: Kell and Fullerton, 1982.

TABLE 14-2
Minimum vehicular volumes for warrant 2

Number of lanes for moving traffic on each approach		*Vehicles per hour on major street Total both approaches*	*Vehicles per hour on higher volume minor street approach (one direction only)*
Major street	*Minor street*		
1	1	750	75
2 or more	1	900	75
2 or more	2 or more	900	100
1	2 or more	750	100

Source: Kell and Fullerton, 1982.

same 8 hours. During each hour, the higher approach volume on the minor street is considered, regardless of its direction. When the 85th percentile speed of major street traffic exceeds 40 mph (64 km/h) or when the intersection lies within the built-up area of an isolated community having a population less than 10,000, the minimum vehicular volume warrant is 70% of the requirement above.

Warrant 2: Interruption of Continuous Traffic

Warrant 2 applies to operating conditions where the traffic volume on a major street is so heavy that traffic on a minor intersecting street suffers excessive delay or hazard in entering or crossing the major street. This warrant is satisfied when for each of any 8 hours of an average day, the traffic volumes given in Table 14-2 existing on the major street and on the higher-volume minor street approach to the intersection and where the signal installation will not seriously disrupt progressive traffic flow. These major street and minor street volumes are for the same 8 hours. During each hour, the higher volume on the minor street is considered, regardless of its direction. When the 85th percentile speed of major-street traffic exceeds 40 mph in either an urban or rural area, or when the intersection lies within the built-up area of an isolated community having a population of less than 10,000, the interruption of continuous traffic warrant is 70% of the requirements in Table 14-2. A reduced volume, similar to that described under warrant 1, can be used in place of those shown in Table 14-2 on higher-speed roads or in smaller communities.

Warrant 3: Minimum Pedestrian Volume

Warrant 3 is satisfied when for each of any 8 hours of an average day, the following traffic volumes exist:

1. On the major street, 600 or more vehicles per hour (vph) enter the intersection (total of both approaches), or 1000 vph or more enter the intersection (total of both approaches) on the major street, where there is a raised median island 4 feet (1.22 meters) or more in width.

2. During the same 8 hours there are 150 or more pedestrians per hour on the highest-volume crosswalk crossing the major street.

When the 85th percentile speed of major-street traffic exceeds 40 mph in either an urban or rural area, or when the intersection lies within the built-up area of an isolated community having a population of less than 10,000, the minimum pedestrian volume warrant is 70% of the requirements above.

Warrant 4: School Crossings

Warrant 4 may be considered as a special case of the pedestrian warrant. The warrant is satisfied for an established school crossing when a gap study shows that the number of adequate gaps in the traffic stream during the period when the children are using the crossing is less than the number of minutes in the same period.

Warrant 5: Progressive Movement

Warrant 5 expresses the desirability of holding traffic in compact platoons. It is satisfied when:

1. On a one-way street or a street that has predominantly unidirectional traffic, the adjacent signals are so far apart that they do not provide the necessary degree of vehicle platooning and speed control.
2. On a two-way street, adjacent signals do not provide the necessary degree of platooning and speed control and the proposed and adjacent signals could constitute a progressive signal system.

The installation of a signal according to this warrant should be based on the 85th percentile speed unless an engineering study indicates that another speed is more desirable. According to this warrant, the installation of a signal should not be considered where the resulting signal spacing would be less than 1000 feet (305 meters).

Warrant 6: Accident Experience

Warrant 6 is satisfied when:

1. An adequate trial of less restrictive remedies with satisfactory observance and enforcement has failed to reduce the accident frequency.
2. Five or more reported accidents of types susceptible to correction by traffic signal control have occurred within a 12-month period, each accident involving personal injury or property damage to an apparent extent of $100 or more.
3. There exists a volume of vehicular traffic not less than 80% of the requirements specified in warrants 1, 2, and 3.
4. The signal installation will not seriously disrupt progressive traffic flow.

Warrant 7: Systems Warrant

Warrant 7 allows both the major and minor streets to be given equal consideration as major routes. It is satisfied when two or more major routes meet at a common intersection and the total existing or immediately projected entering volume is a least 800 vehicles during the peak hour of a typical weekday or each of any 5 hours of a Saturday and/or Sunday.

Warrant 8: Combination of Warrants

In exceptional cases, signals may occasionally be justified when no single warrant is satisfied but where two or more of warrants 1, 2, and 3 are satisfied to the extent of 80% or more of the stated values. Adequate trials of other remedial measures that cause

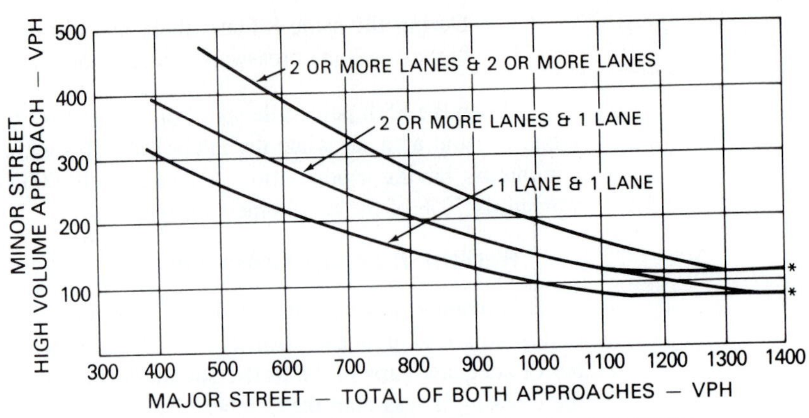

Figure 14-5 Four-hour volume warrant. *Source: Manual of Uniform Traffic Control Devices,* 1988.

less delay and inconvenience to traffic should precede installation of signals under this warrant.

Warrant 9: Four-Hour Volumes

Warrant 9 is satisfied when each of any 4 hours of an average day the plotted points representing the vehicles per hour on the major street (total of both approaches) and the corresponding vehicles per hour on the higher-volume minor-street approach (one direction only) all fall above the curve shown in Figure 14-5 for the existing combination of approach lanes. When the 85th percentile speed of the major-street traffic exceeds 40 miles per hour or when the intersection lies within a built-up area of an isolated community having a population less than 10,000, the 4-hour volume requirement is satisfied when the plotted points referred to fall above the curve in Figure 14-6 for the existing combination of approach lanes.

Warrant 10: Peak-Hour Delay

Warrant 10 is intended for application where traffic conditions are such that for 1 hour of the day minor-street traffic suffers undue delay in entering or crossing the major street. The warrant is satisfied when the conditions given below exist for 1 hour (any four consecutive 15-minute periods) of an average weekday. The peak-hour delay warrant is met when:

1. The total delay experienced by the traffic on one minor-street approach (one direction only) controlled by a stop sign equals or exceeds 4 vehicle-hours for a one lane approach and 5 vehicle-hours for a two-lane approach.
2. The volume on the same minor-street approach (one direction only) equals or exceeds 100 vph for one moving lane of traffic or 150 vph for two moving lanes.
3. The total entering volume during the hour equals or exceeds 800 vph for intersections with four (or more) approaches or 650 vph for intersections with three approaches.

Warrant 11: Peak-Hour Volume

Warrant 11 applies when traffic conditions are such that for 1 hour of the day, minor-street traffic suffers undue traffic delay in entering or crossing the major street. It is satisfied when:

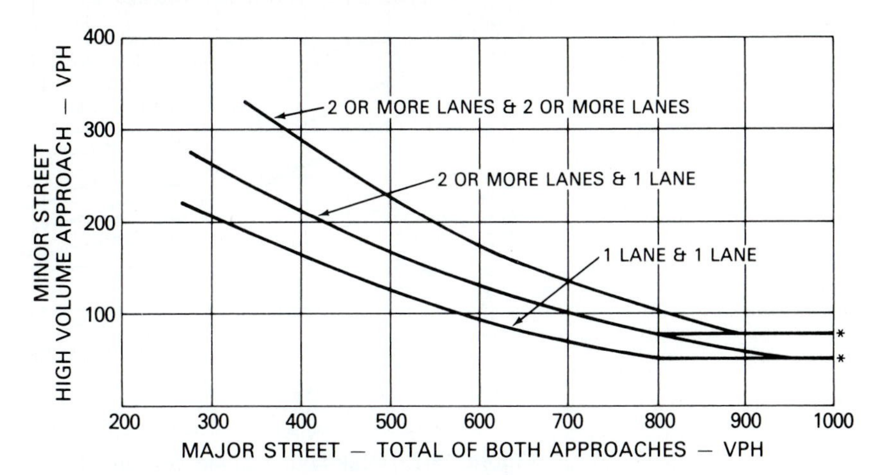

*NOTE: 80 VPH APPLIES AS THE LOWER THRESHOLD VOLUME FOR A MINOR STREET APPROACH WITH TWO OR MORE LANES AND 60 VPH APPLIES AS THE LOWER THRESHOLD VOLUME FOR A MINOR STREET APPROACHING WITH ONE LANE.

Figure 14-6 Four-hour volume warrant for communities of less than 10,000 population or where speeds on major street are above 40 mph. *Source: Manual of Uniform Traffic Control Devices,* 1988.

1. The plotted point representing the vehicles per hour on the major street (total of both approaches) and the corresponding vehicle per hour of the higher volume minor street approach (one direction only) for one hour (any four consecutive 15-minute periods) of an average day falls above the curve in Figure 14-7 for the existing combination of approach lanes.
2. The plotted point referred to falls above the curve in Figure 14-8 for the existing combination of approach lanes if the 85th percentile speed of major street traffic exceeds 40 mph or if the intersection lies within a built-up area of an isolated community having a population less than 10,000.

BASIC DATA REQUIREMENTS FOR WARRANT EVALUATION

The basic data requirements for the evaluation of signal warrants are summarized below. If capacity analysis, signal timing, or other analyses are performed, the data collected may be expanded as needed.

- *Approach volumes* for each hour during 16 consecutive hours of a representative day. The 16 hours selected should contain the greatest percentage of the 24-hour traffic.
- *Traffic movement counts* for each approach, classified by vehicle type (heavy trucks, passenger cars and light trucks, and public-transit vehicles). Counts are needed for each 15-minute period of the two hours in the morning and of the two hours in the afternoon during which total traffic entering the intersection is greatest.
- *Pedestrian volume counts* on each crosswalk during the same periods as the vehicular counts in paragraph 2 of warrant 10 and also during hours of highest pedestrian volume. Where young or elderly persons need special consideration, the pedestrians may be classified by general observation and recorded by age groups as follows:

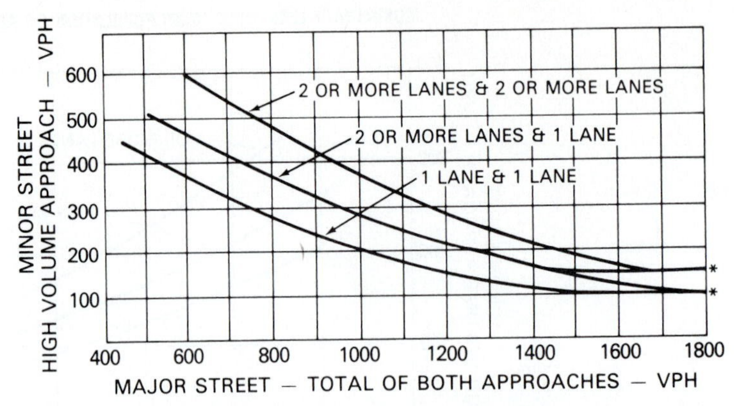

*NOTE: 150 VPH APPLIES AS THE LOWER THRESHOLD VOLUME FOR A MINOR STREET APPROACH WITH TWO OR MORE LANES AND 100 VPH APPLIES AS THE LOWER THRESHOLD VOLUME FOR A MINOR STREET APPROACHING WITH ONE LANE.

Figure 14-7 Peak-hour volume warrant. *Source: Manual of Uniform Traffic Contol Devices,* 1988.

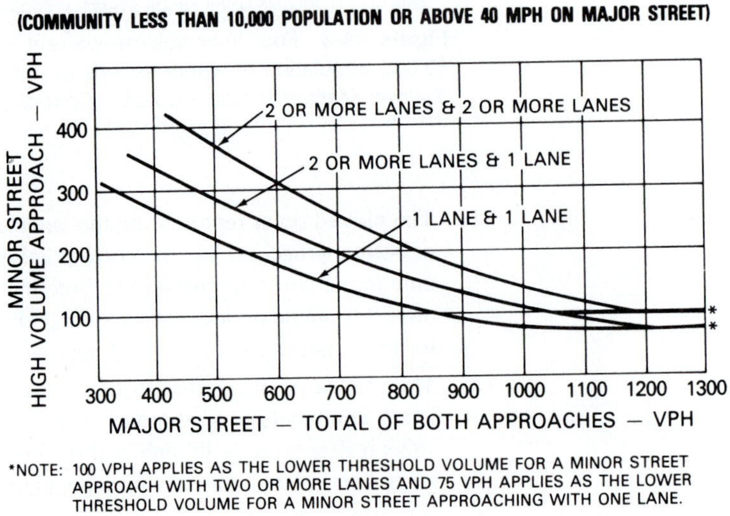

(COMMUNITY LESS THAN 10,000 POPULATION OR ABOVE 40 MPH ON MAJOR STREET)

*NOTE: 100 VPH APPLIES AS THE LOWER THRESHOLD VOLUME FOR A MINOR STREET APPROACH WITH TWO OR MORE LANES AND 75 VPH APPLIES AS THE LOWER THRESHOLD VOLUME FOR A MINOR STREET APPROACHING WITH ONE LANE.

Figure 14-8 Peak-hour volume warrant for communities with less than 10,000 population or speeds over 40 mph on major street. *Source: Manual of Uniform Traffic Control Devices,* 1988.

a. Under 13 years
b. 13 to 60 years
c. Over 60 years

- The *85th percentile speed* of all vehicles on the uncontrolled approaches to the location.
- A *collision diagram* showing accident experience by type, location, direction of movement, severity, time of day, date, and day of week for at least 1 year.
- A *conditions diagram* showing details of the physical layout, including intersection geometrics, channelization, grades, sight-distance restrictions, bus stops and routes, parking conditions, pavement markings, street lighting, driveways, location of nearby railroad crossings, distance to nearest signals, utility poles and fixtures, and adjacent land use.
- *Pedestrian group size and gaps for warrant 4.* The ITE report ''A Program for School Crossing Protection'' describes the procedure for collection data for this warrant (ITE, 1972).

Survey Date ________
Location ________________ Cross walk across ________________

End of Survey (to nearest minute) ______ Start of Survey (to nearest minute) ______ Total Survey time (Minutes) ______			Number of lanes 'N' ______ Roadway Width 'W' ______ Adequate gap time 'G' ______	
Gap Size Seconds	**Number of Gaps Tally**	**Total**	**Multiply by Gap size**	**Computations**
8				
9				G = 1+ [W/35} + 2N
10				
11				G = ________sec
12				
13				
14				T = total survey
15				time X 60
16				
17				T = ________sec
18				
19				
20				D = [(T-t)/T] X 100
21				
22				D = ________%
23				
24				
25				
26				t = total time of all
27				gaps equal to or
28				greater than G
29				
30				
31				
32				
33				
34				
35				
36				
37				
38				
39				
40				
41				
42				
43				
	Totals			

Figure 14-9 Gap size study form for warrant 4. *Source:* Kell and Fullerton, 1982

The pedestrian group size and gap study consists of two field studies: (1) pedestrian group sizes, and (2) vehicular gap sizes. Data for vehicular gaps can be collected and analyzed using the forms in Figures 14-9 and 14-10. The procedure is described below. The adequate gap time between vehicles on the major roadway (G) is estimated using the equation

$$G = R + \frac{W}{3.5} + 2(N - 1)$$

where

G = adequate gap time, seconds
R = reaction time, seconds

Survey date ____________

Crosswalk surveyed ____________

Crosswalk across? ____________

Divided Roadway? Yes No

Curb to curb distance ____________

Group size	Number of rows	Number of groups		Cumulative	Computations
		Tally	Total		
5 or fewer	1				
6 to 10	2				
11 to 15	3				
16 to 20	4				
21 to 25	5				
26 to 30	6				
31 to 35	7				
36 to 40	8				
41 to 45	9				
46 to 50	10				

Time period studied ____________
Number of adequate traffic gaps ____________
Number of minutes in the same period ____________
Is the warrant satisfied? ____________

Figure 14-10 Pedestrian group study form for warrant 4. *Source:* Institute of Transportation Engineers. *A Program for School Crossing Protection: A Recommended Practice*, 3rd Edition, 1972.

W = width of street, feet
N = number of rows of pedestrians crossing in the 85th percentile group

This equation uses a walking time based on 3.5 feet per second. The walking speed may be changed to more accurately reflect observed conditions. R is the perception reaction time: the time required to look both ways, make a decision, and commence the walk. A commonly used value is 3 seconds. The term $2(N - 1)$ represents the pedestrian platoon time. Children are assumed to cross a street in rows of five with an interval of 2 seconds between each row (ITE, 1972). If a reaction time of 3 seconds is used, the equation above can be simplified to

$$G = 1 + \frac{W}{3.5} + 2N$$

Next, the total survey time (T) of the vehicular gap survey is converted from minutes to seconds:

$$T = \text{total survey time} \times 60$$

Finally, the percentage of the study time during which the gaps are of adequate size (D) is calculated:

$$D = \frac{T - t}{T} \times 100$$

where t is the total time of all gaps equal or greater than G.

The data form in Figure 14-10 provides a convenient format for studying pedestrian group size. In the field, the number of groups of pedestrians are recorded by group size. Because schoolchildren are assumed to cross in groups of five, group sizes are incremented by five. The number of adequate traffic gaps is obtained from Figure 14-9 (gap-size study forms). Warrant 4 is satisfied when the number of gaps in the traffic stream during the time children are using the crossing is less than the number of minutes in the same time period.

Agencies often develop forms to assist in the uniform and careful evaluation of warrants. An example from one state agency is shown in Figure 14-11. The first step in evaluating signal warrants is to compare volume and accident data to warrants 1, 2, or 3. Accidents susceptible to correction by traffic signal control are tested against warrant 6. If the requirements of no single one of these warrants are satisfied, the next logical warrant to examine is the combination warrant (8), which requires only 80% of the stated values of two or more of warrants 1, 2, and 3. Warrant 4 is used only when an established school crossing is under study. Warrants 5 and 7 are special warrants used when it is necessary to control arterial or system flow properly. Warrant 5 is based on the speed-distance relationship with adjacent signals. Warrant 7 is used to complete the network of major routes. Warrants 9, 10, and 11 are targeted at special traffic conditions.

SIGNS

The MUTCD prescribes standards for the traffic signing within the right-of-way of all classes of public highways. Traffic signs fall into three broad functional classifications according to use. These include regulatory signs, warning signs, and guide or informational signs. Signs should be used only where warranted by facts and field studies. Signs are essential where special regulations apply at specified places or at specific times only, or where hazards are not self-evident (TCDH). They also give information as to highway routes, directions, destinations, and points of interest. The MUTCD contains the complete, current warrants for a wide range of regulatory, warning and guide signs. The summaries below are presented for convenience only. The MUTCD or the equivalent local manual must be consulted for specific requirements for specific signs.

Stop sign warrants at an intersection require an examination of collision diagrams and a knowledge of operating conditions at the site. A stop sign may be warranted at an intersection where one or more of the following conditions exist:

1. Intersection of a less important road with a main road, where application of the normal right-of-way rule is unduly hazardous
2. Intersection of a county road, city street, or township road with a state highway
3. Street entering a through highway or street
4. Unsignalized intersection in a signalized area

TRAFFIC SIGNAL WARRANTS

CALC ______ DATE ______

CHK ______ DATE ______

DIST CO RTE PM

Major St: ______ Critical Approach Speed ______ mph

Minor St: ______ Critical Approach Speed ______ mph

Critical speed of major street traffic $\geq$ 40 mph ---------- ☐ OR RURAL (R)

In built up area of isolated community of $\leq$ 10,000 pop. ---------- ☐

☐ URBAN (U)

WARRANT 1 – Minimum Vehicular Volume 100% SATISFIED YES ☐ NO ☐ / 80% SATISFIED YES ☐ NO ☐

	MINIMUM REQUIREMENTS (80% SHOWN IN BRACKETS)											Hour
	U	R	U	R								
APPROACH LANES	1		2 or more		/	/	/	/	/	/	/	
Both Apprchs. Major Street	500 (400)	350 (280)	600 (480)	420 (336)								
Highest Apprch. Minor Street*	150 (120)	105 (84)	200 (160)	140 (112)								

* *NOTE: Heavier left turn movement from Major Street included when LT-phasing is proposed* ☐

WARRANT 2 – Interruption of Continuous Traffic 100% SATISFIED YES ☐ NO ☐ / 80% SATISFIED YES ☐ NO ☐

	MINIMUM REQUIREMENTS (80% SHOWN IN BRACKETS)											Hour
	U	R	U	R								
APPROACH LANES	1		2 or more		/	/	/	/	/	/	/	
Both Apprchs. Major Street	750 (600)	525 (420)	900 (720)	630 (504)								
Highest Apprch. Minor Street*	75 (60)	53 (42)	100 (80)	70 (56)								

**NOTE: Heavier left turn movement from Major Street included when LT-phasing is proposed* ☐

WARRANT 3 – Minimum Pedestrian Volume 100% SATISFIED YES ☐ NO ☐ / 80% SATISFIED YES ☐ NO ☐

	MINIMUM REQUIREMENTS (80% SHOWN IN BRACKETS)											Hour
		U	R									
Both Apprchs. Major Street	No Median	600 (480)	420 (336)	/	/	/	/	/	/	/		
Volume	Raised 4' Median	1000 (800)	700 (560)									
Ped's On Highest Volume X-Walk Xing Major Street		150 (120)	105 (84)									

IF MIDBLOCK SIGNAL PROPOSED ☐

MIN. REQUIREMENT	DISTANCE TO NEAREST ESTABLISHED CRWLK.	FULFILLED
150 Feet	N/E ____ ft S/W ____ ft	Yes ☐ No ☐

The satisfaction of a warrant is not necessarily justification for a signal. Delay, congestion, confusion or other evidence of the need for right of way assignment must be shown.

Figure 14-11 Sample signal warrant evaluation form. *Source:* Kell and Fullerton, 1982

TRAFFIC SIGNAL WARRANTS

WARRANT 4 - School Crossings

Not Applicable ☐
See School Crossings Warrant Sheet ☐

WARRANT 5 - Progressive Movement SATISFIED YES ☐ NO ☐

MINIMUM REQUIREMENTS	DISTANCE TO NEAREST SIGNAL	FULFILLED
> 1000 ft	N ______, S ______ ft, E ______ ft, W ______ ft	YES ☐ NO ☐
ON ONE WAY ISOLATED ST. OR ST. WITH ONE WAY TRAFFIC SIGNIFICANCE AND ADJACENT SIGNALS ARE SO FAR APART THAT NECESSARY PLATOONING & SPEED CONTROL WOULD BE LOST		
ON 2-WAY ST. WHERE ADJACENT SIGNALS DO NOT PROVIDE NECESSARY PLATOONING & SPEED CONTROL. PROPOSED SIGNALS COULD CONSTITUTE A PROGRESSIVE SIGNAL SYSTEM		☐ ☐

WARRANT 6 - Accident Experience SATISFIED YES ☐ NO ☐

REQUIREMENT	WARRANT	✓	FULFILLED
ONE WARRANT	WARRANT 1 - MINIMUM VEHICULAR VOLUME		
SATISFIED	OR WARRANT 2 - INTERRUPTION OF CONTINUOUS TRAFFIC		
80%	OR WARRANT 3 - MINIMUM PEDESTRIAN VOLUME		YES ☐ NO ☐
SIGNAL WILL NOT SERIOUSLY DISRUPT PROGRESSIVE TRAFFIC FLOW			☐ ☐
ADEQUATE TRIAL OF LESS RESTRICTIVE REMEDIES HAS FAILED TO REDUCE ACC. FREQ.			☐ ☐
ACC WITHIN A 12 MON. PERIOD SUSCEPTIBLE OF CORR. & INVOLVING INJURY OR > $200 DAMAGE			
MINIMUM REQUIREMENT	NUMBER OF ACCIDENTS		
5 OR MORE*			☐ ☐

** NOTE: Left turn accidents can be included when LT-phasing is proposed*

WARRANT 7 - Systems Warrant SATISFIED YES ☐ NO ☐

MINIMUM VOLUME REQUIREMENT	ENTERING VOLUMES - ALL APPROACHES	✓	FULFILLED
800 VEH/HR	DURING TYPICAL WEEKDAY PEAK HOUR ______ VEH/HR		
	DURING EACH OF ANY 5 HRS OF A SATURDAY AND/OR SUNDAY ______ VEH/HR		YES ☐ NO ☐

CHARACTERISTICS OF MAJOR ROUTES	MAJOR ST	MINOR ST	
HWY SYSTEM SERVING AS PRINCIPLE NETWORK FOR THROUGH TRAFFIC			
CONNECTS AREAS OF PRINCIPLE TRAFFIC GENERATION			
RURAL OR SUBURBAN HWY OUTSIDE OF, ENTERING, OR TRAVERSING A CITY			
HAS SURFACE STREET FWY OR EXPWAY RAMP TERMINALS			
APPEARS AS MAJOR ROUTE ON AN OFFICIAL PLAN			
ANY MAJOR ROUTE CHARACTERISTICS MET, BOTH STS.			☐ ☐

The satisfaction of a warrant is not necessarily justification for a signal. Delay, congestion, confusion or other evidence of the need for right of way assignment must be shown.

Figure 14-11 Continued.

TRAFFIC SIGNAL WARRANTS

WARRANT 8 – Combination of Warrants SATISFIED YES ☐ NO ☐

REQUIREMENT	WARRANT	✓	FULFILLED
TWO WARRANTS	1 - MINIMUM VEHICULAR VOLUME		
SATISFIED	2 - INTERRUPTION OF CONTINUOUS TRAFFIC		
80%	3 - MINIMUM PEDESTRIAN VOLUME		YES ☐ NO ☐

WARRANT 9 – Four Hour Volume SATISFIED* YES ☐ NO ☐

Approach Lanes	One	2 or more	Hour		
Both Approaches , Major Street					
Highest Approaches , Minor Street					

*Refer to Fig. 9-2A (URBAN AREAS) or Figure 9-2B (RURAL AREAS) to determine if this warrant is satisfied.

WARRANT 10 – Peak Hour Delay SATISFIED YES ☐ NO ☐

1. The total delay experienced for traffic on one minor street approach controlled by a STOP sign equals or exceeds four vehicle-hours for a one-lane approach and five vehicle-hours for a two-lane approach; and

   YES ☐ NO ☐

2. The volume on the same minor street approach equals or exceeds 100 vph for one moving lane of traffic or 150 vph for two moving lanes; and

   YES ☐ NO ☐

3. The total entering volume serviced during the hour equals or exceeds 800 vph for intersections with four or more approaches or 650 vph for intersections with three approaches.

   YES ☐ NO ☐

WARRANT 11 – Peak Hour Volume SATISFIED* YES ☐ NO ☐

Approach Lanes	One	2 or more	Hour
Both Approaches , Major Street			
Highest Approaches , Minor Street			

*Refer to Fig. 9-2C (URBAN AREAS) or Figure 9-2D (RURAL AREAS) to determine if this warrant is satisfied.

The satisfaction of a warrant is not necessarily justification for a signal. Delay, congestion, confusion or other evidence of the need for right of way assignment must be shown.

Figure 14-11 Concluded.

5. Unsignalized intersection where a combination of high speed, restricted view, and serious accident records indicate a need for control by the stop sign

Stop signs cannot be erected at intersections where traffic control signals are present, and should not be installed for the sole purpose of controlling the speeds of the motorists.

Warrants for multiway stop control incorporate numerical criteria for accident occurrence, vehicular volumes, and approach speeds. Four-way or all-way stop installations can be used as a safety measure at some locations where the volume on the intersecting roads is approximately equal and the following conditions have been established (MUTCD).

1. Where traffic signals are warranted and urgently needed, the multiway is an interim measure that can be installed quickly to control traffic while arrangements are being made for the signal installation.
2. An accident problem as indicated by five or more reported accidents in a 12-month period of a type susceptible to correction by a multiway stop installation.
3. Minimum traffic volume:
   a. The total vehicular volume entering the intersection from all approaches must average at least 500 vehicles/hour for any 8 hours of an average day.
   b. The combined vehicular and pedestrian volume from the minor street or highway must average at least 200 units/hour for the same 8 hours, with an average delay to minor street vehicular traffic of at least 30 seconds per vehicle during the maximum hour.
   c. When the 85th percentile approach speed of the major street traffic exceeds 40 mph (64 km/h), the minimum vehicular volume warrant is 70% of the foregoing requirements.

Yield sign warrants in the MUTCD recommend that yield signs may be used (MUTCD):

1. On a minor road at the entrance to an intersection when it is necessary to assign right-of-way to the major road, but where a stop is not necessary at all times, and where the safe approach speed on the minor road exceeds 10 mph (16 km/h).
2. On the entrance ramp to an expressway where an adequate acceleration lane is not provided.
3. Within an intersection with a divided highway, where a stop sign is present at the entrance to the first roadway and further control is necessary at the entrance to the second roadway, and where the median width between the two roadways exceeds 30 feet.
4. Where there is a separate or channelized right-turn lane, without an adequate acceleration lane.
5. At any intersection where a special problem exists and where an engineering study indicates the problem to be susceptible to correction by the use of the yield sign.

Warning Signs

Where an engineering and traffic investigation indicates that attention must be called to an actual or potential hazardous condition, warning signs should be installed. Typical locations include:

- Turns, curves, or intersections
- Advance warning for stop signs, signals, or railroad crossings
- Grades, drops, or bumps
- Narrow roadway, bridges, or other points of limited clearance
- Curves or areas where advisory speed is to be indicated

Guide Signs

Adequate signing at intersections to show route directions and destinations includes advance notice of route directions and destinations including advance notice of route junctions and turns, directional route markings at the intersection, and destination signs showing the names of important cities and towns with directions and distances.

REMOVAL OF UNNECESSARY TCDS

Although the MUTCD is generally clear on the warrants to install various signs, particularly multiway stop signs, local citizen or political pressure frequently results in the placement of unnecessary signs. Changes in local conditions may render current signs, markings, or other TCDs unnecessary, redundant, or unduly restrictive. Removal of some signs (stop, yield) may require legal action at the agency level. Other signs, such as parking and regulatory signs, may also require legal approval prior to change or removal. Current practice indicates that sign removals are infrequent unless a MUTCD standard has changed and there is a need to conform to new uniform standards. Even in these cases, conformance through removal is a slow process. A procedure to justify the removal of unneeded traffic signals is described in the report ''A User Guide for the Removal of Not-Needed Traffic Signals'' (JHK, 1980).

EVALUATING NEW INTERSECTIONS AND ROADWAYS

When new intersections and roadways are being planned or where major construction projects will be implemented in the near future, actual traffic volumes cannot be counted and therefore must be estimated. The ITE publication *Trip Generation* (ITE, 1988) may be helpful in deriving trip estimates for this purpose. Additional methods for estimating future traffic volumes are described in Chapters 8 and 9 of this book and in the *Transportation and Traffic Engineering Handbook* (Homburger et al., 1982) and the *Transportation Planning Handbook.* Trip generation estimates are generally expressed in terms of average daily traffic (ADT). Although peak-hour flows can also be estimated, the hourly distribution is not normally available. Consequently, minimum requirements based on estimated ADTs may be obtained from values given in Figure 14-12. These values are based on the assumption that the eight highest hours will each exceed 6.25% of the ADT (equivalent to 500 vph). This can vary by agency (e.g., Texas uses 5.6% of the ADT, which results in ADT values 11% higher than those shown) (Kell and Fullerton, 1982).

EFFECTIVENESS OF TRAFFIC CONTROL DEVICES

The effectiveness of signs and markings or the need for additional control can be measured in a variety of ways. Among the methods that can be used to determine the effectiveness of traffic control devices are:

TRAFFIC SIGNAL WARRANTS

(Based on Estimated Average Daily Traffic—See Note 2)

URBAN ____________ RURAL ____________

	Minimum Requirements EADT			
1. Minimum Vehicular Satisfied ________ Not Satisfied ________	Vehicles per day on major street (total of both approaches)		Vehicles per day on higher-volume minor-street approach (one direction only)	
Number of lanes for moving traffic on each approach				
Major Street / Minor Street	Urban	Rural	Urban	Rural
1 .... / 1	8,000	5,600	2,400	1,680
2 or more .... / 1	9,600	6,720	2,400	1,680
2 or more .... / 2 or more	9,600	6,720	3,200	2,240
1 .... / 2 or more	8,000	5,600	3,200	2,240
2. Interruption of Continuous Traffic Satisfied ________ Not Satisfied ________	Vehicles per day on major street (total of both approaches)		Vehicles per day on higher-volume minor-street approach (one direction only)	
Number of lanes for moving traffic on each approach				
Major Street / Minor Street	Urban	Rural	Urban	Rural
1 .... / 1	12,000	8,400	1,200	850
2 or more .... / 1	14,400	10,080	1,200	850
2 or more .... / 2 or more	14,400	10,080	1,600	1,120
1 .... / 2 or more	12,000	8,400	1,600	1,120
3. Combination Satisfied ________ Not Satisfied ________ No one warrant satisfied but following warrants fulfilled 80% or more ___ ___ (1, 2)	2 Warrants		2 Warrants	

NOTE:

1. Left turn movements from the major street may be included with minor street volumes if a separate signal phase is to be provided for the left-turn movement.
2. To be used only for NEW INTERSECTIONS or other locations where actual traffic volumes cannot be counted.

Figure 14-12 Warrant evaluation for new intersections. *Source:* State of California Traffic Manual.

- Road user compliance studies
- Before-and-after studies of traffic characteristics, accident records, and enforcement records
- Measurements of the device's performance based on its condition
- Analysis of complaints or comments from citizens
- Inappropriate use of a TCD

Driver and pedestrian studies can provide an objective, quantitative evaluation of existing traffic control devices and give guidance for corrective action. Measures of efficiency are usually described by changes in type, frequency, and duration of traffic delays, spot or travel speeds, and increased compliance of drivers or pedestrians to traffic regulations and control devices. Safety evaluations are provided by changes in type and frequency of traffic accidents or conflict. However, other measures of traffic flow and/or safety conditions can be selected as decision-making variables in a before-and-after study.

ROAD USER COMPLIANCE WITH TCDs

Studies are commonly conducted to evaluate driver, pedestrian, or bicyclist compliance with a specific control device or traffic regulation and the effectiveness of traffic control devices. The noncompliance problem appears to be concentrated in specific situations and/or with specific TCDs. These include:

- Exceeding advisory speeds and/or posted speed limits
- Not stopping at stop signs
- Not stopping at right-turn-on-red (RTOR) locations
- Violating the red signal
- Violating active railroad (RR) grade crossing signals
- Violating left-turn-lane signals
- Traveling too fast for conditions

Studies of road-user compliance with stop signs, traffic signals, no-turn restrictions, and right-turn-on-red provisions are discussed specifically in Chapter 15. The techniques presented can be adapted to study compliance of most traffic regulations and control devices: for example, parking regulations, advisory signs, speed limits, and parking regulations. Bicyclist and pedestrian compliance to control devices and regulations may readily be measured by these field procedures as well.

Before-and-After Studies

Before-and-after analyses are commonly used to evaluate the effectiveness of highway or traffic improvements. The criteria for evaluation may be economic or be based on measures of the efficiency and safety of traffic flow through the improved roadway section or intersection. Economic assessments are often expressed as the dollar value of the benefits to the road users, the adjacent property, and the general public as well as the actual costs for making the necessary improvements. These economic evaluations are generally computed on an annual basis. A before-and-after study should be planned as an integral part of the evaluation process for any significant traffic improvement. The design of before-and-after studies is described in Appendix A. Common sources of mistakes made in comparing before-and-after data are (McShane and Roess, 1990):

- Poor choice of time periods for before-and-after data collection
- Inadequate or noncomparable data
- Failure to allow for a stabilizing period for the public to adjust to the change
- Failure to take into account other changes that may affect the situation
- Lack of control data to account for traffic trends
- Failure to account for exposure
- Evaluating a change as significant when in reality the change is due to chance variation

When data are collected for a before–after comparison of traffic movement characteristics such as speed, volume, and delay, time periods should be selected to minimize the influence on the results. For example, if before the installation of a speed zone, a speed study was made in the early afternoons of several weekdays, the after study should be made on similar weekday afternoons. Observations should be spread over a period of days wherever possible. There is less chance of selecting an abnormal period of time if the data for several days are averaged.

Changes in Spot Speeds

Spot speed data may be collected in several ways. Commonly used methods include the use of radar and the enoscope. Methods of collecting spot speed data are discussed in detail in Chapter 3. Once spot speed data have been collected, statistical techniques can be used to test for a change in the average spot speeds following a change such as the installation of a speed limit sign.

This is done as follows:

1. Spot speed studies are conducted before and after the change is made. Conditions for both studies should be as similar as possible.
2. The means (μ_a) and (μ_b) and standard deviations σ_a and σ_b for both studies are calculated independently.
3. A new mean μ is calculated that is the difference between the two estimated means μ_a and μ_b, where $\mu = \mu_a - \mu_b$.
4. A new standard deviation is calculated:

$$\sigma = \sqrt{\frac{\sigma_a^2}{N_a} + \frac{\sigma_b^2}{N_b}}$$

5. The hypothesis in this statistical test is that the distribution has a zero mean, e.g., that there is no difference between mean speed before and mean speed after the change.

$$\mu_a - \mu_b = 0$$

From Table 14-3, the Z corresponding to 95% confidence (in the lower tail of the distribution) is -1.645. Therefore if

$$(\text{Observed Mean} - \text{hypothesized mean}) \div \text{Standard Deviation} < -1.645$$

then the hypothesis is rejected at a 5% level of significance. Sample calculations done on a spreadsheet are shown in Figure 14-13.

TABLE 14-3 Z-values for confidence levels

Confidence (%)	*Z-value*
99	−2.2326
97.5	−1.960
95.0	−1.645
90.0	−1.282

TABLE 14-4 Standard deviations for estimating spot speed sample sizes

Traffic Area	*Highway Type*	*Standard Deviation*
Rural		
	2-lane	5.3 mph
	4-lane	4.2 mph
Intermediate		
	2-lane	5.3 mph
	4-lane	5.3 mph
Urban		
	2-lane	4.8 mph
	4-lane	4.9 mph

Sample sizes for the before and after studies may be calculated as described in Chapter 3. Ideally sample sizes of 50 to 100 vehicles per study should be obtained. The Z test shown above should be used only with sample sizes greater than $N = 30$.

Required samples sizes can be calculated using the equation:

$$1.96 \times \frac{\sigma}{\sqrt{N}} \leq \text{tolerance}$$

Tolerance is the acceptable speed tolerance (e.g., 1.25 mph). The value of 1.96 is based upon a 97.5% confidence level.

The standard deviation for estimating spot speed sample sizes can be selected from Table 14-4. Once the data have been collected, the actual standard deviation should be used in the calculations.

Evaluating Safety Improvements

TCD effectiveness is often evaluated in light of improvements in safety after a device has been installed. In designing before- and-after studies, the period of time considered before and after the improvement must be long enough to observe changes in accident occurrence. For most locations, periods ranging between 3 months and 1 year are selected. The length of the "before" and "after" periods must be the same. Results of before-and-after studies of safety improvements may be confounded by other factors, including changes in exposure (e.g., traffic volumes) and nearby changes that may affect accident occurrence. Evaluating the effectiveness of a single improvement may be further hampered by the fact that roadway improvements may be done in packages. For example, the installation of signals may be accompanied by geometric changes in the intersection and/or improvements on intersection approaches.

In analyzing records for two comparable periods of accident data, it is best to take a full year before and after the improvement. This will help to avoid any difference that might exist because of seasonal fluctuation in traffic patterns. For instance, if a 6-month period is compared immediately preceding and following an improvement that was made in March, the before situation reflects conditions of winter, and the after period reflects summer and autumn traffic conditions. If winter accidents were naturally more frequent

Class boundaries	Class limits	Class midvalues (u)	Frequencies after (f_a)	Before (f_b)
27.5				
	28–29	28.5	0	0
29.5				
	30–31	30.5	1	0
31.5				
	32–33	32.5	2	0
33.5				
	34–35	34.5	14	5
35.5				
	36–37	36.5	7	9
37.5				
	38–39	38.5	20	7
39.5				
	40–41	40.5	38	25
41.5				
	42–43	42.5	29	34
43.5				
	44–45	44.5	35	44
45.5				
	46–47	46.5	15	29
47.5				
	48–49	48.5	12	12
49.5				
	50–51	50.5	9	13
51.5				
	52–53	52.5	4	11
53.5				
	54–55	54.5	0	7
55.5				
	56–57	56.5	0	2
57.5				
Total			186	196

	Before	After
Mean $= \bar{X} = \dfrac{\sum f_i u_i}{\sum f_i}$	42.35	45.13
Standard deviation $= S = \left[\sum f_i u_i^2 - \dfrac{[\sum f_i u_i]^2}{\sum f_i}\right] \times \dfrac{1}{\sum(f-1)}$	4.50	1.23
Difference between means $= (ua - ub) = u$	−2.78	
New std. deviation $\sigma = \sqrt{\dfrac{\sigma_a^2}{N_a} + \dfrac{\sigma_b^2}{N_b}}$	0.34	
$Z = \dfrac{u-0}{\sigma}$	$-8.14 < -1.645$*	

*Difference in speeds with 95% confidence interval

Figure 14-13 Sample analysis of before and after speed studies.

Type of accident	Approach				Total
	EB	WB	SB	NB	
Rear end					
Left turn					
Straight					
Right turn					
Sideswipe					
Fixed object					
Pedestrian or bike					
Right angle					
Other					

Figure 14-14 Summary of accidents by type. *Source:* Box and Oppenlander, 1976.

than summer accidents in the locality concerned, then the "after" phase may include a reduction in accidents not necessarily attributable to the traffic engineering change. If accident statistics are available for only 6 to 8 months prior to the change, then it would be better to select an after period for the same months in the following year. It is generally acceptable to compare periods of several years before and after; however, if a longer time period is used, it is especially important to take changing conditions into account (e.g., trends in car sizes, 55 mph speed limit).

Tabular comparisons of accident frequency should be supplemented by collision diagrams to emphasize the relative frequencies of different types of accidents before and after a specific sign installation. For instance, after the installation of multiway stop signs, collision diagrams might reveal a decrease in right-angle collisions, but an increase in rear-end collisions. Accidents can be separated by a severity classification in making comparisons. Accidents may be tabulated as shown In Figure 14-14 or included on the collision diagram (Figures 14-2).

The normal approximation (or Z) test can also be used to determine if there is a significant change in the number of accidents before and after an improvement. In this application, a value for Z is calculated and used to look up the probability that $Z < Z_1$.

Z is calculated as:

$$Z_1 = (f_a - f_b)/(f_b + f_a)$$

where

Z_1 = test statistic

f_a = number of accidents during the "after" study period

f_b = number of accidents during the "before" study period.

If the calculated value is greater than or equal to the Z value for $p=0.95$, the observed reduction in accident occurrence is significant at the 95% level. If the value is less than the Z-value for $p = 0.95$, the observed accident reduction may not be considered a significant result of improvements made.

TCD CONDITION

Day-to-day traffic operations involve an awareness of the effectiveness of the traffic control devices placed in the field. Studies must be performed and inventories maintained to determine (1) where the devices are and their condition, and (2) how effectively the devices are performing and (3) the maintenance needs of the devices.

Sign Reflectivity

It is desirable periodically to measure the reflectivity of signs. If the reflective properties of a sign when it was new are known, this can be compared with the reflectivity at later times. Reflective properties can be measured by a reflectometer or by the subjective judgment exercised during night inspections. The reflectometer is a sensitive hand-held light meter that emits a beam of light to the sign surface that is reflected back to a sensor unit. The reading is an indication of the existing reflective properties of the sign face. Reflectometers are available but are relatively expensive. Research has been underway for some time to develop an inexpensive simple device that can be used in daylight to measure the reflectivity of a sign at night.

One way to evaluate sign effectiveness (reflectivity) is to observe the signs at night, under both bright and dim headlight illumination. At best, this is an inexact technique since it is dependent on the eyesight of the viewer, the distance and angle from the sign, the judgments of the viewer as to what level of reflectivity is acceptable, and the condition of the headlights of the test vehicle. Another technique suggested by the *Traffic Control Devices Handbook* (FHWA, 1990) is for a two-person crew (driver-observer and recorder) to drive the roads at the posted speed and to evaluate individual signs based on the amount of time that they were visible (legible) to the driver. Signs that are visible for 2 seconds or less should be replaced as soon as possible. Signs that are visible for 3 seconds are considered borderline and should be scheduled for replacement. Signs that are visible for 4 or more seconds are usually considered acceptable.

Feedback from Citizens

Roadway users will often inform agency personnel that certain traffic control devices are damaged, out of order, knocked down, or needed. While some comments from the public are very negative and charged with emotion and others are very calm statements of fact, they are all a valuable source of information as to how the system works. Employees of street and highway organizations, police, and other governmental employees whose duties require that they travel on the roadways should be encouraged to report any damaged or obscured signs at the first opportunity. Many agencies have found that a system for receiving, recording, and responding to complaints is highly desirable. Such a system helps ensure that prompt corrective action is taken when needed and allows for prioritizing other activities. Also, by recording complaints, a permanent record is established that can be reviewed from time to time. These records might form a basis for changing work emphasis areas, reallocating resources, or beginning a training program.

REFERENCES

BOX, P. C. AND J. C. OPPENLANDER (1976). *Manual of Traffic Engineering Studies,* 4th ed., Institute of Transportation Engineers, Washington, DC.

DOUGHTY, J. R. (1982). Traffic signs and markings, in *Transportation and Traffic Engineering Handbook,* W. Homburger, L. Keefer, and W. R. McGrath eds. Prentice Hall, Englewood Cliffs, NJ.

EDWARDS, JOHN D. (1992). *Transportation Planning Handbook*, Prentice Hall, Englewood Cliffs, NJ.

FHWA (1988). *Manual of Uniform Traffic Control Devices*, Federal Highway Administration, Washington, DC.

FHWA (1990). *Traffic Control Devices Handbook*, U.S. Department of Transportation, Federal Highway Administration, Washington DC.

HOMBURGER, W., L. KEEFER, AND W. R. MCGRATH, EDS. (1982). *Transportation and Traffic Engineering Handbook*, Prentice Hall, Englewood Cliffs, NJ.

ITE (1972). *A Program for School Crossing Protection: A Recommended Practice*, 3rd ed. Institute of Transportation Engineers, Washington, DC.

ITE (1988). *Trip Generation*, Institute of Transportation Engineers, Washington DC.

JHK AND ASSOCIATES (1980). *A User Guide for the Removal of Not Needed Traffic Signals*, FHWA-IP-80-1Z, US Department of Transportation, Federal Highway Administration.

KELL, J. H., AND I. J. FULLERTON (1982). *Manual of Traffic Signal Design*, Prentice Hall, Englewood Cliffs, NJ.

MCSHANE, W. AND R. ROESS (1990). *Traffic Engineering*, Prentice Hall, Englewood Cliffs, NJ.

15

Compliance with Traffic Control Device Studies

Donna C. Nelson, Ph.D., P.E.

INTRODUCTION

User disregard of traffic control devices (TCDs) is of increasing concern to those involved in traffic engineering and roadway safety. The level of public obedience is one test of the effectiveness of regulations and devices, the adequacy of publicity and education dealing with these controls, and the level of enforcement (Box, 1984). This chapter focuses on the study of driver and pedestrian compliance with traffic control devices. The specific studies addressed in this chapter include driver compliance with stop signs, traffic signals, no-turn restrictions, and right turn on red. While these studies focus on compliance at intersections, the techniques presented can be adapted to study compliance with most traffic regulations and control devices, at a wide variety of locations. Bicyclist and pedestrian compliance with control devices and regulations may readily be measured by these field procedures as well. Additional background information can be found in FHWA Publication RD-89-103 ''Motorist Compliance with Standard Traffic Control Devices'' (Pietrucha, et al, 1989).

Purpose

Compliance studies are conducted to evaluate driver, pedestrian, or bicyclist compliance with a specific control device or traffic regulation and the effectiveness of traffic control devices.

Applications

Typically, a compliance study is conducted at any roadway or intersection location where information is desired on compliance with traffic regulations or traffic control devices. Studies may be conducted to:

1. Evaluate the effectiveness of traffic control devices.
2. Develop educational programs for drivers, schoolchildren, and the general public.
3. Determine critical locations for selective enforcement efforts.
4. Analyze the effectiveness of traffic improvements through before-and-after studies.

The noncompliance problem appears to be concentrated in specific situations and/or with specific TCDs (Box, 1984). These include:

- Exceeding advisory speeds and/or posted speed limits
- Not stopping at stop signs
- Not stopping at right-turn-on-red (RTOR) locations
- Violating the red signal
- Violating active railroad (RR) grade crossing signals
- Violating left-turn lane signals
- Traveling too fast for conditions

STUDY DESIGN

Location, data collected, sample size, weather conditions, and time of day (week or month) are important factors in the design of road user compliance studies.

Study Locations

Compliance studies may be conducted at any location where there is some indication that a compliance problem exists. Because the establishment of bicycle routes in many communities has produced points of conflict among motor vehicles, bicycles, and pedestrian, evaluation of bicyclist compliance with control devices is sometimes necessary. Studies may be initiated from citizen complaints, observance of a specific traffic problem, or the existence of high accident locations.

Data Needs

The data required for compliance studies are of two types: (1) an inventory and site description, and (2) compliance data. The site diagram and description are very important. During the data analysis, a good description of the site can provide important information in the interpretation of the study results. Inventory and site information can be recorded on the diagram shown in Figure 15-1. The site diagram should be approximately to scale and include all the physical features of the study site (e.g., sidewalks, crosswalks, vegetation, driveways, embankments, signs, traffic signals, markings, roadway shoulders, abutting land uses, bus stops, and other characteristics or conditions that may affect visibility and/or site distances). All streets or highways should be labeled by their official name and/or number. The diagram should indicate estimated measurements for roadway widths, shoulder widths, and lane widths. A checklist of items commonly re-

SITE INVENTORY AND DIAGRAM
Field Sheet

Location:____
Type of Study ____________ Area Type:____________________
Date: ____________________Recorder:_____________________

Indentify on diagram:

Geometry	TCD's and Markings	Area/Intersection Data
1. Lanes, lane width 2. Median, median width 3. Movements by lane 4. Islands, channelization 5. Bus stop locations 6. Objects restricting sight distances 7. Functional classification of roadway	1. Traffic signs before intersection 2. Warning signs/markings within 500 ft. of intersection 3. Speed limit on approach legs 4. Lane designation signs 5. Parking location/prohibitions 6. Roadway edge markings 7. Lane separation markings 8. Crosswalks 9. Other important geometric features	1. Type of Intersection 2. Street/highway names or numbers 3. Directional arrow 4. General area description 5. Traffic volumes

Figure 15-1 Sample site inventory form completed for an intersection study.

quired for a complete site description are shown in Table 15-1. All of these items may not be required for every study.

Compliance Data

Data collected for different types of studies of driver compliance are shown in Table 15-2. These should be modified for specific applications. Conditions of compliance and noncompliance must be well defined before the study begins. Even with well-defined compliance conditions, observations and decisions can be difficult at best. Lack of clearly defined, well-understood definitions can result in inconsistent and unreliable data. For example, four categories of driver action are commonly used for stop sign compliance studies: (1) not stopping, (2) practically stopped, (3) stopped by traffic (forced stop), and (4) voluntary full stop. Categories 1 and 2 represent total and partial noncompliance;

TABLE 15-1
Site Inventory and Study Description Checklist

Study information
- Day and date of study
- Time period of data collection
- General weather conditions
- Investigator's name
- Site location and jurisdiction
- Roadway names/numbers
- Any special conditions

Geometry
- Approach grade
- Number of lanes
- Lane width
- Movements for each lane
- Shoulder width
- Median type and width
- Sight distance restrictions
- Roadway functional classification
- Potentially distracting conditions

Area/intersectional data
- Type of intersection (T, four-way)
- Street/highway names and/or numbers
- Directional arrow (north arrow)
- Intersection control (signal, stop, etc.)
- Signal phasing
- Cycle lengths
- General area description within 0.1 mile (land use type and/or intensity)

Signs and markings
- Traffic signs before intersection
- Warning signs within 500 feet
- Speed limit on approach legs
- Lane designation signs
- Crosswalks
- Roadway edge markings
- Lane separation markings

TABLE 15-2
Data Needed for Compliance Studies

Running the red indication
- Turning movement
- Driver action
- Vehicle type
- Cycle length
- Speed limit
- Approach volume
- Peak hour
- Location

Improper right turn on red
- Queuing conditions
- Type of stop
- Location
- Vehicle type
- Cross street volume
- Approach volume
- Pedestrian paths
- Pedestrian signals

Violations of turn restrictions
- Turning movement
- Type of traffic control
- Vehicle type
- Temporal restrictions
- Approach volume
- Location

Stop sign compliance
- Turning movement
- Queuing conditions
- Type of stop
- Vehicle type
- Cross street volume
- Approach volume
- Pedestrian paths
- Location

TABLE 15-3
Constant Corresponding to Level of Confidence

Constant, K	Confidence Level (%)
1.00	68.3
1.50	86.3
1.64	90.0
1.96	95.0
2.00	95.5
2.50	98.8
2.58	99.0
3.00	99.7

categories 3 and 4 represent compliance. The choice between categories 1 and 2 is at the discretion of the observer and may produce confusion and inconsistencies among data collected by different observers. These and other data definitions are discussed more in the section on data collection.

Sample-Size Requirements

Compliance with a traffic regulation is essentially a "yes–no" effect; therefore, the following equation can be used to calculate the minimum sample size for each observance study site:

$$N = \frac{pqK^2}{E^2}$$

where

N = minimum sample size
p = proportion of drivers or pedestrians that observe the traffic regulation
q = proportion of drivers or pedestrians that do not observe the traffic regulation
K = constant corresponding to the desired confidence level
E = permitted error in the proportion estimate of compliance

The p and q values can be estimated by a preliminary study. The use of 0.5 for p and q provides the most conservative estimate possible of the sample size required. It is, in essence, saying that 50% of the vehicles (or pedestrians) fail to observe the traffic regulation. The sample size required decreases as the value of p decreases or increases from 0.5. The sum of p and q should always be equal to 1.0. The constant K depends on the desired level of confidence. Table 15-3 presents some precalculated values of K based on confidence intervals that represent 1, 2, 3, 4, and 5 standard deviations from the mean. A value of 2.00 for K is often selected for a confidence level of 95.5%, which corresponds to approximately 1 chance in 20 that the proportion of violations found would occur by chance alone. The permitted error E is based on the precision that is required for this estimate. Table 15-4 shows sample size requirements for permitted errors of 5% and 10%, with a confidence level of 90% or 95%. For a more detailed explanation of K, see Appendix C. Box (1984) suggests that samples of 100 are often adequate to indicate compliance with TCDs, except when violations are rare.

TABLE 15-4
Summary of Sample Size Requirements ($p = q = 0.50$)

Permitted Error, E (%)	90% Confidence	95% Confidence
5	270	380
10	70	100

EXAMPLE 15-1

Assume a 20% violation rate ($p = 0.8$, $q = 0.2$), a 5% permitted error, and $K = 2$:

$$N = \frac{(0.8)(0.2)(2)^2}{(0.05)^2} = 256$$

If the violation rate is assumed to be 50% ($p = q = 0.5$), then

$$N = \frac{(0.5)(0.5)(2)^2}{(0.05)^2} = 400$$

Time and Duration of Study

The time and conditions under which compliance studies are conducted can affect results. These studies are usually performed in good weather and under "normal" traffic conditions (e.g., that there are no circumstances present that may affect the results). Samples are taken to cover all applicable periods of the day. Data are commonly collected in 5- or 15-minute intervals for a period that allows the collection of data for more than the minimum sample size. Data should be collected under the conditions which the problem has been observed or is likely to be most evident. For example, traffic delays and accidents generally occur more often during peak traffic periods; therefore, traffic data should be collected during this period. Excessive speeds on some streets may occur only when traffic volumes are light to moderate. Compliance with TCDs in school zones is generally of interest during school hours, specifically when students are traveling to and from school. Compliance studies may be performed in off-peak periods to provide a comparative analysis of the violation problem. If results are to be comparable for before-and-after analysis, similar conditions must exist during both periods of data collection.

Personnel and Equipment Requirements

Compliance studies are most often carried out using manual methods. Observers record road user behavior using tick marks on paper forms or hand-held mechanical counters. One or more observers are used as needed. Because the objective is to count road users violating traffic regulations, the observers must be as inconspicuous as possible. Video cameras may also be useful for the collection of some types of compliance data. If the site allows video cameras to be situated inconspicuously with a clear field of vision, significantly more data can be collected than a person can record manually in real time. Data, however, must still be reduced from the video tapes and entered on paper forms or into a computer program for analysis.

Data Collection Procedures

Typical field sheets and general data collection procedures for several studies are shown below. Figure 15-1 shows a sample site diagram that corresponds to the checklist in Table 15-1 and is used to record the site description and inventory for signalized and unsignalized intersections, respectively. Figures 15-2 to 15-6 are used to record and analyze road user compliance data. These forms should be modified as needed for individual sites and studies. The compliance of each road user to the traffic regulation is recorded by placing a tally mark in the appropriate space on the field sheet. The observations are continued until the desired sample size has been obtained.

Stop sign compliance data are recorded on the form in Figure 15-2. The observer for this study is positioned to see the study vehicles as they arrive at the stop sign, and the cross traffic on the through street. Data are recorded by movement for each driver action: full stop, almost stopped, forced stop, and no stop. A *full stop* is defined as a "complete cessation of movement, however brief." *Nearly stopped* is defined most commonly as "<3 mph." A *forced stop* occurs when the motorist is required to stop because of conflict with cross traffic and pedestrians and *no stop* is commonly defined as ">3 mph." If more than one observer is used, try to make sure that all observers are categorizing vehicles the same way. Cars and trucks (or large vehicles) can be differentiated by entering tick marks in the cells for car or truck.

No-left-turn compliance data may be recorded on the form in Figure 15-3. The observer is positioned to observe the study vehicles as they approach the no-left-turn prohibition location. The number of vehicles passing the study site that do not make an illegal left turn are counted. Through and right-turning trucks may also be recorded.

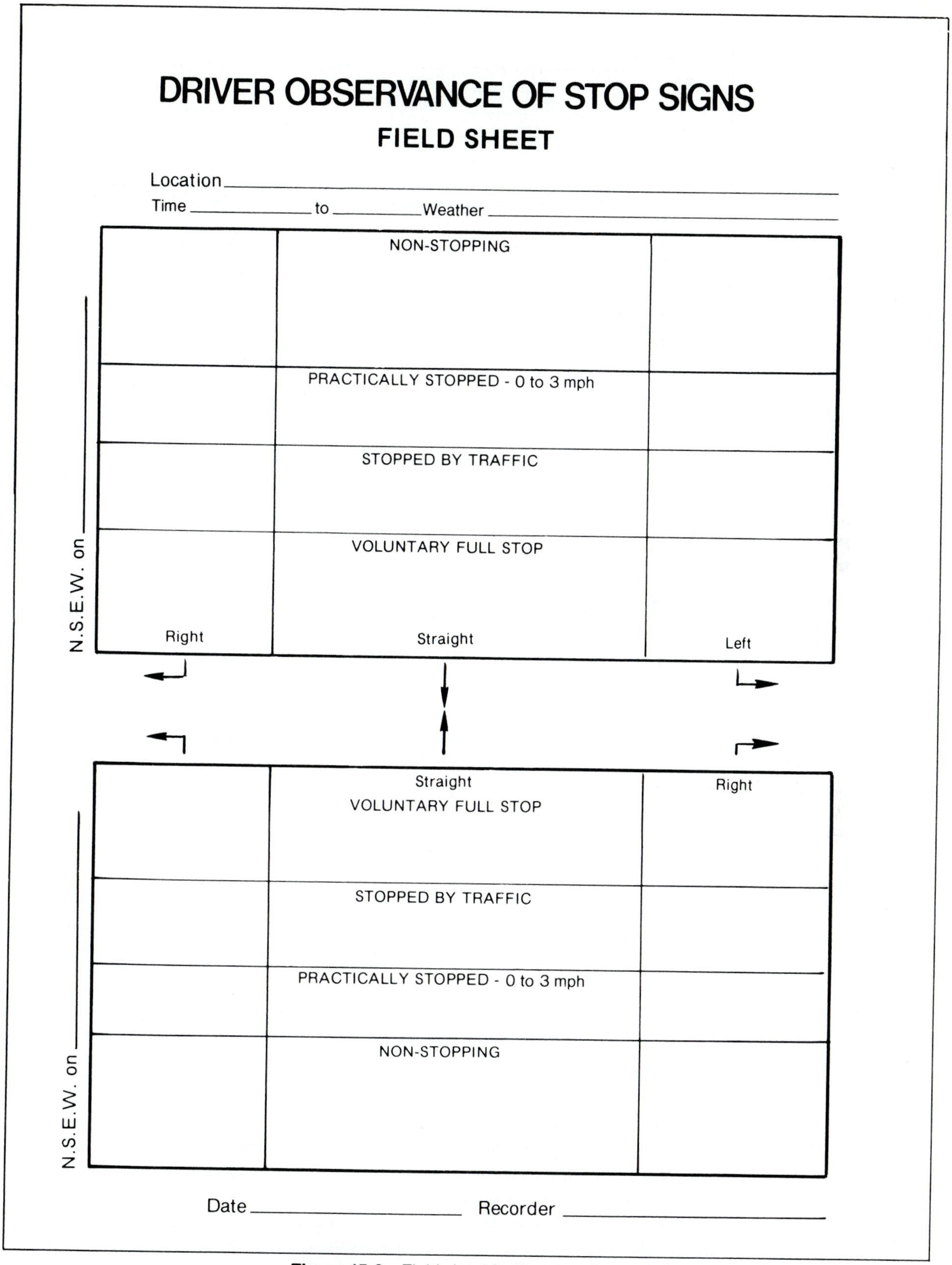

DRIVER OBSERVANCE OF STOP SIGNS

FIELD SHEET

Location ______________________

Time ________ to ________ Weather ________

N.S.E.W. on ________

Right	Straight	Left
	NON-STOPPING	
	PRACTICALLY STOPPED - 0 to 3 mph	
	STOPPED BY TRAFFIC	
	VOLUNTARY FULL STOP	

N.S.E.W. on ________

	Straight	Right
	VOLUNTARY FULL STOP	
	STOPPED BY TRAFFIC	
	PRACTICALLY STOPPED - 0 to 3 mph	
	NON-STOPPING	

Date ________ Recorder ________

Figure 15-2 Field sheet for the stop sign compliance study. *Source:* Box and Oppenlander, 1976.

Location: ______________________________
Direction of Travel: __________ Area Type ______________________
Time: ________ to ______ Weather: ______________________
Date: ______________________ Observer ______________

	LEFT	RIGHT AND THROUGH	TOTAL
CARS			
TRUCKS			

	LEFT	RIGHT AND THROUGH.	TOTAL.
TOTAL CARS			
TOTAL TRUCKS			
TOTAL			

Notes: ______________________________

Figure 15-3 No-left-turn compliance field sheet. *Source:* Adapted from *Motorist Compliance with Standard Traffic Control Devices,* FHWA-RD-89-103.

The *driver compliance with traffic signals* study is concerned particularly with the response of drivers approaching the signal during the clearance interval and red phase. Data may be recorded using the form shown in Figure 15-4. The form has space to record driver actions on four intersection approaches. Driver behavior is tallied by direction of travel as the vehicle leaves the intersection. The observed behavior is related to the traffic signal indication seen by the driver when the vehicle enters the intersection. Entry is usually defined as crossing the near-side curb line. The signal indications are listed on the data form as "green," "yellow after green," "red," and "jumped signal." The forms have space to record driver actions. Care must be taken to consider perception–reaction

DRIVER OBSERVANCE OF TRAFFIC SIGNALS

FIELD SHEET

Location ______________________________

Time __________ to __________ Weather __________

N.S.E.W. on __________

GREEN	YELLOW AFTER GREEN	RED	JUMPED SIGNAL

Left — Straight — Right

N.S.E.W. on __________

	GREEN	YELLOW AFTER GREEN	RED	JUMPED SIGNAL
Left				
Straight				
Right				

N.S.E.W. on __________

	GREEN	YELLOW AFTER GREEN	RED	JUMPED SIGNAL
Left				
Straight				
Right				

N.S.E.W. on __________

GREEN	YELLOW AFTER GREEN	RED	JUMPED SIGNAL

Right — Straight — Left

Date __________ Recorder __________

Figure 15-4 Field sheet for the study of driver compliance with traffic signals. *Source:* Box and Oppenlander, 1976.

times in classifying violations of traffic signals. Signal compliance studies are appropriate during both peak and off-peak hours.

Right turn on red after stop compliance and data on the proper yielding of such drivers to pedestrians and other vehicular traffic can be conducted using the form shown in Figure 15-5. The measure of violation is the observed slowing or brake-light application of through traffic, or interference with pedestrians. Right turn on red may be identified by several conditions: (1) failing to stop completely at the stop line, or before crossing the crosswalks at an urban intersection if pedestrians are present, (2) interfering with a pedestrian in either crosswalk being traversed, or (3) causing sudden slowing of a vehicle on the cross street which has the green indication. The observer should be able to see the study vehicles as they arrive at the traffic signal and the cross traffic on the through street. The data collection form provides space for the following observations. Observe every right-turning vehicle that passes through the intersection during the study. Begin a new data sheet every 15 minutes.

When the signal is green, record the number of study vehicles that:

1. Turned on green or yellow.
2. Stopped on red, waited for green before turning.
3. Stopped on red behind a vehicle waiting for the green indication before turning right. The turn was executed on green.
4. Attempted to turn on red, signal turned to green before turn was executed. The turn was completed on green.

When the signal is red:

1. Determine if study vehicle arrived as a single vehicle or as part of a queue waiting at the signal.
2. Determine if the vehicle made a full stop (a brief cessation of movement), was stopped by vehicular or pedestrian cross traffic, or did not stop at all before entering the intersection.

Pedestrian compliance with traffic signals is recorded on the field sheet (Figure 15-6). Pedestrian behavior is noted as the signal indication on which the person steps from the curb onto the pavement of the intersection. Signal indications are observed for that intersection approach which the pedestrian is crossing and include "green," "yellow," and "red." If signal indications are separately provided for pedestrians, the appropriate indications are "steady don't walk," "flashing don't walk," and "walk." The field sheet is designed to check on crosswalks with a provision for recording diagonal crossings. If a bicycle compliance study is not conducted separately, the compliance of bicyclists can also be recorded by using the letter "B" for bicycle riders and "P" for pedestrians.

DATA ANALYSIS AND SUMMARY

After the field study has been completed, the tally marks are summed to give subtotal and total values. Totals and subtotals are developed for:

1. Each direction of movement
2. Each intersection approach
3. Each compliance category
4. Total number of road users complying
5. Total number of road users not complying

Location: ____________________
Direction of Travel: __________ Area Type ____________
Time: ________ to ________ Weather: ____________
Date: ____________________ Observer ____________

INDICATION	ACTION	CARS	TRUCKS	TOTAL CARS	TOTAL TRUCKS
ON GREEN	Turned on G or Y				
	Stopped on R Waited for G, Turned on G				
	Behind a Waiter (above) Turned on G				
	Attempted to turn on R, turned on G				
ON RED NO QUEUE	Full Stop				
	Stopped by Cross traffic				
	Stopped by Pedestrian Crossing				
	No Stop				
ON RED QUEUE	Full Stop				
	Stopped by Cross Traffic				
	Stopped by Pedestrian Crossing				
Totals					

Notes: ____________________

Figure 15-5 Field sheet for a right turn on red after stop compliance study. *Source:* Adapted from *Motorist Compliance with Standard Traffic Control Devices,* FHWA-RD-89-103.

PEDESTRIAN OBSERVANCE OF TRAFFIC SIGNALS

FIELD SHEET

Location ______

Time ______ to ______ Weather ______

Pedestrians crossing ______ St. on the (N.S.E.W.)______ side ______

of______ St. in ______ direction ______

STEPPED FROM CURB ON	CROSSED STRAIGHT (crosswalk)	TOTAL
RED / WALK		
YELLOW / FLASHING DON'T WALK		
GREEN / STEADY DON'T WALK		
	CROSSED DIAGONALLY	
RED / WALK		
GREEN OR YELLOW / DON'T WALK		
TOTAL		

Date ______ Recorder ______

Figure 15-6 Field sheet for a study of pedestrian compliance with traffic signals. *Source:* Box and Oppenlander, 1976.

Compliance characteristics are summarized as totals for each compliance category. These subtotals and totals are divided by the sample size to yield proportions or percentages. This procedure allows results from other compliance studies (with different sample sizes) to be compared directly. Although summary statistics for compliance studies are usually expressed as proportion or percentage values, other descriptors of central tendency can be developed. Data analysis and statistical summaries are described in detail in Appendix C.

Analysis of Results

To illustrate the analysis of a compliance study, driver compliance with a stop sign was checked on a collector street in an urban area. The following responses were recorded for a sample of 120 vehicles:

1. Voluntary full stop = 66
2. Stopped by traffic = 24
3. Practically stopped = 12
4. Nonstopping = 18
5. Total observed = 120

The percentages of driver compliance are computed by dividing each group by the sample size and multiplying by 100.

1. $(66/120) \times 100 = 55\%$
2. $(24/120) \times 100 = 20\%$
3. $(12/120) \times 100 = 10\%$
4. $(18/120) \times 100 = 15\%$

REFERENCES

Box, P. (1984). Traffic studies, Chapter 17 of *Transportation and Traffic Engineering Handbook*, 5th ed., Institute of Transportation Engineering, Washington, DC, pp. 546–547.

Pietrucha, M., K. Opiela, R. Knoblauch, and K. Cringler (1989). *Motorist Compliance with Standard Traffic Control Devices.* FHWA-RD-89-103, Federal Highway Administration, Washington, D.C., April.

16
Roadway Lighting

L. Ellis King, D. Eng., P.E.

INTRODUCTION

The ability to see is essential for the safe and efficient flow of traffic on our highways. However, in many instances, limitations of the human eye prevent vehicle headlights alone from completely satisfying visual nighttime driving requirements. Fixed roadway lighting supplements vehicle headlights by extending the visibility range both longitudinally and laterally, thus aiding the driver by providing earlier warning of hazards on or near the roadway. Research shows that the nighttime accident rate can be reduced by the provision of adequate roadway lighting (Box, 1971, 1972a, 1972b, 1989; Walker and Roberts, 1976). Lighting defines the roadway geometrics, such as the edge of pavement, curves, and dead ends, and illuminates obstructions in or near the roadway, including channelization islands, bridge piers, and parked cars. Lighting allows the driver to see a pedestrian in the roadway beyond the headlight beam and even before the pedestrian enters the road. It also aids pedestrians by illuminating obstacles on the sidewalk and roadway in their vicinity.

Lighting raises the surrounding brightness level to which the driver's eyes adapt and increases the driver's contrast sensitivity, resulting in an overall improvement in the driver's ability to see. Fixed roadway lighting also contributes to a more pleasant and comfortable night driving environment, which in turn, reduces driver fatigue and improves driver efficiency. Lighting is an aid to police surveillance, and a reduction in street crimes may be experienced following installation of improved street illumination. Auto theft, assault, and vandalism are three of the types of night crimes most frequently cited as being reduced. While the reported impacts of lighting on crime are statistically inconclusive, there are strong indications that the fear of crime is reduced following increases in street lighting and that feelings of safety are higher (Tien, 1979). The negative aspects of lighting include glare, collisions with light poles, initial installation costs, and continuing maintenance and energy costs for the lighting system.

TYPES OF STUDIES

Existing Conditions

In this study, existing conditions are determined at a specific location, along a given route, or in a defined area of the city. Existing facilities are inventoried and accident data are collected. In terms of accidents, the tabulation may yield the percentage of total collisions that occur at night. Another measure is in terms of direct exposure, using the accident rate per million vehicle miles (mvm) or per 100 mvm. The rates may be calculated separately for day and for night accidents. The ratio of these rates, such as the night/day ratio, may be used to identify sites that could benefit from improved lighting.

Studies of existing conditions are comparative in nature. They are first used to determine accident or crime frequency at locations with good lighting. This provides the norm to be used for comparison purposes. If the study concerns crime, it must be focused on the number of total crimes of particular types, and the number and/or percent of these at night since the degree of relative day and night exposure to crime cannot be measured.

Before-and-After Analysis

The potential effect of an improvement in lighting sometimes can be judged by measuring the actual changes in accident or crime conditions for routes or areas that have had an upgrading of illumination. Such studies should be made on a routine basis to justify additional expenditures. The studies may show a need to modify lighting designs if substantial reductions in night accidents and/or crime are not occurring.

DATA COLLECTION

Roadway Grouping

The American National Standard Practice for Roadway Lighting establishes recommended illuminance and luminance levels for the lighting of various types of roadways, pedestrian ways, and bikeways in different areas (IES, 1983). Classifications and recommended illuminance levels are shown in Table 16-1 for roadways and in Table 16-2 for walkways and bikeways. These tables may be used when designing new lighting systems and for evaluating the adequacy of existing systems. Road surface classifications for use with the illuminance recommendations of Table 16-1 are shown in Table 16-3. Any areawide or multiple-route studies of existing lighting conditions should separate the roadways into functional classifications such as those shown in Figure 16-1 (IES, 1983).

Roadway, Pedestrian Walkway, and Bikeway Classifications

- *Freeway:* divided major roadway with full control of access and with no crossings at grade. This definition applies to toll and nontoll roads.
- *Expressway:* divided major roadway for through traffic with partial control of access and generally with interchanges at major crossroads. Expressways for noncommercial traffic within parks and parklike areas are generally known as parkways.

TABLE 16-1
Recommended Maintained Luminance and Illuminance Values for Roadways

(a) Maintained Luminance Values (L_{avg}) (Candelas per Square Meter)[a]

Road and Area Classification	*Average Luminance,* L_{avg}	*Luminance Uniformity*		*Veiling Luminance Ratio (maximum)* L_v to L_{avg}
		L_{avg} to L_{min}	L_{max} to L_{min}	
Freeway class A	0.6	3.5 to 1	6 to 1	0.3:1
Freeway class B	0.4	3.5 to 1	6 to 1	0.3:1
Expressway				
Commercial	1.0	3 to 1	5 to 1	
Intermediate	0.8	3 to 1	5 to 1	0.3:1
Residential	0.6	3.5 to 1	6 to 1	
Major				
Commercial	1.2	3 to 1	5 to 1	
Intermediate	0.9	3 to 1	5 to 1	0.3:1
Residential	0.6	3.5 to 1	6 to 1	
Collector				
Commercial	0.8	3 to 1	5 to 1	
Intermediate	0.6	3.5 to 1	6 to 1	0.4:1
Residential	0.4	4 to 1	8 to 1	
Local				
Commercial	0.6	6 to 1	10 to 1	
Intermediate	0.5	6 to 1	10 to 1	0.4:1
Residential	0.3	6 to 1	10 to 1	

(b) Average Maintained Illuminance Values (E_{avg})[b]

Road and Area Classification	*Pavement Classification*			*Illuminance Uniformity Ratio,* E_{avg} to E_{min}
	R1	*R2 and R3*	*R4*	
Freeway class A	6	9	8	3:1
Freeway class B	4	6	5	
Expressway				
Commercial	10	14	13	
Intermediate	8	12	10	3:1
Residential	6	9	8	
Major				
Commercial	12	17	15	
Intermediate	9	13	11	3:1
Residential	6	9	8	
Collector				
Commercial	8	12	10	
Intermediate	6	9	8	4:1
Residential	4	6	5	
Local				
Commercial	6	9	8	
Intermediate	5	7	6	6:1
Residential	3	4	4	

Notes:

L_v = veiling luminance.

1. These tables do not apply to high mast interchange lighting systems (e.g., mounting heights over 20 meters).

2. The relationship between individual and respective luminance and illuminance values is derived from general conditions for dry paving and straight road sections. This relationship does not apply to averages.

3. For divided highways, where the lighting on one roadway may differ from that on the other, calculations should be made on each roadway independently.

4. For freeways, the recommended values apply to both mainline and ramp roadways.

5. The recommended values shown are meaningful only when designed in conjunction with other elements. The most critical elements are lighting system depreciation, quality, uniformity, luminaire mounting height, luminaire spacing, luminaire selection, traffic conflict area, lighting termination, and alleys.

[a]For approximate values in candelas per square foot, multiply by 0.1.

[b]For approximate values in footcandles, multiply by 0.1.

Source: American National Standard Practice for Roadway Lighting, Illuminating Engineering Society of North America, 1983, IES RP-8-83, page 8, Table 2.

TABLE 16-2
Recommended Average Maintained Illuminance Level for Pedestrian Ways[a] (lux)[b]

Walkway and Bikeway Classification	Minimum Average Horizontal Levels, E_{avg}	Average Vertical Levels for Special Pedestrian Security $(E_{avg})^c$
Sidewalks (roadside) and type A bikeways		
Commercial areas	10	22
Intermediate areas	6	11
Residential areas	2	5
Walkways distant from roadways and type B bikeways		
Walkways, bikeways, and stairways	5	5
Pedestrian tunnels	43	54

[a]Crosswalks traversing roadways in the middle of long blocks and at street intersections should be provided with additional illumination.

[b]For approximate values in footcandles, multiple by 0.1.

[c]For pedestrian identification at a distance. Values at 1.8 meters (6 feet) above walkway.

Source: Illuminating Engineering Society of North America, 1983, RP-8-83, Chapter 16, Table 2, page 9.

TABLE 16-3
Road Surface Classifications

Class	Q_o[a]	Description	Mode of Reflectance
R1	0.10	Portland cement concrete road surface; asphalt road surface with a minimum of 15% of the artificial brightner (e.g., Synopal) aggregates (e.g., labradorite, quartzite)	Mostly diffuse
R2	0.07	Asphalt road surface with an aggregate composed of a minimum 60% gravel (size greater than 10 mm) Asphalt road surface with 10 to 15% artificial brightner in aggregate mix (not normally used in North America)	Mixed (diffuse and specular)
R3	0.07	Asphalt road surface (regular and carpet seal) with dark aggregates (e.g., trap rock, blast furnace slag); rough texture after some months of use (typical highways)	Slightly specular
R4	0.08	Asphalt road surface with very smooth texture	Mostly specular

[a]Q_o, representative mean luminance coefficient.

Source: Illuminating Engineering Society of North America, 1983, RP-8-83, Chapter 16, Table 16-3, page 7.

- *Major:* that part of the roadway system which serves as the principal network for through-traffic flow. The routes connect areas of principal traffic generation and important rural highways entering the city.
- *Collector:* distributor and collector roadways servicing traffic between major and local roadways. These are roadways used mainly for traffic movements within residential, commercial, and industrial areas.
- *Local:* roadways used primarily for direct access to residential, commercial, industrial, or other abutting property. Local roadways do not carry through traffic. Long local roadways will generally be divided into shorter sections by collector roadway systems. Additionally, local streets within residential areas should be grouped into two general types: (1) those serving single- or two-family homes, and (2) those serving apartments or condominiums.
- *Alley:* narrow public ways within a block, generally used for vehicular access to the rear of abutting properties.

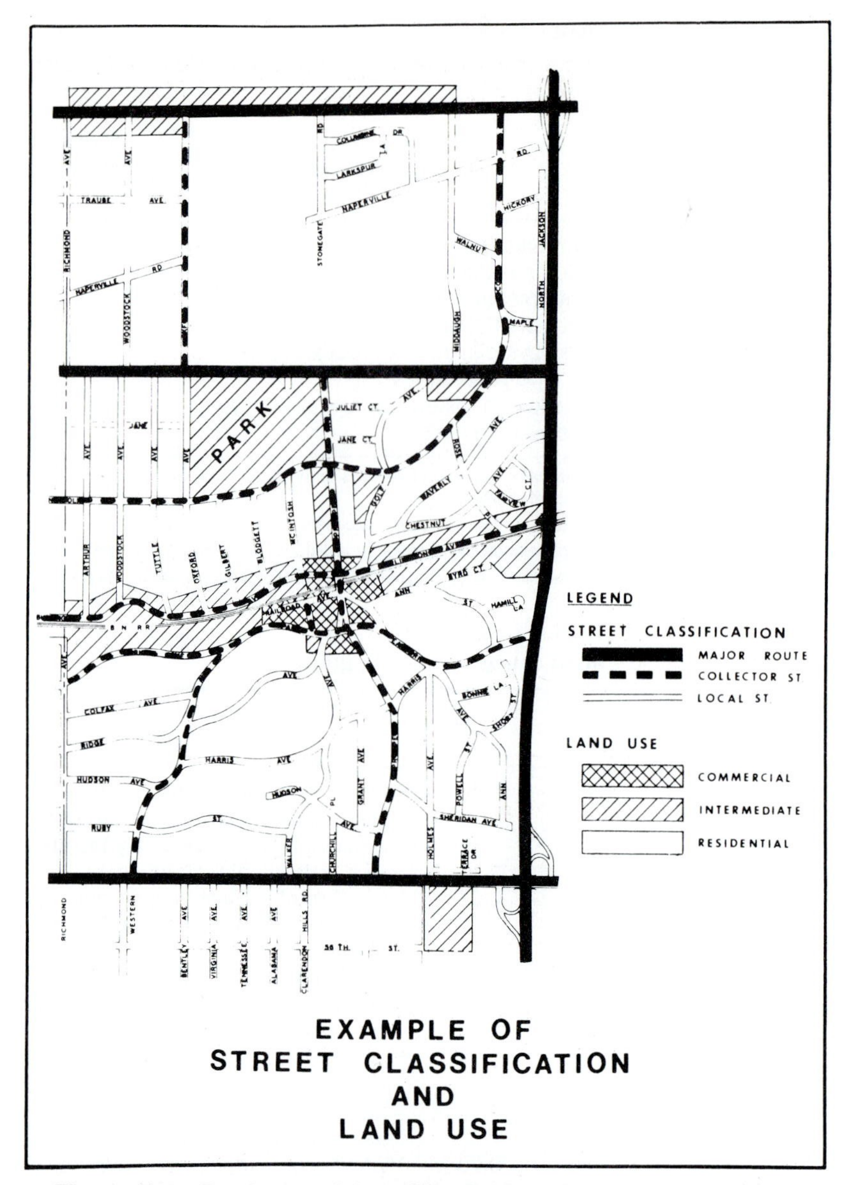

Figure 16-1 Roadway and Area Classification. *Source: American National Practice for Roadway Lighting,* Illuminating Engineering Society of North America, 1983, RP-8-83, Figure 1, page 6.

- *Footway:* paved or otherwise improved areas for pedestrian use, located within public street rights-of-way which also contain roadways for vehicular traffic.
- *Pedestrian walkway:* public walk for pedestrian traffic not necessarily within the right-of-way for a vehicular traffic roadway. Included are skywalks (pedestrian overpasses), subwalks (pedestrian tunnels), walkways giving access to parks or block interiors, and midblock street crossings.
- *Bikeway:* any road, street, path, or way that is specifically designated as being open to bicycle travel, regardless of whether such facilities are designed for the exclusive use of bicycles or are to be shared with other transportation modes. There are two general types of bikeways:

1. *Type A—designated bicycle lane:* portion of roadway or shoulder that has been designated for use by bicyclists. It is distinguished from the portion of the roadway for motor vehicle traffic by a paint stripe, curb, or other similar device.

2. *Type B—bicycle trail:* separate trail or path from which motor vehicles are prohibited and which is for the exclusive use of bicyclists or the shared use of bicyclists and pedestrians. Where such a trail or path forms a part of a highway, it is separated from the roadways for motor vehicle traffic by an open space or barrier.

Area Classifications

- *Commercial:* business or industrial area of a municipality where ordinarily there are many pedestrians during the night hours. This definition applies to densely developed business areas outside, as well as within, the central part of a municipality. The area contains land use that attracts a relatively heavy volume of nighttime vehicular and/or pedestrian traffic on a frequent basis. A commercial area without night pedestrian activity may be classified as a special land use.
- *Intermediate:* those areas of a municipality often characterized by moderate nighttime pedestrian activity such as in blocks having libraries, community recreation centers, large apartment buildings, commercial buildings, or neighborhood retail stores.
- *Residential:* residential development, or a mixture of residential and small commercial establishments, characterized by few pedestrians at night. This definition includes areas with single-family homes, condominiums, and/or small apartment buildings.
- *Problem areas:* certain land uses, such as industrial areas, office parks, commercial parks, and public parks, may be located in any of the foregoing area classifications. The classification selected should be consistent with expected nighttime pedestrian, vehicular, and other related activity. Within any given municipality there should be consistent application of the definitions.

Inventory of Existing Lighting

An inventory of the existing lighting should be made with the results shown on maps and in tabular form. The inventory should include luminaire spacing, mounting height, overhang, and type of luminaire. The data may often be available from the local utility company, contractual records, or the municipal official who is responsible for lighting maintenance. Information regarding street width, sidewalks, median areas, bordering trees, and other vegetation should also be recorded.

When analyzing lighting needs, it is helpful to use the inventory data to prepare maps that show existing lighting conditions. In general, two maps are prepared. One indicates the major traffic routes and, by means of appropriate coding, the average illuminance and/or luminance level of each street. The second map further indicates, by appropriate color coding, the levels found on residential streets. In most municipalities, various neighborhoods are lighted to different levels. Computer programs for calculating roadway illuminance and luminance are readily available, although certain required data, such as mounting height, luminaire overhang, and pavement classification, may require field checks (IES Design Practice Committee, 1981). Field measurements of actual illuminance and luminance levels are highly desirable in order to verify calculated values. Measurement techniques are well established (IES, 1983, 1987). Municipal or utility circuit maps may provide basic data that can aid in making luminance and illuminance calculations. The layout, design, construction, and maintenance of street lighting systems are complex and specialized. Guidance is available from various sources (AASHTO, 1984; IES, 1983, 1987; ITE, 1992; USDOT, 1978).

Accidents

Four general data items are needed for each accident that occurs during the study period. These are the location, severity, type, and whether a day or night occurrence. The location element includes the street name and whether it occurred at a specific intersection or at midblock. In the case of freeway studies, the six desirable location categories include:

1. On mainline between interchanges
2. On mainline within interchanges
3. At ramp exit gore from mainline
4. At ramp entrance to mainline
5. On ramp proper
6. At ramp intersection with crossroad

Accident severity classes include fatal, injury, or property damage only (PDO). In the case of fatal or injury accidents, the number of collisions involving a fatality or an injury should be used, not the number of persons killed or injured. Even when a cost analysis is being performed, it is preferable to assign average unit costs to accidents by class rather than by numbers of persons involved. Due to their low incidence of occurrence, fatal and injury accidents are often combined for analysis purposes.

The primary accident types that are useful in lighting studies are:

1. Vehicle/pedestrian and/or bicycle
2. Vehicle/vehicle
3. Vehicle/fixed object

In specialized studies, other distinctions may be useful, such as separating vehicle accidents into two categories: those involving other moving vehicles and those involving parked vehicles. Fixed-object accidents may be categorized by type of obstacle struck such as light pole, sign post, guardrail, bridge pier, and so on. Pedestrian accidents may also be of particular interest as a special study. The basic tabulation is by day or night, from the light condition indicated by the investigating officer. This item is generally coded on computer printouts and shown on collision diagrams. If not, it will be necessary to determine light conditions directly from the accident reports.

Up to 5% of accidents can be expected to be noted on the accident report as occurring at dusk or dawn. These accidents should not be classified arbitrarily. One option is to omit tabulation of this group. The disadvantages include reduction of sample size and incompleteness in total accident data for comparison with other studies, cross-checking, or use in nonlighting-related studies. If this group is included, there are two methods of placing the dusk or dawn accidents into a day or night category. The simpler method is to assume that all dawn accidents occurred during darkness and all dusk accidents during daylight. The preferred method of grouping is to assume that the time of occurrence is reasonably accurate as shown on the accident report. The local sunrise or sunset time is determined from readily available tables (see, e.g., Table 16-4) and corrected for daylight saving time if it is in effect on the accident date (U.S. Naval Observatory, 1977). If the accident occurred within the period from 15 minutes after sunset to 15 minutes before sunrise, it is classed as a night occurrence, and all accidents during the remaining period are daylight events.

Accident records obtained from computer files may pose some problems. The printouts may code the time of accident occurrence only to the closer hour and some tabulations do not identify exact location, object struck, or other pieces of information that may be vital to an accurate analysis of accidents as related to lighting. Past experience has shown errors of 20 to 60% when comparing automated data versus manual tabulation

TABLE 16-4

Sunrise and Sunset at Winston-Salem, North Carolina (Eastern Standard Time)[a]

Day	Jan.		Feb.		Mar.		Apr.		May		June		July		Aug.		Sept.		Oct.		Nov.		Dec.	
	Rise A.M.	*Set* P.M.	*Rise* A.M.	*Set* P.M.	*Rise* A.M.	*Set* P.M.	*Rise* A.M.	*Set* P.M.	*Rise* A.M.	*Set* P.M.	*Rise* A.M.	*Set* P.M.	*Rise* A.M.	*Set* P.M.	*Rise* A.M.	*Set* P.M.	*Rise* A.M.	*Set* P.M.	*Rise* A.M.	*Set* P.M.	*Rise* A.M.	*Set* P.M.	*Rise* A.M.	*Set* P.M.
1	7 32	5 17	7 22	5 47	6 52	6 16	6 08	6 43	5 28	7 08	5 05	7 32	5 07	7 42	5 28	7 26	5 52	6 49	6 16	6 05	6 43	5 25	7 13	5 07
2	7 32	5 18	7 21	5 49	6 50	6 17	6 06	6 43	5 27	7 09	5 05	7 33	5 08	7 42	5 28	7 25	5 53	6 48	6 17	6 04	6 44	5 24	7 14	5 06
3	7 32	5 19	7 21	5 50	6 49	6 17	6 05	6 44	5 26	7 10	5 05	7 34	5 08	7 42	5 29	7 24	5 54	6 46	6 17	6 02	6 45	5 23	7 15	5 06
4	7 32	5 20	7 20	5 51	6 48	6 18	6 03	6 45	5 25	7 11	5 04	7 34	5 09	7 42	5 30	7 23	5 55	6 45	6 18	6 01	6 46	5 22	7 16	5 06
5	7 32	5 20	7 19	5 52	6 46	6 19	6 02	6 46	5 24	7 11	5 04	7 35	5 09	7 41	5 31	7 22	5 55	6 43	6 19	5 59	6 47	5 21	7 17	5 06
6	7 33	5 21	7 18	5 53	6 45	6 20	6 01	6 47	5 23	7 12	5 04	7 35	5 10	7 41	5 32	7 21	5 56	6 42	6 20	5 58	6 48	5 20	7 18	5 06
7	7 33	5 22	7 17	5 54	6 44	6 21	5 59	6 48	5 22	7 13	5 04	7 36	5 10	7 41	5 32	7 20	5 57	6 40	6 21	5 56	6 49	5 20	7 18	5 06
8	7 32	5 23	7 16	5 55	6 42	6 22	5 58	6 49	5 21	7 14	5 03	7 36	5 11	7 41	5 33	7 19	5 58	6 39	6 22	5 55	6 50	5 19	7 19	5 06
9	7 32	5 24	7 15	5 56	6 41	6 23	5 56	6 49	5 20	7 15	5 03	7 37	5 11	7 41	5 34	7 18	5 58	6 38	6 22	5 54	6 51	5 18	7 20	5 06
10	7 32	5 25	7 14	5 57	6 39	6 24	5 55	6 50	5 19	7 16	5 03	7 37	5 12	7 40	5 35	7 17	5 59	6 36	6 23	5 52	6 52	5 17	7 21	5 07
11	7 32	5 26	7 13	5 58	6 38	6 25	5 54	6 51	5 18	7 17	5 03	7 38	5 12	7 40	5 36	7 16	6 00	6 35	6 24	5 51	6 53	5 16	7 22	5 07
12	7 32	5 27	7 12	5 59	6 37	6 26	5 52	6 52	5 18	7 17	5 03	7 38	5 13	7 39	5 36	7 15	6 01	6 33	6 25	5 49	6 54	5 15	7 22	5 07
13	7 32	5 28	7 11	6 00	6 35	6 26	5 51	6 53	5 17	7 18	5 03	7 39	5 14	7 39	5 37	7 14	6 02	6 32	6 26	5 48	6 55	5 15	7 23	5 07
14	7 32	5 29	7 10	6 01	6 34	6 27	5 49	6 54	5 16	7 19	5 03	7 39	5 14	7 39	5 38	7 13	6 02	6 30	6 27	5 47	6 56	5 14	7 24	5 07
15	7 31	5 30	7 09	6 02	6 32	6 28	5 48	6 54	5 15	7 20	5 03	7 40	5 15	7 38	5 39	7 11	6 03	6 29	6 28	5 45	6 57	5 13	7 24	5 08
16	7 31	5 31	7 08	6 03	6 31	6 29	5 47	6 55	5 14	7 21	5 03	7 40	5 16	7 38	5 40	7 10	6 04	6 27	6 28	5 44	6 58	5 13	7 25	5 08
17	7 31	5 32	7 07	6 04	6 29	6 30	5 45	6 56	5 13	7 21	5 03	7 40	5 16	7 37	5 40	7 09	6 05	6 26	6 29	5 43	6 59	5 12	7 26	5 08
18	7 30	5 33	7 05	6 05	6 28	6 31	5 44	6 57	5 13	7 22	5 03	7 40	5 17	7 37	5 41	7 08	6 05	6 24	6 30	5 42	7 00	5 11	7 26	5 09
19	7 30	5 34	7 04	6 06	6 27	6 32	5 43	6 58	5 12	7 23	5 04	7 41	5 18	7 36	5 42	7 06	6 06	6 23	6 31	5 40	7 02	5 11	7 27	5 09
20	7 30	5 35	7 03	6 07	6 25	6 32	5 42	6 59	5 11	7 24	5 04	7 41	5 19	7 35	5 43	7 05	6 07	6 21	6 32	5 39	7 03	5 10	7 28	5 10
21	7 29	5 36	7 02	6 08	6 24	6 33	5 40	7 00	5 11	7 25	5 04	7 41	5 19	7 35	5 44	7 04	6 08	6 20	6 33	5 38	7 04	5 10	7 28	5 10
22	7 29	5 37	7 01	6 09	6 22	6 34	5 39	7 00	5 10	7 25	5 04	7 41	5 20	7 34	5 44	7 03	6 09	6 18	6 34	5 37	7 05	5 09	7 29	5 11
23	7 28	5 38	6 59	6 10	6 21	6 35	5 38	7 01	5 09	7 26	5 04	7 42	5 21	7 34	5 45	7 01	6 09	6 17	6 35	5 35	7 05	5 09	7 29	5 11
24	7 28	5 39	6 58	6 11	6 19	6 36	5 37	7 02	5 09	7 27	5 05	7 42	5 21	7 33	5 46	7 00	6 10	6 15	6 36	5 34	7 06	5 09	7 29	5 12
25	7 27	5 40	6 57	6 12	6 18	6 37	5 35	7 03	5 08	7 28	5 05	7 42	5 22	7 32	5 47	6 59	6 11	6 14	6 37	5 33	7 07	5 08	7 30	5 12
26	7 26	5 41	6 56	6 13	6 16	6 38	5 34	7 04	5 08	7 28	5 05	7 42	5 23	7 31	5 48	6 57	6 12	6 12	6 38	5 32	7 08	5 08	7 30	5 13
27	7 26	5 42	6 54	6 14	6 15	6 38	5 33	7 05	5 07	7 29	5 06	7 42	5 24	7 31	5 48	6 56	6 12	6 11	6 39	5 31	7 09	5 07	7 31	5 13
28	7 25	5 43	6 53	6 15	6 13	6 39	5 32	7 06	5 07	7 30	5 06	7 42	5 24	7 30	5 49	6 55	6 13	6 09	6 40	5 30	7 10	5 07	7 31	5 14
29	7 24	5 44	6 53	6 16	6 12	6 40	5 31	7 06	5 06	7 30	5 06	7 42	5 25	7 29	5 50	6 53	6 14	6 08	6 41	5 28	7 11	5 07	7 31	5 15
30	7 24	5 45			6 11	6 41	5 30	7 07	5 06	7 31	5 07	7 42	5 26	7 28	5 51	6 52	6 15	6 06	6 41	5 27	7 12	5 07	7 32	5 16
31	7 23	5 46			6 09	6 42			5 06	7 32			5 27	7 27	5 51	6 50			6 42	5 26			7 32	5 16

[a]Add 1 hour for Daylight Saving Time if and when in use.

Source: Gale Research Co., Detroit, MI.

from original accident report forms (Box, 1971). If there are serious doubts regarding the validity of the computer printout data, the original accident report forms should be used. Suggestions for locating and possibly correcting erroneous accident data are given in Chapter 11.

The effect of lighting on roadway accidents is tenuous and easily masked by other variables. Controls in variability can be achieved by careful selection of the analysis method, provided that an adequate accident sample size is available. In an existing condition study, it is generally desirable to use 3 to 4 years of data. In a before-and-after study, 2 years of data are desirable during each period. Such periods minimize chance findings. To account for any time effects, it is advisable to include a set of unimproved control sites, having essentially similar characteristics as the improved sites, in the after period. More details regarding the design of before-and-after experiments are given in Appendix A.

The decision criteria shown in Figure A-3 of Appendix A may be used to determine whether a change in the number of accidents recorded during a before-and-after study is significant. These criteria assume that accidents are Poisson-distributed and also recognize that the before count is a random variable. To illustrate the use of Figure A-3, consider an urban roadway that received a lighting improvement. A total of 12 nighttime accidents were recorded during the 2-year before period prior to the improvement, and three occurred during the 2-year after period following the improvement. This represents a decrease of (12 − 3)/12 = 75%, which is greater than the figure value of 67% for a 95% confidence level. Therefore, it may be concluded with 95% certainty that the reduction in accidents was not due to chance alone. If the number of accidents in the before period had been considerably larger, say 26 occurrences, the same conclusion would be reached if the number of accidents in the after period were reduced to 13 or fewer, since the 26/13 = 50% change is greater than the figure value of 46%.

Traffic Volumes

Night traffic volume data are generally gathered by use of automatic-type traffic counters. Counts should be selected so as to represent traffic during various seasons of the year. When averaged, January, April, July, and October counts yield a good yearly estimate. Count data should also be selected to represent different types of routes, such as:

1. Industrial streets, which may show very low night volumes
2. Primary business streets, which may show high night volumes on days of evening shopping and low night volumes on days of early closing
3. Boulevards or parkways, which may show high night volumes in summer and low volumes in winter

In general, other types of routes show little daily variation. Volume counts are tabulated for both the night period and for the full 24 hours.

A typical computation of night traffic volume at a single location is made as follows:

- Sunset: 1905 + 15 minutes = 1920
- Sunrise: 0548 − 15 minutes = 0533
- Night traffic = (1920 to 0533) = 4200
- 24-hour count = 17,000
- Night percentage 4200/17,000 = 25%

The night percentages are computed separately for each seasonal count at the location. These percentage are then averaged to provide the typical annual percentage of night traffic by type of route. This step is repeated for each route.

Night traffic percentage by type of route may be applied individually when conducting lighting studies. However, the type of abutting land use typically changes from point-to-point along major traffic routes. Portions of a given route may carry predominantly industrial-type traffic; other portions may carry shopper traffic; and still others, social-recreational traffic. To simplify the accident studies, a citywide major route night volume percentage should be used.

A separate residential night volume percentage may easily be computed by averaging four seasonal counts for one typical street in each of several different neighborhoods:

1. Single-family—low cost
2. Single-family—medium cost
3. Single-family—high cost
4. Multiple-family—low rent
5. Multiple-family—medium rent

Past experience indicates a consistent range of 23 to 27% of total vehicle miles are traveled at night and the use of an average value of 25% for the night portion of the 24-hour volume appears warranted (Box, 1971). If it is desired to determine the actual percentage of night traffic at a given location, hourly tabulations of volume are needed for all 365 days of the year and should include the volumes in both directions of travel. These hourly data are generally available only from a limited number of automatic recording stations along freeways or other high-volume facilities.

ANALYSIS

Control for Normal Variation

No two routes have identical traffic, geometric, or operational characteristics. Comparative studies of similar routes are valid only to the degree that similarity actually exists. Accident rates per mvm may change with volume, and therefore miles of travel do not truly represent exposure. Thus a simple comparison of night accident rates among different routes, even of the same functional class, cannot be expected to yield meaningful results. The ideal method of comparison is to test a facility against itself, with the basic variable being day or night condition. This can be done in two ways: by calculation of the percentage of total night accidents that occurred along the route or by calculation of the *N/D* ratio of rates (night accident rate per mvm divided by day accident rate per mvm). At intersections, the percentage of total accidents occurring at night is the usual comparison; however, if the *N/D* ratio is used, it should be based on millions of vehicles entering the intersection.

Use of Night Percentage

The percentage of accidents at night ($\%N$) is directly calculated, as shown below:

$$\%N = \frac{100A_N}{A_N + A_D} \qquad (16\text{-}1)$$

where A_N is the number of night accidents and A_D is the number of day accidents. Since approximately 25% of vehicle travel is at night, a condition of $N = 25\%$ would typically represent a night accident rate at night equal to the day rate. A night accident percentage in excess of the night traffic percentage may be evidence of inadequate illumination.

A major study of lighted urban freeways found an average of 32% of accidents occurring at night, while the unlighted freeway average was 44% (Box, 1971). Similar findings have been reached in a before-and-after study of major routes (Box, 1989). Based on such research, a rule of thumb is sometimes applied, with night proportions up to 30 or 35% representing reasonable safety, while greater percentages may indicate a need for lighting improvement.

Ratio of Rates

When the actual vehicle miles of day and night travel are known, the *N/D* rate ratio can be calculated directly by dividing the night accident rate by the day accident rate. The study of freeways found an average *N/D* accident rate ratio of 1.4 for lighted routes and 2.4 for unlighted routes, while the major route study *N/D* accident rate ratio was 1.35 during the before study and 0.87 in the after study. The ratio of rates can also be calculated directly if the percentage of night travel is known or estimated. In such a case, data on vehicle-miles of travel are not required. This method, which greatly simplifies data gathering, uses the following equation:

$$R = \frac{A_N(1 - P)}{A_D P} \tag{16-2}$$

where

R = *N/D* ratio of accident rates
P = percent of travel at night
A_N = number of night accidents
A_D = number of day accidents

For instance, if P is taken as 25%, the equation simplifies to

$$R = \frac{3A_N}{A_D} \tag{16-3}$$

SCHEDULING IMPROVEMENTS

A cost-effective lighting installation or improvement program requires use of both hazard identification (such as high night percentage of accidents or high *N/D* rate ratio) and the annual number of night accidents (usually averaged over the last 2 years). The added cost of providing lighting must also be considered. Table 16-5 shows how two alternative lighting projects could be compared. Project A is evidently more effective, although a simple ranking using night percentage would rank project B as first.

The *N/D* rate ratio can also be used to rank improvements, where the expected number of night accidents (E_N) is given by the following formula, using previously defined variables:

$$E_N = \frac{RA_DP}{1 - P} \tag{16-4}$$

TABLE 16-5
Cost Comparison for Two Lighting Improvement Projects

	Project A	*Project B*
1. Existing night percentage of accidents	40%	50%
2. Expected night percentage of accidents[a]	30%	30%
3. Annual number of existing total accidents	100	40
4. Annual number of existing night accidents (No. 1 × No. 3)	40	20
5. Annual number of expected night accidents (No. 2 × No. 3)	30 (30%)	12 (30%)
6. Annual accident reduction by lighting (No. 4 × No. 5)	10	8
7. Added annual lighting cost[b]	$3000	$3000
8. Added cost per reduced accident	$300	$370

[a]Based on assumed local experience for routes provided with recommended illumination level.

[b]From assumed lighting design calculations of amortized capital cost, plus annual maintenance and energy charges, minus existing lighting costs (if any).

The E_N value is subtracted from the existing number of night accidents, and the annual cost per reduced accident is calculated as in the preceding example. A separate calculation is made for each project, and the rankings then follow in simple order, starting with the lowest annual cost per reduced accident.

Since pedestrian accidents are especially susceptible to reduction by improved lighting, this class of accident can be singled out for separate analysis. Priorities can be established along major routes by directly calculating the number of accidents per mile and total number at each intersection. Sound engineering judgment must be used to prevent one or two unusual accidents from creating unrealistic programming. Priority schedules for residential lighting may be established by comparing crime percentages (night percentage of total daily) for the various areas or by using the total number of night crimes. The nature of the crimes may also be a consideration.

REFERENCES

AASHTO (1984). *An Informational Guide for Roadway Lighting,* American Association of State Highway and Transportation Officials, Washington, DC.

Box, P. C. (1971). Relationship between illumination and freeway accidents, *Illuminating Engineering,* May–June.

Box, P. C. (1972a). *Comparison of Accidents and Illumination,* Highway Research Record 416, Highway Research Board, Washington, DC, p. 10.

Box, P. C. (1972b). *Freeway Accidents and Illumination,* Highway Research Record 416, Highway Research Board, Washington, DC.

Box, P. C. (1989). *Major Road Accident Reduction by Illumination,* Transportation Research Record 1247, Transportation Research Board, Washington, DC.

IES (1983). *American National Standard Practice for Roadway Lighting,* ANSI/IES RP-8-1983, Illuminating Engineering Society of North America, New York.

IES (1987). *IES Lighting Handbook,* 1987 Application Volume, Section 14, *Roadway Lighting,* Illuminating Engineering Society of North America, New York.

IES Design Practice Committee (1981). Available lighting computer programs: a compendium and a survey, *LD&A,* March, p. 35.

ITE (1992). *Traffic Engineering Handbook,* 4th ed., Institute of Transportation Engineers and Prentice Hall, Englewood Cliffs, NJ.

Tien, J. M. (1979). Lighting's impact on crime, *LD&A,* December, p. 20.

USDOT (1978). *Roadway Lighting Handbook,* Implementation Package 78-15, Federal Highway Administration, Washington, DC, December.

U.S. Naval Observatory, Nautical Almanac Office (1977). *Sunrise and Sunset Tables for Key Cities and Weather Stations of the U.S.,* Gale Research Co., Detroit, MI (out-of-print).

Walker, F. W., and S. E. Roberts (1976). *Evaluation of Lighting On Accident Frequency at Highway Intersections,* Transportation Research Record 562, Transportation Research Board, Washington, DC, p. 73.

17

Public Transportation Studies

Joseph E. Hummer, Ph.D., P.E.

INTRODUCTION

Public transportation studies provide operators with the information they need to make intelligent choices about services. Studies provide the numbers and trends which indicate that changes in operations are needed. Studies indicate whether patrons have responded to changes. Studies also provide data for comparisons between agencies, to measure the quality of service and to provide the basis for funding decisions. It is important that public transportation studies be conducted properly. A poorly designed or executed study may be worse than no study because the numbers may provide operators with extra confidence in a mistaken course of action. In addition, studies may consume large quantities of scarce funds. A staff of over 30 full-time data collection personnel (referred to as "checkers" by public transportation professionals) is not unusual in a large agency, and automatic data collection equipment can cost tens of thousands of dollars per bus or railcar.

In this chapter public transportation studies are introduced to novice public transportation professionals, checkers, and students. Experienced professionals can use this chapter as a quick reference. Agencies or companies spending large sums to conduct important studies should refer to one of the comprehensive manuals produced by the Urban Mass Transportation Administration (UMTA) (1985a,b) before proceeding. Note that in 1991, UMTA became the Federal Transit Administration (FTA). The chapter applies primarily to studies of fixed-route bus transportation. Many of the techniques discussed also apply for other fixed-route modes. Demand-responsive (i.e., paratransit) services have different purposes and vastly different ways of operating than do fixed-route services, so their study techniques are generally different. Those interested in studies of paratransit operations are referred to a manual produced for UMTA (Dornan and Middendorf, 1983).

The studies discussed in this chapter are primarily those which are useful for operators studying their own services. Studies are also conducted to meet the needs of outside funding sources, investors, the media, and the public. In the United States, agencies that receive federal funding conduct studies to provide Section 15 data to the FTA. Operators conduct those studies under very rigid sets of rules (see UMTA, 1985c, for Section 15 rules) that are not repeated in this chapter. This chapter also focuses on the conduct of the study rather than the establishment of data collection programs or the manipulation of study results. Much has been written on both of those topics, and each could occupy an entire manual. Finally, the chapter is limited to studies requiring actual field data collection and does not include any discussion of the collection and analysis of routine management information. Counting revenue, counting transfers, conducting an inventory of equipment and parts, collecting personnel data, compiling cost information, and other continuous management information efforts are outside the scope of this chapter.

CHOICE OF STUDY AND METHOD

Problem Identification

The type of public transportation study to be conducted depends on the problem facing the operator. Operators could have problems planning future services, scheduling current services, evaluating current services, controlling current services, or reporting to funding sources and investors. The first task when conducting a public transportation study is, as for other studies, to identify the problem to be solved.

Data Items Needed

For each of the common problems facing operators, there are one or more data items that they could collect during a study. The second task in conducting a public transportation study is to select the type of data to be collected. Table 17-1 shows the most common data items collected in public transportation studies and their related uses. Boardings are the most common data item for public transportation studies because they can be applied to the widest range of problems. Analysts must make sure that the data items identified for study are relevant to the problem. In addition, analysts should identify a proven solution methodology that requires the data item before spending money on data collection.

TABLE 17-1
Common Public Transportation Study Data Items and Uses

Data Item	*Uses*
Load at peak point or other key point	Scheduling, planning
Running time and delay	Scheduling, planning
Schedule adherence at specified points	Scheduling, evaluation, control
Boardings	Scheduling, evaluation, planning, reporting
Distribution of boardings by fare category	Planning, marketing
Boarding and alighting by stop	Planning
Passenger miles	Evaluation, reporting
Passenger characteristics and attitudes	Planning, marketing
Passenger origin and destination pattern along route	Planning, marketing

Source: Based on UMTA, 1985b.

Usually, the selection of a data item for study is straightforward. There may be only one relevant data item. If several data items are relevant to the problem, it may be possible to collect them simultaneously. Analysts must take the time to understand the problem thoroughly, however, or they may overlook a relevant data item. For example, analysts evaluating advertisements for public transportation services will usually collect boardings, and may be interested in boardings by fare category, as measures of behavior. Since advertising usually has the objective of changing attitudes and awareness as well as behavior, analysts should also collect data on passenger attitudes and awareness in this case (Everett et al., 1987). Data items selected for study must be closely related to the problem or to the objectives of the program being evaluated.

Data Items to Be Collected

After identifying the data items needed, the next task in conducting a public transportation study is to decide which data items are to be collected in the field. Usually, the data items identified as needed to solve the problem will be the same as those to be collected. Two distinct possibilities arise when data items that solve the problem are not the same as the items to be collected. First, the data items identified to solve the problem could be available from some previous study or continuous collection effort. Use extreme caution before reusing old data, since conditions in public transportation change relatively quickly. Data items that need updating each season or each year for which older data are not reliable include (UMTA, 1985b):

- Peak point load
- Boardings
- Passenger miles
- Schedule adherence
- Running time

Second, the data item identified may be approximated accurately from an auxiliary data item that is much cheaper to collect. Table 17-2 lists examples of common data items that can be accurately inferred from easier-to-collect auxiliary data items. Furth and McCollom (1987) provide equations for determining an estimate of the desired data item from the auxiliary data item, for determining the accuracy of the estimate, and for determining the sample size necessary to derive a conversion factor between the auxiliary and inferred data items.

To reduce the need for "ad hoc" data collection efforts, some agencies develop continuous data collection programs. These programs often consist of a baseline phase and a monitoring phase (UMTA, 1985b). The baseline phase provides a comprehensive one-time "snapshot" of system operations. Data items collected simultaneously during the baseline phase provide all the data necessary for long-term planning and other infre-

TABLE 17-2
Auxiliary Data Items

Auxiliary Data Item	*Inferred Data Item*
Peak point load or revenue	Boardings
Boardings, peak point load, or revenue	Passenger miles
Peak point load	True maximum load[a]
Load at point near peak point	Peak point load or true maximum load
Revenue	Peak point load, true maximum load

[a]True maximum load is the maximum number of passengers observed on a trip, while peak point load is the number of passengers observed on board where the load is expected to be highest.
Source: UMTA, 1985b.

quent major efforts, provide conversion factors for auxiliary items, and provide the beginning points for trend analyses. During the monitoring phase, key data items are updated as needed, using conversion factors whenever possible. During the monitoring phase, operators should also make periodic checks to reveal problems that require attention.

Choosing a Study Method

Following the identification of data items to be collected, the analyst chooses a study method. There is a surprisingly small number of study methods in widespread use. Most studies are performed by checkers in the vehicle (ride checks), by checkers stationed along the route (point checks), by drivers, or with automatic data collection equipment. Analysts use surveys during some studies. Table 17-3 shows the study methods available for the data items listed in Table 17-1. The next section of this chapter describes each of these major methods in greater detail.

For some data items, there is no choice of study method. For other items, there are several choices. The prime determinants when a choice is presented are availability of resources, measurement errors, and cost. Availability of resources means that questions such as the following must be addressed:

- *Driver studies:* Do driver work rules permit data collection?
- *Point and ride checks:* Are sufficient numbers of trained checkers available? If not, can checkers be hired and trained?
- *Studies with automatic data collection equipment:* Is the equipment available? If not, can the funds be procured and allocated to purchase and maintain the equipment?
- *Surveys:* Is the expertise to write, administer, and analyze a survey available?

If the resources are not available to conduct the study using a particular method, analysts must explore alternative methods, change the data item to be collected, or abandon the study.

Measurement errors occur with every study method. For driver studies, increasing passenger loads and heavy traffic are major factors in causing error. For point checks, many standees and vehicles with tinted windows lead to measurement error. For ride checks, simultaneous boarding and alighting, two or more doors in use, a multitude of

TABLE 17-3
Study Methods Used to Collect Common Data Items

Data Item	*Study Method*
Load at peak point or other key point	Point check, ride check
Running time and delay	Ride check, trail car
Schedule adherence at specified points	Point check, ride check, trail car
Boardings	Driver study, ride check
Distribution of boardings by fare category	Driver study, ride check
Boarding and alighting by stop	Automatic data collection equipment, ride check
Passenger miles	Automatic data collection equipment, ride check
Passenger characteristics and attitudes	Survey
Passenger origin and destination pattern along route	Special ride check, survey, inferred from automatic data collection equipment, point check, ride check

Source: UMTA, 1985b.

fare categories, and crowded vehicles cause error. Automatic data collection equipment can fail systematically (due to large crowds, passengers with large packages blocking sensors, etc.) or can fail from mechanical or electrical problems. Finally, surveys are extremely sensitive in the choice of sample, format, wording of questions, and many other ways, as described in Appendix B. When considering a study method, analysts must consider any problem particular to that method and the data item needed. For a boarding count, which is the most common type of public transportation study, automatic data collection equipment will generally provide the most accurate data. The ride check method will be close to automatic data collection equipment in accuracy, and point check data will be less accurate (UMTA, 1985b; Deibel and Zumwalt, 1984).

Cost is more important than measurement error in choosing a study method. This is because all the methods in Table 17-3 provide reasonably accurate data and because increased sample sizes can reduce the effects of many measurement errors. The choice of a study method is easy when costs are considered, since driver studies are generally the least expensive study method and point checks are the next least expensive study method. Thus UMTA (1985b) recommends:

1. Automatic data collection systems, if available
2. Driver studies, for items that cannot be collected automatically
3. Point checks, for items that cannot be collected automatically or by drivers
4. Ride checks only for items that cannot be collected automatically, by drivers, or by point checks

A survey is a relatively expensive and risky last resort when none of the other common methods can deliver the needed data items. There are a few circumstances when ride checks are less expensive than point checks, such as when many simultaneous point checks are needed along a route or when headways are long.

STUDY METHODS

Driver Studies

Having drivers count boarding passengers or boarding passengers by fare category is a very efficient and accurate study method. The method is efficient because no labor costs for checkers are incurred. Driver counts are accurate because they are kept simple, because drivers are monitoring boarding passengers anyway, and because drivers are familiar with the various fare categories being recorded. For example, the Chicago Transit Authority reports that its continuous count of boardings by fare category is 98% accurate (Foote and Hancox, 1989).

Driver studies are impossible on unmanned vehicles and when passengers can board the vehicle through more than one door. In these cases, transit agencies must use other study methods.

The driver can keep the count on a mechanical counter or, as in the case of the Chicago Transit Authority, the count board may be integrated into an electronic farebox. Usually the count board presents no more than six keys to the driver so some fare categories may have to be recorded together (UMTA, 1985a). The driver could write the total count or the count per fare category on a summary form like that given in Figure 17-1 at the end of each trip, or the driver could enter the code for "end of trip" into an electronic count board. Appendix G provides blank data collection forms suitable for copying.

Statistics produced from driver studies include the mean number of passengers per trip and the proportions of passengers by fare category. The analysis of statistics produced in public transportation studies is discussed later in the chapter.

BOARDING COUNT FIELD SHEET

ROUTE ______________________ BLOCK NUMBER ______________________

DAY __________ DATE __________ WEATHER ______________________

OBSERVER ______________________

Route Segment		Boarding Passengers					
From	To	Full Fare	Reduced Fare	Transfer	Full + Transfer	Reduced + Transfer	All Passes

Figure 17-1 Boardings by fare category field sheet. *Source:* UMTA, 1985a.

Point Checks

Checkers conduct point checks while standing along a route, usually at a stop, watching boarding, alighting, and on-vehicle activity. Point checks are efficient if the point has moderate to high levels of public transportation activity because the checker will be able to collect information from many different vehicles. Point checks are also efficient if both directions of a street can be monitored by one checker. Point checks are reasonably accurate, although not as accurate as ride checks or automatic data collection equipment. Operators should not consider point checks when the checker is particularly vulnerable to crime or foul weather.

Point checks primarily provide load counts and schedule adherence. Figure 17-2 provides a convenient data form for these primary items. Checkers will need an accurate watch to record schedule adherence. Secondary purposes of point check studies include:

- Boarding and alighting counts at the particular point
- Deceleration, stop, and acceleration times at the particular point

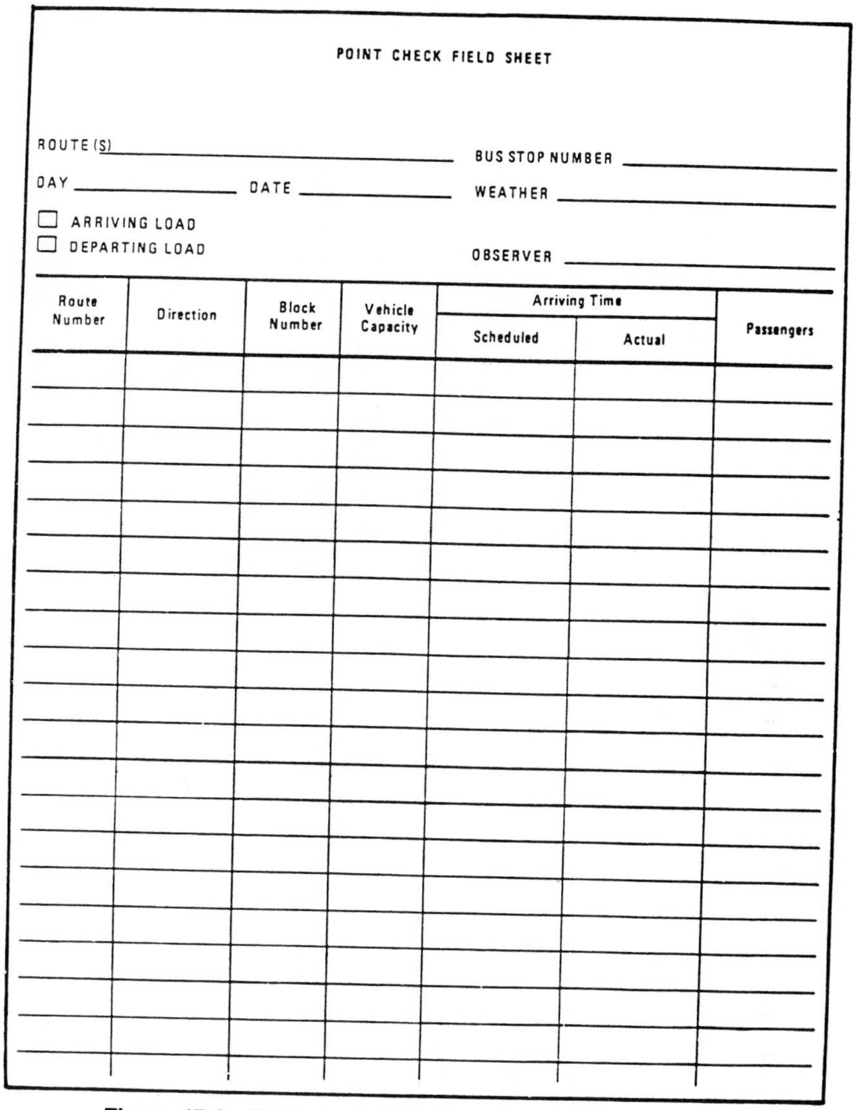

POINT CHECK FIELD SHEET

ROUTE (S) ______ BUS STOP NUMBER ______

DAY ______ DATE ______ WEATHER ______

☐ ARRIVING LOAD

☐ DEPARTING LOAD

OBSERVER ______

Route Number	Direction	Block Number	Vehicle Capacity	Arriving Time		Passengers
				Scheduled	Actual	

Figure 17-2 Point check field sheet. *Source:* UMTA, 1985a.

- Data on passenger behavior, such as arrival times, wait times, directions of travel before or after transit trip, and modes of travel before or after transit trip at the particular point

Operators can alter the data form in Figure 17-2 to include columns for these secondary elements if they are of interest. Appendix G contains blank data collection forms suitable for copying.

During most point checks, the checker estimates the load while standing outside the vehicle. An efficient technique is to count the passengers if the load is light, count empty seats if the load is moderate, and count standees if the load is heavy (UMTA, 1985a). Checkers should know the seating and standing capacities of all vehicles subject to data collection. Checkers may have to board the vehicle for a quick count if tinted glass, heavy loads, obstructed views, or other factors preclude a count from the roadside. Drivers should be made aware that a study is under way if boarding is anticipated. Load counts are difficult to conduct for both directions on a wide street. However, recording schedule adherence on both directions of a street is not a problem at any point where the view is not obstructed.

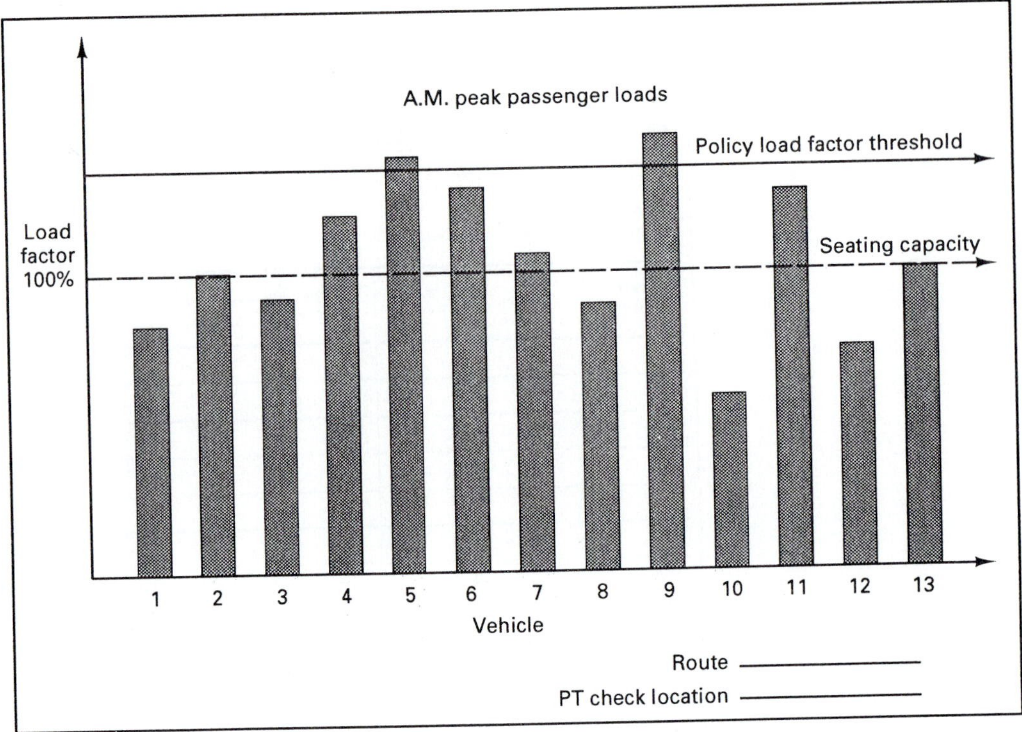

Figure 17-3 Vehicle load factor profile. *Source:* UMTA, 1985a.

Operators usually conduct point checks at the peak load point of a route (i.e., the point where the maximum number of passengers is expected). Other key points along a route may be of interest, such as transfer points and points where vehicles may turn around short of the end of the route. It is very efficient to conduct a point check where several routes overlap, even if that point does not happen to be a peak load or key point for each route. Sometimes, operators conduct several point checks simultaneously on one route. Furth (1989) provides a methodology for estimating the passenger origin and destination pattern along a route from several simultaneous point checks and previous ride check or survey data.

Office personnel usually enter data into a computer manually from point check forms. However, the Toronto Transit Commission reports that optical scanning of special point check forms is quicker (although not necessarily more accurate) than manual entry (Pratt, 1991). Optical scanning may be useful for other surveys as well.

Standard point check data provide several useful plots and statistics. Figures 17-3 and 17-4 show two of the more useful plots with load data and with schedule adherence and load data, respectively. The vertical axis in Figure 17-3 is the *load factor* in percentage of seating capacity, rather than the actual number of passengers, to remove the effects of different sizes of vehicles. In a section later in this chapter the analysis of relevant statistics such as the mean load and the proportion of vehicles on schedule is discussed.

Ride Checks

A checker conducts a ride check by riding in the vehicle of interest as it covers its route. Ride checks provide more accurate data than point checks and are comparable in accuracy to automatic data collection equipment under most typical loading conditions for many data items. Ride checks are usually much more expensive than driver studies and point checks because a checker is gathering data on only one trip at a time. In addition,

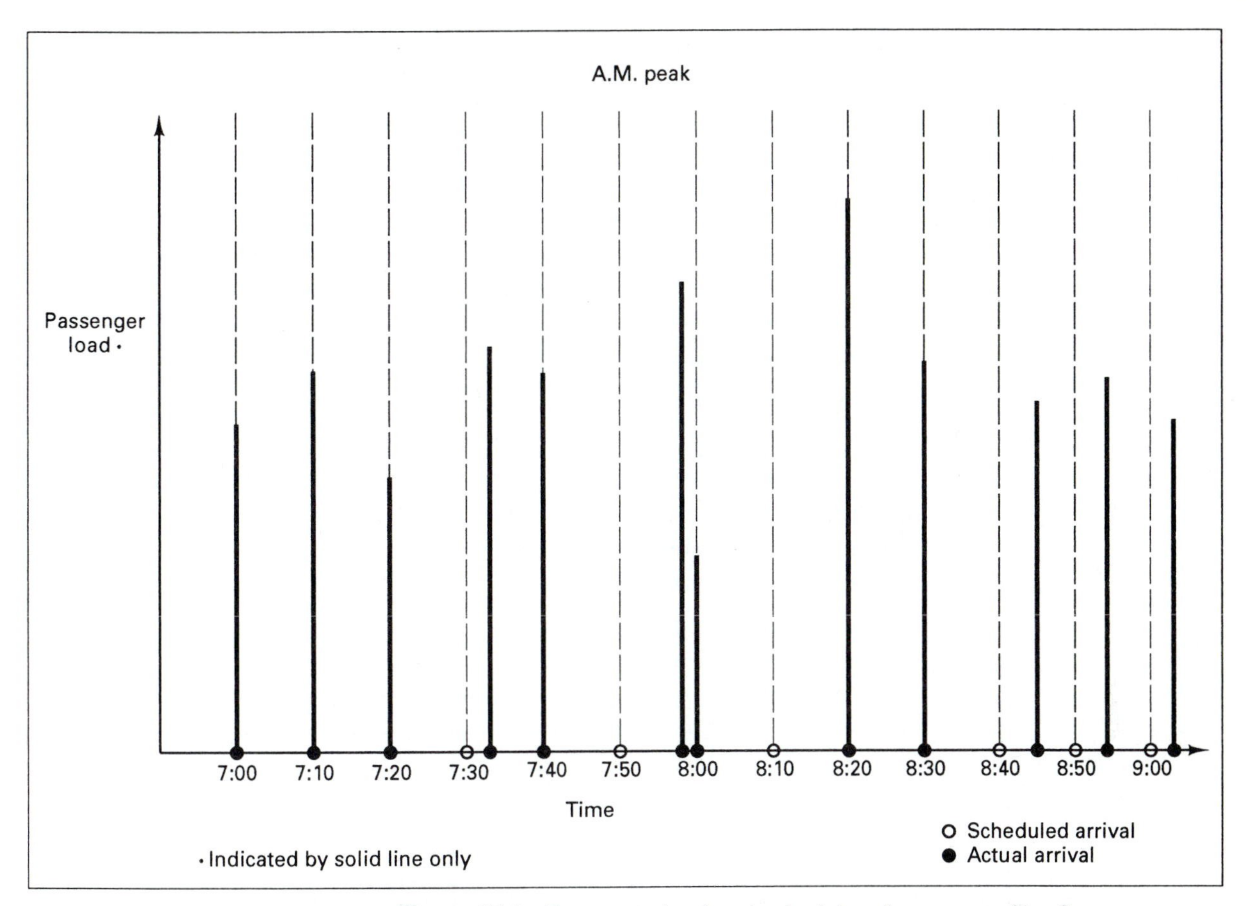

Figure 17-4 Passenger load and schedule adherence profile. *Source:* UMTA, 1985a.

ride checkers lose considerable time finding or waiting for the assigned vehicle and traveling between assignments.

Ride checks are flexible and provide several useful data items. Ride checks most often provide boarding and alighting volume and the vehicle arrival time at each stop and at other key points. Figure 17-5 shows a form suitable for recording these basic items. A checker must wear an accurate watch to record arrival times. Programs are available that allow data to be recorded on a hand-held or portable computer and later transferred to a personal computer for analysis (see e.g., Barnes and Urbanik, 1990).

Ride checks sometimes provide more detailed data on vehicle movements than simply arrival time at stops. Ride checkers can record deceleration times, dwell times at stops, acceleration times, and durations and causes of delays other than stops for passengers. Figure 17-6 shows a form for these speed and delay items. If they are busy, checkers do not need to compute column 7 for delay time in the field. Column 2 for the time at control points allows schedule adherence and various speed measures to be calculated. A ride checker using a form such as the one shown in Figure 17-6 can rarely collect data on passenger boarding and alighting at the same time.

A checker in an automobile trailing the transit vehicle of interest can also collect the speed and delay items on the form shown in Figure 17-6. The trail car method is advantageous because the checker is (perhaps) out of sight of the transit vehicle driver, so no bias is introduced by changed driver habits. However, the trail car method is more expensive than the ride check method because it always requires an automobile and under heavy traffic conditions may require a second checker. The trail car method also does not allow a full view of the causes of delays. Because the disadvantages of the trail car

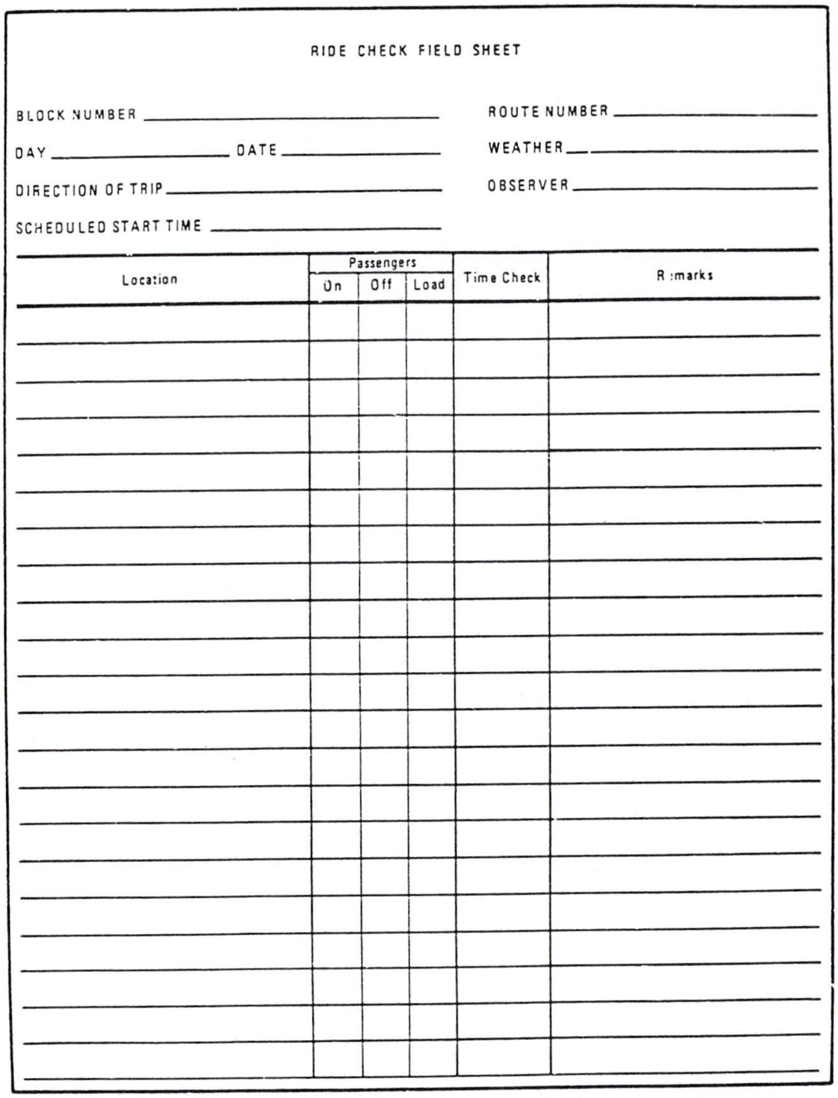

RIDE CHECK FIELD SHEET

BLOCK NUMBER ______________ ROUTE NUMBER ______________

DAY ____________ DATE ____________ WEATHER ______________

DIRECTION OF TRIP ______________ OBSERVER ______________

SCHEDULED START TIME ______________

Location	Passengers			Time Check	Remarks
	On	Off	Load		

Figure 17-5 Ride check field sheet. *Source:* UMTA, 1985a.

method rarely outweigh the advantages, operators use the ride check method more often for speed and delay data items.

Ride checks can also provide many of the data items that driver studies and point checks provide. Checkers can count boardings by fare category during a ride check. Load counts and boarding counts per trip can be made directly during a ride check or can easily be derived from a ride checker's boarding and alighting by stop data. Ride checkers usually occupy the seat directly behind the driver. Boarding and alighting from the front door, the causes of delay, and the type of fare paid are all best observed from that seat. Boarding and/or alighting through a rear door, with many standees, may require a change in position or may mean that the checker stands to conduct the counts. In rare instances, more than one ride checker may be needed on a vehicle. Checkers must be familiar with the causes of delay or the possible fare categories, depending on the data items being collected.

Ride check data are analyzed in several ways. Figure 17-7 shows how boarding and alighting by stop data are often analyzed. An alternative to the bar graph of the load profile in Figure 17-7 is to vary the width of the bars by the distance between stops. It is also common with boarding and alighting by stop data to calculate the passenger miles

PUBLIC TRANSPORTATION VEHICLE DELAY FIELD SHEET

Day:
Date:
Weather:
Route:
Direction:
Vehicle Type:

Observer:
Method:
Trip Number:
Trip Start Time:
Trip End Time:

(1) Location	(2) Time at control point	(3) Time slower than walking speed	(4) Stop time	(5) Time faster than walking speed	(6) Delay cause	(7) (5) – (3) Delay time (sec.)
					TOTAL DELAY TIME, SECONDS	

Symbols for delay cause: P = passenger loading, S = traffic signal, SS = stop sign, PK = parked cars, DP = double parked, PED = pedestrians, RT = right turns, LT = left turns, T = general congestion, KT = intentionally killed time, O = other (explain).

Remarks:

Figure 17-6 Public transportation vehicle delay field sheet.

per trip. The passenger load between two stops is multiplied by the distance between the stops; the number of passenger miles per trip is then the sum of these quantities for the entire route. Operators examine transit vehicle time and delay data by producing a space and time plot as in Figure 17-8 or a bar graph with magnitudes of delay by type as in Figure 17-9. Useful statistics from ride check data include the mean number of passengers boarding at stop x, mean passenger miles per trip, mean delay at signal y, mean overall travel speed and the proportion of delay due to turning vehicles. In a later section the analysis of typical statistics from public transportation studies is discussed.

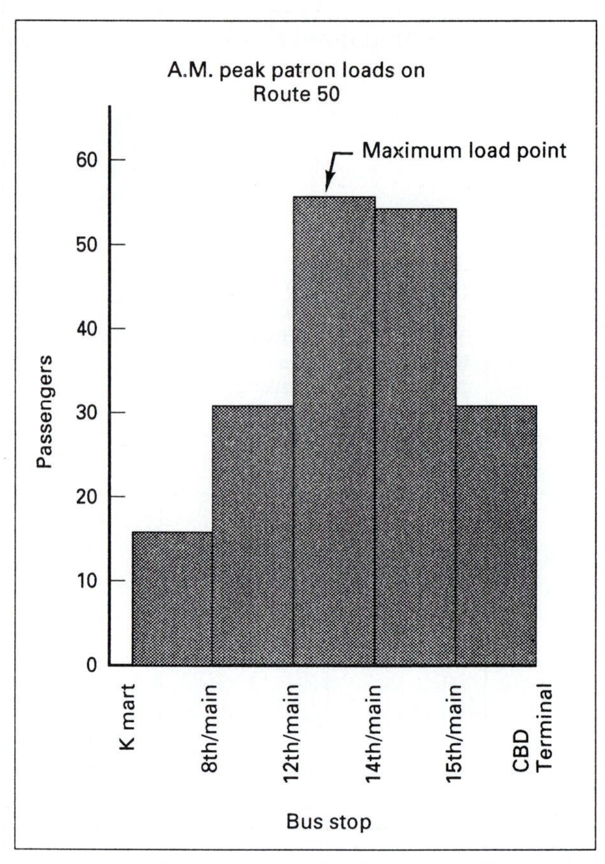

Figure 17-7 Route load profile. *Source:* UMTA, 1985a.

Passenger Origin and Destination

Operators can use ride check data to estimate passenger origin and destination patterns by stop on a route (i.e., not "true" door-to-door origins and destinations) at least four different ways, including:

1. A checker stands at the rear of the vehicle and notes on a seating chart the stop where each passenger boards and alights (Simon and Furth, 1985).
2. A checker distributes a card to each passenger as he or she boards. The card is coded with the originating stop. The checker collects the cards as passengers alight (Simon and Furth, 1985).
3. A ride checker (or automatic data collection equipment) provides boarding and alighting data by stop. Then a relatively simple algorithm provided by Simon and Furth (1985) estimates the origin and destination pattern.
4. The analyst uses point check load data and a minimum of ride check boarding and alighting by stop data in the algorithm from method 3 (see Furth, 1989).

Each of the four ride check methods of determining origin and destination patterns by stop on a route has advantages and disadvantages. The first two methods involve more direct measurement, so they are generally more accurate but more expensive than methods 3 and 4. Method 1 is appropriate only on trips with no standees and limited seating turnover. Method 2 has a high response rate but may cause initial confusion among passengers. The checker must be able to explain quickly, possibly in more than one language, the purpose of the card and what the passenger is to do with it (Simon and Furth,

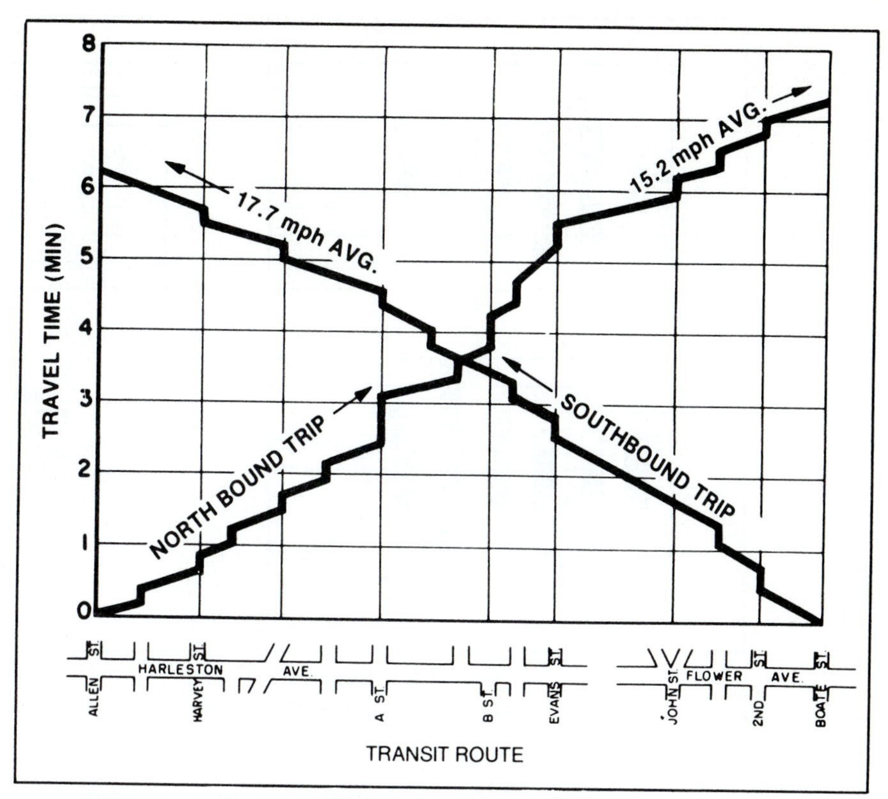

Figure 17-8 Public transportation vehicle time–space diagram. *Source:* Box and Oppenlander, 1976.

1985). If analysts can tolerate somewhat less accuracy and have the needed ride check or point check data on hand, methods 3 and 4 are the best ways to estimate origin and destination patterns.

Operators also use surveys to estimate origin and destination patterns, particularly when they need "true" door-to-door origins and destinations. In a later section onboard surveys are discussed, and survey methods are described in greater detail in Appendix B.

Automatic Data Collection

Automatic data collection equipment includes data acquisition, recording, transfer, and reporting components, as shown in Figure 17-10. In most installations, a series of photoelectric beams near the door or pressure-sensitive pads in the stairs provides boarding and alighting counts by stop. Computer algorithms decipher the activity recorded by the sensors and decide, for example, whether a sequence of signals represents a boarding or alighting passenger. The equipment records travel distances (through the vehicle odometer) and times to index the counts and to provide schedule adherence and similar data items.

Some automatic data collection systems include additional features. *Signposts,* one of the more common extras, are radio transmitters mounted on the roadside along a route that provide the on-board processing unit with location information as the vehicle passes. Signposts eliminate the need for difficult decisions about which boarding and alighting data belong to which stop. Without signposts these decisions often cannot be made, which results in discarded data. Two signposts per route, located at the ends of the route, may be justified on the basis of higher-quality data offsetting the higher capital costs

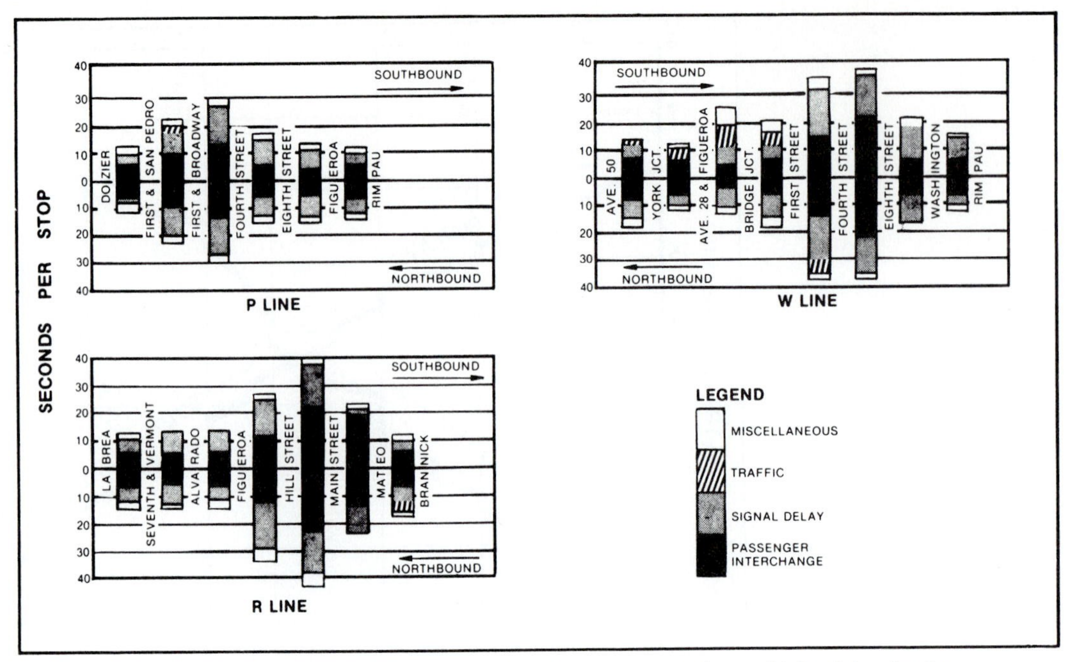

Figure 17-9 Summary of public transportation vehicle delay by type and location. *Source:* Box and Oppenlander, 1976.

(Deibel and Zumwalt, 1984). Signposts have other applications besides data collection that may help justify their costs as well. Sensors to record when the vehicle door is open and keyboards that allow extensive manual inputs to the data record are other features of some automatic data collection systems.

Automatic data collection equipment is expensive, so it is installed on only a fraction of the vehicles in a fleet. Deibel and Zumwalt (1984) concluded that a useful, valid, reliable database can be maintained if about 10% of the vehicle fleet is equipped for automatic data collection. Scheduling an equipped vehicle to a trip on which data are to be collected is challenging due to a variety of problems, including vehicle servicing and incompatible vehicle sizes. For example, operators cannot schedule a data collection vehicle with a capacity of 40 passengers on a route that needs 60-passenger vehicles. However, equipping 10% of the fleet provides an adequate margin for most of these problems.

Automatic data collection equipment produces much more raw data than manual study methods. The data processing and report generating capabilities of the system must be sufficient to handle this load. With fast microcomputers and other recent advances in computer technology, though, this is more an administrative than a technical challenge. The reports themselves are no different from the types of reports produced with manual methods and described elsewhere in this chapter.

Data collected automatically are usually more accurate than data collected manually, but bias can be a problem. Automatic systems tend to undercount for several reasons, including mechanical or electrical failures, environmental factors, and passengers blocking sensors. If bias is suspected, the operators can compare the automatic count to a careful, trusted manual count. If the operators find a bias, they can repair the system or correct for the bias mathematically during data processing.

The costs of automatic data collection systems vary widely with time, system specifications, and other factors. Generally, costs are well over $10,000 (U.S. 1991 dollars) per vehicle. Because the costs vary so widely, it is impossible to state exactly which public transportation agencies would find the equipment cost-effective and which would not. However, experience and cost analyses indicate that:

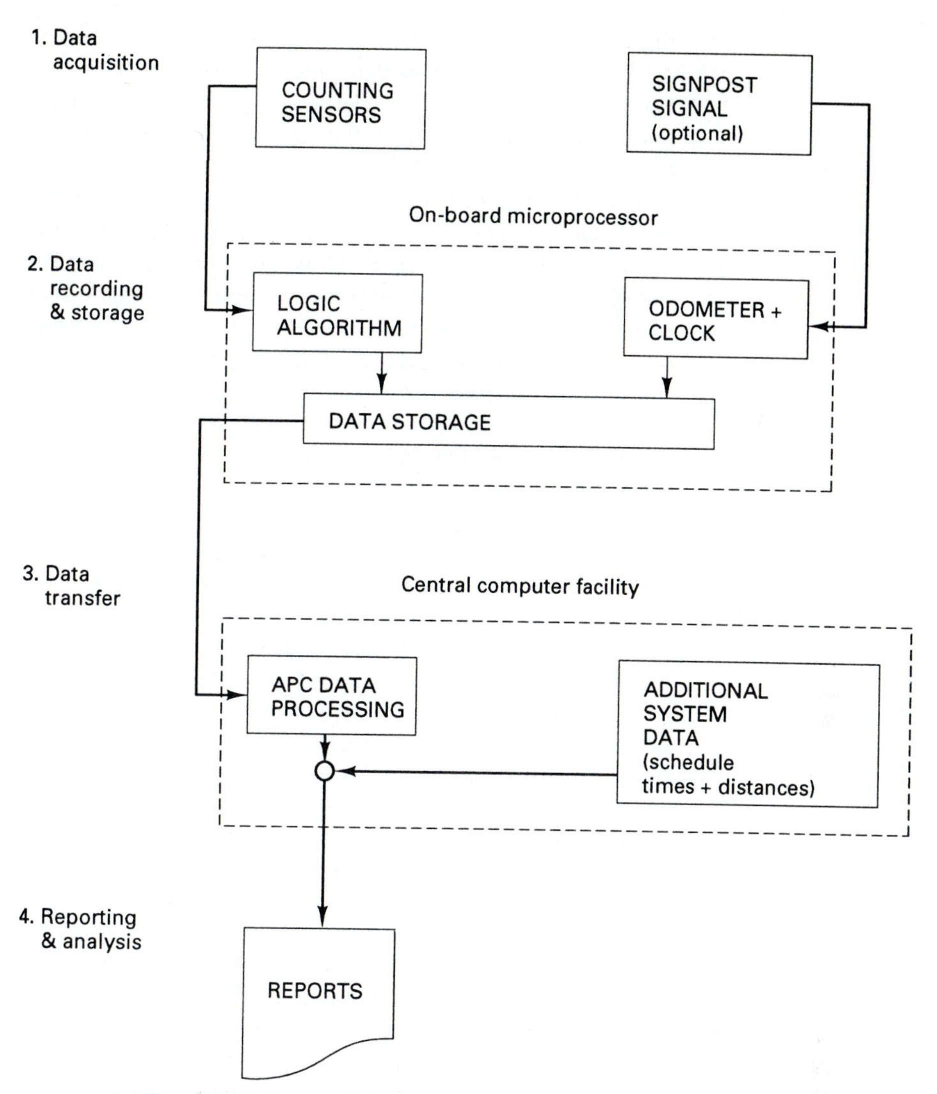

Figure 17-10 Components of an automatic data collection system. *Source:* UMTA, 1985b.

- For larger agencies (over 400 vehicles) the equipment will usually be cost-effective.
- For midsized agencies (100 to 400 vehicles) the equipment may be cost-effective depending on data needs.
- For small agencies (under 100 vehicles) the equipment will usually not be cost-effective.

When the availability of funds, the data accuracy needed, the data turnaround time, and other factors are considered with the cost-effectiveness, deciding whether or not to invest in automatic data collection equipment becomes very complex.

Surveys

Surveys provide information on passenger demographic traits, door-to-door origins and destinations, travel desires, attitudes, future plans, and other characteristics. Operators cannot collect these data items with the other study methods presented in this chapter. Surveys are relatively expensive to develop and conduct and are probably less accurate than the other study methods discussed in this chapter. Therefore, if operators can collect a data item with a study method other than a survey, they should do so. Analysts conduct many surveys for public transportation agencies by telephone or by visiting respondents' homes. Appendix B discusses these types of surveys. In this section we highlight the features that make on-board surveys distinct from other surveys. The sampling and analysis techniques for on-board surveys do not differ much from other study methods discussed in this chapter. Those techniques are presented later in the chapter.

Analysts generally use two types of on-board surveys: hand-in and mail-in. For both survey types, survey workers distribute the forms with a quick word of explanation to passengers upon boarding. Hand-in surveys are collected from passengers as they alight. Hand-in surveys are short, consisting of only a few questions, because there is limited time for passengers to complete them. Passengers answer hand-in surveys exclusively on the vehicle, so only information that passengers remember readily can be asked for. Hand-in surveys are printed on a card that can be written on and are distributed with pencils.

Passengers may complete mail-in surveys on the vehicle or elsewhere. Thus they do not need to be as short as hand-in surveys. They can pose more difficult questions, since passengers have time to ponder questions and seek assistance for answers. Mail-in surveys are printed on ordinary paper and do not have to be distributed with pencils. The forms can be stamped and addressed on the reverse side, or a stamped and addressed envelope can be provided with each form.

The choice of survey type depends primarily on the number of questions that the operator must ask. Shorter surveys are generally better, but on-board surveys can be longer than postcard size if the questions provide needed data items. Response rate is another factor that influences the choice between hand-in and mail-in surveys. Hand-in surveys have higher response rates, which could approach 90% usable returns, while mail-in surveys generally achieve 30 to 60% usable responses. However, a high response rate does not necessarily mean an unbiased survey (Brog and Meyburg, 1981; Doxsey, 1983). Finally, if passengers are on board for a very short trip or if there is a significant number of standees, severe biases will occur with hand-in surveys.

An intriguing option that combines the best features of hand-in and mail-in surveys is the two-part survey (Stopher, 1985). A two-part survey consists of a short hand-in form and a longer mail-in form that are handed out together. The hand-in form asks for very basic information on passenger characteristics and should have a high return rate. The mail-in form asks the detailed questions that are truly of interest to the survey sponsor as well as the same questions asked in the hand-in form. The analyst can compare the characteristics of passengers who returned both parts to the characteristics of the passengers who only returned the hand-in part, and will thereby gain insight into biases in the sample that answered the important questions.

Bamford et al. (1984) described a useful innovation for hand-in surveys. Instead of distributing pencils with a hand-in survey, they created a survey form on which passengers could rub off responses with a fingernail or coin, similar to the instant lottery tickets sold in some states. They reported a high response rate and that the eight-question survey took only 30 to 60 seconds to complete. The main disadvantage they noted was the need to precode all possible answers, so a survey with detailed questions could not use the format.

Stopher (1985) offers several excellent suggestions for conducting on-board surveys. First, he recommends that a count be maintained of the number of survey forms distributed for each trip. This count is essential for measuring the response rate. Second, Stopher argues that drivers should never be asked to distribute survey forms. Distribution of forms by drivers may bias questions about the driving, may not allow the control of forms noted above to be maintained, and may not provide passengers with the assistance they need to complete the survey. Next, he states that separate forms should be available for each predominant language in the survey area. A quick greeting by the checker distributing survey forms should be enough to indicate which language a particular passenger can respond in. Stopher also advises that boxes should be placed at each exit door for the return of forms, with large signs posted nearby to draw attention to the boxes. The checker distributing survey forms may be busy with boarding passengers and may not be available to collect every completed form. Finally, Stopher urges that every fare-paying passenger be given a survey form, since there are many problems with sampling a subset of passengers on a vehicle.

STATISTICAL ANALYSIS

In this section sampling, sample sizes, and means and proportions are discussed. These issues apply to all the study methods presented above.

Sampling

A sample is necessary for almost every public transportation study because the cost of a complete census is prohibitive. Public transportation agencies face increasing financial pressure, especially for expenditures perceived as nonessential, such as data collection, and cannot tolerate waste. Fortunately, the ability to generalize about an entire population from a sample allows useful data to be collected and analyzed for a reasonable cost.

After determining the study method to be used, the analyst conducting a public transportation study must determine the sampling unit. The sampling unit is the entity that represents one data point. For most public transportation studies, the analyst chooses the trip, one vehicle in revenue service traveling from the start to the end of a route, as the sampling unit. For example, in determining whether boardings have risen due to the introduction of a new fare pass, driver studies are to be conducted. Each driver participating in the study reports the number of passengers on each trip. From these data the analyst can calculate mean boardings per trip. In turn, multiplying this mean by the total number of trips will produce the desired quantity for comparison.

Once a sampling unit is chosen, the analyst devises a sampling plan. Historically, analysts have used simple random samples in most public transportation studies. With a simple random sample, each unit in the entire population of units has an equal chance of being selected for measurement. The sample size and analysis formulas for simple random samples are straightforward. However, the plan sometimes causes inefficient use of resources. A simple random sample of trips measured by ride checks, for example, may mean that checkers will be scrambling all over the route network to be on their assigned trips. A simple random sample of passengers measured using an on-board survey may mean distributing survey forms to only a few passengers on each vehicle.

For more efficiency, analysts can conduct many studies using stratified or multistage sampling plans. Stratified plans require the analyst to divide the population of units into several identifiable groups and then sample randomly from within those groups. For example, analysts could distinguish radial routes from crosstown routes and could sample trips from those groups. Another common stratification is by time of day: one could draw separately from A.M. peak, P.M. peak, between-peak, evening, and overnight "owl" services. Multistage plans are more common and usually more efficient than stratified plans. Most multistage plans require (1) the random selection of routes from the entire system of routes, then (2) the random selection of directions of travel from the selected routes, and finally, (3) the random selection of time periods from the selected route/direction combinations. With typical multistage plans, point checkers can stay in place throughout a peak period, for example, or survey distributors can give a form to every passenger boarding a vehicle. Stratified and multistage plans introduce more order into the data collection process, assure a more accurate estimate of the statistic of interest, and can result in a reduced sample size. The sample size and analysis formulas for stratified and multistage sampling plans are more complex than for simple random samples. Refer to Cochran (1977) and other sampling texts for details. Analysts contemplating stratified or multistage sampling plans should seek advice from a professional statistician.

Sample Sizes

Once analysts have planned the sample, they can calculate the necessary sample size. For estimating mean values with a simple random sample, one can determine the number of units to be sampled, *n*, using

$$n = \left(t \frac{v}{d}\right)^2 \tag{17-1}$$

where

t = a constant that depends on the confidence level needed in the estimate
v = coefficient of variation (the standard deviation divided by the mean)
d = tolerance on the estimate

Analysts can conservatively assume values for t as 1.8 with 90% percent confidence and 2.2 with 95% confidence, based on UMTA (1985b). A tolerance of 0.03 with a 90% confidence level means that 90% of samples of the size n will have mean values that fall within 3% of the estimated mean. Obviously, advanced knowledge of the coefficient of variation is necessary to estimate a sample size. Table 17-4 therefore provides coefficients of variation for common data items if estimates are not available locally. Tolerances recommended by UMTA (1985b) range from 10 to 30% for most common data items.

Many data items collected in public transportation studies are proportions, such as the proportion of passengers paying special student fares and the proportion of trips more than 5 minutes behind schedule. For estimating a proportion using a simple random sample, analysts can estimate the number of units n to sample using

$$n = \frac{t^2 p(1 - p)}{d^2} \tag{17-2}$$

where

t = a constant related to the confidence level
p = prior estimate of the proportion of interest
d = tolerance on the estimate

Values of t^2 in use with equation (17-2) are generally 3 for 90% confidence and 4 for 95% confidence. If no prior estimate of p is available, analysts can use a conservative value of $p = 0.5$. In public transportation studies, a common value of tolerance for estimating proportions is 0.1. For values of p near zero or 1, analysts need asymmetric tolerance ranges (see UMTA, 1985b).

Selecting a Sample

After analysts determine needed sample size they select the units to sample. They can use random number tables and computer programs that generate random numbers. Stratified and multistage sampling plans will lighten the burden of this task compared with simple

TABLE 17-4
Default Coefficients of Variation for Key Data Items

Data Item	Time Period	Route Classification	Default Value
Load	Peak	Load < 35 pass./trip	0.50
	Peak	Load > 35 pass./trip	0.35
	Off-peak	Load < 35 pass./trip	0.60
	Off-peak	Load 35–55 pass./trip	0.45
	Off-peak	Load > 55 pass./trip	0.35
	Evening	All	0.75
	Owl[a]	All	1.00
	Sat., 7 A.M.–6 P.M.	All	0.60
	Sat., 6 P.M.–1 A.M.	All	0.75
	Sun., 7 A.M.–1 A.M.	All	0.75
Boardings, passenger-miles	Peak	Peak load < 35 pass./trip	0.42
	Peak	Peak load > 35 pass./trip	0.35
	Off-peak	Peak load < 35 pass./trip	0.45
	Off-peak	Peak load = 35–55 pass./trip	0.40
	Off-peak	Peak load > 55 pass./trip	0.35
	Evening	All	0.73
	Owl[a]	All	0.80
	Sat., 7 A.M.–6 P.M.	All	0.45
	Sat., 6 P.M.–1 A.M.	All	0.73
	Sun., 7 A.M.–1 A.M.	All	0.73
Running time	All	Short (< 20 min)	0.16
	All	Long (> 20 min)	0.10

[a]Owl default values are the same for weekdays and weekends.
Source: UMTA, 1985b.

random sampling plans, in which every unit must be labeled. Operators schedule checkers to study the selected units while considering their work rules, the need for breaks, and travel times between study locations.

Foul weather and special events may force analysts to drop certain units from the sample. If analysts want an estimate of annual ridership, for example, they should not have to alter the sample plan due to foul weather and special events. However, if analysts need an estimate of schedule adherence for scheduling purposes, they should delete trips from the sample that are affected by foul weather or special events.

Estimating Mean Values and Proportions

In earlier sections of this chapter graphs were presented that were useful for presenting the results of studies. Along with a graph of key results, many public transportation studies also require an estimate of a mean value and knowledge of the tolerance associated with the estimate. The best estimate of the mean value for a data item in a population is the mean value from a sample. The tolerance associated with the estimated mean value is found to be

$$d = t\frac{v}{n^{0.5}} \tag{17-3}$$

where the variables are as defined for equation (17-1). One can compute the coefficient of variation, v, for equation (17-3) from the sample (i.e., sample standard deviation divided by the sample mean). Values for t depend on the confidence level and the sample size. For 90 and 95% confidence levels and large samples (more than 30 units), analysts should use t values of 1.7 and 2.0, respectively. Most statistics texts contain tables of t-distribution values for smaller samples and other levels of confidence.

Analysts can estimate the tolerance on a proportion from a sample using

$$d = t\left[\frac{p(1 - p)}{n}\right]^{0.5} \tag{17-4}$$

where the variables are as defined for equation (17-2). The sample provides the value of p, and the value of t is as described for equation (17-3). Presenting estimated means or proportions and the tolerances on those estimates graphically is usually of great assistance to decision makers viewing the study results.

REFERENCES

BAMFORD, C. G., R. J. CARRICK AND R. MACDONALD (1984). Public transport surveys: A new effective technique of data collection, *Traffic Engineering and Control,* Vol. 25, June.

BARNES, K. E. AND T. URBANIK II (1990). *Automated Transit Ridership Data Collection Software Development and User's Manual,* UMTA/TX-91/1087-3F, Texas State Department of Highways and Public Transportation, Austin, September.

BOX, P. C. AND J. C. OPPENLANDER (1976). *Manual of Traffic Engineering Studies,* 4th ed., Institute of Transportation Engineers, Washington, DC.

BROG, W. AND A. H. MEYBURG (1981). *Consideration of Nonresponse Effects in Large-Scale Mobility Surveys,* Transportation Research Record 807, Transportation Research Board, Washington, DC.

COCHRAN, W. G. (1977). *Sampling Techniques,* 3rd ed., Wiley, New York.

DEIBEL, L. E. AND B. ZUMWALT (1984). *Modular Approach to On-Board Automatic Data Collection Systems,* National Cooperative Transit Research and Development Program Report 9, Transportation Research Board, Washington, DC.

DORNAN, D. AND D. MIDDENDORF (1983). *Planning Services for Transportation-Handicapped People: A Data Collection Manual,* DOT-I-83-40, Urban Mass Transportation Administration, Washington, DC, August.

DOXSEY, L. (1983). *Respondent Trip Frequency Bias in On-Board Surveys,* Transportation Research Record 944, Transportation Research Board, Washington, DC.

EVERETT, P. B., R. W. SHOEMAKER AND W. E. DOSCHER (1987). *Public Transportation Marketing Evaluation Manual, Final Report,* DOT-T-88-06, Urban Mass Transportation Administration, Washington, DC, October.

FOOTE, P. J. AND W. A. HANCOX (1989). *Producing Section 15 Service-Consumed Data: Challenge for Large Transit Authorities,* Transportation Research Record 1209, Transportation Research Board, Washington, DC.

FURTH, P. G. (1989). *Updating Ride Checks with Multiple Point Checks,* Transportation Research Record 1209, Transportation Research Board, Washington, DC.

FURTH, P. G., AND B. MCCOLLOM (1987). *Using Conversion Factors to Lower Transit Data Collection Costs,* Transportation Research Record 1144, Transportation Research Board, Washington, DC.

PRATT, R. H. (1991). *Collection and Application of Ridership Data on Rapid Transit Systems,* Synthesis of Transit Practice 16, National Cooperative Transit Research and Development Program, Transportation Research Board, Washington, DC, September.

SIMON, J., AND P. G. FURTH (1985). Generating a bus route O–D matrix from on-off data, *Journal of Transportation Engineering,* Vol. 111, No. 6, November.

STOPHER, P. R. (1985). The design and execution of on-board bus surveys: some case studies, *New Survey Methods in Transport,* 2nd International Conference, VNU Science Press, Utrecht, The Netherlands.

UMTA (1985a). *Review of Transit Data Collection Techniques: Final Report,* DOT-I-85-26, Urban Mass Transportation Administration, Washington, DC, March.

UMTA (1985b). *Transit Data Collection Design Manual, Final Report,* DOT-I-85-38, Urban Mass Transportation Administration, Washington, DC, June.

UMTA (1985c). *Revenue Based Sampling Procedures for Obtaining Fixed Route Bus Operating Data Required under the Section 15 Reporting System,* Circular UMTA-C-2710.4, Urban Mass Transportation Administration, Washington, DC.

18

Goods Movement Studies

Marsha D. Anderson

INTRODUCTION

Goods movement, the distribution of raw materials and finished products, is most frequently handled by truck, train, and airplane. Some commodities move by ship or pipeline, but these modes tend to be less visible to the average transportation planner. Incorporating freight planning into general transportation planning activities is becoming increasingly important as the regional and local road networks become more congested, funds to expand physical capacity continue to decrease, air quality concerns heighten, and healthy economic activity is essential.

Goods movement is an enormous enterprise. Intercity and intracity freight movement cost $96.4 billion in 1988 (U.S. Dept. of Commerce, 1988). Approximately 22,000 establishments were involved in the intercity component, employing almost 1.4 million people. More than $20 billion in user fees (U.S. DOT, various years), largely in the form of road taxes, have been collected in each of the last five years. Intercity freight movement was handled by more than 4.9 million trucks, according to the 1987 Census of Transportation, Truck Inventory and Use Survey. Truck miles driven each year continues to rise at 5% per year (De Gasperis, 1991). It has been estimated that in 1990 more than 5 billion tons of freight were picked up or delivered in urban areas, taking more than 35 billion vehicle stops and more than 50 billion miles of travel (US DOT, 1976). There are a great many views represented with respect to freight movement: those of shippers and carriers, public agencies such as departments of transportation at all levels, planners and engineers, public officials, and the general motoring public. The role that each plays when any study is being conducted will inspire the study design and therefore the results. The question of whether one is looking for short-term, practical solutions to existing problems or projections for long-term planning purposes will also influence the types of studies conducted. Investigations of freight movement may lead to changes in operating practices, zoning or development regulations, design standards, fees, tariffs, or taxes. It may also influence the timing of public or private projects and the dollars expended. In this chapter we address the process of collecting data, the impacts of route characteristics, loading/unloading studies, vehicle weight studies, and hazardous materials movement.

DATA COLLECTION

The goals and objectives of a study will dictate the types of data and quantity needed for drawing reasonable conclusions. Documenting the facilities, equipment, and activities related to freight movement occurs at many levels. At the most macroscopic level is the Census of Transportation, conducted by the U.S. Department of Commerce, Bureau of the Census. The eight major transportation groups are railroad transportation, local and interurban passenger transit, trucking and warehousing, U.S. Postal Service, water transportation, transportation by air, pipelines (except natural gas), and transportation services. Those reported in the Census are trucking and warehousing, water transportation, and transportation services. The data categories include number of establishments, employees, revenues, and expenses.

These surveys are done every 5 years and are supplemented by other works, such as the Truck Inventory and Use Survey (TIUS). In this document one can find statistics on the number of trucks, miles driven, length and weight, revenues, range of operations, products carried, and equipment. Another product of the Census Bureau, the Commodity Transportation Survey, provides information at a macroscopic level for the development of indices and measures of flows. To supplement the TIUS, the Federal Highway Administration, Federal Railroad Administration, and Office of the Secretary of the U.S. Department of Transportation have developed a second instrument, the Nationwide Truck Activity and Commodity Survey (NTACS), which focuses on the volume of commodities by highway and rail modes, relationships between commodities and equipment, trucks, and multimodal flows. The first data files became available in late 1991.

There are a variety of other sources and data sets available at the national or regional level. Figure 18-1 presents the major sources. Highly specialized databases exist; for example, the Agricultural Marketing Service and Office of Transportation of the U.S. Department of Agriculture (USDA) collect information such as fruit and vegetable shipment data. Included are the volume of each commodity unloaded, freight rates, and the availability of truck service to the shippers. The USDA supplies operating cost information for both fleets and owner-operators. The Census of Agriculture also provides useful information on the magnitude, by type of commodity to be moved.

Two surveys conducted by the Arizona Department of Transportation produced information on commodities shipped, carrier type, tonnage, origins and destinations, routes used, and issues and concerns. These are the Freight Movement Survey (FMS) and the Highway Carrier Attitude Survey (CAS). Twenty-one hundred carriers were surveyed, with a total response of approximately 25% (Carey, et al.,1988), requiring three mailings, select telephone follow-ups and in person meetings. The Arizona Freight Network Analysis Decision Support System (Radwan, et al., 1988) was developed as a result of incorporating the survey findings into the planning process and is represented by Figure 18-2. The major steps in the process include modal decisions, forecasting and simulations, and management and strategic planning capabilities.

The Chicago Area Transportation Study (CATS) frequently conducts freight movement studies for planning updates. A 1990 survey identified mobility issues and operations-related problems for the carriers. The main output was the location of Hot

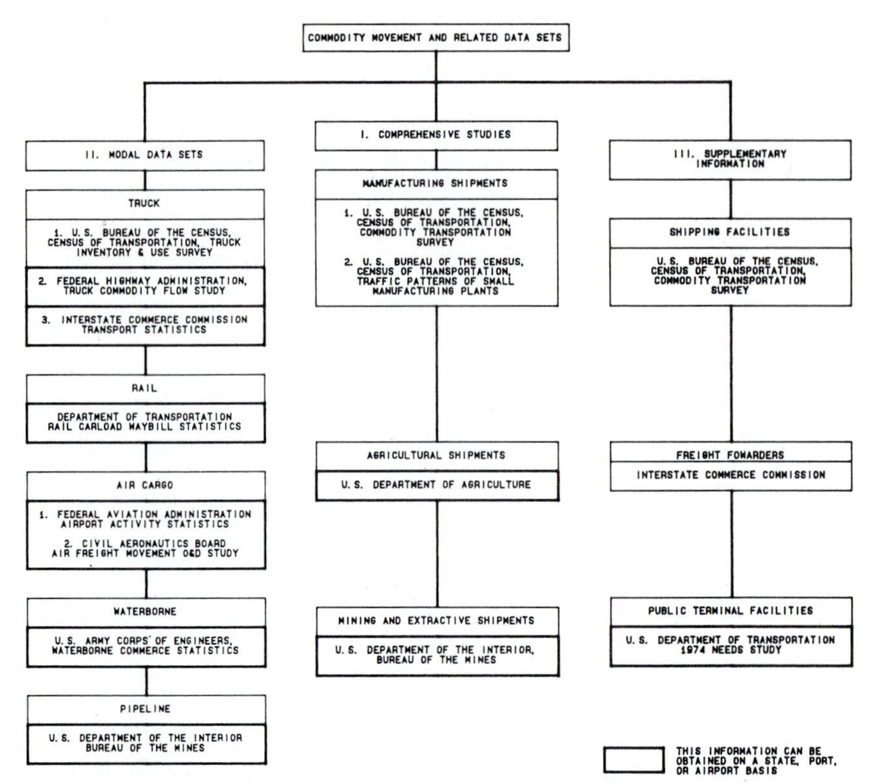

Figure 18-1 National and regional commodity movement data sets. *Source:* Bolger, F. and H. Bruck. Urban Goods Movement Projects and Data Sources, prepared for the U.S. Department of Transportation, Office of Systems Analysis and Information, Washington, D.C., 1973.

Spots impeding movement and service. It also helped in the generation of a wish list for physical improvements and regulatory change. The questionnaire asked for problem locations and roadway elements such as signs, signals, lane widths, turning bays, speed limits, merges, ramps, bridges, grade crossings, and terminal access.

Survey of Shippers

The Commodity Transportation Survey, conducted in 1983 by the Census Bureau, provides data on the distribution of products by class and shipper group. A sample of manufacturing establishments and a sample of their shipping documents is used to develop the database. The data elements in the survey include weight shipped by Standard Industrial Classification (SIC), state of origin, destination, and means of transportation.

The Annual Survey of Manufacturers, along with Current Industrial Reports (CIR) and other publications by the Census Bureau, provide measures of total sales, market share, quantity and cost of materials consumed, production hours and transportation modes, and location information. Approximately 55,000 of 370,000 sites are included in the survey process. Reports generated include a series by industry, by location of manufacturers, and geographic area statistics. The CIR are typically commodity reports.

Survey of Goods Movement Companies

A broad perspective is presented on goods movement companies as a part of the Census of Transportation via the survey of transportation industries. The data collected include physical location of operations, type of organization, operating revenues, payroll, and employment. The business classifications are based on the Standard Industrial Codes

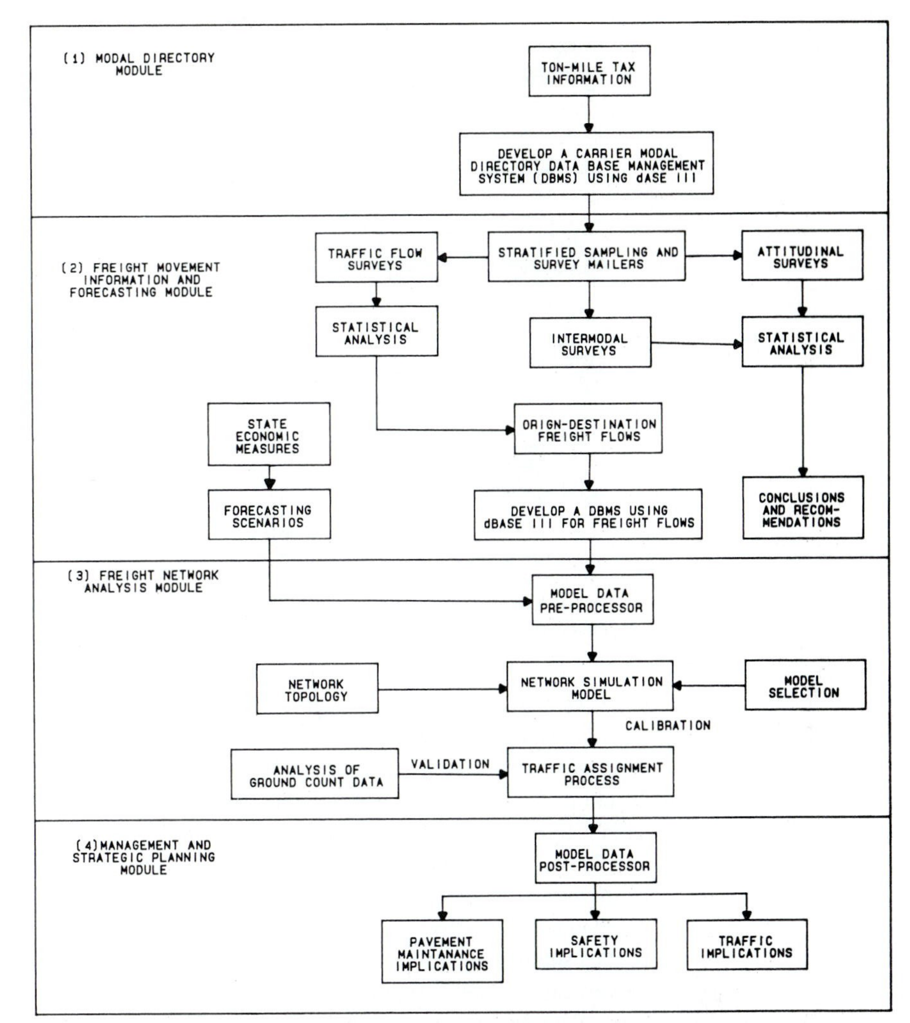

Figure 18-2 Arizona freight network analysis decision support system modules and tasks. *Source:* Radwan, A. Essam, et. al., A Decision Support System for Freight Transport in Arizona, prepared for the Arizona Department of Transportation, Transportation Research Record #1179, 1988.

(SICs), and for this survey the major groups included are motor freight transportation and warehousing (SIC 42), water transportation (SIC 44), and transportation services (SIC 47). Based on the most recent survey, done in 1987, it was found that there were more than 102,000 businesses providing motor freight transportation and warehousing. Water transportation of freight was served directly by more than 2300 firms. In 1988, the total operating revenues for trucking services were estimated at more than $100 billion (SIC category 421). This was based on a fleet of more than 240,000 trucks, 550,000 truck-tractors, and 1151 full and semitrailers.

Existing Data Bases

The Truck Inventory and Use Survey is conducted by the Census Bureau every 5 years for years ending in 2 and 7. The data are then summarized for comparison with previous years, and presented in detailed tabular form for the current survey year. For transportation planning applications, the following are among the data included: major use of vehicle, body type, annual miles, range of operation, mileage outside base-of-operation

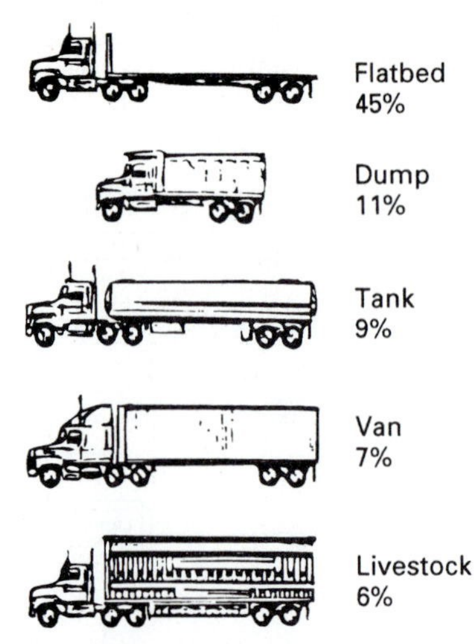
Flatbed
45%
Dump
11%
Tank
9%
Van
7%
Livestock
6%
Between 1982 and 1987, the number of large Wyoming registered trucks engaged in commercial activities decreased by 4,000; this is a 15% reduction from the 27,000 vehicles registered in 1982.

Figure 18-3 Wyoming large trucks by body type. *Source:* American Trucking Associations and Wyoming Trucking Associations, Trucking in Wyoming, 1987.

state, vehicle size, average weight, total length, and products carried (including hazardous material breakdown).

Numerous documents are published with extensive data, such as the American Trucking Association's financial and operating statistics, transport statistics (U.S. Interstate Commerce Commission), containership databank (National Magazine Company and Transmodal Industries Research), and vehicle/pavement interaction data (Society of Automotive Engineers). Affiliate state trucking association organizations also publish detailed data on activities in their regions, such as largest manufacturing facilities, numbers of vehicles by type and taxes paid, agricultural production, and magnitude of freight movement by mode. Figure 18-3 presents an example of the information presented on truck fleet mixes.

In Canada, a major effort was undertaken in the late 1970s to develop a significant database. The report includes a discussion of the sampling procedures and public databases employed (Tee Consulting Services, 1979). Main elements of the survey were body type and gross vehicle weight, fuel type, main use of vehicle, and an origin–destination query including establishment type serviced, and the load factor for a truck–trailer at each stop.

Several efforts have resulted in the development of databases relating truck trip generation to land use. Data collection was done in many cities in the United States, with the results reported in proportion to acreage or 10,000 square feet of development. Some of the findings are summarized in Figure 18-4.

Special activity centers have trucking needs that may be atypical, and standard trip generation rates or weight expectations may not apply. A study done by the Texas Transportation Institute (Middleton, 1990) used manual counts and portable weigh-in-motion scales to assess the characteristics of truck traffic for timber, produce, grain, beef cattle, limestone, and sand/gravel. Interviews were used as a follow-up to quantify the typical conditions.

Many studies were conducted in the 1980s during deregulation to establish and assess truck size and weight limits. The prevalent characteristics in the United States are 80,000 lb maximum gross vehicle weight (GVW), 20,000 lb for single axles and 34,000 lb for tandem axles. Maximum lengths range from 55 to 60 feet for single trailers to 65 feet for twin trailers. In Canada, the average maximum GVW is 116,000 lb, with almost 21,000 lb for single axles and 39,000 for tandem axles (Staley, 1981). Length maxima range from 67 to 75.3 feet. In Europe, a wide range of sizes and weights are permitted.

Truck Stop Generation as Related to Land Use							
Type of Establishment	Daily Truck Stops per 100,000 Square Feet of Floor Area						
	Source of Data (Reference No.)					Range	Mean
	1	2	3	4	5		
Office	2.0	2.2	2.4	1.5	2.0	1.5-2.4	2.0
Bank	-	-	-	-	3.0	-	3.0
Brokers	-	-	-	-	3.0	-	7.0
Credit	1.8	-	-	-	3.0	1.8-3.0	2.4
Insurance Co.	-	-	-	-	1.0	-	1.0
Real Estate	-	-	-	-	2.0	-	2.0
Trade	-	-	-	-	7.0	-	7.0
Wholesale	-	-	-	-	8.0	-	8.0
General Retail	-	-	-	-	2.0	-	2.0
Misc. Retail	2.5	-	-	-	1.0	1.0-2.5	1.8
Apparel	1.8-6.7	-	-	-	4.0	1.8-6.7	4.5
Department Store	1.5-3.7	-	-	2.7	-	1.5-3.7	2.6
Furniture	1.9	-	-	-	6.0	1.9-6.0	3.9
Restaurant/Bar	27.-61.	-	-	-	32.0	27.0-61.0	36.0
Drug Store	37.0	-	-	-	-	-	37.0
Food	-	-	-	-	42.0	-	42.0
Variety Store	1.8	-	-	-	-	-	1.8
Jewelry Store	1.1-3.8	-	-	-	-	1.1-3.8	2.5
Barber Shop	2.2	-	-	-	-	-	2.2
Photo Studio	6.9	-	-	-	-	-	6.9
Services	-	-	-	-	7.0	-	7.0
Hotel	2.0	-	-	-	0.4	0.4-2.0	1.2
Institution	1.0	-	-	-	-	-	1.0
Personal	-	-	-	-	12.0	-	12.0
Business	-	-	-	-	18.0	-	18.0
Parking	-	-	-	-	0.3	-	0.3
Amusement	-	-	-	-	4.0	-	4.0
Administration	-	-	-	-	4.0	-	4.0

Figure 18-4 Truck trip generation by land use. *Source:* Dallas CBD Goods and Services Distribution Project, 1980.

Figure 18-5 presents the specified values for some of the countries surveyed. The range of GVW maximum weights ranged from 16 to 44 metric tons.

Water carrier data are available from the Corps of Engineers and the *Journal of Commerce* as well as the Census Bureau. Analysis of rail shipments is possible using waybill samples available through the Interstate Commerce Commission, Public Use Waybill File. Items represented in these files include tonnage by commodity, carloads by

LIMITS OF EUROPEAN MOTOR VEHICLE SIZES AND WEIGHTS

COUNTRY	LENGTH		AXLE LOAD			
	TRUCK SEMI TRAILER	OTHER COMBI-NATIONS	SINGLE	TANDEM	TRIDEM	RANGE
	meters	meters	metric tons	metric tons	metric tons	metric ton
Austria	16.0	18.0	10.0	16.0	16.0	16.0-38.0
Belgium	15.5	16.0	10.0 12.0	18.0 20.0	--	19.0-44.0
Denmark	16.0	18.5	10.0	16.0	24.0	16.0-38.0
France	15.5	16.0	13.0	varies	varies	13.0-40.0
Germany (East)	16.5	18 22	10.0 11	11 16 18	16.5 24.0	varies
Greece	15.0	18.0	10.0 13.0	20.0	--	19.0-30.0
Italy	15.5	18.0	12.0	17.0 20.0	--	18.0-44.0
Yugoslavia	16.5	18.0	10.0	16.0	24.0	22.0-40.0

Figure 18-5 Size and weight limits for trucks in selected countries in Europe. *Source: Limits of Motor Vehicle Sizes and Weights* prepared for the International Road Federation, 1989.

type of rail car, intermodal trailer movements, origins and terminations by commodity and state, average length of haul, freight rates by commodity, and size of shipment.

A California database (Teal, 1988), TASAS (Traffic Accident Surveillance and Analysis System), provides details on all accidents on the state highway system that generate a police report. It includes truck-related accidents but does not include information on lane closures or time duration of incident. A study was conducted to supplement that database using a random sampling of 13 days' worth of events, with 239 truck-related incidents. Data elements included type of accident, location, time, duration, and number of lanes closed.

For studies involving air cargo, data are available from the Air Transport Association of America, the Civil Aeronautics Board, the Federal Aviation Commission, and some of the data service bureaus. Pipeline information is obtainable through the Department of Energy, Interstate Commerce Commission, and the Petroleum Publishing Company.

ROUTE CHARACTERISTICS

Rail and truck movement of freight are often affected by similar factors, and one area where this is true is in the routing of vehicles. For rail shipments, the ownership of certain segments of the rail network can influence routing. For truck shipments, the designated truck networks, or restricted segments of a road network, exert some influence. In Chicago, a recent study showed that trucks were as much as 28% of the expressway traffic stream (Chigago Area Transportation Study, 1991). In Atlanta, that number is often as high as 30%; in some parts of the I-75 corridor that percentage is, at times, as much as 50%. On arterials, the percentages are less, but with a high, in Chicago, of 20%.

In the last ten years, changes in laws and regulations regarding the maximum weight and length specifications issued at the national level, and adopted by many states, has resulted in some use of double and triple trailers. Concern for the infrastructure, roadway capacity, truck productivity, route continuity, and safety of the motoring public led to the designation of a National Truck Network. The access rules for local roadways greatly expands the network. In some states the impact of route designations is simulated to determine the impact on the overall road network. Such simulation studies may be used to depict the rate at which nonessential truck traffic is discouraged from using streets in congested areas, particularly in peak hours. With growing concern for air quality, such analyses may rise in importance and help answer such questions as: Will the increase in truck vehicle miles traveled (vmt) create more or less pollution than the added congestion expected without specified truck routes?

When defining a study, it is necessary to understand the universe of candidate vehicles. The Observable Truck Characteristics Study (Hanscom, 1981) includes gross vehicle weight, front axle weight, maximum axle weight, empty versus loaded, empty weight by vehicle type, payload, length, width of load, number of trailers, cab type, and overall truck type. Topology, geometrics, and truck attributes in an area must be assessed simultaneously when drawing conclusions about operating characteristics with respect to freight movement.

In Vancouver, truck route rules are applied only to vehicles with more than two axles or greater than 32,000 lb GVW (Swan, 1979). A test was run to compare the existing route network to a highly concentrated scheme of four north-south and four east-west arterials. The evaluation looked at the additional cost to the trucking industry in terms of driving time increases, vehicle miles traveled, and noise impact. The data collection effort was an origin/destination study, from which travel patterns were derived. For coding purposes the study area was divided up into 70 zones. The classifications used for the rest of the coding detailed the body types, GVW, fuel type, use of vehicle, land use classifications, and commodity classifications. The Standard International Trade Classification (SITC) of the United Nations are the groupings for motor transport traffic in Canada. A computer model was used to test the original alternatives and subsequent variations, using all-or-nothing assignments.

To determine the suitability of a particular roadway to handle trucks of different sizes, field observations may be made. A study conducted for the Federal Highway Administration (Hummel, et al., 1988) involved more than 1100 trucks. The characteristics inventoried included number of legs in each approach, average lane widths by approach, width of lane from which turn was made, average curb radius, signalization at intersection, and phasing of signal, where appropriate. The data were then stratified based on truck type (semi 40-foot, 45-foot, 48-foot, and double 28-foot). Turn times were measured as well as encroachments and conflicts caused by truck maneuvers.

Wisconsin applied the Caltrans turning template software, together with overhead photographing and ground-level phototriangulation, to determine the adequacy of various ramps and road segments to handle truck traffic. Computer simulations were run for each study site and compared to field observations. Truck-related factors examined included angle of turn, radius of turn, tractor wheelbase, trailer length, trailer width, and axle width. Standard roadway geometrics were examined.

In Nashville, Tennessee (Stammer, et al., 1985) visual overlays, sensitivity analyses by computer modeling, and sketch planning methods were used to evaluate several system alternatives. Factored vehicle classification counts provided the means for developing zone-level truck trip-end productions. The Tennessee Computerized Road Information Management System (TRIMS) provided all of the geometric data, and a simple noise prediction model was employed in the study. Base maps were overlaid with transparencies shaded for truck-sensitive areas based on three criteria: truck terminal locations, congested areas, and high truck–automobile accident location.

Modeling similar to that done for road networks, can be done for railroads, with ownership being added as one of the critical variables for each link in the system. The impact of rail line abandonments and changes in ownership can be evaluated by modeling freight flows and alternative path choices. The intermodal implications and highway issues are tied directly to the railroad component of the commodity movement.

LOADING AND UNLOADING STUDIES

The movement of trucks through urban areas is essential for the pickup and delivery service provided, but this same movement is a component of the urban congestion. Further, much of the downtown freight loading and unloading is handled curbside, adding yet another dimension to the importance of adequate facilities planning. At the local level, data collection is usually undertaken in response to the identification of goods movement problems, which can be grouped as presented in Figure 18-6, which cites operational, economic, and external impact areas for downtown, terminal, and network applications.

To service businesses, on- and off-street loading spaces must be designated. How many, of what type, and where they can be located must be determined. It may be necessary to impose time restrictions, and specifying those time ranges should be based on the local objectives, supported by an appropriate data collection effort. Truck trip generation rates, the result of one type of study, are used to derive the number and type of spaces. As shown previously, these rates will vary by land use and category of establishment.

A major data collection effort is under way through the Institute of Transportation Engineer's Technical Council Committee 6A-48. The objective is to gather input on the loading dock requirements for new development from four groups: consultants/architects, motor carriers, cities/counties, and developers/businesses. Under each category a series of questions are being posed: where pickups and deliveries are accommodated, desirable design elements (vertical clearance, dock height, grades, etc.) maneuvering space and orientation, actual number and mix of spaces, provision of ancillary equipment (dock levelers), queuing space, dumpster/compactor requirements, dock requirements by land use type, and variance opportunities.

The pickup and delivery activities include at least the following steps: finding the appropriate parcel if it is for a delivery, locating the precise address for the interaction, receiving the item if it is a pickup, identifying the responsible person, walking between the truck and the location, and completing the documentation. If the delivery schedule is not located conveniently, the dwell time can be excessive and there can be a measurable impact on the adjacent road network. The types of vehicles used for urban deliveries includes everything from private passenger cars, taxis, pickup trucks, and so on, through tractor–trailer combinations. According to a CBD study in Dallas, Texas (Christiansen, 1979), the majority of delivery vehicles are single-unit trucks (40%), with vans accounting for an additional 27%, and passenger cars providing another 18%. Tractor–trailers were only 3% of the delivery vehicle population.

Sample data collection forms are included as Figures 18-7 through 18-9 for individual trucks, single loading docks, or curbside pickup and delivery operations. Studies of single-truck operations are conducted to determine such factors as trip-length characteristics and dwell times by land use type. The investigation of an existing loading dock allows the derivation of the number of trucks per day, peak period and peak-period percentage, and dwell times, all for a given land use. Curbside studies also allow the determination of space requirements by land use.

PRELIMINARY DATA COLLECTION FOR USE IN IDENTIFICATION OF LOCAL URBAN GOODS MOVEMENT PROBLEMS

Problem Area Classification		
Downtown or High Density Distribution Centers	Interface/Terminals	System/Network
I. Operational	I. Operational	I. Operational
A. Queuing counts at destination related to capacity of receiving and shipping facility	A. Volume Counts	A. Traffic Counts with detailed vehicle classification
B. Accident Data	B. Street use in area	B. Speed and delay
C. Curbside usage	C. Measures of total activity time, number of trucks, cargo hauled	1. System segments 2. Rail at grade crossings 3. Moveable bridges
D. Inventory of loading zones	D. Access route examined - time runs	C. Accident Data
E. Illegal stopping to load on streets	E. Ease of access to loading dock	D. Access problems (low bridges, steep grades, load zoned roads, restricted access by time of day.)
II. Economic	II. Economic	II. Economic
Survey of shippers, receivers and operators	Survey of terminal operators	Survey of carriers operators
		A. Truck B. Rail C. Water D. Air
III. External Impacts	III. External Impacts	III. External Impact Monitoring

Figure 18-6 Data needs for identifying urban goods movement problems. *Source:* Gendell, David S., et.al., Urban Goods Movement Considerations in Urban Transportation Planning Studies prepared for the U.S. Department of Transporation, Engineering Foundation Conference, 1974.

A. Compute ranges of pollution using volume data	A. Aerial photos of terminal and surrounding area	A. Air B. Noise C. Water
B. Monitoring	B. Monitoring	
1. Air 2. Noise		

LOG DATE

Truck Mileage Reading at Start of First Trip________(miles)
Truck Mileage Reading at End of Last Trip________(miles)

Stop Number	Arrival and Departure Time at Each Stop		Location of Each Stop by Zone	Type of Business or Activity at each Stop	Type and Quantity(units) of Freight Handled at Each Stop		Load Factor for Truck or Trailer after Each Stop
Start of 1st Trip							
1st Stop							
2nd Stop							
3rd							
4th							
5th							
6th							
7th							
8th							
9th							
10th							
11th							
12th							

Figure 18-7 Urban truck daily log form. *Source:* Transportation Consulting MDA, Inc., d/b/a Street Smarts, 1992.

Past studies indicate that most pickup and delivery trucking activities occur between 6 A.M. and 6 P.M. (Christiansen, 1979). Activities tend to peak between 10 A.M. and 2 P.M. (Habib, 1980), with dwell times ranging from 11.5 to almost 20 minutes. Some types of establishments have different patterns. For example, restaurants experience an early morning operation, with food and supplies arriving prior to the start of the regular day. There are selected seasonal variations in freight movement, including the response to commercial requirements for the winter holiday seasons; there are also agricultural peaks which move throughout the country.

Calculating loading dock requirements includes the number of expected truck arrivals per day and an estimate of the peaking characteristics. Based on the processing time anticipated or measured, a service rate can be calculated. Applying queuing equations, one can determine the number of docks needed to achieve the desired level

LOADING DOCK OPERATIONS LOG FORM

LOCATION: ____________________ DATE: ____________

SIZE OF BLDG: ____________ SQ. FT. OCCUPIED SPACE ____________

LAND USE ____________________________

		TYPE OF VEHICLE			COMMODITY		
COMPANY NAME	TIME IN	VAN	SUT	T-T	DELIVERED	PICKED-UP	TIME OUT

Figure 18-8 Loading dock operations log form. *Source:* Transportation Consulting MDA, Inc., d/b/a Street Smarts, 1992.

of operation (see chapter 19 for further discussion). Queuing equations also allow for testing assumptions, such as the impact on waiting time if one less dock is provided. The final step in this process is the prediction of the mix of the vehicle types at, and therefore the design requirements for, the loading dock space. The Dallas study provides a reasonable estimate by land use type, as shown in Figure 18-10.

VEHICLE WEIGHTS

Data on truck weights are collected for many purposes, including pavement design, revenue estimates, motor carrier enforcement, highway cost allocation, and other planning and engineering activities. The vehicle weights are reported by motor freight companies as part of their reporting requirements; the recipient varies by state but is frequently the motor vehicle department through the registration process. Roadside weight checking is conducted with either permanent or portable scales, usually as part of an enforcement program. Following the changes in trucking regulations in the 1980s and the resulting changes to fleet mixes, truck size and weight studies are often conducted to evaluate the impact on types of trucks being used, pavement, and geometric requirements and changes in industry efficiencies.

In Wisconsin, the Department of Transportation used statistical criteria to locate their weigh stations (Gardner, 1983). The Federal Highway Administration suggests considerations for site selection to include ADT volumes, percent trucks, percent trucks by type, percent trucks by commodity, seasonal areas, interstate versus intrastate trips, land use characteristics as related to trip origins and destinations and site suitability, and nearby alternative routes (Winfrey, 1976).

Static Measurements

The use of static scales is required for the certification of truck weights in the case of a cited violation. The accuracy of these scales, when calibrated, makes the data defensible, if necessary, in a court of law. Typical permanent stations are often costly to construct, equip, and operate. From a transportation network perspective, having to bring the truck to a complete stop often causes backup and safety problems, as a weigh station may be processing many thousands of trucks in a single day. The data collected at permanent

P.U.D. OPERATIONS SURVEY

A. LOCATION

____

B.
DATE______________________________________

C. TIME OF ARRIVAL______________ TIME OF DEPARTURE__________

D. PARKING MODE

Vehicle Parked In
1. Travel Lane______________
2. Non-Travel Lane Illegally________
3. Non-Travel Lane Legally________
4. Other______________

E. ACTIVITY RECORDED

1.
Location______________________________

Operation______________________________

Commodity__________ No. of Parcels______ Total Weight____________

NOTE: Continue below only if additional establishments are visited

2.
Location______________________________

Operation______________________________

Commodity______ No. of Parcels_________ Total Weight__________

Figure 18-9 Curbside inventory and pickup and delivery operations form.
Source: Transportation Consulting MDA, Inc., d/b/a Street Smarts, 1992.

weigh stations is often biased, due to general knowledge of their location and operation. Oversized and/or overweight vehicles are suspected of circumventing the stations.

Weigh-in-Motion

Weigh-in-motion (WIM) is used in some locations to determine if a truck is traveling within a reasonable range of the legal limits as adopted by a local government. The data collected may include all or some of the following: gross vehicle weight, axle weight, and

LOADING SPACE SIZE DISTRIBUTION FOR VARIOUS LAND USES

Land Use Category	Percentage of Space Sizes		
	55'	35'	20'-25'
Office	--	40%	Balance
Retail and Personal Services	--	60%	Balance
Retail, if over 60,000 sq.ft.	25%	25%	Balance
Commercial/Industrial	--	40%	Balance
Hotel/Motel	1 space	75%-1	Balance
Food and Beverage Services	--	40%	Balance

Figure 18-10 Loading space size distribution for various land uses. *Source:* Walters, Carol A., 1980. CBD Dallas: A Case Study in Development of Urban Goods Movement Regulations prepared for the City of Dallas, Texas, Office of Transportation Programs, Dallas, Texas.

tandem axle weight. These allow the application of the inner and outer bridge formulas. In the United States, the Department of Transportation or other responsible agency monitors the data collected, and together with an enforcement agency, is further charged with penalizing violators.

The weigh-in-motion scales are often found at permanent truck weigh station sites, but are also used by some agencies in a roving mode. The forest products industry is experimenting with an automated, on-site weigh-scale program to allow for better truck utilization and assurance that safety codes are being met prior to dispatch of the load onto the roadway. Other weight data are collected using techniques built into existing bridge components (Halkyard, 1991). Florida, Kentucky, and New Mexico participated in a correlation study to compare static weighings to the bridge weigh-in-motion study. The accuracy of the weights collected by a WIM device are affected by many factors. The characteristics that affect the accuracy include roadway factors, such as longitudinal profile, transverse profile, grade, cross slope, and curvature; vehicular factors, such as speed, acceleration, axle configuration, body type, suspension system, tires, load, load shifts, aerodynamic characteristics, and center of gravity; and environmental factors, such as wind, temperature, and ice (ITE Technical Committee 5D-6, 1986, p. 7). The final report of the ITE Technical Committee on Truck Weighing-in-Motion presents the findings of many test studies conducted around the United States with respect to the factors listed above. In addition, scale calibration is a significant factor in WIM readings.

HAZARDOUS MATERIALS

Many basic commodities come under the heading of hazardous materials when being transported. Gasoline for automobiles must be delivered to your local filling station. Chemicals to develop your film and clean your pool or exterminate your house must move from factory to end user. Moving these materials in a manner to minimize risk is essential, yet often controversial. There is a unique code notation in the Standard Transportation Commodity Codes (STCC) which indicates hazardous materials, allowing for the development of databases. The Office of Technology Assessment published a summary of

ESTIMATED TRANSPORTATION OF HAZARDOUS MATERIAL BY MODE IN 1982

MODE	NUMBER OF VEHICLES/VESSELS USED FOR HAZARDOUS MATERIALS	TONS TRANSPORTED	TON MILES
Truck	337,000 dry freight or flat bed	927 million	93.6 billion
Rail	130,000 cargo tanks	73 million	53.0 billion
Water-borne	4,909 tanker barges	549 million	638.5 billion
Air	3,772 commercial planes	.285 million	459.0 million
TOTAL		1.5 billion	784.0 billion

Figure 18-11 Estimated transportation of hazardous materials by mode in 1982. *Source:* OTA, Transportation of Hazardous Materials U.S. Congress, Office of Technology Assessment, OTA SET 304, Washington, DC, US Government Printing Office, July 1986.

commodity-based information by mode. Figure 18-11 shows the magnitude of hazardous moving, by mode.

Shipments should be routed to optimize transit time and reliability, minimize accident probability, and minimize the impact on people and property in the event of an accident. These goals can sometimes be in conflict. Computerized files exist for analyzing rail movement of hazardous materials. This is done by combining network data, Federal Railroad Administration accident/incident files, and demographics near system links. To determine the primary and backup routes for moving hazardous materials, a detailed study was conducted (McMillen et al., 1984). Existing conditions were inventoried, similar to a typical traffic impact study inventory; however, more in-depth information was gathered on accidents (by location, type, and vehicles involved) and surrounding land uses. The areas to be avoided as much as possible include hospitals, schools, historical areas, and parks. Proximity to public safety agencies and emergency medical services is desirable.

A risk assessment is the next step in the process. Data from the Materials Transportation Board and local accident records are valuable for the development of the route criteria. With a first goal of minimizing the risk of an incident, the next goal is to minimize the impact of an incident should one occur. Plans need to be in place for detection and notification, responding to the incident with a hierarchy of players specified, as well as a site protection and traffic control plan. Regulations addressing the movement of hazardous materials often contain guidelines for increasing route safety. Those factors include lowering speed limits in critical areas, designating special truck lanes in critical areas, use of heavy-duty guiderails on bridges, catch basins to retain spills, and mandatory stops during severe inclement weather.

REFERENCES

American Trucking Associations, Wyoming Trucking Associations (1987). *Trucking in Wyoming* ATA/WTA, Burlington, MA.

Bolger, F., and H. Bruck (1973). *Urban Goods Movement Projects and Data Sources,* U.S. Department of Transportation, Office of Systems Analysis and Information, Washington, DC.

Carey, D., H. Mahmassani, and G. Toft (1988). *Air Freight Usage Patterns of Technology-Based Industries,* Transportation Research Record 1179, Transportation Research Board, Washington, DC, pp. 33–39.

Chicago Area Transportation Study, 1991. *Operation GreenLight, Freight Movements and Urban Congestion in the Chicago Area,* Chicago, March, pp. 9–10.

Christiansen, D. (1979). *Urban Transportation Planning for Goods and Services,* Federal Highway Administration, Washington, DC.

Christiansen, D. (1980). *Dallas CBD Goods and Services Distribution Project,* City of Dallas, Texas, Office of Transportation Programs.

de Gasperis, S. Presentation to MIT Students/Faculty, UPS Efficiency/Urban Goods Movement, 1991, Cambridge, MA.

Gardner, W. D. (1983). *Truck Weight Study Sampling Plan in Wisconsin,* Transportation Research Record 920, Washington, DC, pp. 12–18.

Gendell, D. S., et al. (1974). *Urban Goods Movement Considerations in Urban Transportation Planning Studies,* FHWA 32-01-23 RFP 397, U.S. Department of Transportation, Engineering Foundation Conference, 1974.

Habib, P. (1980). *Transportation System Management Options for Downtown Curbside Pickup and Delivery of Freight,* Transportation Research Record 758, Transportation Research Board, Washington, DC, pp. 63–69.

Halkyard, T. (1991). *Field Trials of Low Cost Bridge Weigh-in Motion,* Federal Highway Administration, McLean, VA, June.

Hanscom, F. (1981). *The Effect of Truck Size and Weight on Accident Experience and Traffic Operations,* Vol. II, *Traffic Operations,* FHWA/RD-80-136, U.S. Department of Transportation, Federation Highway Administration, Washington, DC.

Hummer, J., C. Zegeer, and F. Hanscom (1988). *Effects of Turns by Larger Trucks at Urban Intersections,* Transportation Research Record 1195, Transportation Research Board, Washington, DC, pp. 64–74.

International Road Federation (1989). *Limits of Motor Vehicle Sizes and Weights,* Washington, DC.

ITE Technical Committee 5D-6 (1986). *Trucking Weighing-in Motion,* Informational Report, Institute of Transportation Engineers, Washington, DC, p. 7.

McMillen, R., M. Anderson, and C. Cerbone (1984). Traffic and transportation analysis: hazardous waste disposal facility, *National Conference on Hazardous Wastes and Environmental Emergencies,* Houston, TX.

Middleton, D. (1990). *Results of Special-Use Truck Data Collection,* FHWA/TX-420-3F, U.S. Department of Transportation, Federal Highway Administration, Washington, DC.

OTA (1986). *Transportation of Hazardous Materials,* OTA SET 304, U.S. Congress, Office of Technology Assessment, Washington, DC, July.

Radwan, A. E., J. Cochran, and M. Farris (1988). *A Decision Support System for Freight Transport in Arizona,* Transportation Research Record 1179, Transportation Research Board, Washington, DC, pp. 23–30.

Radwan, A. E., M. Rahman, and S. Kalevela (1988). *Freight Flow and Attitudinal Survey for Arizona,* Transportation Research Record 1179, Transportation Research Board, Washington, DC, pp. 16–22.

Staley, R. (1981). *Foreign Truck Size and Weight Limits,* American Trucking Associations, Research and Economics Division, Washington, DC.

Stammer, R., C. Wright, and J. Donaldson (1985). *Conducting Truck Routing Studies from a New Perspective,* Transportation Research Record 1038, Transportation Research Board, Washington, DC, pp. 59–63.

Swan Wooster Engineering, Co. Ltd. (1979). *Evaluation of Urban Trucking Rationalization in Vancouver: Phases 1 and 2,* Vol. 5, Urban Transportation Research Branch of Canadian Surface Transportation Administration, Montreal, p. 2–15.

TEAL, R. (1988). *Estimating the Full Economic Costs of Truck Incidents on Urban Freeways*, AAA Foundation for Traffic Safety, Falls Church, VA.

TEE CONSULTING SERVICES, INC. (1979). *Framework for Urban Goods Movement Information in Canada*, Vol. 9, Urban Goods Movement Report Series, Urban Transportation Research Branch, Transport Canada, Montreal, September.

TRANSPORTATION CONSULTING, MDA, Inc., d/b/a Street Smarts, Duluth, Georgia (1992).

U.S. DEPARTMENT OF COMMERCE, Bureau of the Census (1988). *1988 Motor Freight Transportation and Warehousing Survey*, U.S. Department of Commerce, Washington, DC.

USDOT (assorted years). *Federal Transportation Financial Statistics*, U.S. Department of Transportation, Office of Economics, Washington, DC.

USDOT (1976). *Urban Goods Movement Input to National Transportation Plan, Final Report*, U.S. Department of Transportation, Office of the Secretary, Washington, DC, March.

WALTERS, C. A. (1980). *CBD Dallas: A Case Study in Development of Urban Goods Movement Regulations*, City of Dallas, Texas, Office of Transportation Programs, Dallas.

WINFREY, R., P. D. HOWELL, AND P. M. KENT (1976). *Truck Traffic Volume and Weight Data for 1971 and Their Evaluation*, Federal Highway Administration, Washington, DC, December.

19 Queuing Studies

H. Douglas Robertson, Ph.D., P.E.

INTRODUCTION

A queue forms when the demand on a facility exceeds its capacity or the headway for arriving vehicles is less than the time it takes to service them at a specific location. Queues may be either stopped or slow moving, but in both cases, the result is delay. Queuing occurs in many everyday situations: at store checkout counters, drive-through windows, ticket counters, restaurants, and entertainment events. Queuing is present in all transportation modes and affects motorists and pedestrians. On streets and highways the queuing process is found at intersections, incident sites (accidents and disabled vehicles), bottlenecks, parking and drive-through facilities, and toll plazas.

This chapter focuses on how to obtain the input data required for examining queuing processes. It does not present a comprehensive description of queuing analysis methods and techniques. Readers can find an excellent treatment of queuing analysis and simulation in Chapters 12 and 13 of *Traffic Flow Fundamentals* (May, 1990). Additional references on the analysis of queuing applications are given at the end of the chapter. The chapter addresses queuing analysis input data as they relate to freeway incidents, bottlenecks, work zones, toll facilities, and parking and drive-through facilities. Queuing at intersections is discussed in Chapter 5. The collection of queuing analysis input data is similar to performing volume counts. This chapter addresses data collection issues peculiar to queuing studies. Refer to Chapters 2 and 13 for the particulars of counting traffic or pedestrians.

INPUT REQUIREMENTS FOR QUEUING ANALYSIS

The data required to analyze a queuing process depend on the type of analysis anticipated and the nature of the data available to support that analysis. In every case, something must be known about (1) the characteristics of traffic arriving at the point of constraint where queuing begins, and (2) the nature of the service being provided at the point of constraint that produces queuing.

Levels of Analysis

Queuing analysis may be undertaken at two different levels of detail. The macroscopic level may be used where the arrival and service patterns are considered to be continuous or follow a probability distribution. The microscopic level is used where the arrival and service patterns are considered to be discrete or may be simulated as discrete arrivals and departures. In simple terms, macroscopic analysis deals with the characteristics of groups of vehicles; microscopic analysis deals with the characteristics of individual vehicles.

Analysis Approaches

Queuing analysis is classified as either deterministic or stochastic, depending on the type of data available and what is known about those data. Analysts select deterministic queuing analysis when they know the arrival and service times of each vehicle. Stochastic queuing analysis is selected when either the arrival or the service distributions are probabilistic (i.e., the exact arrival and/or service time of each vehicle is unknown). The flowchart shown in Figure 19-1 depicts a decision process for selecting either an analytical or simulation approach based on whether the data are deterministic or stochastic. In Figure 19-1, the traffic intensity (ρ) is defined as the mean arrival rate (λ) divided by the mean service rate (μ) for each lane. See May (1990) for a complete discussion of these analysis techniques.

DATA COLLECTION

A queuing analysis requires the following input data elements:

- Arrival times or rates
- Service times or rates
- Arrival distribution
- Service distribution
- Queuing conditions

Arrival and service data and queuing conditions are collected in the field. Analysts derive arrival and service distributions from the field data. Each of the data elements above is defined in the following sections, and procedures for collecting each type of field data within that element are described.

Arrival Data

Arrival data may also be referred to as *demand* or *input data*. Arrival data may be in the form of individual vehicle arrival times or expressed as either a flow rate, such as vehicles per hour, or as a time headway, such as seconds per vehicle.

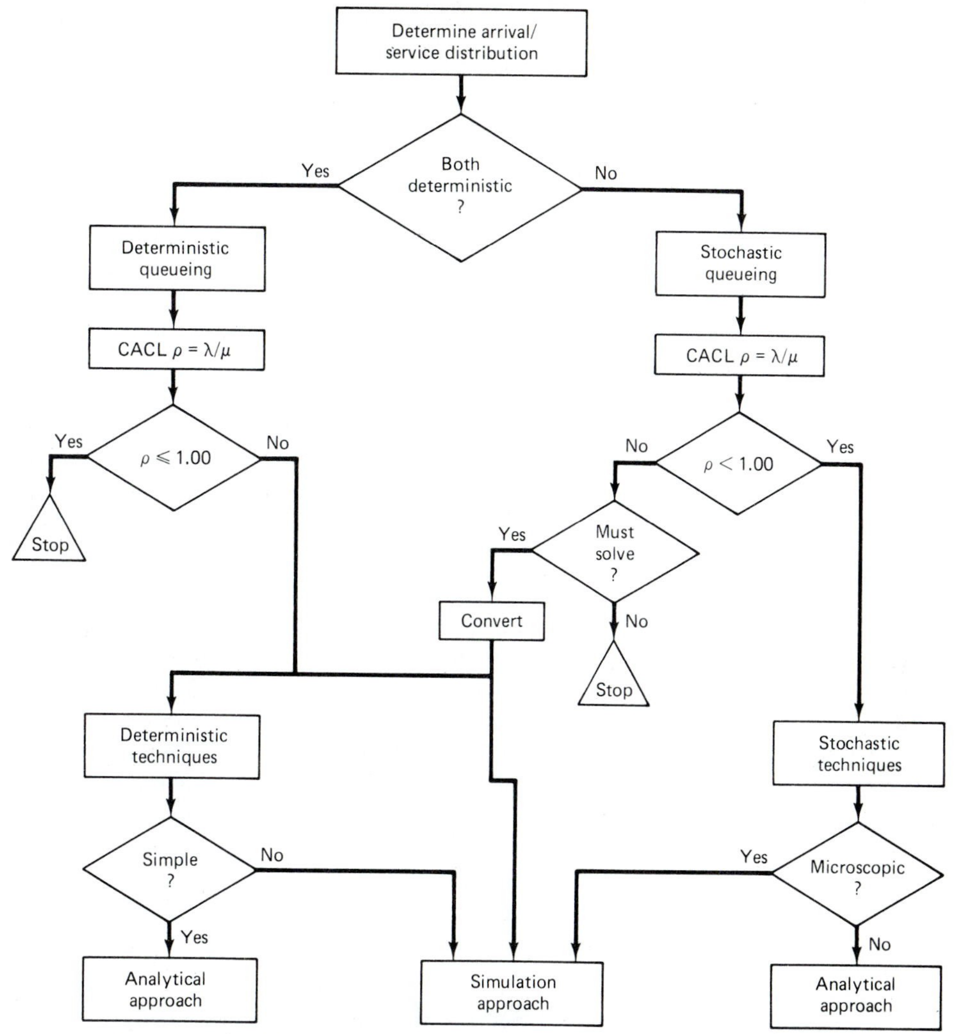

Figure 19-1 Flowchart of queuing analysis approaches. *Source:* May, A. D., *Traffic Flow Fundamentals,* Prentice-Hall, Inc., Englewood Cliffs, NJ, 1990, p. 361. Reprinted by permission.

Service Data

Service data may also be referred to as *capacity, departure,* or *output data.* Service data may be in the form of individual vehicle departure times or may be expressed as a flow rate, such as vehicles per hour, or as a time headway, such as seconds per vehicle.

Equipment and Personnel

Arrival and service data may be collected either manually with observers using count boards or hand-held computers, or automatically with detectors in the travel lanes or video recorders. The equipment and personnel required for collecting arrival and service rates for use in the macroscopic level of analysis are the same as described in Chapter 2. Arrival and service rates are taken directly or are derived from volume counts.

Data collection is somewhat easier for service rates, because the departure point is fixed. The number of observers required will depend on the number of approach lanes being studied and the volume level of traffic. In the cases of bottlenecks and incidents, the service point usually has fewer lanes than the approaches. Toll plazas, however, may require additional observers. If the microscopic level of analysis is used, observers normally record both the arrival time and the departure time of each vehicle.

Field Procedures

If the level of analysis is to be macroscopic, the procedures for collecting arrival and service data are essentially the same as described in Chapter 2 for performing traffic volume counts. Vehicles are counted, either manually or with automated equipment, as they pass a point on the roadway (arrival data) or as they depart the front of the queue (service data) for a given time interval. The study period should begin prior to the start of queuing and continue until the queue dissipates. The count interval should be either 5 or 15 minutes. Hourly counts may be used but are not recommended. Shorter count intervals increase the accuracy of the queuing analysis. Study teams should collect arrival and service data at the same time.

Arrival data should be collected just upstream of the end of the longest expected queue. Traffic at that point should be traveling at a normal speed for prevailing conditions, unaffected by the queuing situation forming downstream. If the study team expects long queues, the observers may have to move the data collection point upstream as queuing occurs. At toll plazas the data collection point should be upstream of where lanes are added that lead into toll collection booths.

Service data should be collected at the front of the queue. The observer should count the departing vehicle as it begins to accelerate to normal speed. Again, the number of observers required depends on the number of open lanes or service bays at the departure point. One person should be able to count up to six lanes unassisted. Automated data collection using the equipment and techniques described in Chapter 2 is particularly effective and efficient at fixed service points. For the microscopic level of analysis, the observer must be positioned to identify a specific vehicle and track it from the time it approaches the end of the queue until it is serviced (i.e., departs the front of the queue). This generally means selection of a high vantage point and the use of binoculars.

Arrival and Service Data Distributions

Analysts use standard statistical distribution fitting techniques to derive arrival and service data distributions from the arrival and service data. Most traffic arrivals follow a random distribution unless upstream traffic signals create platoons. Service rates are generally uniform in bottleneck situations, exhibit small variances at toll booths and parking facility exits, and may show larger variances at drive-through facilities. If the arrivals and departures are constant and uniform or their distributions have constant mean values, a deterministic queuing approach is taken. If either the arrival or service distributions are probabilistic, analysts must use a stochastic analysis approach. Probability distributions common to traffic situations include the random (Poisson), Erlang, and the generalized distributions. Equations have been developed to calculate performance measures for the various combinations of arrival and service data distributions. These distributions are also used to generate microscopic input data in simulation models (May, 1990).

Queue Conditions

The last element of field data needed for queuing analysis is a description of the site under study. The items of interest include:

- Number of approach lanes or paths
- Number of service points, open lanes, or paths
- Queue discipline

The most common system of queue discipline in traffic applications is "first in, first out." In other words, the first to arrive at the head of the queue is the first to depart. Other queue disciplines include "first in, last out" and "served in random order." The observer should note in some detail the type of constriction that is causing traffic to break down and form queues. Observers should also record the environment, weather, and pavement conditions at the time of data collection. This information may explain unusual or unexpected characteristics of the data once they are reduced and examined.

Sampling Considerations

The required sample size depends on how widely the arrival and service rates vary over time. If the rates are uniform throughout the study period of interest, a sample of 30 to 50 vehicles is adequate. If either the arrival rates or the service rates vary during the study period, the study needs larger samples to maintain the desired level of accuracy. In many cases the mean arrival or service rate over the study period will provide for an adequate and meaningful analysis. However, the arrival and/or service rates may change significantly one or more times during the data collection period. For this reason, 5-minute count intervals are recommended so that such changes may be detected and accounted for in the analysis.

QUEUING DATA APPLICATIONS

Traffic streams break down when the rate of arrivals exceeds the capacity of the roadway due to a permanent or temporary constriction (or reduction in number of lanes) at a point along the roadway. The constriction may not always be physical but may result from traffic slowing to observe an activity along the side of the roadway or in opposing lanes. This phenomenon is commonly referred to as *rubbernecking*. Figure 19-2 illustrates the effects of a breakdown in terms of the formation of queues. In the following sections, situations where queuing studies may be applicable and the implications for data collection are discussed briefly.

Queuing at Permanent Bottlenecks

Data collection at permanent constrictions such as bridges and underpasses is relatively straightforward. The location is fixed and the times when traffic stream breakdowns occur and queues form are usually known. The procedures described earlier in the chapter may be employed to record arrivals and departures for use in queuing analysis. Additional information on queuing at bottlenecks may be found in Banks (1990).

Queuing at Accident/Incident Sites

Not all queuing situations are easily studied in the field. Accidents or disabled vehicles that block lanes and cause queuing are difficult to study in the field because they occur at different places without forewarning. In urban areas where incident occurrence is high, a data collection team could be sent to an incident site to collect arrival/departure data for some portion of the queuing period. However, this approach is expensive and seldom

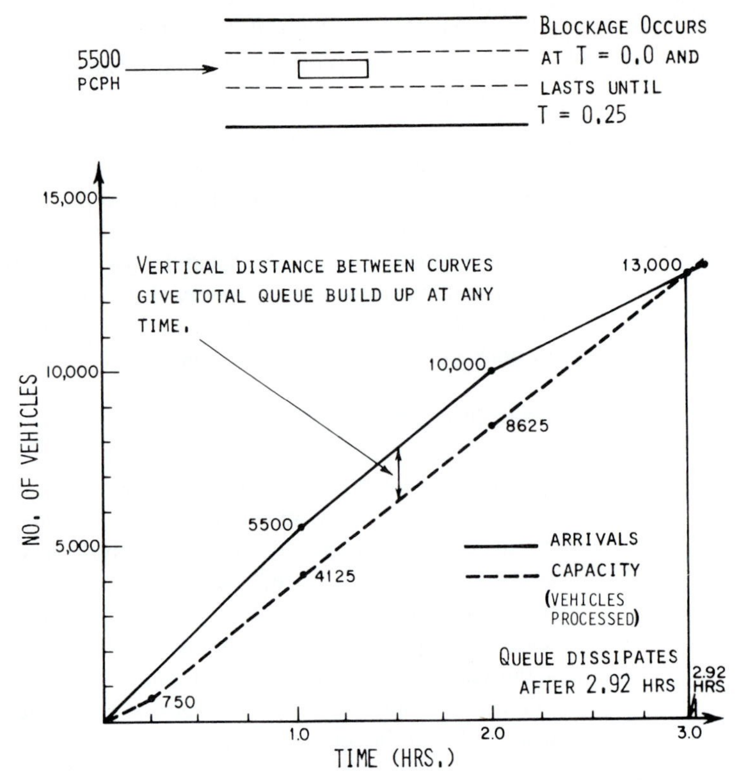

Figure 19-2 Effects of breakdown. *Source:* TRB, 1985.

practical. Instrumented roadways, such as those with freeway management systems, may afford the opportunity to collect queuing data, depending on where in the system the incident occurs.

In analyzing the effects of queuing due to accidents or incidents, it may be sufficient to estimate arrival rates from local volume counts taken on another day in the vicinity of the incident and during the same time period. Service rates may be determined from having observed constricted lane operations under similar accident or incident conditions at another time and place. When this approach is taken, caution should be exercised that the level of accuracy is stated clearly together with the study results.

Queuing at Work Zones

While work zones are temporary, when and where they occur and under what conditions is known beforehand. Thus field data may be collected while they are in operation. The study of queuing at work zones allows reasonably accurate estimates of queue length and delay to be calculated for similar types of operations and conditions. These estimates are useful in determining the optimal times, durations, and configurations of lane closures and work zone operations. Arrival counts are made upstream of the first warning of the presence of a work zone where traffic is flowing at normal speed. Observers obtain service data where the normal lane configuration resumes. Figure 19-3 shows a sample calculation for the queuing analysis of a work zone on an urban freeway.

Queuing at Fixed Service Sites

Fixed service sites include toll plazas, parking facility entrances and exits, and drive-through facilities. Field data collection is relatively easy at these locations. These locations are also more amenable to automated data collection techniques than are the work zone or incident site situations. The results of queuing studies at fixed service sites are

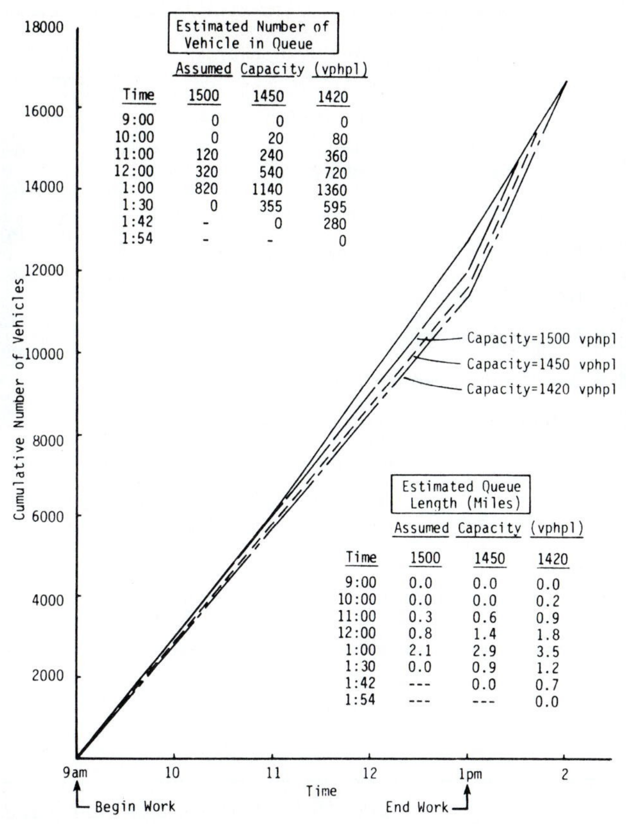

Figure 19-3 Sample calculation: queue analysis for a work zone. *Source:* Dudek, 1982.

used to design new facilities or improve operations at existing facilities. The results are very reliable for similar types of sites and conditions.

QUEUING MODELS

A number of computer models have been developed to assist in performing queuing analyses. Three of them are discussed briefly below. The models discussed here all operate with standard input data collected using the methods described above. The software is available from the Center for Microcomputers in Transportation (1992).

dQUEUE

The dQUEUE program is a dynamic queuing analysis model based on the Monte Carlo simulation technique and traffic flow theory (Lin and Hoang, 1990). It was developed by the Florida Department of Transportation and used to evaluate traffic delays at toll facilities. The program runs on IBM PC-, XT-, or AT-compatible microcomputers with DOS 2.1 or higher and 384K of RAM. Figure 19-4 shows the data input screen. The program graphically animates each vehicle's movement as it approaches the toll plaza. Analysts can observe queues as they form and dissipate. Figure 19-5 shows the animation display screen. The following measures of effectiveness (MOEs) are calculated for each simulation run:

- Average delay per vehicle
- Total delay

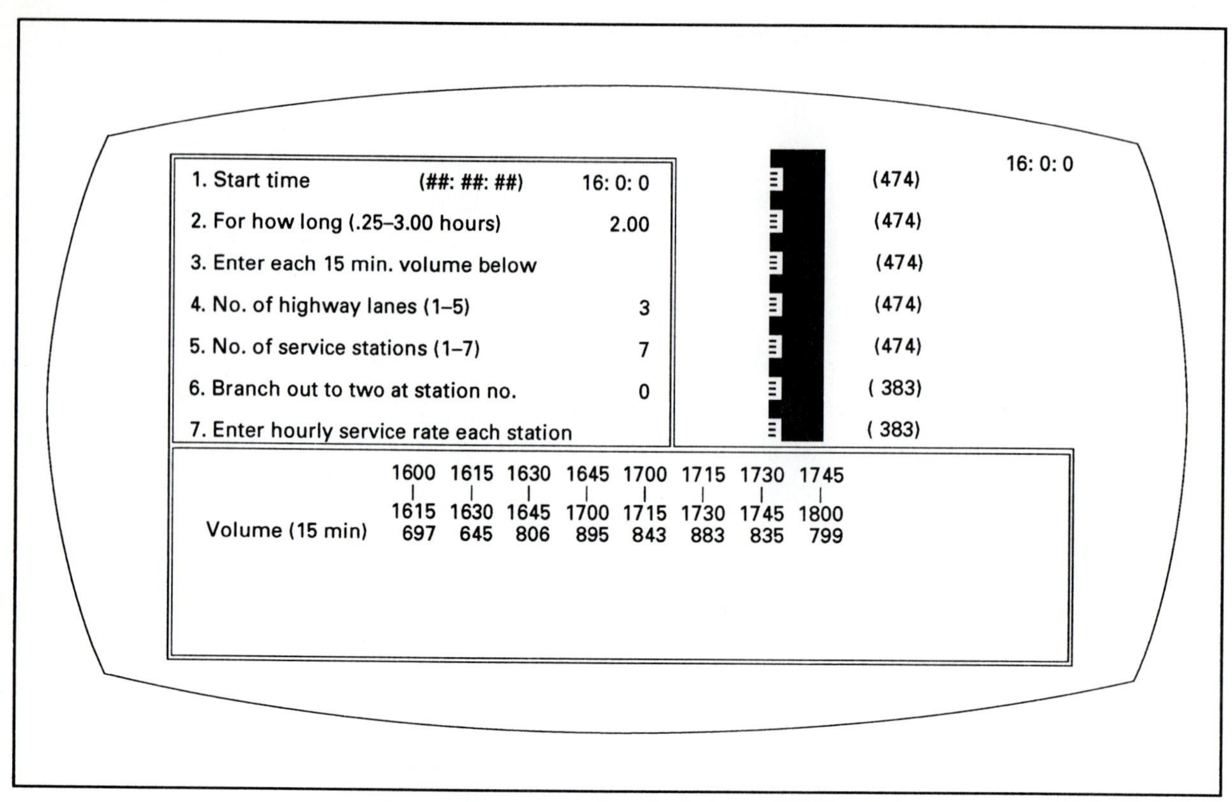

Figure 19-4 dQUEUE data input screen. *Source:* Lin and Hoang, 1990.

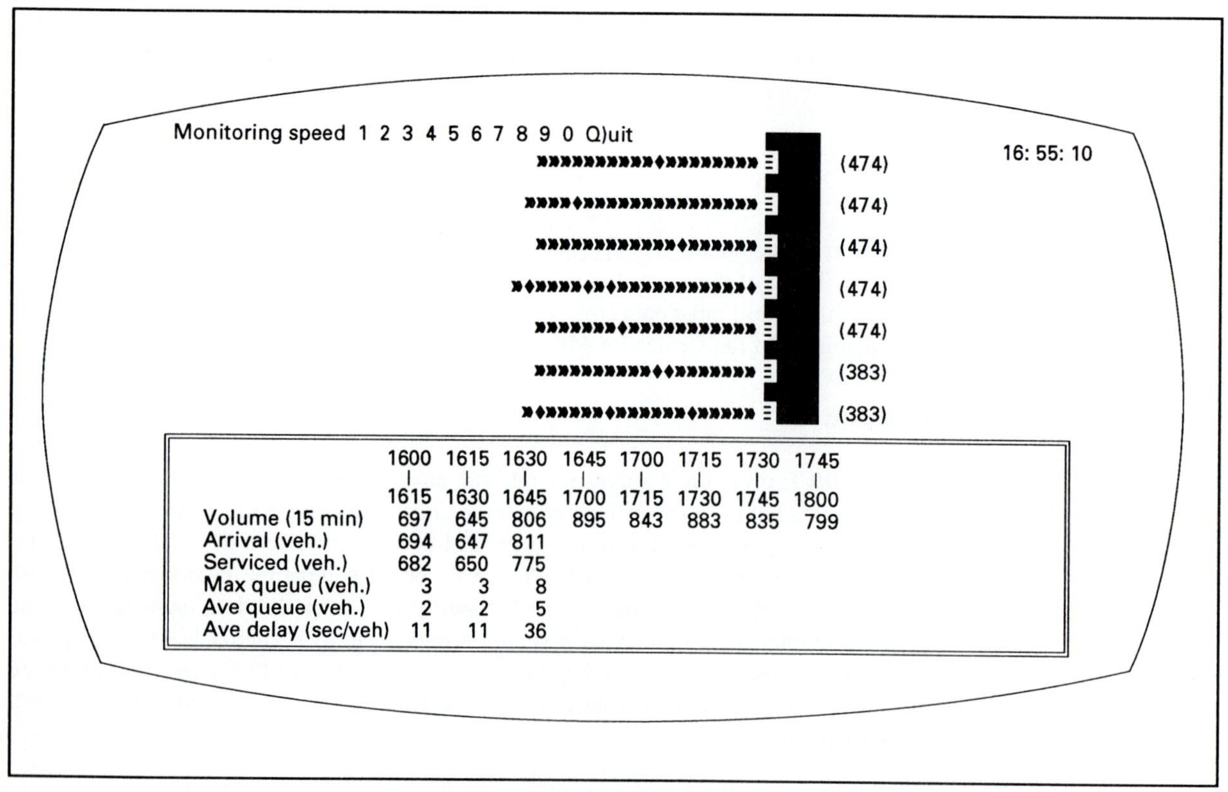

Figure 19-5 dQUEUE system animation display. *Source:* Lin and Hoang, 1990.

- Average queue length
- Maximum queue length
- Level of service

This model is designed to provide realistic delay estimates that are often overstated when using the queuing theory approach and understated when using the input/output model. A 2-hour queuing simulation can be run in about 3 minutes using a microcomputer with an 80386 processor.

QUEWZ

The Texas Transportation Institute developed the QUEWZ model as a tool for planning and scheduling freeway work zone operations (Krammes et al., 1987; Memmott and Dudek, 1982). The model analyzes traffic flow through freeway work zones and estimates the lengths of queues and road user costs attributed to work zone operation. Input and output data are shown in Table 19-1.

The model will run on an IBM PC with MS DOS 2.0 or higher. Figure 19-6 shows a summary of example problem outputs (Memmott and Dudek, 1982).

DELAY

The DELAY software package uses an interactive Lotus 1–2–3 spreadsheet to estimate traffic delay and congestion and assess the trade-offs in cost-effectiveness among alternative measures (Morales, 1989). Analysts can use it to estimate the impact of planned freeway work zone lane closures and the consequences of freeway incidents. Flow rates and incident durations constitute the input data. The program computes total delay, time to return to normal flow, and maximum queue length. A graph of the delay condition is also produced. Figure 19-7 shows a sample printout. The program runs on any IBM-compatible microcomputer with 128K or more of memory.

TABLE 19-1
Input and Output Data for QUEWZ

Input data
- Required
  - Lane closure strategy
  - Total number of lanes
  - Number of open lanes through work zone
  - Length of closure
  - Time of lane closure and work zone activity
  - Actual traffic volumes by hour
- Optional
  - Factor to update cost calculations
  - Percentage trucks
  - Speeds and volumes for speed–volume curve
  - Capacity estimate risk reduction factor or work zone capacity
  - Problem description

Output data
  - Vehicle capacity
  - Average speed through work zone by hour
  - Hourly user costs
  - Daily user costs
  - If a queue develops, average length of queue each hour

Source: Memmott and Dudek, 1982, p. 4.

*** SUMMARY OF EXAMPLE PROBLEMS ***

PROB NO	TOTAL NUMBER OF LANES		NUMBER OF OPEN LANES THRU WZ		LENGTH OF WORK ZONE (MILES)	NORMAL CAPACITY EACH DIRECTION (VPH)	RESTRICTED CAPACITY WORK ZONE INACTIVITY HOURS (VPH)		RESTRICTED CAPACITY WORK ZONE ACTIVITY HOURS (VPH)		HOURS OF RESTRICTED CAPACITY		HOURS OF WORK ZONE ACTIVITY		LONGEST EST QUEUE LENGTH (MILES)		TOTAL ADD. DAILY USER COSTS DUE TO LANE CLOSURE ($)
	INB	OUTB	INB	OUTB			INB	OUTB	INB	OUTB	BEG	END	BEG	END	INB	OUTB	
1	2	2	1	2	1.00	4000.	1800.		1332,		8	16	9	15	1.9	0.0	17647.
2	2	2	1	1	1.00	4000.	1800.	1800.	1354.	1354.	8	16	9	15	1.7	2.9	35112.
3	2	2	1	2	1.00	4000.	1800.		1650.		0	23	9	15	1.0	0.0	11214.
4	2	2	1	1	1.00	4000.	1800.	1800.	1354.	1354	0	23	9	15	1.7	3.7	78343.
5	3	3	2	3	1.00	6000.	3600		2983.		8	16	9	15	0.0	0.0	546
6	3	3	1	3	1.00	6000.	1800.		1127.		8	16	9	15	3.6	0.0	64108.
7	3	3	2	3	1.00	6000.	3600.		2983.		0	23	9	15	0.0	0.0	847.
8	3	3	1	3	1.00	6000.	1800.		1127.		0	23	9	15	4.1	0.0	120878.
9	4	4	4	3	1.00	8000.		5400.		4577.	0	23	9	15	0.0	0.0	368.
10	4	4	4	2	1.00	8000.		3600.		2968.	0	23	9	15	0.0	0.0	986.
11	4	4	4	1	1.00	8000.		1800.		1200.	0	23	9	15	0.0	3.2	101485.
12	5	5	4	5	1.00	10000.	7200.		6200.		0	23	9	15	0.0	0.0	214.
13	5	5	3	5	1.00	10000.	5400.		4500.		0	23	9	15	0.0	0.0	436.
14	5	5	2	5	1.00	10000.	3600.		2745.		0	23	9	15	0.0	0.0	1126.
15	5	5	1	5	1.00	10000.	1800		1200.		0	23	9	15	1.7	0.0	81736.
16	6	6	5	6	1.00	12000.	9000.		8250.		9	15	9	15	0.0	0.0	58.
17	PROBLEM NOT PROCESSED																
18	6	6	3	6	1.00	12000.	5400.		4500.		9	15	9	15	0.0	0.0	217.
19	6	6	2	6	1.00	12000.	3600.		2800.		9	15	9	15	0.0	0.0	551.
20	6	6	1	6	1.00	12000.	1800.		1200.		9	15	9	15	0.8	0.0	27495.

Figure 19-6 Summary of example problems. *Source:* Memmott and Dudek, 1982.

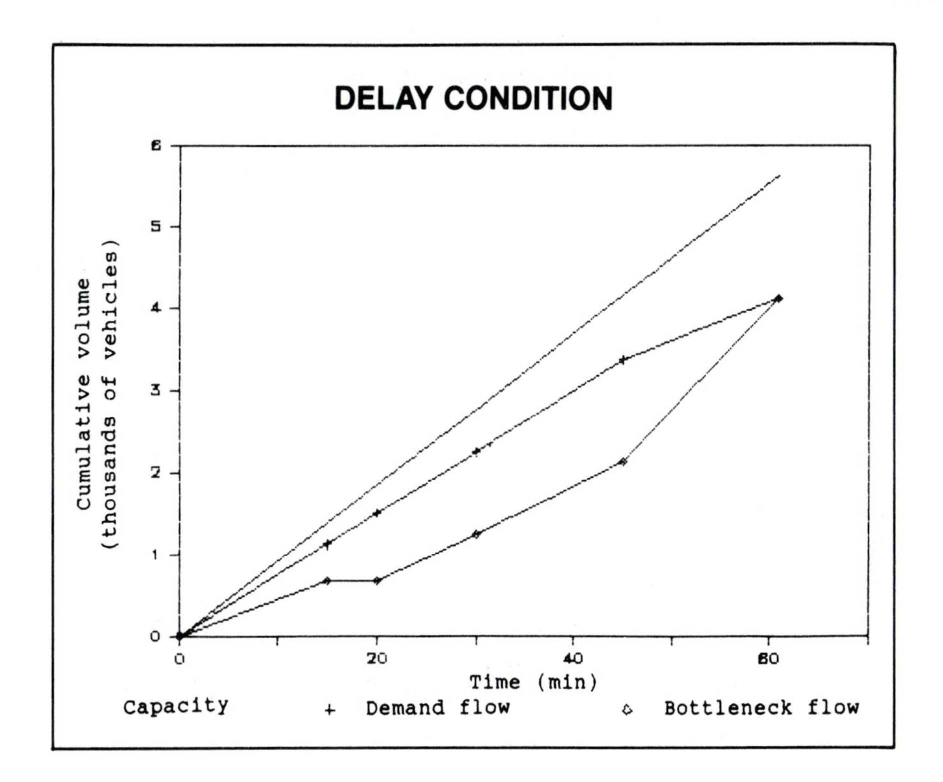

A Method for Calculating Delay, Time, and Queue for Trade-Off Analyses

Any Place, U.S.A.

	Number of Lanes:	3
Capacity flow rate of the facility, veh/hr	S1 =	5550
Initial demand flow rate, veh/hr	S2 =	4500
Initial bottleneck flow rate, veh/hr	S3 =	2700
Adjusted bottleneck flow rate, veh/hour	S4 =	3500
Revised demand flow rate, veh/hr	S5 =	2800
Incident duration until first change, min.	T1 =	15
Duration of total closure, min.	T2 =	5
Incident duration under adjusted flow, min.	T3 =	10
Elapsed time under initial demand, min.	T4 =	45
RESULTS:	Total Delay, veh-hrs =	572.5
	Time to Normal Flow (TNF), min =	60.9
	Maximum extent of queue, veh =	1242
	Maximum length of queue, miles =	2.35

Figure 19-7 DELAY sample printout: 15-minute detection and arrival time. *Source:* Morales, 1989.

REFERENCES

BANKS, J. H. (1990). *Flow Processes at a Freeway Bottleneck,* Transportation Research Record 1287, Transportation Research Board, Washington, DC, pp. 20–28.

CENTER FOR MICROCOMPUTERS IN TRANSPORTATION (1992). *McTrans Catalog,* University of Florida, Gainesville, FL, June, pp. 25, 35.

DUDEK, C. L. AND S. H. RICHARDS (1982). *Traffic Capacity Through Urban Freeway Work Zones in Texas,* Transportation Research Record 869, Transportation Research Board, Washington, DC, Figure 6-13.

KRAMMES, R. A., C. L. DUDEK, AND J. L. MEMMOTT (1987). *Computer Model for Evaluating and Scheduling Freeway Work-Zone Lane Closures,* Transportation Research Record 1148, Transportation Research Board, Washington, DC, pp. 18–24.

LIN, J., AND HOANG, L. (1990). *dQUEUE Version 1.200, User's Guide,* Florida Department of Transportation, Tallahassee, FL, October.

MAY, A. D. (1990). *Traffic Flow Fundamentals,* Prentice Hall, Englewood Cliffs, NJ, pp. 338–375.

MEMMOTT, J. L. AND C. L. DUDEK, C.L. (1982). *A Model to Calculate the Road User Costs at Work Zones (QUEWZ),* FHWA/TX-83/20+292-1, Federal Highway Administration, Washington, DC, September.

MORALES, J. M.(1989). *Analytical Procedures for Estimating Freeway Traffic Congestion,* Transportation Research Circular 344, Transportation Research Board, Washington, DC, January, pp. 38–46.

TRB (1985). *Highway Capacity Manual,* Special Report 209, Transportation Research Board, Washington, DC, pp. 6-7 to 6-14.

A

Experiment Design

Joseph E. Hummer, Ph.D., P.E.

INTRODUCTION

Experiments are comparisons between two or more conditions that are manipulated by, or are under the control of, the experimenter. Experiments are conducted in a systematic and scientific manner so that inferences can be drawn about general populations from measurements taken on samples from the population. For example, one could consider a series of two spot speed studies to determine the effects of a change in the speed limit on the 85th percentile speed to be an experiment. However, a spot speed study conducted to determine the 85th percentile vehicle speed on a highway would not be an experiment since the analyst makes no comparison and does not manipulate the conditions during the study.

Experiments are one of the major means by which transportation engineers gain an understanding of the transportation systems they design and operate. Experiments are also one of the major uses of the data collection methods described in this manual. However, a poorly designed experiment can provide invalid results (i.e., results from which no inferences outside the entities actually measured can be made) or can cost more than it should. This appendix provides a brief overview of the facets of experiment design that are used most often by transportation engineers, so that these outcomes may be avoided. General concepts and terms are presented first, followed by discussions of several types of experiments.

Experiment design reaches far beyond the relatively simple concepts provided in this appendix. Readers seeking details on more advanced topics are encouraged to consult texts such as Cochran and Cox (1957), Anderson and McLean (1974), Hicks (1982), and Montgomery (1984). The advice of professional statisticians is essential for the design of advanced experiments with significant costs.

GENERAL CONCEPTS

Statisticians use several important terms in discussions of experiment design. *Units* or *subjects* are the entities that are selected for participation in an experiment, *measures of effectiveness* (MOEs) are the traits that are measured during an experiment, *factors* are the variables being manipulated in an experiment, and *levels* are the particular states or conditions of factors. In the spot speed experiment mentioned above, the units were the vehicles, the MOE was vehicle speed, the factor was the speed limit, and the levels were the values of the old and new speed limits. Other terms important during discussions of experiment design include *treatments* and *replications*. A *treatment* is a particular combination of levels of various factors. For example, an experiment on observance of traffic control devices had two factors (sign type and sign size), each at two levels (sign types were "stop" and "yield" and sign sizes were small and large). Thus there were four treatments: small stop sign, small yield sign, large stop sign, and large yield sign. The number of *replications* is the number of units to which each treatment is applied. In the experiment on observance of traffic control devices, if there were 28 different intersection approaches (units) tested, and the experimenter applied the treatments evenly, seven replications were made.

The process of experiment design begins with recognition of a problem. Next, the engineer must determine whether an experiment is needed to solve the problem. This determination is often made with a very rough estimate in mind of the cost of an experiment. Better estimates of the cost of an experiment will be available after the design is complete. After the design is complete, the engineer can still reasonably decide not to proceed with the experiment due to excessive cost. The central tasks of experiment design are to define the MOE(s), list the units to be studied, and determine a treatment for each unit. The experiment design is complete when an analysis plan is developed.

One other important task of the experiment designer is to specify the inference space of the experiment. Experiment results should apply outside the narrow range of the particular units measured. "How widely do the results apply?" is the important question that faces the designer. The answer to the question depends primarily on the number of factors and levels studied and the population of units from which a sample is drawn. If the experiment draws units from a vast population and includes a wide range of factors and levels, a large inference space is likely.

Engineers often use the results of an experiment to accept or reject a hypothesis about the relative effects of the treatments applied. Typically, two hypotheses are constructed: (1) that the mean value of the MOE associated with one treatment is not different from the mean value associated with the other treatment, and (2) that the mean values are different. Statisticians call the former the null hypothesis. Rejecting the null hypothesis when it should be accepted (i.e., concluding that there was a difference when there really was not) is called a *type I error. Confidence levels*, such as the "95% confidence level" used so often, refer to the probability that a type I error will not be made. Experimenters make *type II errors* when they accept a null hypothesis that should have been rejected (i.e., conclude that there was not a difference when there really was). Most experimenters estimate sample sizes (i.e., the number of replications) and construct confidence intervals based on a low probability of type I errors, because those are usually more critical. Most type II errors merely preserve the status quo. Sample-size formulas are available for experiment design that minimize both type I and type II errors.

The experiment designer must address several potential problems, including the presence of factors that affect the MOE but are not accounted for in the experiment, results that are unrelated to the project objectives, treatments for one unit that affect the responses of other units (i.e., treatments are not independent), and others. Many experiment designs fail due to nonrandom assignment of treatments to units. Nonrandom assignment of treatments to units is sometimes difficult to overcome and could lead to biased results in a number of ways. Nonrandom assignment of treatments to units might occur, for example, in an experiment to evaluate the effect of an educational program for persons convicted of reckless driving on future accident rates. If the program is offered on a voluntary basis during the experiment, the volunteers may have some unmeasured characteristic, such as anxious spouses, that distinguish them from the general population of reckless drivers. That characteristic may influence the experiment results; a lower future accident rate may be due to the anxious spouses rather than the educational program. Most experimenters using nonrandom procedures to assign treatments to units acknowledge a lower degree of confidence in their results than if units are chosen for treatment at random.

A random number table or computer program that generates random numbers is useful in ensuring that units to be given a particular treatment are randomly selected. For example, if the experiment designer is to select a sample of 30 buses from a fleet of 500, she could assign each bus a number from 001 to 500. Then she could list the first 30 unique three-digit numbers between 001 and 500 in a column of random numbers. The buses corresponding to the numbers on this list would be treated in the experiment. The requirement that the numbers be unique is necessary to ensure that the same bus is not sampled twice. This requirement does not unduly influence the results from samples of large populations.

SIMPLE COMPARISONS

The simplest experiment designs involve the comparison of two levels of a single factor, or two treatments. The speed limit example mentioned earlier is an example of this simple experiment. This design requires two random samples from the population of units. The experimenter applies one treatment to one sample and the other treatment (or no treatment) to the other sample. With small samples of fewer than 30 replications the *t*-test may be an appropriate analysis method for comparing the mean values of the samples. To use the *t*-test on small samples, analysts must assume that the distribution of the MOE value in the population is normal (bell-shaped). If that assumption is poor, the *t*-test outcome may be misleading. With larger samples, no such normality assumption is necessary to compare mean values using *Z*-tests. *F*-tests are appropriate for testing differences in variances with normally-distributed samples. *T*-tests and other useful statistical analysis methods are described in Appendix C.

Experimental comparisons between more than two levels of one factor are common. Such comparisons are designed in a very similar way to comparisons between two treatments described above. One-way analysis of variance (ANOVA) is a convenient and usually valid method for comparing more than two levels of one factor. ANOVA is valid only if the data meet certain conditions. Fortunately, tests are available to see whether a data set meets those conditions. The results of a one-way analysis of variance will show whether there are significant differences in the MOEs among the levels of the factor tested. A means test, such as Newman–Keuls, Duncan's multiple range, or Tukey's, may then be used to demonstrate the levels that were significantly different from other levels (Anderson and McLean, 1974).

Intersection	Number of conflicts observed		d = Student − Teacher
	Student	Teacher	
1	13	17	−4
2	26	30	−4
3	26	21	5
4	36	36	0
5	8	11	−3
6	17	28	−11
7	53	58	−5
8	34	33	1
9	28	22	6
10	9	14	−5
11	19	13	6
12	13	8	5
13	15	19	−4
14	29	14	15
15	34	34	0

Total d = 2
Average d = 0.133
Std. Dev. d = 6.413

To test null hypothesis that Average d = 0, compute t:

t = Average d/(Std. Dev. $d/\sqrt{n}$) = $0.133/(6.413/\sqrt{15})$
$t = 0.08$

For 14 degrees of freedom, t (0.025 level) = 2.145. Since t computed is less than t (0.025 level) accept null hypothesis at 95 percent level.

Figure A-1 Paired comparison experiment data and analysis.

PAIRED COMPARISONS

One common variation on the simple experiment design described above for comparing two treatments is called a *paired comparison.* Paired comparisons are made by blocking, which is forming groups of units that have similar characteristics. After the blocks are formed, the experimenter chooses two units from each block for study. One unit receives one treatment, while the second unit receives the second treatment. As usual, units must respond independently to treatments. Experimenters analyze the differences in MOEs for each pair using the t-test for small samples or the Z-test for large samples. Use of paired comparisons or blocking helps filter out variations among units, allowing a more focused analysis of the effect of the treatment itself (Bhattacharyya and Johnson, 1977).

An example of an effective paired comparison design is in testing the accuracy of traffic conflict data collectors after training. A newly trained collector and a teacher together spend one-half hour at each of 15 different intersection approaches, and independently estimate the number of conflicts observed. Each intersection approach represents a block generating a pair of measurements: one for the student and one for the teacher. The hypothesis tested is that there is no difference between the student's and the teacher's estimated numbers of conflicts at approaches. Figure A-1 provides sample data and calculations that show that the hypothesis is accepted. The experiment design is effective because the variation from different intersections is removed. By contrast, it would be possible with random matching of treatments and units for the student to observe more higher-volume intersections than the teacher.

BEFORE-AND-AFTER EXPERIMENTS

The most common form of experiment in transportation engineering, the before-and-after experiment, is a special type of paired comparison. In a before-and-after experiment, measurements are taken at the same location twice: once before a change and once afterward. The treatments are "not changed" and "changed," the units are points in time and space (i.e., a location at a particular time), and the blocks are the locations.

Before-and-after experiments are attractive for statistical and practical reasons. Statistically, a before-and-after experiment allows a paired comparison to be performed, removing from consideration variation between locations. Practically, before-and-after experiments can be conducted during improvement programs and require measurements at fewer locations than other experiments. Before-and-after experiments are easily understood by engineers and nontechnical readers and make intuitive sense.

Drawbacks to Before-and-After Experiments

Engineers thinking of conducting a before-and-after experiment must consider seven serious drawbacks of that experiment type. First, a before-and-after experiment may require a longer time between the decision to conduct an experiment and the achievement of a conclusion than in other types of experiments. The engineer conducting a before-and-after experiment must wait through both the before and the after periods. Second, before-and-after experiments are very difficult to design while treatments are being implemented or after treatments have been implemented. It is difficult to obtain data for the before period later (or "post hoc") from routine sources. Third, units may not react instantaneously to a treatment. Drivers encountering a traffic control device soon after its placement, for example, may not know how to react and may exhibit unusual behaviors that bias any experiment data being collected. If the experimenter allows a few weeks to elapse before collecting data, the novelty effect may be much smaller. Fourth, units may react to the treatment in an unstable or random fashion (termed *instability*).

People often confuse the fifth and sixth serious drawbacks to before-and-after experiments. History refers to changes in MOE values through the before and after periods caused by factors other than the treatment (Council et al., 1980). For instance, an experiment measuring the number of fatal accidents on rural freeways in the United States through the 1980s could be affected by the change of speed limit on many of those roads in 1987 and 1988. Experiments that compare a before measurement from one season of the year to an after measurement from another season are likely to suffer from a history bias. *Maturation,* the sixth drawback, refers to trends in MOE values with time (Council et al., 1980). For instance, fatal accident rates have been falling in many developed countries for years due to a number of factors. That trend can affect experiments measuring those rates.

Statisticians call the seventh serious drawback to before-and-after experiments regression to the mean (Council et al., 1980). *Regression to the mean* refers to the tendency for a fluctuating characteristic of an entity to return to a typical value in the time period after an extraordinary value has been observed. Engineers have observed this tendency in many databases. Table A-1 shows regression to the mean in a typical accident database. Note that in the second year all groups of roadway sections regressed toward the overall mean of approximately 0.7 accident per section.

Regression to the mean affects before-and-after experiments whenever experimental units are chosen on the basis of a high or low MOE. For instance, treating the highest-

TABLE A-1
Accident Data Illustrating Regression to the Mean

Number of Sections in Group	*Accidents per Section*[a] *First Year*	*Second Year*	*Change (%)*
12,859	0	0.404	Increase
4,457	1	0.832	−16.8
1,884	2	1.301	−35.0
791	3	1.841	−38.6
374	4	2.361	−41.0
160	5	3.206	−35.9
95	6	3.695	−38.4
62	7	4.968	−29.0
33	8	4.818	−39.8
14	9	6.930	−23.0
33	≥10[b]	10.39	−22.0

[a]Overall first- and second-year average accidents per section is approximately 0.7.
[b]Average = 13.33.
Source: Hauer and Persaud, 1982.

accident locations in a jurisdiction with a program of intense traffic law enforcement and evaluating the results on accident rates with a before-and-after experiment is a classic candidate for regression to the mean bias. The high-accident locations would probably have experienced a decline in accidents in the after period anyway, regardless of the treatment.

Overcoming Before-and-After Drawbacks

Some engineers doubt that a before-and-after experiment is ever valid, because so many previous efforts have suffered from one or more of the drawbacks described above. This is an unfortunate perception because a before-and-after experiment does produce valid results economically under certain circumstances if the drawbacks are considered and (if possible) treated. In this section we describe ways to overcome the drawbacks described previously. If a serious drawback is identified that cannot be treated, another experiment type should be used.

The first drawback listed above, that before-and-after experiments require more time than other experiments, is not easy to overcome. Often, analysts and their sponsors must accept the longer time required for the before-and-after experiment. If the longer time cannot be tolerated but an analyst still wants to perform a before-and-after experiment, he or she can change MOEs and use a surrogate measure. For example, if an analyst cannot wait years for the outcome of a before-and-after study using accidents, traffic conflicts (see Chapter 12) may be an acceptable alternative that will deliver results within months.

There is no simple way to overcome the difficulty in constructing post-hoc experiments. Good recordkeeping and inventories may help an analyst assemble data on before periods during or after the after periods. However, even the best records are likely to be missing key elements occasionally, and an analyst cannot simply measure a missing data element in the field in a post-hoc analysis. The decision to conduct a before-and-after experiment should be made prior to the before period.

The third drawback to before-and-after designs mentioned previously is of concern when adaptation to the treatment by the units is not instantaneous. In this situation, experimenters use warm-up periods to allow the unit to adapt. For example, experimenters should wait for one to several months after installing a traffic control device to collect after data on motorist reaction. Experimenters should also take care not to unwittingly use data from the period when the treatment was being constructed. Experimenters interested in the ''novelty'' effect of a treatment do not use warm-up periods.

Designers overcome the fourth drawback, instability, by using a sample of sufficient size and by using statistical analysis methods to draw conclusions from the data. The main point to remember regarding instability is that a count of before events is a random variable, rather than a constant, and must therefore be treated with statistical techniques.

Designers can deal with history and maturation, the fifth and sixth drawbacks to a before-and-after experiment, by shortening the time span of the study. Designers changing from 3-year before and after periods to 1-month periods greatly reduce the chance of large history or maturation biases. Again, it is helpful to use a surrogate measure that can be collected quickly. Experimenters also attempt to overcome history and maturation by literature reviews that identify and quantify biases. Manual adjustments for possible history and maturation biases rely heavily on judgment and are often open to debate, however.

Overcoming Regression to the Mean

Regression to the mean was the last of the serious drawbacks to before-and-after experiments discussed previously. Experimenters can avoid it while selecting units for treatment or can overcome it during analysis. During unit selection, experimenters can avoid regression to the mean by randomly choosing units for treatment from the entire list of units. For instance, in evaluating the effects of a program of school zone speed enforcement, every school zone in the study area must be a candidate for enforcement and data collection. The major problem with this approach is that the units most in need of treatment (i.e., school zones with higher speeds) are not necessarily chosen for treatment. The experiment results do not show whether enforcement in higher-speed zones is effective.

Experimenters can overcome regression to the mean bias during analysis if the experimental data are Poisson distributed. Data are Poisson distributed if they meet certain conditions, including (Steen, 1982):

1. They are structured as a number of events that happen during a particular interval of time (or space or other dimension).
2. The underlying rate of event occurrence does not change through the time studied.
3. The occurrence or nonoccurrence of an event in one time segment is unrelated to the probability of event occurrence in subsequent time segments.
4. More than one occurrence in a very short time is unlikely.

Traffic accident and traffic conflict are examples of data that are usually Poisson-distributed.

For data that are Poisson distributed, Hauer and Persaud (1982) provided a useful method to adjust the number of events experienced in the before period for regression to the mean. The expected number of after events at a location that had k events before is estimated by the number of events that occurred in the before period at locations with $k + 1$ events divided by the number of locations with k events in the before period. For example, consider in Table A-1 the 1884 roadway sections with two accidents each in the before period. An estimate of accidents in the after period adjusted for regression to the mean is $(3 \times 791)/1884 = 1.3$ accidents per section. If an experimental treatment had been applied to these sections and records had indicated an average of 1.5 accidents per section in the after period, a naive analyst would have concluded that the treatment had been effective in reducing accidents when actually, the treatment had probably been harmful.

The major problems with this adjustment for regression to the mean bias are that it only applies to Poisson-distributed data and that it requires a large database. Hauer and

Persaud (1982) analyzed an accident data set from 82 sites and reported that estimation using the adjustment method above was very poor. However, when the data were examined cumulatively (i.e., the analysts examined sections with four or more accidents together instead of analyzing sections with four accidents separately from sections with five accidents, six accidents, etc.), the adjustment method produced more accurate results than did unadjusted before-and-after comparisons.

Hauer and Lovell (1986) have demonstrated another procedure for removing regression to the mean bias during the analysis of before-and-after experiments. The procedure involves relatively advanced statistical techniques that are beyond the scope of this appendix but should be less data intensive than the adjustment described above.

Analyzing a Before-and-After Experiment

Experimenters analyze a before-and-after design like a standard paired comparison test if the data are normally distributed and mean values are of interest. This is the case in many experiments, such as studies of travel times due to revised signal timing plans, speeds in response to enforcement programs, and transit vehicle ridership in response to fare changes. The difference between the mean value of the MOE in the before period and the mean value of the MOE in the after period is analyzed using *t*-tests for small samples or Z-tests for large samples (see Appendix C). The analyst assumes that the difference observed in MOEs between the periods is due entirely to the treatment unless she adjusts for history, maturation, or other known biases.

For before-and-after experiments with accident, traffic conflict, or other data that are Poisson distributed, the usual analysis methods are inappropriate. Before-and-after experiment data that are Poisson-distributed should be analyzed using the *modified binomial test.* Figure A-2 shows criteria for rejecting the null hypothesis (that the before and after rates are the same) based on the modified binomial test. An example use of Figure A-2 follows.

EXAMPLE A-1

Observers noted 14 traffic conflicts at an intersection controlled only by stop signs during a typical weekday. Flashing red signals were then installed to supplement the stop signs. After a sufficient time elapsed for driver acclimation to the signals, observers recorded seven traffic conflicts on a typical weekday. Were the signals effective in reducing traffic conflicts at the 95% confidence level?

Solution: Figure A-2 shows that for 14 before events an analyst needs a 60% reduction to reject the null hypothesis. Since a 50% reduction was experienced (i.e., $(7/14) \times 100\%$) the analyst cannot reject the null hypothesis at the 95% level.

BEFORE-AND-AFTER WITH CONTROL EXPERIMENTS

Before-and-after with control is the strongest experiment design commonly used in transportation engineering. Before-and-after with control experiments overcome three of the major drawbacks of simple before-and-after experiments: history, maturation, and regression to the mean. With a sound statistical design to counter instability and an adequate warm-up period to let units adjust to the treatment, this experiment design is difficult to discredit.

Experimenters select control units at random from the population of units when treatment units are selected. Experimenters measure control units during the before and after periods like treatment units, but do not alter control units during the experiment.

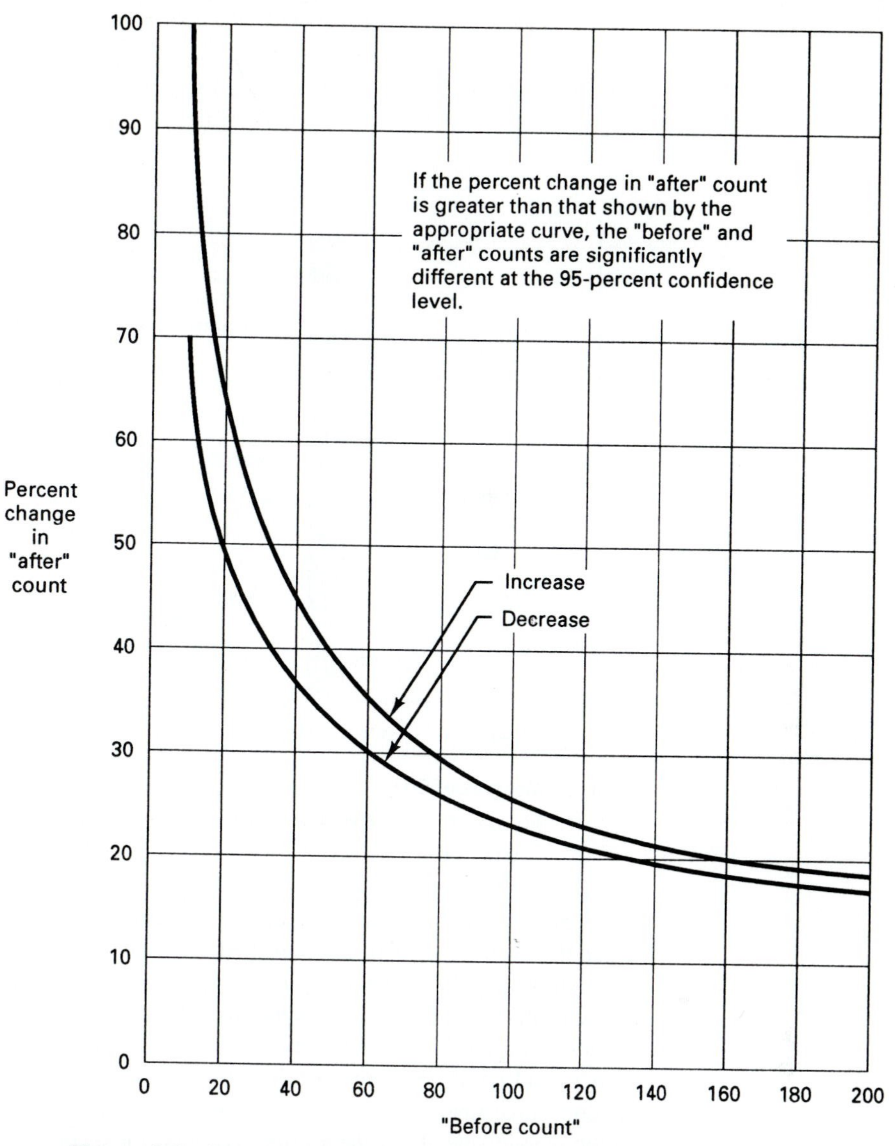

Figure A-2 Rejection criteria for Poisson-distributed data based on the binomial test. *Source:* Weed, R. M., "Revised Decision Criteria for Before-and-After Analyses," *Transportation Research Record 1068,* Transportation Research Board, National Research Council, Washington, DC, 1986, p. 11.

Figure A-3 illustrates the control unit concept. Mean values from a before-and-after with control experiment could be tested with a *t*-test for small samples or a *Z*-test for large samples. A typical null hypothesis is that the mean difference between the experimental units before and after treatment is the same as the mean difference between the control units in the before and after periods.

Before-and-after with control experiments are strong but require more resources than do other experiment types. Like a simple before-and-after experiment, the before-and-after with control experiment may require a relatively long time for data collection and the randomly chosen units may be widely dispersed. However, with control units the amount of data collection will increase over a simple before-and-after experiment. In addition, it is essential that a before-and-after with control experiment be set up early, since control units must be chosen randomly from all possible units.

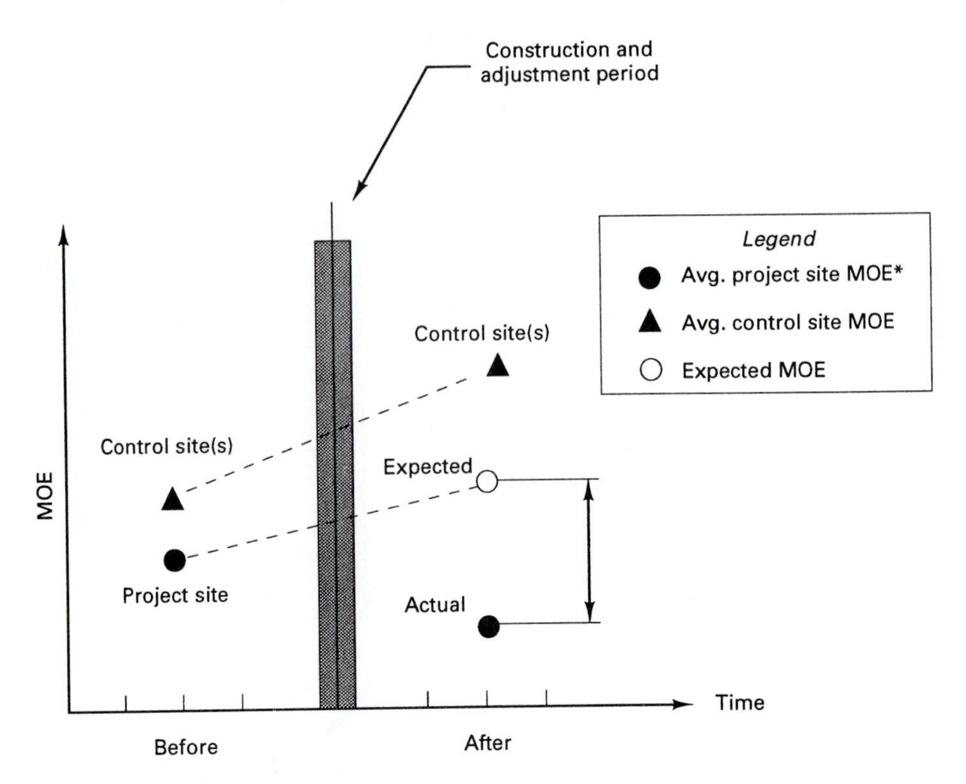

*MOE = Measure of Effectiveness

Figure A-3 Before-and-after experiment with control sites. *Source:* FHWA, 1981.

Some people believe that a before-and-after with control experiment is not feasible when the treatment is to be implemented across an entire jurisdiction or agency. However, Council et al. (1980) suggest that an experimenter can take advantage of the fact that implementation of a treatment is not instantaneous, due to budgeting and personnel constraints, to establish control units. Suppose that guardrails are to be improved over a 2-year period on all freeways in a jurisdiction. An experimenter may be able to select at random some freeway sections to be improved in the first year and some to be improved in the second year. Then the sections selected for second-year improvement can serve as control units during the first year. The key to this plan is random control and treatment unit selection. If the project engineer does not permit random selection, no true "control" is established and a bias may affect the results.

A before-and-after with control experiment may be politically unpopular and open to liability questions if high-accident or high-priority units are untreated. If an experimenter is faced with this criticism, Council et al. (1980) suggest that the treatment is still unproven (that is why an experiment is needed), so withholding it is not necessarily harmful. In addition, the experimenter may promote the idea of an experiment with a promise that high-priority units not treated during the experiment will be treated immediately after the final set of experiment measurements, regardless of the experiment outcome.

BEFORE-AND-AFTER WITH COMPARISON EXPERIMENTS

Before-and-after with comparison experiments are a compromise between the drawbacks of the simple before-and-after experiment and the rigid rules of the before-and-after with control experiment. As with control units, experimenters measure comparison units during the before and after periods and do not alter comparison units during the experiment. However, experimenters select comparison units deliberately rather than at random. This is an important difference.

Comparison units may help overcome the history, maturation, and regression to the mean drawbacks of the simple before-and-after experiment. The ability of comparison units to overcome these drawbacks depends on how closely they compare to experimental units, and proving mathematically that the drawbacks are alleviated is impossible. To alleviate history biases, the same events must affect the experimental and comparison units. To alleviate maturation biases, the same trends must affect both. To alleviate regression to the mean, the experimental and comparison units must have the same group mean of the characteristic being measured in the before period and overall. This is the most difficult constraint on comparison units and means that there is little chance of finding units to compare to a set of high-priority units scheduled for treatment. An experimenter with a large database can employ an adjustment method such as Hauer and Persaud's (presented earlier for simple before-and-after experiments) to estimate the direction and magnitude of regression to the mean. However, most experimenters simply try to choose comparison units similar enough to experimental units so readers and sponsors believe that the regression effects are similar.

Like the other before-and-after experiment designs, a before-and-after with comparison experiment requires a relatively long time until results are available, a warm-up period, and a statistical procedure to account for instability. Comparison units mean more data collection effort than for a simple before-and-after experiment. Comparison units may not be as scattered as control units.

Locating enough high-quality comparison units is a challenging task in many experiments. The number of comparison units needed is estimated from standard statistical sample-size formulas. A common method for choosing comparison units is to choose a unit for treatment and then use the unit next to it for comparison. For example, in evaluating a new traffic control device at minor intersections, an experimenter randomly selects the intersection of Main Street and First Street for treatment. He would then use the next minor intersection along Main Street (i.e., Main Street and Second Street) as a comparison unit. He should not use this method if operations at the sites are dependent, however. If drivers can see the traffic control device at Main and First from Main and Second, for instance, the latter is probably not an appropriate comparison unit. Comparison units are often much easier to locate when experiments are designed prior to or during the before period rather than during or after the after period. Reassembling data from several prior years to locate adequate comparison units is extremely difficult, even in agencies with outstanding recordkeeping systems.

FACTORIAL DESIGNS

The experiment designs described earlier in this appendix primarily involved comparisons of one treatment to another. These methods work well for determining the effects of a few treatments, but the sample sizes required for valid comparisons of more than a few treatments quickly prove onerous. Fortunately, factorial experiment designs are available to allow engineers to make valid comparisons between many factors and levels using minimum samples.

In addition to efficiency, one of the prime benefits of many factorial designs is the ability to estimate the effects of interactions between factors. The interaction is the combined effect of two or more factors. For example, consider an experiment examining the relationship of traffic volume and the presence of a turn bay to delay. The experimenter found that both factors working alone (i.e., the main effects) were related to delay, but that information was not very useful. She also found that the interaction between volume and the presence of a turn bay was related to delay. That information was very useful: She

TABLE A-2
ANOVA Results from a Four-Factor Experiment

Factor or Interaction[a]	*Degrees of Freedom*	*Sum of Squares*	*F Value*	*Significance Probability*
L	1	3.3	14.2	0.0004
T	3	81.6	116.6	0.0001
P	3	272.2	389.5	0.0001
S	2	218.3	467.7	0.0001
L*T	3	0.3	0.4	0.7272
L*P	3	1.4	2.0	0.1244
L*S	2	1.8	3.8	0.0271
T*P	9	3.1	1.5	0.1799
T*S	6	6.9	5.0	0.0004
P*S	6	0.4	0.3	0.9418
Error	57	13.3		
Total	95	589.9		

[a]The notation L*T, for example, means the interaction between the left-turn volume factor (L) and the through volume factor (T).

Source: Hummer et al., 1989.

TABLE A-3
Results of Means Tests on the Main Effects of the Four-Factor Experiment

Factor	*Level*	*Number of Observations*	*Mean Delay (seconds per vehicle)*	*Levels of Same Factor That Were Not Significantly Different at 0.05 Level Using Student–Newman–Keuls Test*
P	None	24	11.7	—
	Left dir.	24	9.1	—
	Opposite dir.	24	9.5	—
	Both dirs.	24	7.0	—
L	140	48	9.2	—
	230	48	9.5	—
T	400	24	8.4	—
	600	24	8.7	—
	800	24	9.6	—
	1000	24	10.8	—
S	Permissive	32	7.2	—
	Perm.-pro.	32	10.4	Pro.-perm.
	Pro.-perm.	32	10.4	Perm.-pro.

Source: Hummer et al., 1989.

then knew that there were some levels of volume where turn bays would be more beneficial and could recommend designs based on the breakpoint in the relationship. Two-way interactions often provide the most useful information from an experiment. Three-way interactions can be useful but are sometimes difficult to interpret. Interactions among four or more factors are not generally useful in transportation engineering experiments.

Analysis of variance (ANOVA) is the usual method of analyzing factorial experiments. Experimenters must verify several assumptions about their data before trusting ANOVA results. Experimenters must also specify a model of the relationship between the factors, the interactions, and the MOE to use ANOVA. Most standard computer statistical packages contain ANOVA and quickly generate estimates of the significance levels of any number of main effects and interactions. Table A-2 shows ANOVA results from an experiment with four factors, where the model included only main effects and two-way interactions. Analysts use means tests to find the particular levels of significant factors that deviate from other levels. Table A-3 illustrates the application of a means test to the main effects of the four-factor experiment.

Levels of factor "traffic control" \ Levels of factor "enforcement"	Unit numbers			
	None	Two patrols per hour	Four patrols per hour	Contin-uous patrols
None	31 19	2 28	18 5	23 11
Signs only	6 22	32 21	1 26	4 27
Markings only	3 7	10 12	8 15	13 16
Signs and markings	30 14	29 24	17 25	20 9

Figure A-4 Example of a completely randomized factorial experiment design.

Completely randomized factorial designs have one or more observations in every cell of the matrix of factors and levels. Figure A-4 illustrates such a design with two factors, each at four levels, and two replications. The units, numbered 1 through 32, were assigned at random to a cell in the matrix. Completely randomized factorial designs must be at least partially replicated (i.e., at least a few cells must have more than one observation) to allow the significance of the main effects and all the interactions to be estimated without additional assumptions.

Many types of advanced factorial designs are available. *Block designs* utilize groups of similar units, as described in the section "Paired Comparisons" above, to remove extraneous sources of variation efficiently. Block designs are useful when treating some units one day and other units another day, for example. The possible effect of the day treated can then be separated from the effect of the treatment. Within each block, experimenters assign units to a treatment at random. Experimenters use *nested designs* when one or more factors are embedded within other factors so that it is impossible to obtain an interaction between the factors. For example, nested designs are convenient when an experimenter identifies several jurisdictions and then selects for treatment several households within each of those jurisdictions. The experimenter can separate the effect of the jurisdiction from the effect of the household (Anderson and McLean, 1974).

Other advanced factorial designs called *fractional factorials* allow the experimenter to gain full information on main effects and partial information on interactions while running only a portion of a replication. Fractional factorials are very difficult to design but can be very efficient, especially when treatments are expensive. Experimenters often use fractional factorials in exploratory work to determine which variables should be examined in depth. Figure A-5 shows the design of an experiment with eight factors, each at two levels, that would have required 256 units for a full replication. The experimenters designed a one-quarter fractional factorial experiment that required only 64 units (labeled "run number" in Figure A-5) and provided complete information on all eight main effects and all interactions between two factors. *Latin square* and *Graeco-Latin square* designs are very specialized fractional factorial designs that have been applied successfully in transportation engineering experiments. Designers interested in block, nested,

Run Number	Variable Combination	Run Number	Variable Combination
1	I	33	BCDEF
2	AB	34	ACDEF
3	CD	35	BEF
4	ABCD	36	AEF
5	GH	37	BCDEFGH
6	ABGH	38	ACDEFGH
7	CDGH	39	BEFGH
8	ABCDGH	40	AEFGH
9	DE	41	BCF
10	ABDE	42	ACF
11	CE	43	BDF
12	ABCE	44	ADF
13	DEGH	45	BCFGH
14	ABDEGH	46	ACFGH
15	CEGH	47	BDFGH
16	ABCEGH	48	ADFGH
17	FH	49	BCDEH
18	ABFH	50	ACDEH
19	CDFH	51	BEH
20	ABCDFH	52	AEH
21	FG	53	BCDEG
22	ABFG	54	ACDEG
23	CDFG	55	BEG
24	ABCDFG	56	AEG
25	DEFH	57	BCH
26	ABDEFH	58	ACH
27	CEFH	59	BDH
28	ABCEFH	60	ADH
29	DEFG	61	BCG
30	ABDEFG	62	ACG
31	CEFG	63	BDG
32	ABCEFG	64	ADG

Letter Key

Letter	Variable	Units	Level With No Letter	Level With A Letter
A	Through volume	VPH	300	700
B	Opposing Volume	VPH	300	700
C	Left-turn volume	VPH	20	50
D	Right-turn volume	VPH	20	50
E	Mean vehicle speed	MPH	35	50
F	Upstream signal distance	Miles	0.5	1.5
G	Downstream signal distance	Miles	0.5	1.5
H	Bypass lane present	—	No	Yes
I	Variables A through H all at "no letter" level			

Figure A-5 Example one-quarter fraction factorial experiment design with eight factors. *Source:* Bruce, E. L. and J. E. Hummer, "Delay Alleviated by Left-Turn Bypass Lanes," *Transportation Research Record 1299,* Transportation Research Board, National Research Council, Washington, DC, 1991.

fractional, or other advanced factorial designs are encouraged to consult a professional statistician or a text on statistical experiment design such as by Cochran and Cox (1957), Anderson and McLean (1974), Hicks (1982), or Montgomery (1984).

REFERENCES

ANDERSON, V. L., AND R. A. MCLEAN (1974). *Design of Experiments, A Realistic Approach,* Marcel Dekker, New York.

BHATTACHARYYA, G. W., AND R. A. JOHNSON (1977). *Statistical Concepts and Methods,* Wiley, New York.

BRUCE, E. L., AND J. E. HUMMER (1991). *Delay Alleviated by Left-Turn Bypass Lanes,* Transportation Research Record 1299, Transportation Research Board, Washington, DC.

COCHRAN, W. G., AND G. M. COX (1957). *Experimental Designs,* Wiley, New York.

COUNCIL, F. M., ET AL. (1980). *Accident Research Manual,* FHWA/RD-80/016, U.S. Department of Transportation, Federal Highway Administration, Washington, DC, February.

FHWA (1981). *Highway Safety Evaluation Procedural Guide,* U.S. Department of Transportation, Federal Highway Administration, Washington, DC.

HAUER, E., AND J. LOVELL (1986). *New Directions for Learning about the Safety Effect of Measures,* Transportation Research Record 1068, Transportation Research Board, Washington, DC.

HAUER, E., AND B. PERSAUD (1982). *Common Bias in Before-and-After Accidents Comparisons and Its Elimination,* Transportation Research Record 905, Transportation Research Board, Washington, DC.

HICKS, C. R. (1982). *Fundamental Concepts in the Design of Experiments,* Holt, Rinehart and Winston, New York.

HUMMER, J. E., R. E. MONTGOMERY AND K. C. SINHA (1989). *An Evaluation of Leading versus Lagging Left Turn Signal Phasing,* FHWA/IN/JHRP-89/17, Federal Highway Administration and Indiana Department of Transportation, Indianapolis, IN, December.

MONTGOMERY, D. C. (1984). *The Design and Analysis of Experiments,* 2nd ed., Wiley, New York.

STEEN, F. H. (1982). *Elements of Probability and Mathematical Statistics,* Duxbury Press, Boston.

WEED, R. M. (1986). *Revised Decision Criteria for Before-and-After Analyses,* Transportation Research Record 1068, Transportation Research Board, Washington, DC.

B
Survey Design

Joseph E. Hummer, Ph.D., P.E.

INTRODUCTION

Transportation engineers have used surveys for many years, particularly in transportation planning. However, many engineers assigned to design and conduct surveys have not had training or experience in survey preparation. Those engineers are often surprised to learn that even small surveys are very difficult to design and can fail if designed poorly. This appendix serves as a starting point for beginners and provides a reference for those already experienced in designing surveys. There is no substitute for experience in conducting surveys. An engineer inexperienced in surveys who is conducting a survey that is expensive or important should seek help from an experienced professional.

Surveys are essentially systematic, structured conversations. Surveys range from written questionnaires where each respondent answers the same questions with the same set of possible answers to rambling interviews held together by a central theme. Most survey results are coded and analyzed using standard statistical techniques (discussed in Appendix C).

Articulating Study Objectives

Before analysts decide to conduct a survey they must carefully define the study problem and articulate the study objectives. If possible, the study objectives should be written clearly in one short paragraph. Clear objectives are important since many subsequent decisions depend on them. If analysts cannot clearly articulate study objectives, they should suspend further work on the study.

Why Survey?

Surveys are an effective study technique for several reasons. If conducted properly, engineers can collect a large and valid data set relatively cheaply. Surveys have flexible formats and allow several types of answers. Surveys are often the only way to gather feedback from users of a system if data on market responses (i.e., revenue, passenger volume, etc.) are not available. Surveys are also the primary way to measure attitudes and determine why people act a certain way. Finally, despite recent abuses, such as sales personnel posing as interviewers, people generally try hard to answer well-designed survey questions, especially on topics of wide interest, such as transportation.

A survey may not be the most effective source of data for a particular study, for two good reasons. First, an analyst should not conduct a survey if the study objectives can be achieved by some other data collection method, such as direct measurement. A survey only provides the estimates and opinions of imperfect people. Surveys may be conducted to supplement data from other sources, of course. Second, analysts should not conduct a survey unless adequate resources are available to do the job properly. Skipping the steps of proper survey design, such as a pretest of the survey instrument, undermines the validity and reliability of the survey.

METHODS

Survey data are collected from the verbal or written responses of a sample of people. Common methods of collecting written responses include a mail-out and mail-back form, a hand-out and mail-back form, and a hand-out and hand-back form. Common methods of collecting verbal responses include telephone interviews and personal interviews. Personal interviews are conducted at the respondent's home, at the respondent's workplace, at the survey analyst's office, at some public place, at an event, or almost anywhere. There are also hybrid methods that combine the foregoing techniques. Each method has advantages and disadvantages.

Table B-1 shows some of the relative advantages of three common survey methods: personal interviews, telephone interviews, and mail-out and mail-back questionnaires. The major advantage of the personal interview method are the ability to probe responses in depth and the high probability of obtaining responses from the desired sample. By contrast, mail-out and mail-back questionnaires offer lower costs and minimize interviewer bias. Telephone interviews usually take less time than the other methods shown in Table B-1 but do not allow the presentation of visual stimuli. Telephone interviews generally have traits that are between the extremes of the personal interview and the mail-out and mail-back questionnaire, making them an attractive choice for many surveys. Interview surveys have improved in recent years with advances in computers. Interviewers using hand-held computers in the field or personal computers in the office can immediately code and store responses. Computers also help interviewers follow complex "skips" where some respondents are not asked some questions.

Hand-out and hand-back or hand-out and mail-back questionnaires are common in transportation studies. They have many of the advantages and disadvantages of mail-out and mail-back questionnaires but can be less costly. Costs are especially low if workers (bus drivers, for example) distribute and/or collect questionnaires while performing their regular duties. Hand-out and hand-back or hand-out and mail-back surveys also have higher response rates than mail-out and mail-back surveys.

The choice of a survey method depends on the type of questions and the sample anticipated. Typically, a designer tentatively chooses a method in the early stages of

TABLE B-1
Comparison of Major Survey Methods

Criterion	*Personal Interview*	*Telephone Interview*	*Mail-Out and Mail-Back Questionnaire*
Data collection costs	High	Medium	Low
Data collection time required	Medium	Low	High
Sample size for a given budget	Small	Medium	Large
Data quantity per respondent	High	Medium	Low
Reaches widely dispersed sample	No	Maybe	Yes
Reaches special locations	Yes	Maybe	No
Interaction with respondents	Yes	Yes	No
Degree of interviewer bias	High	Medium	None
Severity of nonresponse bias	Low	Low	High
Presentation of visual stimuli	Yes	No	Maybe
Instruction complexity possible	High	Medium	Low
Respondent can consult records	No	No	Yes
Respondent anonymity possible	No	Yes	Yes
Field worker training required	Yes	Yes	No

Sources: Alreck, P. A. and Settle, R. B., *The Survey Research Handbook,* Richard D. Irwin, Inc., Homewood, IL, 1985; Sudman, S. and Bradburn, N. M., *Asking Questions,* Jossey-Bass Publications, San Francisco, 1982.

survey development and then reevaluates his choice while drafting questions and a selecting a sample.

SAMPLE SELECTION

A major part of the work of survey design is the selection of a sample of people who will respond to the survey. The sample is intended to represent the population of interest. Sampling is necessary because it is usually too costly and too slow to survey the entire population of interest. Several common sampling techniques are discussed here. Cochran (1977), Lerman and Manski (1976), and others provide more details and descriptions of other techniques.

Random Sampling

With *random sampling* each member of the population of interest has some established probability of being selected as part of the sample to be surveyed. Random sampling is common and has several advantages over nonrandom sampling, which is discussed later. Random sampling requires the analyst to regard the *population* under study as a collection of *sampling units.* The population under study can be quite specific and must be related to study objectives. For example, an engineer might limit the population for a survey on the environmental impacts of a new highway to residents living and employees working within 1000 feet of the highway right-of-way. Sampling units are the nonoverlapping entities in the population that will be surveyed. In most surveys, sampling units are individual people, although transportation surveys have also used families, neighborhood groups, and other entities. Once the designer establishes the sampling unit, she must obtain a complete list of sampling units in the population, called a *frame.* A reasonably accurate and complete frame is necessary for most random sampling techniques. Compiling a frame is often very difficult because directories, tax rolls, and other common framing materials are often out of date or error-prone.

Simple Random Sampling

With this method, each sampling unit in the population has an equal chance of being chosen. The analyst chooses units to be sampled by assigning a number to each unit in the population and then drawing random numbers. Simple random sampling can be accomplished *with replacement,* when sampling units are eligible to be drawn more than once, or *without replacement,* when units cannot be drawn a second time. Some formulas for analyzing samples with replacement are simpler, but for most surveys there is no difference in the analysis. Therefore, to remove the possibility that a person is asked the same questions twice, sampling without replacement is standard.

Simple random sampling has a few advantages over other sampling methods, but for most surveys it is inferior. Simple random sampling is easy to understand, is commonly used (making it easier to compare results with other surveys), and allows a designer to select a sample with no population data besides a frame. However, simple random sampling is usually inconvenient, since survey personnel must track down widely dispersed units. Worse still, simple random sampling is almost always less precise than sampling methods that utilize more information about the population. With the same size sample, other sampling schemes will usually produce better estimates of means and proportions than will simple random sampling.

Table B-2 provides formulas for estimating mean values, variances, and needed sample sizes for a simple random sample. The sample-size formula for continuous (not proportional) data requires an estimate of the variance of the mean for the entire population or a prior estimate of the coefficient of variation, which is the variance divided by the mean. Analysts can obtain this prior estimate from an educated guess, the results of a prior survey, or a pilot study of the current population. The sample-size formulas are based on the desire to estimate the mean value of some characteristic within a specified tolerance from the true mean value for the entire population. For example, an analyst may want to estimate mean household income within $1000 of the true mean.

Stratified Random Sampling

In this scheme, analysts divide the population into nonoverlapping groups of units called *strata.* A simple random sample is then drawn from each stratum. Estimates of means, proportions, and other desired statistics are based on the samples in the strata. Analysts should select strata so that the characteristic of interest in the survey differs between them. For example, in a survey on the travel habits of households, it may be desirable to stratify by the number of automobiles owned by a household since auto ownership is strongly related to travel patterns.

Stratified random sampling almost always provides more precise estimates of a quantity than does simple random sampling. This is especially true for populations where the quantity to be estimated is similar within strata but differs greatly between strata. Stratified random sampling is also advantageous because estimates for the strata can be made. In addition, Cochran (1977) points out that stratified random sampling may be convenient, such as during surveys to be administered by field offices, where each office can survey a stratum. The main disadvantage of stratified random sampling in comparison to simple random sampling is the extra work needed to select a reasonable strata. The analyst must know the variable(s) on which the strata are based before stratification. The analyst must also know the number of units in each stratum before producing estimates, but if a complete frame is available, this is not a major obstacle.

The formulas for sample-size estimation and for estimating means and variances are more complex for stratified random sampling than for simple random sampling and are not reproduced in this appendix. Consult Cochran (1977) or another statistic or sampling text for those formulas.

Cluster Sampling

This random sampling scheme is best explained through an example. Instead of choosing individual people as the sampling unit for a travel survey of the residents of a town, the analyst chooses neighborhoods as the sampling unit. Then the individual

TABLE B-2
Simple Random Sampling Formulas

Quantity to be Estimated	*Formula*[a]	*Variable Definitions*
	Continuous Data	
Mean	$\bar{y} = \dfrac{\sum_{i=1}^{n} y_i}{n}$	y: Individual observation i: Observation number n: Total number of observations
Variance of mean	$s^2 = \dfrac{\sum_{i=1}^{n} (y_i - \bar{y})^2}{(n - 1)}$	
Sample size	$n = \left(\dfrac{tS}{d}\right)^2$ or $n = \left(\dfrac{tS}{r\bar{Y}}\right)^2$	t: A constant corresponding to the desired level of confidence, α percent (i.e., the statistic for the normal distribution; see Table 12-2) S: Standard deviation of the population mean $\bar{Y}$: Population mean d: $\lvert \bar{Y} - \bar{y} \rvert$ should not be greater than d more than $(100 - \alpha)$ percent of the time r: $\lvert \bar{Y} - \bar{y} \rvert$ should not be greater than $r\bar{Y}$ more than $(100 - \alpha)$ percent of the time
	Proportional Data	
Proportion	$p = \dfrac{a}{n}$	a: Number of units in sample answering "yes"
Variance of proportion	$\text{var}(p) = \dfrac{pq}{n - 1}$	q: Proportion of units in sample answering "no"
Sample size	$n = \dfrac{t^2pq}{d^2}$ or $n = \dfrac{t^2q}{r^2p}$	P: Proportion of units answering "yes" in population d: $\lvert P - p \rvert$ should not be greater than d more than $(100 - \alpha)$ percent of the time r: $\lvert P - p \rvert$ should not be greater than rP more than $(100 - \alpha)$ percent of the time

[a]Formulas for sampling with replacement or for sampling without replacement when less than 5% of the units in the population are sampled.
Source: Cochran, 1977.

people are known as the sampling *subunits*, and each subunit in a sampled neighborhood would be asked to respond to the survey. This sampling plan is known as *cluster sampling* because the sampling unit (a neighborhood) consists of a cluster of subunits. Cluster sampling is a very popular way to conduct surveys in transportation and many other fields.

Lower cost is the main advantage of cluster sampling over other sampling methods. The travel survey mentioned above saves money by sending survey personnel to a relatively small number of neighborhoods rather than to dwellings scattered across town. In addition, cluster sampling does not require a frame of all the subunits in the population. The travel survey mentioned above only needed lists of neighborhoods in the town and of residents of the sampled neighborhoods. The major disadvantage of cluster sampling is that a convenient unit and subunit must be available. Cluster sampling is impossible if reasonable clusters cannot be defined. Another disadvantage of cluster sampling, cited by Fink and Kosecoff (1985), is a set of relatively complex formulas for estimating sample sizes, means, and variances. Those formulas differ depending on the exact cluster sampling technique used and are not reproduced in this appendix.

Nonrandom Sampling

Nonrandom sampling refers to schemes wherein each unit does not have an established probability of being selected. Engineers usually draw nonrandom samples from the most readily accessible units in the population in a haphazard way. An example of a nonrandom sampling plan would be an engineer surveying pedestrian habits in an area by stopping the first 100 people walking along a particular sidewalk. Most surveys by transportation engineers use nonrandom sampling.

Nonrandom samples are appropriate for many studies but have several disadvantages. Nonrandom samples have an advantage over random samples in that they are much more convenient for the designer. There is no need to draw a random sample. Indeed, if one of the steps involved in drawing a random sample cannot be performed, a nonrandom sample may still be sufficient to achieve project objectives. The major disadvantage of nonrandom samples is that one cannot prove the validity of the findings mathematically. With nonrandom sampling, the sampling error (the error in the quantity being estimated attributed to the fact that one particular sample was drawn of the many possible samples in the entire population) remains unknown, so engineers cannot easily generalize their findings from the sample to the entire population. A common mistake made by engineers conducting surveys with nonrandom samples is to use the analysis techniques of random sampling and thereby mislead readers of their reports. Another disadvantage of nonrandom sampling is that there is no theoretically sound way to estimate the needed sample size. Finally, survey workers may be susceptible to bias in the selection of a nonrandom sample.

COMPOSING QUESTIONS

Writing good survey questions is very difficult. Consequently, many surveys have questions with no feasible responses, questions with too many feasible responses, too many questions, vague instructions, and other flaws. The best sampling plan and the most exhaustive statistical analysis are meaningless unless valid and reliable responses are obtained from the persons sampled. This section provides guidance in composing questions so that major pitfalls may be avoided.

Number of Questions

Several factors affect the length of a survey. First, the study objectives will affect the survey length. Complex and multiple objectives require longer surveys, while simple objectives require short surveys. Designers should include as many questions as needed to accomplish study objectives and should make sure that each question relates to the objectives. One way to ensure that each question is related is to state in writing its unique contribution to accomplishing the objectives. Second, the topic affects the survey length. Respondents will tolerate longer surveys on topics that are interesting or important to them. Sudman and Bradburn (1982) advise that on topics important to respondents, home interviews can last up to $1\frac{1}{2}$ hours and mail-back questionnaires can be up to 16 pages long. Mail-back questionnaires on topics unimportant to the respondent should be limited to two to four pages. Third, survey length depends on the format chosen. Analysts can ask more questions per respondent with personal or telephone interviews than with mail-back surveys. Finally, there is a trade-off with a fixed budget between the number of questions to be asked and the sample size, especially for personal and telephone interviews.

Question Order

Research has shown that responses to a question can differ depending on placement of the question in relation to other questions (Alreck and Settle, 1985). There are no rules on question order, but several general suggestions for obtaining higher-quality responses include:

1. Place easier questions early in the survey to avoid discouraging respondents.
2. Place questions that may be threatening (i.e., embarrassing or intimidating) to respondents in the middle of the survey (Sudman and Bradburn, 1982).
3. Place demographic questions at the end of the survey.
4. Avoid *response set bias* (a series of questions in a row with the same response).

In addition, designers should move questions likely to bias responses to later questions. For example, do not follow a question on traffic congestion with a question asking for "the worst suburban problems of our time" with "traffic congestion" as one possible response.

Some surveys contain skips, where certain respondents are directed to skip certain questions. If a survey contains skips, designers must ensure that questions pertaining to all respondents are not skipped. Provide clear directions for skips, especially for mail-back surveys when respondents cannot ask a survey worker for help.

Begin a survey with an introduction explaining who is conducting the survey, the purpose of the survey, the subject matter to be covered by the survey, and information on the respondent's rights and privileges. After the last question has been asked, leave time or space for respondent comments. The comments may prove useful and the opportunity to make a comment helps the respondent feel that her opinion is important. Finally, the respondent should always be thanked and may be offered a summary of the survey results.

Types of Questions

There are two basic types of questions. *Closed questions* present a fixed set of alternate answers to the respondent. *Open questions* allow the respondent to create his own answer. A skilled interviewer can use open questions effectively, but for most surveys closed questions are preferred for several reasons. First, answers to a closed question are often ready for analysis, whereas answers to an open question must be quantified or classified before analysis. Second, closed questions allow respondents a common frame of reference (Converse and Presser, 1986). For example, a question asking for a list of retail establishments respondents visited the previous day may be confusing if left open, because respondents are unsure whether hair care, shoe repair, and other services are "retail." Third, closed questions aid respondent recall. Fourth, Converse and Presser (1986) point out that open questions are more vulnerable to question order and other biases. Fifth, closed questions reduce the amount of writing required of respondents or interviewers, which is especially important if the survey is conducted in moving vehicles or other difficult locations. Finally, closed questions are less taxing mentally, so respondents will provide higher-quality answers with a higher response rate (Alreck and Settle, 1985).

Closed questions have a few disadvantages relative to open questions, but these disadvantages are easily mitigated. Closed questions are more difficult to create because a list of response possibilities must be established. However, asking an open question during pretests may reveal a list of possibilities to be offered later with a closed question. Some have also criticized closed questions because they do not allow respondents to say why they chose a particular response. To mitigate this, closed questions may be followed with probing open questions (Converse and Presser, 1986).

Closed questions take many forms. ''Yes or no'' questions are common and simple to score. However, because ''yes or no'' questions are absolute, respondents may feel more intimidated by them and are more likely to misinterpret them (Fink and Kosecoff, 1985). Distribute likely ''yes'' and ''no'' responses randomly on the survey form so that respondents do not discern a pattern.

Checklist questions provide respondents with a list of possible responses, from which they may choose one or more items. The keys to constructing unbiased checklist questions include (Alreck and Settle, 1985):

1. The list must include all possible answers.
2. The items must be mutually exclusive.
3. There should be more variation between items than within individual items.

''No opinion'' or ''don't know'' options should be offered on most checklists (and with most ''yes or no'' questions) to include all possible answers and to avoid bias (Converse and Presser, 1986).

Closed survey questions may provide *scales* to the respondents so that they can choose positions on a continuous spectrum of possibilities. Scales are easier to construct than checklists since designers must simply fix the extremes and establish an increment on the scale. In comparison to other forms of closed questions, scales may save space on a written survey or save time in an interview. Once a scale has been presented to or learned by the respondent, it may not have to be repeated while different questions are asked. Scales also lead to more possibilities for numerical analysis than ''yes or no'' and checklist questions which are usually analyzed only as proportions (Alreck and Settle, 1985).

Among the most common types of scales are Likert, ordinal, and ranking. *Likert scales* measure the degree to which respondents agree with a statement, and usually run from 1 = ''strongly agree'' through 3 = ''neutral'' or ''no opinion'' to 5 = ''strongly disagree.'' Likert scales are very common but there are several good reasons for caution in their use. Respondents generally want to agree with the analyst more often than disagree, so biased responses are more likely with Likert scales (Converse and Presser, 1986). Also, Likert scales sometimes confound extreme and intense views on an issue. For example, a person ''strongly agreeing'' with a statement that ''More highways must be built to solve traffic congestion'' may be expressing a strong desire to have two highways built (an intense view) or a weaker desire to have 100 highways built (an extreme view). Rewording questions can help overcome this problem with Likert scales.

In an *ordinal scale,* items are in some logical sequence. For example, for a question on when people prefer to perform errands, a convenient (and oversimplified) ordinal scale may be 1 = ''before work,'' 2 = ''during lunch,'' and 3 = ''after work.'' Ordinal scales provide good data on where items are relative to other items, but provide less information for the analysis than do checklist questions. *Ranking scales* require the respondent to put a list of items in order using some criterion. Ranking scales are useful for simulating real-life situations when choices must be made. However, data on the absolute standing of each item on the list are unavailable from a ranking scale. Fink and Kosecoff (1985) and other references on conducting surveys provide more details on the above-mentioned scales and descriptions of other useful scales.

JOB CATEGORIES

A. MANUFACTURING OF TRANSPORTATION EQUIPMENT
B. OTHER MANUFACTURING
C. AGRICULTURE, FORESTRY AND FISHERY
D. MINING
E. BUSINESS SERVICES AND REPAIR SERVICES
F. PROFESSIONAL AND RELATED SERVICES
G. WHOLESALE OR RETAIL TRADE
H. FINANCE, REAL ESTATE OR INSURANCE
I. TRANSPORTATION, COMMUNICATIONS, UTILITIES
J. CONSTRUCTION
K. ENTERTAINMENT OR RECREATION SERVICES
L. GOVERNMENT
M. OTHER (Please Describe)

OCCUPATION TYPES

A. PROFESSIONAL OR TECHNICAL
B. FARMER OR FARM MANAGER
C. FARM LABORER OR FARM FOREMAN
D. OTHER LABORER
E. MANAGER, OFFICIAL, OWNER OF A BUSINESS
F. CLERICAL AND SIMILAR WORKERS
G. SALES
H. CRAFTSMAN OR FOREMAN AND SIMILAR WORKERS
I. EQUIPMENT OPERATOR OR MOTOR VEHICLE OPERATOR
J. PRIVATE HOUSEHOLD WORKER (MAID, BUTLER, ETC.)
K. OTHER SERVICE WORKER
L. MILITARY
M. OTHER (Please Describe)

Figure B-1 U.S. Census Bureau standard job categories and occupation types. *Source:* Sheskin, I. M. and Stopher, P. R., "Pilot Testing of Alternative Administrative Procedures and Survey Instruments," *Transportation Research Record 886,* Transportation Research Board, National Research Council, Washington, DC, 1982.

General Tips

Regardless of question format, in every effort to compose questions designers should consider some general tips. The main rule on composing questions is to keep the study objectives in mind. Every question should help satisfy those objectives. Effective questions should also:

- Be as short as possible.
- Use standard demographic questions and terms whenever possible (Sudman and Bradburn, 1982). For example, standard U.S. Census Bureau occupational category definitions are available (Figure B-1) that will allow easy comparisons of results to other surveys, among other advantages. In fact, borrowing successful questions from other analysts (after obtaining permission and promising to give proper credit to the author) is an acceptable, inexpensive way to construct many surveys quickly.
- Avoid technical jargon and slang.
- Use words simple enough to be understood by the least sophisticated respondent. A survey of the U.S. general public should not exceed a grade 7 to 9 reading

level. For comparison, this chapter is written for a college-educated audience at a grade 14 to 16 reading level. Many word processing and grammar checking computer programs judge the reading level required to understand text.

- Define terms carefully. Respondents sometimes misinterpret very simple terms. The word "you," for example, can mean the respondent or the respondent's family, depending on the content of previous questions and other factors (Sudman and Bradburn, 1982).
- Avoid controversial and other words that evoke strong responses. Respondents associate such words as *crisis, liberal,* and *traffic congestion* with strong images that can bias the response to an otherwise mundane question.
- Avoid racial, ethnic, sexist, or otherwise biased language. Look for the presence of biased language during pretests (discussed later).
- Avoid leading questions (i.e., "How clear is it to you that mass transit saves fuel . . . ?").
- Avoid loaded questions. Loaded questions usually associate one aspect of the issue under question with some desirable trait. Alreck and Settle (1985) provided the following example of a loaded question: "Do you advocate a lower speed limit to save human lives?"
- Avoid double negatives (i.e., "Do you agree or disagree that highways should *not. . . ?").*
- Avoid "double-barreled" questions. In double-barreled questions, respondents may answer one part one way and another part another way. Double-barreled questions usually result from trying to do too much in one question and are treated by creating separate questions from the parts.
- Be very careful about prompting respondents' recall. Mentioning an example of a possible answer during the question may bias the response in favor of the item mentioned.
- Responses to threatening questions may be biased. The number of respondents claiming to use seat belts, ride public transportation, and perform other socially desirable actions will probably be overestimated unless special techniques (see Sudman and Bradburn, 1982, for example) are employed.
- Avoid hypothetical questions. If hypothetical questions cannot be avoided, append them onto one or more questions on real-life experiences. Then use additional questions to find out what respondents were thinking while answering the hypothetical question (Converse and Presser, 1986). For example, a designer may want to ask whether a respondent would use an electronic map in her automobile that shows traffic congestion. The designer might lead the hypothetical question by a question on whether the respondent listens to traffic reports on the radio and might follow the hypothetical question with a checklist of the desirable features of an electronic map.
- Ask respondents to recall events in the most immediate time frame feasible. The choice of an appropriate time frame avoids *telescoping,* the tendency to underestimate the passage of time since an event occurred. Longer timeframes are possible for events that were very important to respondents.

Constructing high-quality survey questions is time consuming. Even after several stages of writing and revising, it is likely that some questions will prove flawed during administration of the survey. Thus surveys should ask for information that is central to the study objectives at least twice, in different formats. This redundancy is tolerable considering the risk in putting survey objectives on the line in a single question.

EXAMPLE SURVEY FORM

The example survey form in Figure B-2 illustrates several interesting traits in survey design. The survey purpose was to collect origin and destination, demographic, and other information about current bus patrons. Survey workers distributed the form on buses. Passengers could complete the form on the bus and return it to the worker, or passengers could return the completed form by mail. A cover sheet introduced the survey and promised a pass for free bus rides for returned survey forms. The survey primarily required respondents to check boxes, which is appropriate for on-board completion. However, questions 4 and 6 are open, requiring respondents to write addresses, and require fairly advanced reading comprehension. The survey designers were aware of those limitations. Take-home surveys were therefore also given to a limited sample of respondents.

Notice the skips in questions 2, 3, and 5. The use of special graphics such as bold arrows and dash-dot boxes help guide respondents through those difficult areas. Other features to note include:

- "Other" responses are provided for.
- Simple language is used throughout (and the survey worker on the bus provided help to weak readers).
- Precise directions are given with several questions.
- Demographic questions are placed at the end of the form.
- Respondent comments are encouraged.

The designers could have improved the survey form by eliminating the coding spaces. Also, many on-board surveys suffer from biased samples, since those who ride more often will receive more forms. To adjust for that bias, a question on how often the respondent rides the bus is usually included in all on-board surveys (see Doxsey, 1983).

PROTECTING RESPONDENTS

The United States and other countries have firmly established the right to privacy. Competent adults generally cannot be forced to answer any survey except a periodic census. Therefore, survey designers must take precautions to avoid violating respondent rights. In addition, survey personnel must treat respondents with respect and dignity. The relatively high response rates enjoyed by some surveys may result from goodwill established with respondents from previous surveys.

The most common precaution undertaken to protect respondent rights is informing the respondent of the purpose and method of the survey before administration. In the United States, federal guidelines specify that each respondent receives (Fink and Kosecoff, 1985):

1. A description of the survey procedures and an explanation of the purpose of each procedure
2. A description of risks and benefits
3. An offer to answer any inquiries
4. An offer to withdraw from the survey at any time without prejudice

For many surveys, describing the survey purposes in too much detail could bias the eventual answers. Since respondents do not need detailed descriptions to make rational decisions about whether to participate, general descriptions oriented to what the respondent needs to know are provided. The description of risks and benefits is crucial. Respondents must understand the schedule of compensation for participation and survey

000293 METROPOLITAN DADE COUNTY: ON-BOARD TRANSIT SURVEY

METRO-DADE

PLEASE HELP US IMPROVE THE BUS SYSTEM BY FILLING OUT AND RETURNING THIS SHORT SURVEY FORM DURING YOUR BUS RIDE. EVEN IF YOU HAVE ALREADY FILLED OUT ONE OF THESE FORMS, PLEASE COMPLETE THIS ONE ALSO. IF YOU HAVE ANY QUESTIONS, OR IF YOU NEED HELP IN FILLING OUT THIS FORM, THE SURVEY PERSON WILL BE GLAD TO HELP YOU.

1. HOW LONG DID YOU WAIT FOR THIS BUS AT THE BUS STOP? ______ MINUTES

2. DID YOU USE A TRANSFER WHEN YOU GOT ON THIS BUS? YES ☐ NO ☐

IF YES, WHAT TYPE OF TRANSFER WAS IT?
- BROWARD COUNTY TRANSFER ☐
- MONEY PLUS TRANSFER ☐
- TRANSFER ☐

FROM WHICH BUS DID YOU TRANSFER?

WHERE DID YOU GET OFF THAT BUS? PLEASE GIVE NEAREST STREET CORNER ______

IF NO, HOW MUCH DID YOU PAY WHEN YOU GOT ON THIS BUS?
- 25¢ STUDENT ☐
- 25¢ SENIOR CITIZEN ☐
- 25¢ HANDICAPPED ☐
- 50¢ FARE ☐
- 75¢ EXPRESS ☐
- FREE PASS ☐

3. HOW DID YOU GET TO THE BUS STOP WHERE YOU GOT ON THIS BUS? CHECK ONLY ONE PLEASE
- WALKED ☐
- BY CAR: DROPPED OFF ☐ / PARKED NEAR THE STOP ☐
- TAXI OR JITNEY ☐
- BICYCLE, MOTORCYCLE, OR MOPED ☐
- ANOTHER BUS ☐
- OTHER (PLEASE DESCRIBE) ______

IF YOU CAME ON ANOTHER BUS, HOW DID YOU GET TO THE STOP WHERE YOU GOT ON THAT BUS?
- BY CAR: DROPPED OFF ☐ / PARKED NEAR THE STOP ☐
- WALKED ☐
- TAXI OR JITNEY ☐
- BICYCLE, MOTORCYCLE, OR MOPED ☐
- OTHER (PLEASE DESCRIBE) ______

HOW LONG DID YOU WAIT FOR THAT BUS? ______ MINUTES

WHEN YOU TRANSFERRED DID YOU WALK MORE THAN ONE BLOCK TO ANOTHER BUS STOP? NO ☐ YES ☐ IF YES, HOW FAR DID YOU WALK? ______ BLOCKS

4. A) WHERE DID YOU START OUT FROM BEFORE GETTING ON THIS BUS? CHECK ONE ONLY PLEASE

FROM
- HOME ☐
- WORK ☐
- SCHOOL ☐
- RESTAURANT ☐
- AIRPORT ☐
- SHOPPING ☐
- A FRIEND/RELATIVE'S HOME ☐
- HEALTH CARE (DOCTOR/DENTIST/HOSPITAL/ETC.) ☐
- HOTEL/MOTEL ☐
- RECREATION (BEACH/PARK/THEATER/MOVIES/ETC.) ☐
- OTHER PLEASE DESCRIBE

B) WHERE IS THAT? PLEASE TELL US THE ADDRESS, OR NEAREST STREET CORNER, OR THE BUILDING NAME OR PLACE NAME

STREET/CORNER/BUILDING ______

CITY/TOWN/VILLAGE ______ ZIP ______

C) HOW FAR IS THAT FROM THE BUS STOP? ______ BLOCKS

5. WILL YOU TRANSFER TO ANOTHER BUS WHEN YOU GET OFF THIS BUS? NO ☐ YES ☐

IF YES, WHAT BUS WILL YOU USE? ______

WHERE WILL YOU GET THAT BUS? PLEASE GIVE NEAREST STREET CORNER OR PLACE NAME

STREET CORNER/PLACE ______

6. A) WHERE ARE YOU GOING ON THIS BUS RIDE? CHECK ONE ONLY PLEASE

TO
- HOME ☐
- WORK ☐
- SCHOOL ☐
- RESTAURANT ☐
- HOTEL/MOTEL ☐
- AIRPORT ☐
- SHOPPING ☐
- A FRIEND/RELATIVE'S HOME ☐
- HEALTH CARE (DOCTOR/DENTIST/HOSPITAL/ETC.) ☐
- RECREATION (BEACH/PARK/THEATER/MOVIES/SPORTS/ETC.) ☐
- OTHER PLEASE DESCRIBE

B) WHERE IS THAT? [illegible]

STREET/CORNER/BUILDING ______

CITY/TOWN/VILLAGE ______ ZIP ______

C) HOW WILL YOU GET THERE FROM THE BUS STOP?
- WALK ☐
- BY CAR: PARKED NEAR THE STOP ☐ / PICKED UP ☐
- TAXI OR JITNEY ☐
- BICYCLE, MOTORCYCLE, MOPED ☐
- OTHER PLEASE DESCRIBE

7. FOR THIS BUS RIDE, COULD YOU HAVE USED A CAR INSTEAD OF THE BUS?
- YES, AS A DRIVER OR PASSENGER ☐
- YES, AS A DRIVER ☐
- YES, AS A PASSENGER ☐
- NO, NOT AT ALL ☐

8. HOW DO YOU LEARN ABOUT MTA (METROBUS) SERVICES? CHECK ALL THAT APPLY PLEASE
- DAILY NEWSPAPER ☐
- WEEKLY NEWSPAPER ☐
- COMMUNITY NEWSPAPER ☐
- RADIO ☐
- TELEVISION ☐
- MTA BY PHONE ☐
- MTA LITERATURE ☐
- BUS DRIVER ☐
- WORD OF MOUTH ☐
- OTHER (PLEASE DESCRIBE) ______

9. ARE YOU FEMALE ☐ OR MALE ☐

10. IS YOUR AGE: UNDER 12 ☐ 12-17 ☐ 18-34 ☐ 35-64 ☐ 65-69 ☐ 70 OR OVER ☐

11. DO YOU HAVE A DRIVER'S LICENSE? NO ☐ YES ☐

12. HOW MANY MONTHS OF THE YEAR DO YOU SPEND IN THE MIAMI AREA?
- MORE THAN 6 MONTHS ☐
- 3-6 MONTHS ☐
- 1-3 MONTHS ☐
- LESS THAN 1 MONTH ☐

IF LESS THAN 1 MONTH, HOW MANY TIMES HAVE YOU VISITED THIS AREA IN THE PAST 5 YEARS?
- ONCE (THIS TIME) ☐
- 2-3 TIMES ☐
- MORE THAN 3 TIMES ☐

PLEASE TELL US YOUR LOCAL ADDRESS, SO THAT WE CAN SEND YOU A FREE BUS PASS FOR FREE RIDES ON METROBUS FOR HELPING US IN THIS SURVEY.

NAME ______
STREET ADDRESS ______
CITY/TOWN/VILLAGE ______ ZIP ______
IF YOU ARE A VISITOR, UNTIL WHAT DATE WILL YOU BE AT THIS ADDRESS? MONTH ______ DAY ______

DO YOU WISH TO BE ON A MAILING LIST TO GET INFORMATION ABOUT THE MTA METROBUS SYSTEM? NO ☐ YES ☐

THANK YOU FOR YOUR HELP

WE WELCOME ANY COMMENTS YOU MAY HAVE ON BUS SERVICE OR THIS SURVEY, AND INVITE YOU TO USE THE SPACE BELOW FOR YOUR COMMENTS.

NO POSTAGE NECESSARY IF MAILED IN THE UNITED STATES

BUSINESS REPLY MAIL

POSTAGE WILL BE PAID BY ADDRESSEE

OFFICE OF TRANSPORTATION ADMINISTRATION
METRO-DADE TRANSIT SURVEY
44 WEST FLAGLER ST., 9th FLOOR
MIAMI, FLORIDA 33130

a this is form a

Figure B-2 Example on-board bus survey form. *Source:* Sheskin and Stopher, 1982.

workers must inform them of the slightest possibility of discomfort or embarrassment during questions. Methods for asking sensitive questions without embarrassing or threatening respondents are available (Sudman and Bradburn, 1982). Respondents are also entitled to an accurate estimate of how long the survey will last.

Although it is usually a good idea, it is not always necessary to identify the survey sponsor. Some sponsors inspire such strong images that biased responses or high nonresponse rates are likely. If a controversial sponsor must be identified, it is better to do so late in the survey to avoid as much bias as possible (Alreck and Settle, 1985).

Survey designers should guarantee to keep responses confidential whenever possible. If a name, an address, a telephone number, a driver license number, citizenship status, or other personal data must be gathered, the respondent must be assured that those data will be used only to achieve survey objectives. Readers should never be able to identify individual respondents from published survey results. Also, survey designers should not require respondents to sign the survey form.

Many funding sources, government agencies, and universities have established procedures for reviewing surveys before they are conducted. These reviews protect respondent rights and/or ensure that the study objectives cannot be achieved by other means. For example, the U.S. Office of Management and Budget (OMB) reviews surveys of 10 or more people conducted with federal funds. Wise designers submit their surveys for review early. Universities typically require several weeks, while the OMB requires an average of 60 to 75 days but not more than 90 days.

TRAINING INTERVIEWERS

Interviewer training is crucial to the success of interview surveys. Interviewers can waste a well-written question if they ask incorrectly or if they record the responses inaccurately. The goal for training interviewers is to produce results that are consistent between interviewers and consistent over time for individual interviewers.

Hiring interviewers can be difficult for engineers untrained in personnel matters. Judge potential interviewers first for integrity and honesty, and second for any previous interviewing skills or technical knowledge. Several authors recommend that, if possible, interviewers should have demographic characteristics similar to those of the respondent population (Fink and Kosecoff, 1985; Alreck and Settle, 1985). Professional survey and market research firms may be able to provide skilled interviewers for a lower cost than agencies will incur to hire and train their own interviewers.

Once hired, survey designers should introduce interviewers to the survey method, questions, and answer form. If several interviewers are being trained, a lecture format may be appropriate for this part of the training. Interviewers need to learn "how" to ask questions and "why" the questions are being asked. For each question, the survey designer needs to tell the interviewers the extent to which the "script" must be followed. The training schedule should allow plenty of time for interviewers to ask questions of the survey designer.

After a general orientation to the survey, interviewers should practice asking questions and recording answers. Interviewers can practice on each other, on the survey designer, or on a sample of respondents during pretesting. Survey designers can make sure that interviewers are sufficiently trained, or can spot weaknesses in interview technique and suggest corrective actions, by observing an interview of a respondent during a pretest.

PRETESTS

During pretesting, survey conditions are simulated to scrutinize the survey method and questions. If the survey is found lacking during the pretest, as most are, the designer can revise it before data collection. Pretests are necessary for all surveys. Some authors insist that an analyst without enough resources to pretest should not conduct a survey (Sudman and Bradburn, 1982).

While every facet of survey design can benefit from pretesting, pretests are especially useful during two particular stages in the process of developing a survey. First, pretests are useful while beginning to compose questions. The results of pretests during this time may prompt wholesale changes, such as revising the format, dropping an entire sequence of questions, or adopting different types of questions. Second, pretests are useful after questions have been fully developed, in a "dress rehearsal." The designer can reword questions, add closed question choices, reform scales, and delete individual questions. However, larger changes are rarely warranted at this later stage. Pretesting data recording forms, data reduction methods, and data analysis techniques (with made-up data, if necessary) is strongly encouraged during the second stage of pretesting. Many surveys of all sizes have faltered after data collection when analysts found themselves with sets of inappropriate or unusable responses.

Pretest a respondent sample that is similar to the actual sample, especially during the "dress rehearsal." Although drafts of questions should be circulated for comment among colleagues, pretesting on colleagues, employees, students, relatives, or other groups unlike the sample of respondents may not be helpful. Informing respondents that they are part of a pretest is useful for pretests during early stages of survey development if detailed probes of the responses to questions are then performed. When respondents are not informed that a pretest is being conducted, the pretest is more realistic, but it will not generate as much information that could be useful for diagnosing problems.

The interviewers who will conduct the survey should also conduct the pretest. The survey designer should intensely debrief the interviewers about each respondent during a pretest. The literature recommends pretest sample sizes from 20 to 75.

SURVEY ADMINISTRATION

Once a survey instrument has been pretested, a few final details must be attended to before data are collected. Written surveys must be printed, assembled, and mailed. For effective written surveys, Alreck and Settle (1985) recommend that designers avoid bright-colored (pink, blue, or yellow) paper. White, off-white, beige, and many pastel colors are fine for mailing. Mail-back surveys achieve highest response rates with first-class stamps and lowest response rates with bulk mail permits, although the latter is less expensive per response received. Assemble mail-out and mail-back surveys with a cover letter explaining the purpose of the survey, asking for participation, detailing how completed surveys are to be mailed back, and providing other information.

The issue of respondent compensation must be addressed before conducting the survey. Most transportation surveys do not compensate respondents. Compensation introduces an additional expense that may restrict sample size and may bias results. However, designers should consider compensation for longer surveys or when respondents perform difficult tasks. Compensation shows appreciation, indicates goodwill, gains attention, or creates an obligation to respond (Alreck and Settle, 1985).

Respondents do not have to be compensated with cash. Designers can reduce the paperwork burden of providing cash (receipts, tax statements, etc.) by providing small gifts that respondents will value. The on-board bus survey discussed earlier in the appendix (Figure B-2) promised each respondent a pass for free rides and is a good example of such a small, valued, noncash gift.

Before telephone interviewing begins, designers should establish procedures for handling the possible outcomes from a call. Table B-3 lists most of the possible outcomes from an attempted telephone call and possible responses to those outcomes. A response should be decided on beforehand for each possible outcome. Interviewers should keep a log of telephone calls so that respondents not reached with the original call are retried and so that respondents are not called twice.

TABLE B-3
Telephone Call Outcomes and Interviewer Options

Possible call outcomes

1. Call answered by qualified respondent.
2. Call answered by unqualified respondent.
3. Number proves to be wrong location or subscriber.
4. Call is not answered after six to 10 rings.
5. Busy signal is received.
6. Call reaches answering machine or answering service.
7. Number has been changed and no new number is given.
8. Number is not in service and no new number is given.

Interviewer options

1. Interview respondent.
2. Ask if a qualified respondent is present.
3. Leave message asking respondent to return call.
4. Terminate and place call to next number.
5. Terminate and call back in a few minutes.
6. Terminate and call back at different day and time.

Source: Alreck and Settle, 1985.

Survey designers need to ensure that the survey is being answered by the respondents intended. For example, a survey on the effects of raised medians should be distributed to adjacent business proprietors, customers, and through traffic drivers, among others. Before interviews, the qualifications of potential respondents need to be checked with a brief question or two. Unqualified persons should be thanked and dismissed. On written surveys, designers can ask for respondent qualification at the beginning, directing unqualified respondents to discard the survey or pass it to a qualified respondent. Analysts can also check respondent qualification on a returned written survey and discard returns from unqualified respondents.

Because certain members of a household tend to answer the telephone at certain times of the day, some survey designers use a response matrix like that in Figure B-3 to avoid biased telephone interview samples. The interviewer asks the person who answers the telephone one or two questions about household composition, consults the response matrix, and requests the selected household member to come to the telephone for the interview. To ensure that each adult household member has an equal chance of being interviewed, designers create several versions of the matrix with different entries. Then interviewers select a version of the matrix at random before each call.

Survey designers should monitor interviewers regularly to reduce bias. If direct observation of interviewers is impossible, the survey designer should recontact a small subsample of respondents for comments on the interview. Mostly, survey designers should watch the tendency of interviewers to speed delivery of questions. Respondents need time to formulate answers, so survey designers must insist that interviewers maintain a slow, constant pace during every interview. Survey designers should also remind interviewers not to register emotions as respondents answer questions. Even subtle movements or noises by interviewers, such as a nod of the head or a brief "OK," can bias later responses. Analysts should compare survey results from different interviewers for major discrepancies that may indicate improper technique or fraud.

The number of responses received from mail-back surveys should be noted each day. That number will usually follow a skewed distribution with a long "tail." Most responses are received within a few weeks, but a few will dribble in months after distribution. Analysts should set a cutoff date for returns once the long tail of the distribution has been reached. Any surveys received after the cutoff date should not be analyzed (Alreck and Settle, 1985). Analysts should scan all returned surveys or interview answer sheets for missing sections, intentionally misleading data, and other obvious errors be-

1. Sequence number ________
2. Telephone number ________

3. First, could you tell me how many people 18 or older live there?

0 → Terminate call

4. How many of these are female?

	1	2	3	4	5	6 or more specify ______
0	Sole male	Younger male	2nd youngest male	2nd youngest male	2nd youngest male	4th oldest male
1	Sole female	Sole male	Older male	2nd youngest male	2nd youngest male	3rd oldest male
2	X	Younger female	Younger female	Older male	2nd youngest male	2nd oldest male
3	X	X	2nd youngest female	Youngest female	Older male	Oldest male
4	X	X	X	2nd youngest male	Youngest female	Youngest female
5	X	X	X	X	2nd youngest female	2nd youngest female
6 or more specify: ______	X	X	X	X	X	4th oldest female

Tell the person on the phone which member of the household is the selected repondent and attempt an interview.

5. Description of final result: ____

____

____

____

____

Figure B-3 Example telephone interview response matrix. *Source:* Groves, R. M. and Kahn, R. L., *Surveys by Telephone,* Academic Press, Inc., 1979.

fore coding. Analysts can correct defective forms, if it is possible to recontact the respondent without undue bias, or can discard forms that cannot be corrected. Answers are then coded into a computer file for analysis.

SOURCES OF ERROR

Analyze survey data, like most other transportation study data, using statistical processes described in Appendix C. The major difference between survey data and other transportation study data is in the different types of errors expected with survey data. Cochran (1977) identified four different types of error that are important for surveys:

1. Sampling error
2. Coding and reduction errors
3. Biased responses
4. Nonresponse

Sampling error is due to the selection of a sample from the subject population. Most samples from a population will provide results that differ from other samples. Analysts can control the amount of sampling error when they use accepted procedures for random sampling.

Analysts can detect and correct most *coding and reduction* before analysis. Thorough training of coders, close contact between analysts and coders, and frequent double-checks of coded and reduced data are needed.

Biased responses are very difficult to detect in surveys. However, once detected, there are mathematical procedures to estimate the effects of biases on survey results (Cochran 1977). In addition, Brog et al. (1982) and others have explored typical biases in transportation surveys and have made suggestions for overcoming some of them. A thorough pretest is one of the best ways to avoid biased responses.

Nonresponse is a problem in surveys when the portion of the desired sample that does not respond differs from the portion of the desired sample that does respond and a bias of unknown size is introduced. Correcting for nonresponse requires time and effort, so smaller surveys may suffer more from nonresponse errors. Response rates from 20 to 70% are typical for transportation surveys. However, there is no standard response rate that should cause concern for analysts. A 20% response rate where the respondents are similar to the nonrespondents is superior to a 70% response rate where the respondents differ from the nonrespondents.

The best way to treat nonresponse is through prevention: Designers should consider response rate during every stage of survey development. A classification scheme for nonresponse developed by Cochran (1977) is helpful for designers. The classes include:

1. *Noncoverage:* People in the sample are not given the opportunity to respond to the survey.
2. *Unable to answer:* People are reached but do not have sufficient information to answer the question.
3. *Unwilling to answer:* People are reached but refuse to respond.
4. *Not-at-home:* Survey workers try but cannot reach some people in the sample.

Designers can prevent many noncoverage problems by using an up-to-date frame (list of units in the population). Most ''unable to answer'' problems should be detected during pretesting. If respondents are frequently unable to answer a question, the designer can change the format of, reword, or discard the question. Skillful introductions, proper inducements, and polished questions will reduce the number of people unwilling to answer. Finally, survey workers should repeatedly attempt to contact not-at-home nonrespondents. A procedure for determining the optimum number of callback attempts is available from Cochran (1977).

Stopher and Sheskin (1982) describe ways to detect and treat nonresponse bias in transportation surveys. Usually, analysts can judge the extent of nonresponse bias after a special effort to call or visit a small number of nonrespondents. Unbiased nonresponse can be corrected by distributing the survey to more people without violating the original sampling plan. Analysts can correct biased nonresponse mathematically using the different characteristics of the respondent and nonrespondent subsamples.

REFERENCES

ALRECK, P. A., AND R. B. SETTLE (1985). *The Survey Research Handbook*, Richard D. Irwin, Homewood, IL.

BROG, W., E. ERL., A. H. MEYBURG, AND M. J. WERMUTH (1982). *Problems of Nonreported Trips in Surveys of Nonhome Activity Patterns,* Transportation Research Record 891, Transportation Research Board, National Research Council, Washington, DC.

COCHRAN, W. G. (1977). *Sampling Techniques,* 3rd ed., Wiley, New York.

CONVERSE, J. M., AND S. PRESSER (1986). *Survey Questions: Handcrafting the Standardized Questionnaire,* Sage Publications, Newbury Park, CA.

DOXSEY, L. B. (1983). *Respondent Trip Frequency Bias in On-Board Surveys,* Transportation Research Record 944, Transportation Research Board, National Research Council, Washington, DC.

FINK, A., AND J. KOSECOFF (1985). *How to Conduct Surveys: A Step-by-Step Guide,* Sage Publications, Beverly Hills, CA.

GROVES, R. M., AND R. L. KAHN (1979). *Surveys by Telephone,* Academic Press, New York.

LERMAN, S. R., AND C. F. MANSKI (1976). *Alternative Sampling Procedures for Calibrating Disaggregate Choice Models,* Transportation Research Record 592, Transportation Research Board, National Research Council, Washington, DC.

SHESKIN, I. M., AND P. R. STOPHER (1982). *Pilot Testing of Alternative Administrative Procedures and Survey Instruments,* Transportation Research Record 886, Transportation Research Board, National Research Council, Washington, DC.

STOPHER, P. R., AND I. M. SHESKIN (1982). *Method for Determining and Reducing Nonresponse Bias,* Transportation Research Record 886, Transportation Research Board, National Research Council, Washington, DC.

SUDMAN, S., AND N. M. BRADBURN (1982). *Asking Questions,* Jossey-Bass Publications, San Francisco.

C

Statistical Analyses

L. Ellis King, D. Eng., P.E.

INTRODUCTION

After field data have been collected in a traffic engineering study, the information is arranged or tabulated for visual inspection and analysis. Then the application of appropriate statistical techniques aids in making the proper and most effective evaluation of the study results. To ensure complete and accurate knowledge of existing traffic conditions, the field study must be designed and performed properly with due consideration for the following statistical analysis. If engineers collect and analyze data with appropriate statistical procedures, they can avoid inaccurate and improper interpretations of the traffic situation. Success in determining traffic improvements depends greatly on the reliability and correct interpretation of the information that describes the traffic problem.

The arrangement of data into a convenient form is covered in the section "Data Reduction," and the subject of statistical analyses is discussed in the sections "Descriptive Statistics" and "Statistical Inference." "Descriptive Statistics" is concerned with summarizing the data collected in a traffic engineering study, while "Statistical Inference" describes the procedures for the development of statistical estimates and the testing of statistical hypotheses. The final section of this chapter describes some of the calculation aids that are available.

This appendix describes very basic statistical analyses used by transportation engineers (Greenshields, 1978; Taylor and Young, 1988). When they anticipate complex analyses, engineers should seek the advice of a professional statistician prior to data

TABLE C-1
Random Spot Speeds on Urban Arterial (mph)

49	35	37	48	52	50	43	46	41	50
43	46	45	47	44	48	42	35	53	47
46	45	45	41	40	41	39	44	52	42
40	46	53	45	48	48	47	52	49	49
44	45	40	46	45	55	51	42	46	45
47	45	44	48	41	48	46	44	49	44
49	41	38	51	54	42				

TABLE C-2
Spot Speeds on Urban Arterial in Increasing Order (mph)

35	35	37	38	39	40	40	40	41	41
41	41	41	42	42	42	42	43	43	44
44	44	44	44	44	45	45	45	45	45
45	45	45	46	46	46	46	46	46	46
47	47	47	47	48	48	48	48	48	48
49	49	49	49	49	50	50	51	51	52
52	52	53	53	54	55				

collection. Costly and frustrating analysis mistakes can be avoided with thorough planning and good professional advice beforehand.

DATA REDUCTION

When data are arranged systematically according to the frequency of occurrence for different size classifications, such as the grouping of speeds in a spot speed study, the resulting tabulation may be shown as a frequency distribution or histogram. If the collection of information is based on the time of occurrence, such as the passings of vehicles in a traffic volume study, the array of values may be treated as a time series or temporal distribution. Data that have been recorded according to geographic location, such as the location of accidents on a street map, may be plotted as a spatial distribution. These distributions are commonly used in summarizing and presenting the results of transportation engineering studies.

Tables and Graphs

Data from transportation studies generally can be presented in tables or graphs. Appendix D provides suggestions on the format and appearance of tables and figures, while this discussion is focused on the content of tables and figures. Use of a microcomputer makes data manipulation a relatively easy task, and numerous tables may be generated that present various aspects of the data. The simplest table is a straight-line listing of the data in the order in which they were recorded. Table C-1 shows such a listing for 66 spot speeds collected on an urban arterial with a posted speed limit of 50 mph. The data are of little value when presented in this form and are more useful when ordered from lowest to highest value, as shown in Table C-2. The extreme values and groups of data are now apparent.

Frequency Distribution

Analysts can condense and summarize the large number of data points shown in Tables C-1 and C-2 by developing a frequency table. The construction of a frequency table first requires the selection of the group or class size. If too few or too many groups are

TABLE C-3
Frequency Distribution for Spot Speed Data

Class Boundaries	Class Interval	Class Midpoint	Class Frequency	Relative Frequency	Cumulative Frequency Number	Cumulative Frequency Relative
32.5						
	33–34	33.5	0	0		
34.5					0	0
	35–36	35.5	2	0.030		
36.5					2	0.030
	37–38	37.5	2	0.030		
38.5					4	0.060
	39–40	39.5	4	0.060		
40.5					8	0.121
	41–42	41.5	9	0.136		
42.5					17	0.257
	43–44	43.5	8	0.121		
44.5					25	0.378
	45–46	45.5	15	0.227		
46.5					40	0.605
	47–48	47.5	10	0.152		
48.5					50	0.757
	49–50	49.5	7	0.106		
50.5					57	0.863
	51–52	51.5	5	0.076		
52.5					62	0.939
	53–54	53.5	3	0.046		
54.5					65	0.985
	55–56	55.5	1	0.015		
56.5					66	1.000
	57–58	57.5	0	0		
58.5						
			66	1.000		

selected, detail is lost in the data reduction. The appropriate number of classes generally ranges from 10 to 20. After the field data have been collected and tabulated, the range in measurements is determined by subtracting the lowest from the highest values. The range is then divided by 10 and 20 to estimate, respectively, the maximum and minimum class sizes that are reasonable for the observed data. A convenient class size is selected within these minimum and maximum values. After the analyst determines the class size, he selects class intervals to define completely the actual sample values that are contained in each class. These limits are written to the same precision as the original data. Table C-3 shows a frequency table for the spot speed data in Tables C-1 and C-2.

After the class intervals have been recorded on the frequency table, each field-recorded vehicle speed is placed in the appropriate class interval. Summing the number of vehicles in each class intervals gives the frequency of occurrences for each of the intervals selected in the spot speed study. The resulting table of occurrence in the various class intervals is known as a *frequency distribution,* and the sum of the class frequencies is equal to the sample size or total number of field observations.

A *relative frequency distribution* is obtained by dividing the total number of vehicles in each class by the total number of vehicles in the sample; the relative frequencies must total 1.0. Relative frequency distributions provide a more convenient format for data summaries because the user does not need to refer to the sample size. The use of relative frequencies expressed either as proportions or percentages, also permits direct comparison of the results obtained from different studies with varying sample sizes. Table C-3 shows both frequency and relative frequency.

The *cumulative frequency distribution* provides a listing of the total number of observations that are less than or greater than a specified value. A cumulative frequency distribution is developed beginning with the class at either the bottom or the top of the frequency table. The class frequencies, either actual or relative, are summed in the selected direction with respect to the class boundaries. The class boundaries are the most extreme values included in a given class and are calculated to $\frac{1}{2}$ unit of greater precision than the original data. Table C-3 shows the cumulative frequency distribution developed from the frequency distribution discussed previously. If the frequency summation is from the small to the large values of the study variable, as in Table C-3, the analyst matches the higher class boundary with the corresponding cumulative frequency. The analyst selects the lower class boundary for this matching when he develops the cumulative frequency distribution from the large to the small measurements. The information contained in a frequency distribution table may also be presented graphically as a histogram, frequency diagram, or cumulative frequency diagram. When plotted for the complete range of observations, these diagrams provide an opportunity to evaluate the data shape with regard to conformance to a recognized distribution, degree of symmetry, and presence of irregularities.

Analysts construct a *histogram* by plotting a diagram with class boundaries on the horizontal axis and the corresponding frequency or relative frequency of occurrence on the vertical axis. The resultant rectangle represents the number of observations within each class interval. Figure C-1 is a histogram for the classes of spot speeds shown in Table C-3. Although actual or relative frequencies may be selected, the use of relative frequencies provides a diagram that has general application and facilitates comparisons since it eliminates the influence of sample size.

The *frequency diagram* is constructed by plotting a graph with the midpoint of each class on the horizontal axis and the corresponding frequency or relative frequency of the class on the vertical axis. Figure C-2 shows a frequency diagram for the classes of spot speeds from Table C-3. The points are plotted and then connected by straight lines. The diagram is closed by connecting the extreme points with the next class midpoints that have a frequency of zero. As with the histogram, actual or relative frequencies may be selected for plotting.

The *cumulative frequency diagram* is constructed by plotting a graph with a class boundary of each class on the horizontal axis and the corresponding cumulative frequency of the class on the vertical axis. The higher class boundary is matched with the corresponding cumulative frequency when the frequency summation is from the small to the large values of the study variable, and the lower class boundary is selected for this matching if the cumulative frequency is summed from the large to the small observations. A smooth curve is used to connect the plotted points, and extended to the two extreme class boundaries with cumulative frequencies of 0 and 1. On cumulative frequency diagrams, analysts generally express class frequency on a relative basis. Figure C-3 shows a cumulative frequency diagram of the spot speed data from Table C-3.

Time Series Distribution

Data that are observed and tabulated with respect to the exact time of occurrence provide a *time series distribution*. The time interval for recording data is selected to accommodate the purpose of the study and may range from a few seconds to several years. For example, vehicular volumes are tabulated in 5-minute intervals for the determination of peaking characteristics, while traffic accident statistics are frequently summarized on an annual basis. Figure C-4 illustrates time series distributions for traffic volumes by hour, day, and month. Observations that are collected in short-time intervals may be combined to produce summary totals for longer time series. For example, 5-minute volume counts can be summed to provide hourly values, which are then summed to produce the daily total. This summation can be extended to the largest time interval desired.

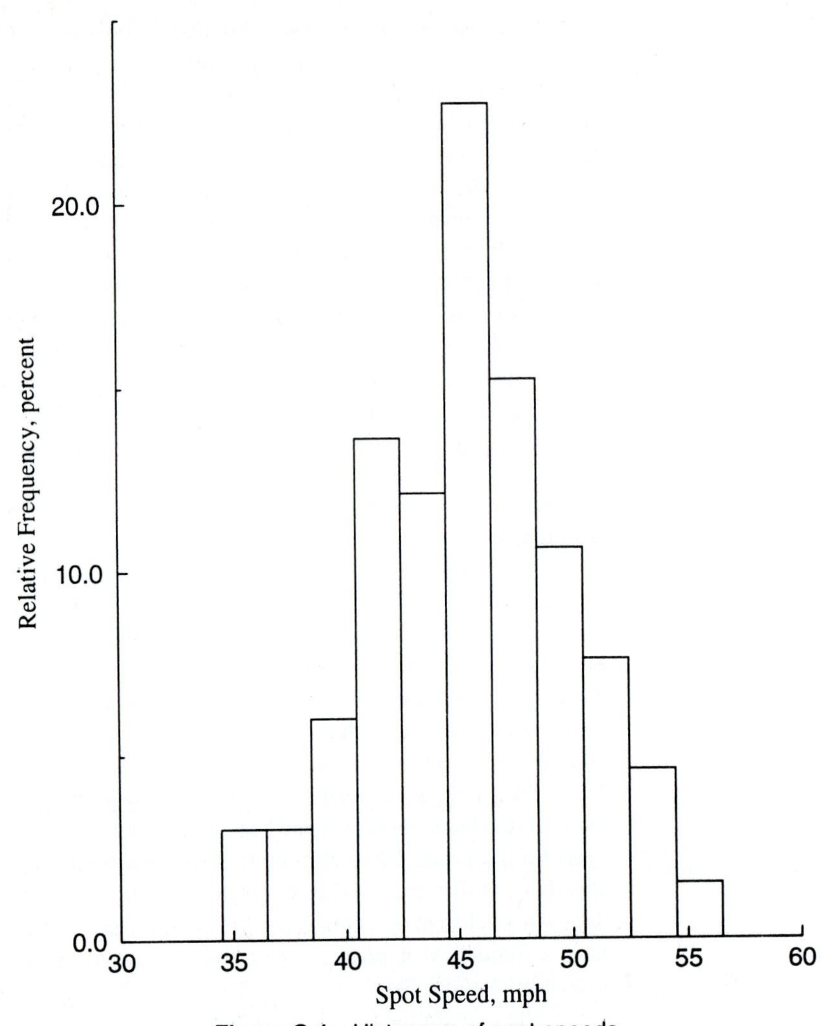

Figure C-1 Histogram of spot speeds.

Spatial Distribution

Traffic information is often presented with reference to the geographical location where the event occurred or is present. For example, vehicular volumes, travel speeds and delays, traffic accidents, trip desires, parking supplies and demands, public transportation usage, and inventories of traffic control devices and regulations are often summarized as spatial distributions. Spot maps, route logs, and area-wide maps are commonly used to illustrate data distributions according to geographical location. The spot map of traffic accident locations shown in Figure C-5 is an example of a spatial distribution.

DESCRIPTIVE STATISTICS

The function of *descriptive statistics* is to describe a collection of data by a few representative values. Descriptive statistics also permit the efficient evaluation and analysis of the traffic problem of interest. Descriptive statistics involve the central tendency, variability, and shape of the data.

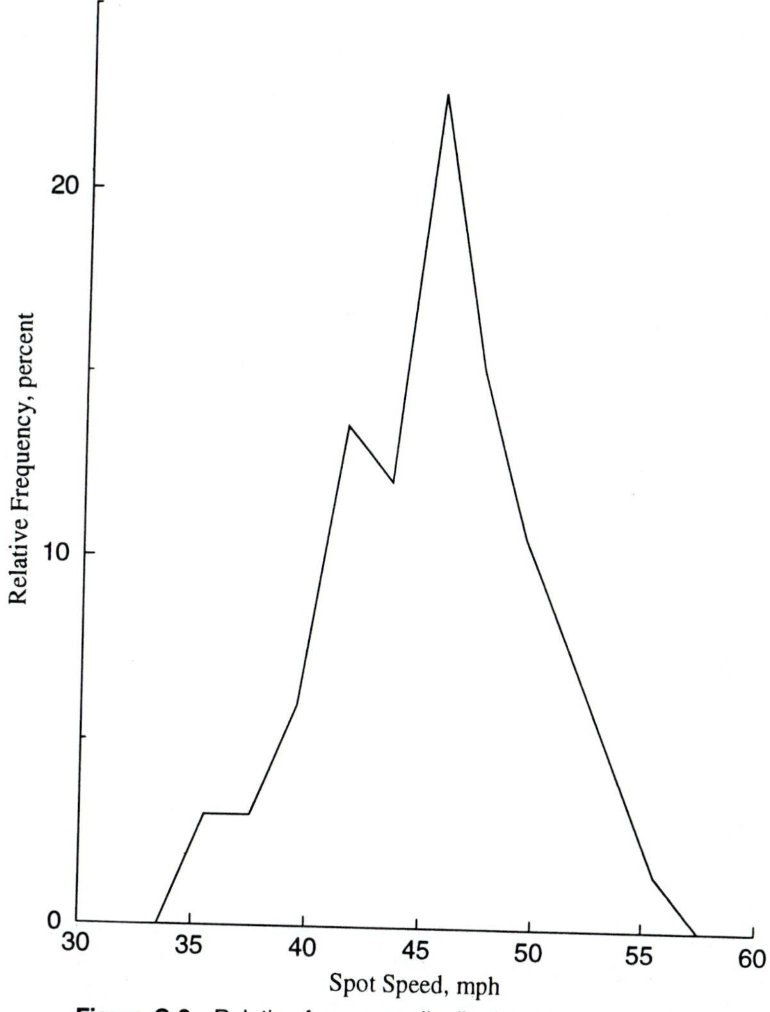

Figure C-2 Relative frequency distribution of spot speeds.

Central Tendency

Various measures of an "average" are available to describe the *central tendency* of a set of observations. *Average* is a broad term that generally includes many measures of central tendency. Commonly used measures of central tendency for summarizing the results of transportation engineering studies include the:

1. Arithmetic mean
2. Median
3. Mode

An average is generally located near the center of the data distribution, as illustrated for the frequency distribution of spot speeds in Figure C-6. Depending on the shape of the distribution, the various measures of central tendency may or may not coincide.

Arithmetic Mean

The *arithmetic mean* is the most common measure of central tendency and is obtained by dividing the sum of the individual observations by the total number of observations. The general expression for the mean unclassed data is

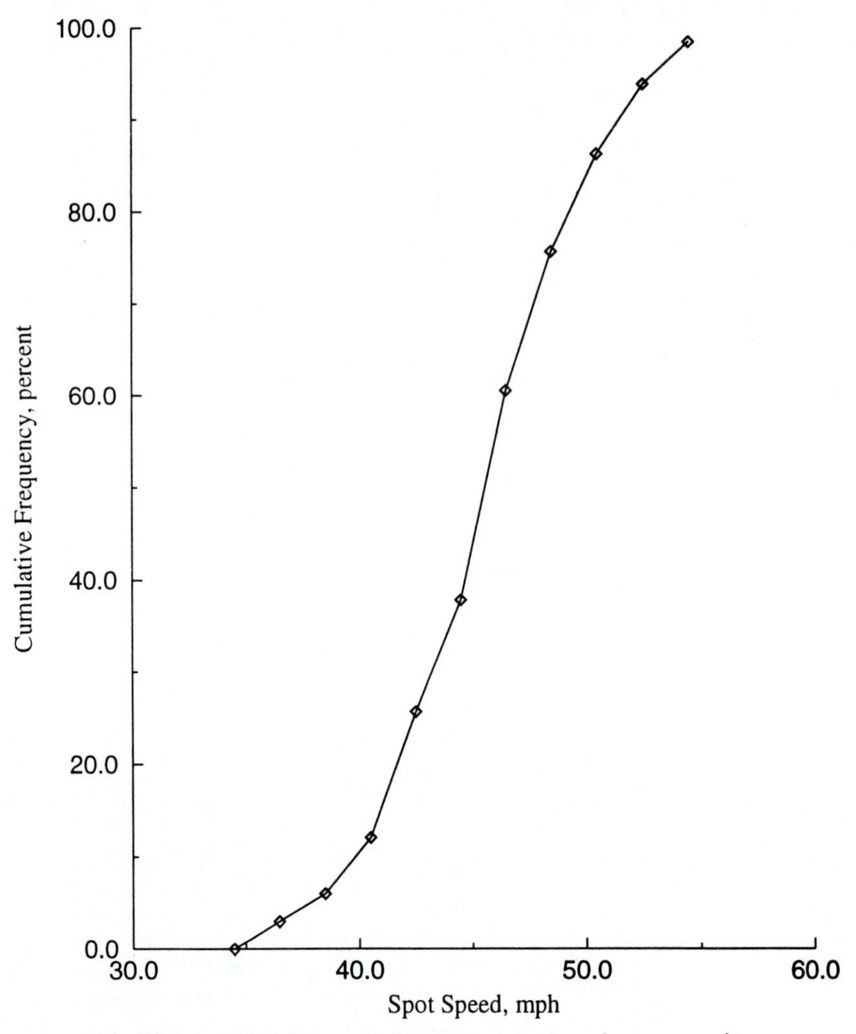

Figure C-3 Cumulative frequency distribution of spot speeds.

$$\overline{X} = \frac{\Sigma x_i}{N} \tag{C-1}$$

where

$\overline{X}$ = arithmetic mean
Σx_i = sum of individual observations
N = total number of observations

If the measurements have been placed into classes as shown in Table C-3, analysts using the following relationship to compute the arithmetic mean:

$$\overline{X} = \frac{\Sigma f_i u_i}{\Sigma f_i} \tag{C-2}$$

where

$\overline{X}$ = arithmetic mean
$\Sigma f_i u_i$ = sum of products of frequency and class midvalue for each class
Σf_i = sum of frequencies for all classes

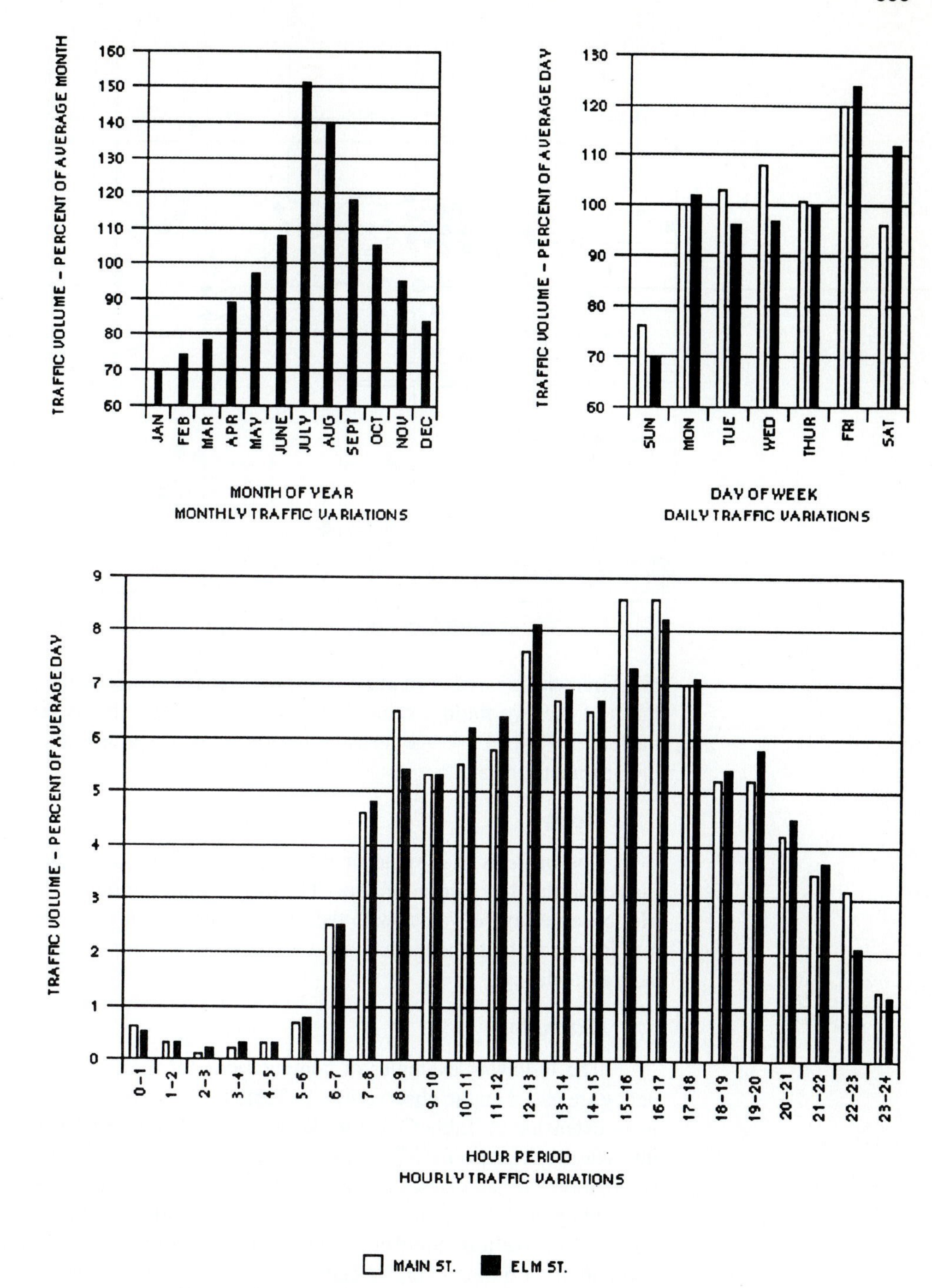

Figure C-4 Time series distributions for a traffic volume study.

Most procedures for determining the minimum sample size in various transportation engineering studies are based on the mean as the desired measure of central tendency. In statistical inference, the mean is generally the most efficient estimator of the average value for the population characteristics being studied.

Median

The *median* represents the middle value in a series of measurements that have been ranked in order of magnitude and divides the measurements into two equal parts. When the number of observations is odd, the median is the middle value in the list of ranked

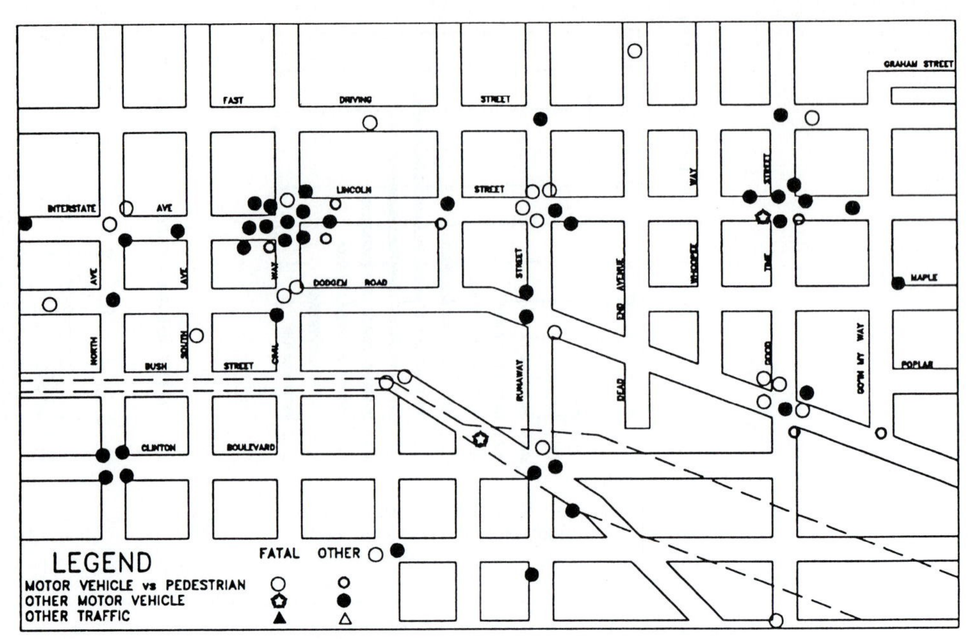

Figure C-5 Spatial distribution for traffic accident data.

measurements. For an even number of ranked measurements, the median is the arithmetic mean of the two middle values. The 50th percentile value is equal to the median. The median is a useful average measure because it is less affected by extreme values than is the arithmetic mean.

Mode

The *mode* is that value or that class midpoint that occurs with the greatest frequency in the distribution of data. The mode is not a reliable ''average'' for small samples because several values with the same frequency can occur by chance. As the sample size increases, both the median and the mode become more meaningful as an estimate of central tendency.

Examples

The following examples of computations for the various measures of central tendency are based on the observations in Table C-2 for unclassed data and Table C-4, which is an extension of Table C-3, for classed data. When speed measurements are grouped into classes, equation (C-2) is used to calculate the mean speed. From Table C-4, the summation over all classes is obtained for the product of the class midpoint and the corresponding class frequency. This value is then divided by the sum of frequencies for all classes (i.e., the total number of observations) to provide the arithmetic mean, which is 3006/66 or 45.5 mph. This procedure is similar for unclassed data, except that in accordance with equation (C-1), the individual measurements are summed in the numerator, which is then divided by the total number of observations, again giving a mean speed of 45.5 mph.

The median or 50th percentile value is found for classed data by a linear interpolation across the class in which the middle value occurs. In the spot speed example, this value is the mean of the 33rd and 34th frequencies and lies in the class with boundaries of 44.5 and 46.5. A linear interpolation for the location of 33.5 in this class is accomplished by calculating:

$$\text{median} = 44.5 + \left(46.5 - 44.5 \times \frac{33.5 - 25}{40 - 25}\right) = 44.5 + 1.1 = 45.6 \text{ mph}$$

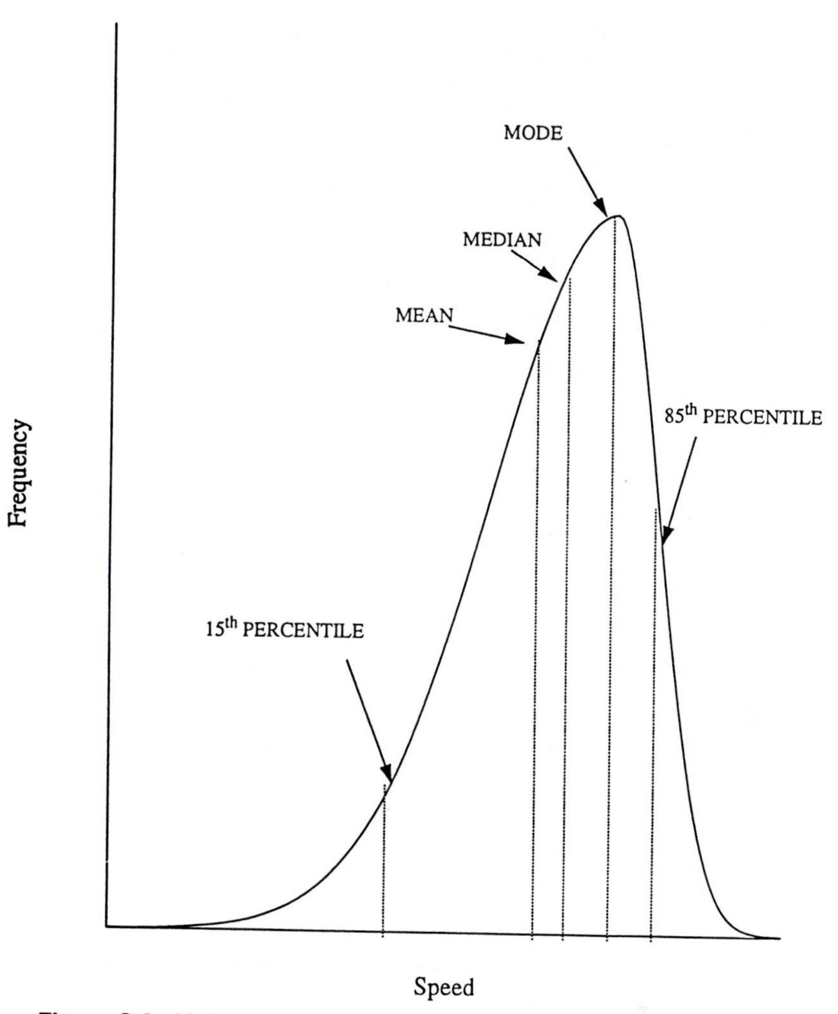

Figure C-6 Various average measures and percentiles for spot speed data.

The values of 25 and 40 are, respectively, the cumulative frequencies at the class boundaries of 44.5 and 46.5 mph. For the unclassed data of Table C-2, the median is the arithmetic mean of the two middle values and is calculated as

$$\text{median} = \frac{45 + 46}{2} = 45.5 \text{ mph}$$

The mode is the class midpoint for that class with the largest frequency of occurrence and is equal to 45 mph from Table C-4. For the unclassed data of Table C-2, the most frequently occurring value or mode is 45 mph.

Variability

A statistic describing the variability, dispersion, or spread of sample data is also valuable. The following two measures of variability are widely applied in the summary of transportation engineering data:

1. Range
2. Standard deviation

TABLE C-4
Summary Calculations for Classed Spot Speed Data

Class Boundaries	Class Interval	Class Midpoint u_1	Class Frequency f_1	f_1u_1	$f_1u_1^2$
32.5					
	33–34	33.5	0	0	0
34.5					
	35–36	35.5	2	71	2,521
36.5					
	37–38	37.5	2	75	2,813
38.5					
	39–40	39.5	4	158	6,241
40.5					
	41–42	41.5	9	374	15,500
42.5					
	43–44	43.5	8	348	15,138
44.5					
	45–46	45.5	15	683	31,054
46.5					
	47–48	47.5	10	475	22,563
48.5					
	49–50	49.5	7	347	17,152
50.5					
	51–52	51.5	5	258	13,261
52.5					
	53–54	53.5	3	161	8,587
54.5					
	55–56	55.5	1	56	3,080
56.5					
	57–58	57.5	0	0	0
58.5					
			66	3,006	137,910

Although the range is easier to compute, the standard deviation is a more reliable measure of data variability. The *range* is the interval between the smallest and the largest observations. The range is markedly dependent on the size of the sample and is greatly influenced by outliers or erroneous measurements. This summary statistic is readily interpretable and is often used with small samples of 10 or fewer observations. As the sample size increases, the variability within the sample may increase while the range remains essentially unchanged. Due to this phenomenon, the range should not be used in any comparative evaluation, such as before-and-after analysis, that involves samples of different sizes.

The more important measure of variability is the *standard deviation,* which is the positive square root of the variance. The variance is the sum of squares of the deviations of the observations less one. The expression for sample standard deviation is written in the following form for unclassed data:

$$s = \sqrt{\frac{\Sigma(X_i - \overline{X})^2}{N - 1}} \qquad \text{(C-3)}$$

where

s = standard deviation
$\overline{X}$ = arithmetic mean

X_i = individual observation
N = total number of observations

The $N - 1$ term is related to the statistical concept of degrees of freedom. The N observations and variance have $N - 1$ degrees of freedom since the deviations of the N observations must sum to zero and only $N - 1$ deviations can be used to determine the remaining one.

The following equation for standard deviation is applicable to classed data:

$$s = \sqrt{\frac{\Sigma f_i u_i^2 - \dfrac{(\Sigma f_i u_i)^2}{\Sigma f_i}}{\Sigma f_i - 1}} \tag{C-4}$$

where

s = standard deviation
u_i = class midpoint for ith class
f_i = frequency of ith class

The standard deviation increases in value as the observations become dispersed at greater distances from the mean. Frequency distributions are shown in Figure C-7 for data from two spot speed studies with different dispersions. The standard deviation for the data with large dispersion would be greater than the standard deviation for the data with small dispersion. The standard deviation summary statistic provides a reliable variability measure that has direct application in various statistical inferences.

If the shape of the data is approximately a bell-shaped curve or the normal probability distribution curve, multiples of the standard deviation can be expressed on either side of the mean and represent limits that contain various percentages of the total values in a selected sample. Limits of ± 1, 2, and 3 standard deviations about the mean contain, respectively, 68.3, 95.5, and 99.7% of the total observations. These relationships are shown in Figure C-8.

The following illustrative calculations for range and standard deviation are based on the spot speed data shown in Tables C-2 and C-4. The range for the classified data is the difference between the upper class interval for the highest speed group and the lower class interval for the lowest speed group and is equal to (56 − 35) or 21 mph. For unclassed data, the difference between the maximum and minimum observation is equal to (55 − 35) or 20 mph, the exact value. The application of equation (C-4) permits the computation of the standard deviation for classed data. The numerator contains the following class summations, which are tabulated in the "total" row of Table C-4:

1. Product of class frequency and square of corresponding class midpoint
2. Square of product of class frequency and corresponding class midpoint
3. Class frequency

The denominator is equal to the sum of the class frequencies less one. The standard deviation is then determined by taking the positive square root of the value that results from the indicated algebraic operations on the various summations in the numerator and the denominator of equation (C-4). The summary figures for the spot speed example are $[(137{,}910 - (3006)^2/66]/(66 - 1)$, and the square root of this expression gives a value of 3.9 mph as the standard deviation. The standard deviation for the unclassed data in Table C-2, calculated according to equation (C-3), using actual measurements rather than class midpoints and class frequencies, gives a value of 4.4 mph. The lack of agreement

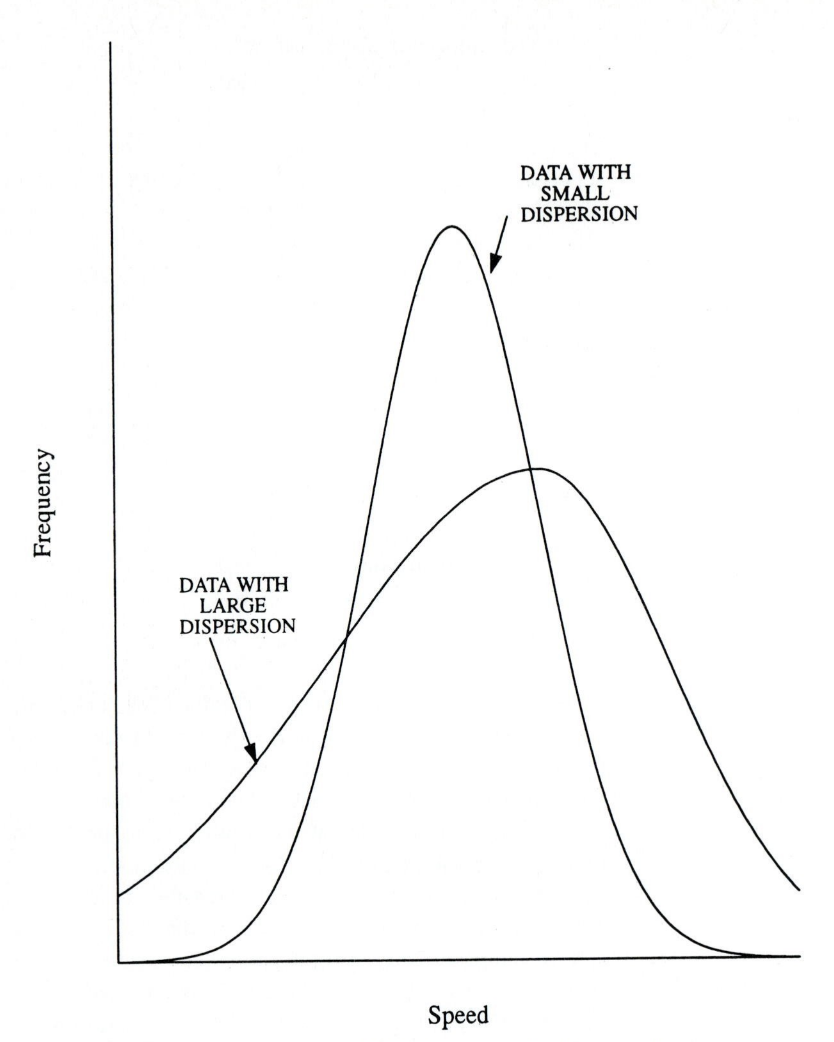

Figure C-7 Two spot speed distributions with different dispersions.

between the two calculated values is an indication of the loss in accuracy that may occur when data are grouped.

STATISTICAL INFERENCE

Various techniques of statistical inference provide valuable tools for the interpretation of the results of traffic engineering studies. Statistical inference permits the generalization of sample findings into statements about the population or universe from which the sample was taken. Probability considerations are involved in the development of statistical inferences. Several useful procedures for statistical inference are presented below in sections on estimation and on significance testing. Additional information on these two subjects is given in readily available textbooks on statistics (Walpole and Myers, 1985; Kennedy and Neville, 1986).

Estimation

When sample measurements of vehicle speeds, volumes, occupancy, and other characteristics are made, the tabulated results consist of a number of different values and a single value is often selected from the array to represent the data. This representative

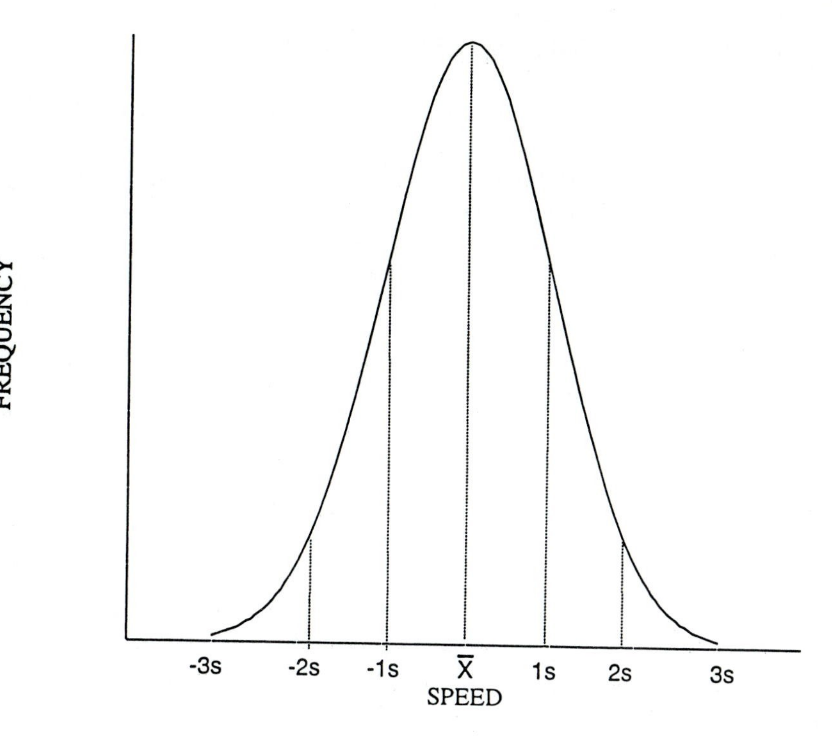

$\bar{X}$ = Mean
s = Standard Deviation
68.3% of all values are within ± 1s.
95.5% of all values are within ± 2s.
99.7% of all values are within ± 3s.

Figure C-8 Percentage variations of sample values about the mean for a normal distribution.

value, or average, is generally computed as the mean, the median, or the mode. This average value is then generally taken not only to represent the sample of traffic observations, but also to represent the entire population of values from which the sample was taken. The results of many studies are reported in the form of single-value or point estimates.

The accuracy of a sample depends on two factors. The first factor is chance, which is the probable amount of error due to change in the average of any sample and which can be estimated. This section describes how to calculate this estimate. The second factor affecting the accuracy of the sample is the method of sampling. The sample represents only the population from which it was drawn. In a speed study, if only the fastest speeds were recorded, the average would be representative of only the faster element of traffic on that street. To obtain a truly unbiased or representative sample of speeds in a spot speed study, vehicles must be selected at random in such a way that each vehicle has an equal chance of being included in the sample.

The requirements for a representative sample are:

1. The sample must be selected without bias.
2. The components of the sample must be completely independent of one another.
3. There should be no underlying differences between areas from which the data are selected.
4. Conditions must be the same for all items constituting the sample.

Reliability of the Sample

Assuming that the sample had been chosen in an unbiased manner, it is possible to calculate the degree of error that may be due to chance by calculating a confidence interval estimate. An *interval estimate* of the population mean is useful in reporting results of various transportation studies and is expressed by the following inequality:

$$\bar{X} - \frac{t_\alpha S}{\sqrt{N}} < \mu < \bar{X} + \frac{t_\alpha S}{\sqrt{N}} \tag{C-5}$$

where

μ = mean of the population
$\bar{X}$ = mean of the sample
s = standard deviation of the sample
t_α = statistic of the t distribution for $(N - 1)$ degrees of freedom and the probability defined by the subscript α
N = total number of observations
α = 1.0 − confidence coefficient

The expression $(s/\sqrt{N})$ in this inequality is defined as the standard error of the estimate for the mean. Table C-5 presents typical values of the t statistic for various degrees of freedom and for selected levels of α. The degrees of freedom are defined as the sample size less one, and the selection of a confidence coefficient is generally confined to the probability levels of 0.90, 0.95, and 0.99. The corresponding values of α are 0.10, 0.05, and 0.01 and are listed as column headings in Table C-5. A confidence coefficient of 95% provides an acceptable interval estimate for most purposes. The terms $(\bar{X} - t_\alpha s/\sqrt{N})$ and $(\bar{X} + t_\alpha s/\sqrt{N})$ define, respectively, lower and upper limits of the confidence interval. The interpretation of the probability statement for an interval estimate is that the confidence interval contains the population value with a probability that is equal to the confidence coefficient.

The following example illustrates the development of an interval estimate of the mean speed for the classed data that are presented in Table C-3. Sample values of mean and standard deviation were previously calculated as 45.5 and 3.9 mph, respectively. The term $(t_\alpha s/\sqrt{N})$ is now calculated as 2.000 $(3.9/\sqrt{66})$ or 0.96 mph for a confidence coefficient of 95%. This amount is subtracted from and added to the sample mean to determine, respectively, lower and upper limits of the interval estimate, which is written as $44.5 < \mu < 46.4$. That is, the population mean of the spot speeds is expected to be contained in the interval from 44.5 to 46.4 mph with a confidence that is equal to 95%. For the unclassified data with a mean of 45.5 and standard deviation of 4.4, the term $(t_\alpha s/\sqrt{N})$ calculated as 2.000 $(4.4/\sqrt{66})$ or 1.1. This results in lower and upper limits of 44.4 and 46.6, respectively.

This interval is the measure of precision of the sample, assuming that the sample selection is random and unbiased. If the sample error is too large, more observations should be obtained, making certain, of course, that conditions have not changed since the original observations were made.

When a study is made in which the result is expressed as a proportion or a percentage, the precision of the estimate is indicated by the following inequality:

$$p - t_\alpha \sqrt{\frac{pq}{N}} < \Phi < p + t_\alpha \sqrt{\frac{pq}{N}} \tag{C-6}$$

TABLE C-5
***t* Values**

Degrees of Freedom	Level of α		
	0.10	0.05	0.01
1	6.314	12.706	63.657
2	2.920	4.303	9.925
3	2.353	3.182	5.841
4	2.132	2.776	4.604
5	2.015	2.571	4.032
6	1.943	2.447	3.707
7	1.895	2.365	3.499
8	1.860	2.306	3.355
9	1.833	2.262	3.250
10	1.812	2.228	3.169
11	1.796	2.201	3.106
12	1.782	2.179	3.055
13	1.771	2.160	3.012
14	1.761	2.145	2.977
15	1.753	2.131	2.947
16	1.746	2.120	2.921
17	1.740	2.110	2.898
18	1.734	2.101	2.878
19	1.729	2.093	2.861
20	1.725	2.086	2.845
21	1.721	2.080	2.831
22	1.717	2.074	2.819
23	1.714	2.069	2.807
24	1.711	2.064	2.797
25	1.708	2.060	2.787
26	1.706	2.056	2.779
27	1.703	2.052	2.771
28	1.701	2.048	2.763
29	1.699	2.045	2.756
30	1.697	2.042	2.750
40	1.684	2.021	2.704
60	1.671	2.000	2.660
120	1.658	1.980	2.617
∞	1.645	1.960	2.576

where

Φ = proportion of the population
p = proportion of the sample
$q = 1.0 - p$
t = statistic of the t distribution for $(N - 1)$ degrees of freedom and the probability defined by the subscript α
N = number of observations
$\alpha = 1.0 -$ confidence coefficient

The term $(\sqrt{pq/N})$ is defined as the standard error of the estimate for the proportion. Table C-5 provides typical values of the t statistic for the various degrees of freedom and for selected levels of α. The determination of the sample precision for proportions or percentages is very similar in procedure and interpretation to the situation for the mean of a sample discussed previously.

As an example of a percentage estimate, 400 drivers were randomly sampled in regard to their location of employment and 120, or 30%, reported that they worked downtown. The term $(t\ \sqrt{pq/N})$ is calculated as 1.960 times the square root of (0.3)(0.7)/400 or 4.5% as the precision of the percentage estimate for a confidence coefficient of 95%. That is, the estimate of 30% is subject to a possible sampling error of ±4.5%, and the upper and lower limits of the percentage estimate are expressed as $25.5 < \phi < 34.5$ with a confidence of 95%.

Significance Testing

Analysts are often concerned whether the difference in average values between two sets of sample data is statistically significant or merely due to chance variations that result from sampling. Significance testing is a valuable way to address those concerns. For instance, certain improvements might reduce the average parking duration from 40 minutes to 30 minutes. The analyst must decide whether this reduction is really significant or is due to chance alone, and significance testing can help in that decision.

If two samples of data are taken from the same population, there will probably be a difference between the averages of these two samples. This difference would be due to chance alone. A greater number of observations probably provides a smaller chance error, and hence a smaller difference between the study averages, because the same population is being sampled. However, if two studies are made at different locations where conditions are not identical, the two populations will have different mean values. This difference, added to the chance variation between the two studies, equals the total observed difference between the means of the two studies.

Because differences between pairs of sample means from a given population occur only due to chance error, these differences are subject to the laws of probability and follow a normal curve. Any difference that is large enough to fall at an extreme point on this curve is not within the realm of chance error and represents a significant difference. The *significance test* for equality or inequality of the means of two populations with unequal variances is based on samples of 30 or more observations for each population. A test statistic for this situation is determined by the following formula:

$$t = \frac{\overline{X}_1 - \overline{X}_2}{\sqrt{\frac{{s_1}^2}{N_1} + \frac{{s_2}^2}{N_2}}} \tag{C-7}$$

where

t = statistic of the t distribution
$\overline{X}_1$ = mean of first sample
$\overline{X}_2$ = mean of second sample
s_1 = standard deviation of first sample
s_2 = standard deviation of second sample
N_1 = number of observations in first sample
N_2 = number of observations in second sample

The computed value of t is then compared with the critical t value (t_c), as obtained from Table C-5, to determine the significance of the difference between the two sample means. The value of t_c is selected in accordance with the specified level of significance (α). The value of 0.05, corresponding to 95% confidence, is often chosen as the level of significance, although an α of 0.01 to 0.10 is within the proper range for most evaluations of transportation data.

If the computed value (either positive or negative) of t is greater than t_c, the difference between the two means is considered significant and not due just to chance variations alone. The difference between the two means is defined as nonsignificant and due to chance variation alone when the calculated t value (either positive or negative) is less than the critical t value.

Analysts perform significance testing for the difference between two proportions or percentages in a similar manner except that they use the following equation to compute the t statistic:

$$t = \frac{p_1 - p_2}{\sqrt{p_0 q_0\left(\frac{1}{N_1} + \frac{1}{N_2}\right)}} \tag{C-8}$$

where

t = statistic of the t distribution
$p_0 = (p_1 N_1 + p_2 N_2) / (N_1 + N_2)$
p_1 = proportion observed in first sample
p_2 = proportion observed in second sample
N_1 = number of observations in first sample
N_2 = number of observations in second sample
$q_0 = 1.0 - p_0$

Either proportional or percentage values may be used in equation (C-8) for p and q.

The following example is presented to illustrate significance testing. Under old parking regulations, a study showed that 185, or 28.5%, of 648 vehicles were parked overtime. After new parking regulations were adopted, a similar study revealed that 119 of 512 vehicles or 23.2% were parked overtime. The weighted average, p_0, of the two percentages is first calculated as $[(28.5 \times 648) + (23.2 \times 512)]/(648 + 512)$ or 26.2%, and q_0 is equal to $(100.0 - 26.2)$ or 73.8%. The various values are now inserted in equation (C-8) to determine the calculated value of t as $(28.5 - 23.2)$ divided by the square root of $(26.2 \times 73.8)(1/648 + 1/512)$ or 2.038. The critical t value is obtained from Table C-5 as 1.960 for a significance level of 5%. Because t is larger than t_c, the difference between the two percentages is significant, and the new parking regulations appear to be effective in reducing overtime parking.

Nonparametric Tests

Significance testing requires large samples selected from a normal population and requires numerical data. However, not all transportation study data meet these requirements. When this is the case, various nonparametric significance tests are available for evaluating the study results. These distribution-free tests are particularly convenient when the sample size is small and/or the data are qualitative rather than numerical.

The following example illustrates the use of a nonparametric rank-order correlation test to see whether the speeds of vehicles on two freeways with greatly differing traffic volumes are independent of each other. The average vehicle speeds shown in Table C-6 were recorded at seven randomly selected locations on each freeway. The speeds for each freeway have been ranked from 1 for the highest to 7 for the lowest. The differences between the ranks have been calculated, squared, and totaled. The test statistic is computed as follows:

$$r_s = 1 - \frac{6\Sigma d_i^2}{N(N^2 - 1)} \tag{C-9}$$

TABLE C-6
Vehicle Speeds on Two Freeways

Average Speeds		Ranking			
Freeway 1	*Freeway 2*	*Freeway 1, X*	*Freeway 2, Y*	*Difference, d = X − Y*	d^2
63	61	3	3	0	0
59	62	4	2	2	4
67	59	1	4	−3	9
55	58	6	5	1	1
65	68	2	1	1	1
51	56	7	6	1	1
56	53	5	7	−2	4
					20

TABLE C-7
Critical r_s Values

	Significance Level		
N	*0.100*	*0.050*	*0.010*
5	0.900	1.000	1.000
6	0.829	0.886	1.000
7	0.714	0.786	0.929
8	0.643	0.738	0.881
9	0.600	0.700	0.833
10	0.564	0.648	0.794
11	0.536	0.618	0.818
12	0.497	0.591	0.780
13	0.475	0.566	0.745
14	0.457	0.545	0.716
15	0.441	0.525	0.689
16	0.425	0.507	0.666
17	0.412	0.490	0.645
18	0.399	0.476	0.625
19	0.388	0.462	0.608
20	0.377	0.450	0.591
21	0.368	0.438	0.576
22	0.359	0.428	0.562
23	0.351	0.418	0.549
24	0.343	0.409	0.537
25	0.336	0.400	0.526
26	0.329	0.392	0.515
27	0.323	0.385	0.505
28	0.317	0.377	0.496
29	0.311	0.370	0.487
30	0.305	0.364	0.478

where

r_s = Spearman rank-order correlation coefficient
Σd_i = summation of the differences
N = number of measurement pairs

Critical values for r_s are given in Table C-7. Any calculated value greater than the critical value in absolute value is a statistically significant indication of independence.

Substituting the example data in Table C-6 in equation (C-9) gives a coefficient of

Histogram of SPEED			N = 66	
Midpoint	Count			
33.5	0			
35.5	2	**		
37.5	2	**		
39.5	4	****		
41.5	9	*********		
43.5	8	********		
45.5	15	***************		
47.5	10	**********		
49.5	7	*******		
51.5	5	*****		
53.5	3	***		
55.5	1	*		
57.5	0			
	N	MEAN	MEDIAN	STDEV
SPEED	66	45.5	45.5	4.43
	MIN	MAX		
SPEED	35	55		

Figure C-9 Example output from statistical analysis program.

$$r_s = 1 - \frac{(6)(20)}{7(7^2 - 1)} = 1 - 0.36 = 0.64$$

The critical r_s value is obtained from Table C-7 as 0.786 for a significance level of 5% and $N = 7$. Since the calculated r_s is less than the critical value, it may be concluded that speeds on the two freeways are dependent (i.e., are related to each other).

Numerous nonparametric tests are available for use when transportation data are not suitable for the classical statistical tests and only one has been presented here. Complete explanations of the procedures may be found in Siegel and Castellan (1988), and illustrative examples are given in Taylor and Young (1988).

CALCULATION AIDS

Data analysis and evaluation may be greatly facilitated by the use of calculators and computers. Simple scientific calculators have built-in programs for calculating descriptive statistics such as means, standard deviations, and variances, while more advanced calculators can be programmed to perform more complex statistical tests. Notebook-sized, battery-powered microcomputers may be used both to collect and analyze data at field locations and/or store data for later office analysis. Desktop personal microcomputers are available to most engineers and numerous statistical packages are available for use on these. Many of these packages have been developed specifically for engineering use and include graphical capabilities. Specialized microcomputer programs for analyzing the results of transportation studies are available from commercial sources, government agencies, and user groups. Figure C-9 shows a simple frequency diagram and descriptive statistics for the spot speed study data of Table C-1 that was produced by a typical statistical analysis program and desktop microcomputer/printer.

REFERENCES

GREENSHIELDS, B. D., AND F. M. WEIDA (1978). *Statistics with Applications to Highway Traffic Analyses*, 2nd ed., Eno Foundation for Transportation, Inc., Westport, CT, revised by D. L. Gerlough and M. J. Huber.

KENNEDY, J. B., AND A. M. NEVILLE (1986). *Basic Statistical Methods for Engineers and Scientists*, 3rd ed., Harper & Row, New York.

SIEGEL, S., AND N. J. CASTELLAN, JR. (1988). *Nonparametric Statistics for the Behavioral Sciences*, 2nd ed., McGraw-Hill, New York.

TAYLOR, M. A. P., AND W. YOUNG (1988). *Traffic Analysis: New Technology and New Solutions*, Hargreen Publishing Company, North Melbourne, Australia.

WALPOLE, R. E., AND R. H. MYERS (1985). *Probability and Statistics for Engineers and Scientists*, 3rd ed., Macmillan, New York.

D

Data Presentation Techniques

Donna C. Nelson, Ph.D., P.E.

INTRODUCTION

Data presentation is an essential part of most engineering reports and oral presentations. Illustrations (in the form of tables, figures, drawings, and photographs) are efficient, powerful tools for the presentation of data and results. However, illustrations must be prepared carefully to ensure that the intended message is conveyed. Poorly designed illustrations can confuse, distort, and fail to communicate the relevant information.

In recent years, a wide variety of microcomputer-based tools have come into common use. Professional-looking graphics can be created using spreadsheets, graphics managers and presentation software, paint programs, computer-aided drafting (CAD), and programs that allow the integration of scanned video images with computer-generated graphics. These new tools have made the creation of report graphics fast and easy. This has also shifted the burden of producing final graphics and tables from the graphics staff to the traffic engineering staff. Speed and ease of use have also encouraged the tendency to graph everything, a practice that can be ineffective and confusing.

DESIGN OF ILLUSTRATIONS

Illustrations are used to convey or to clarify information. Page layout, type of illustration, content, and skill of execution all contribute to the success or failure of an illustration as a communication tool. Illustrations must be designed for the audience that will view them

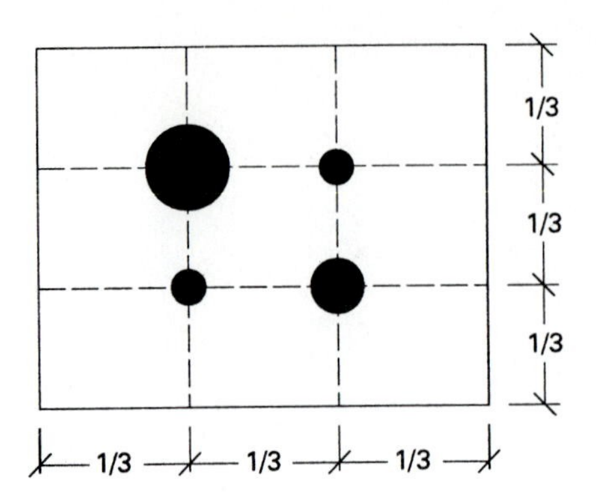

Figure D-1 Relative weight by location.

and for the circumstances under which they will be used. This section includes some general guidelines for the design of illustrations.

Focus and Attention

Eye movement and memory recall studies suggest that some positions on a page are more important than others. For example, the eye usually focuses first at the upper left corner of the page, then moves around the illustration in steps toward the lower right corner, where the eye exits the page. Research suggests that it is easier for a reader to recall the first item seen and to a lesser extent, the last item seen. The items viewed between the first and last items are more difficult to remember. Research also suggest that fewer than 10 to 15 items are retained in the viewers short-term memory. The illustrations in Figures D-1 to D-5 illustrate some of these concepts.

There are a variety of ways to determine the location of focal points on a page. A simple technique is to divide the page into thirds in both directions as shown in Figure D-1. The intersections of the dividing lines are focus points. The size of the circles placed over locations 1 through 4 show the relative visual strength of these locations. The upper left location is the strongest visual position because the eye appears to spend more time focused in this area. An item placed here is the most likely to be remembered. The lower right intersection is the second strongest visual position and, as the last item viewed, is also likely to be remembered. Items placed away from these locations tend to serve as background information. They support items in areas of strength but are less likely to be remembered.

The *visual weight* of items or areas on an illustration is also important. Visual weight is an illusive combination of position, shape, color, contrast, and meaning. If there is more than one object on the page, the eye tends to focus on the visually heaviest object first. The eye either stays focused on this point or continually returns to it from other items (Figure D-2). If all the items on a page have the same visual weight, there is little for the eye to focus on. With no point to focus on, the viewer may become bored and the eye may move off the illustration (Figure D-3). Items placed within the focus areas (as shown in Figure D-1) draw the viewer's attention.

The shape of the items can also direct the viewer's focus around the page (Figure D-4). Items shaped like arrows that point to the lower right corner tend to direct the interest of the observer off the page. Placing a shape that either stops the flow or points back onto the page helps the viewer focus on the page and on the message (Figure D-5).

Composition

The structure of an illustration can be compared to that of written paragraph. In a paragraph, the first sentence informs the reader what the paragraph is about: telling the reader what to expect. This is analogous to the first item of focus on the illustration. The main body of the paragraph develops the idea presented in the first sentence. Similarly, the main body of the illustration serves as the background, supporting the main message of

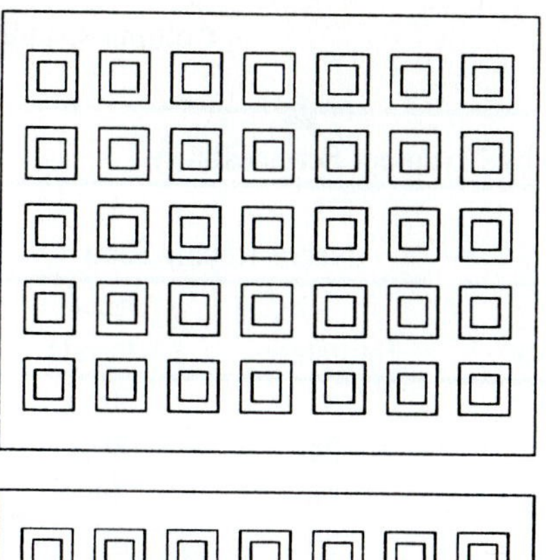

Figure D-2 This figure has a strong focal point.

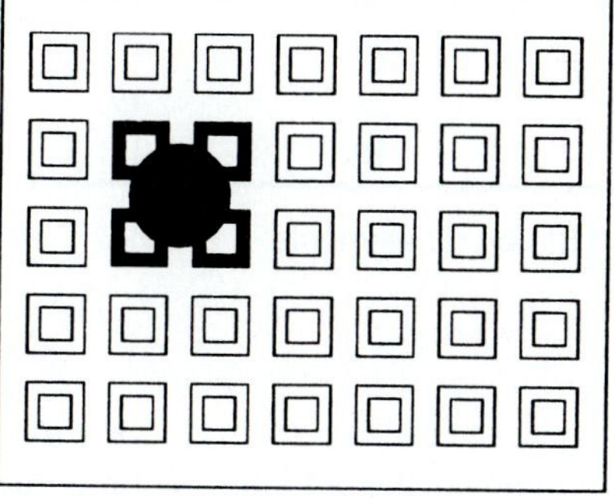

Figure D-3 This figure lacks a focal point.

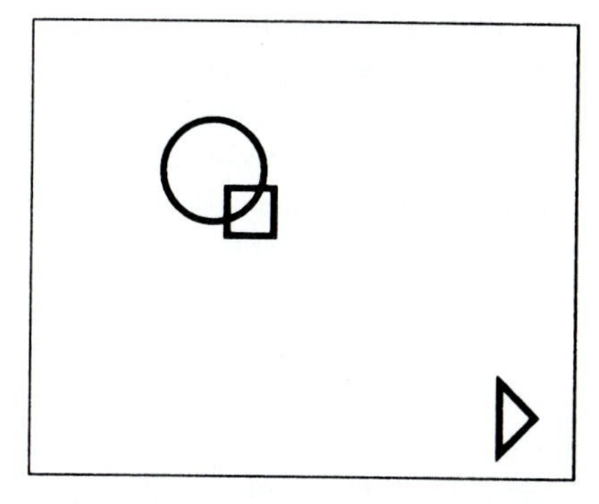

Figure D-4 The shape of objects directs the eye away from the illustration.

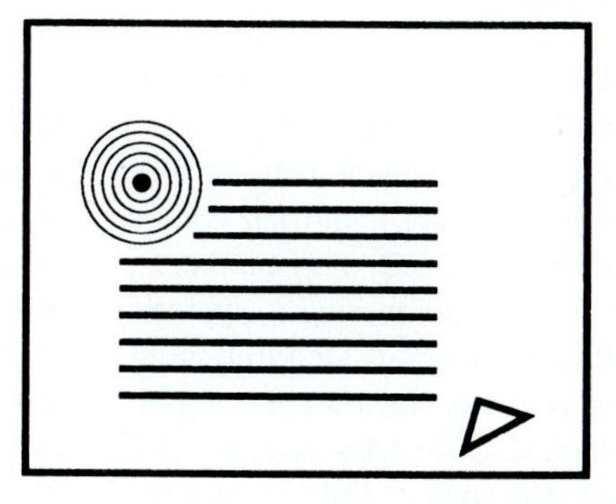

Figure D-5 Visual weight and objects can be used to balance an illustration.

the visual. Points of interest are established by the objects of heavier visual weight. The final sentence reinforces, concludes, and tells the reader what the paragraph was about. For an illustration, this usually corresponds to the item at the lower right area of focus. This is the last thing the viewer sees before leaving the illustration.

TABLES

Formal tables serve two basic purposes: to summarize information or data discussed in the text and to compile reference data. The purpose of the table guides how it is designed and where it is placed in a report. Summary tables save space and enhance

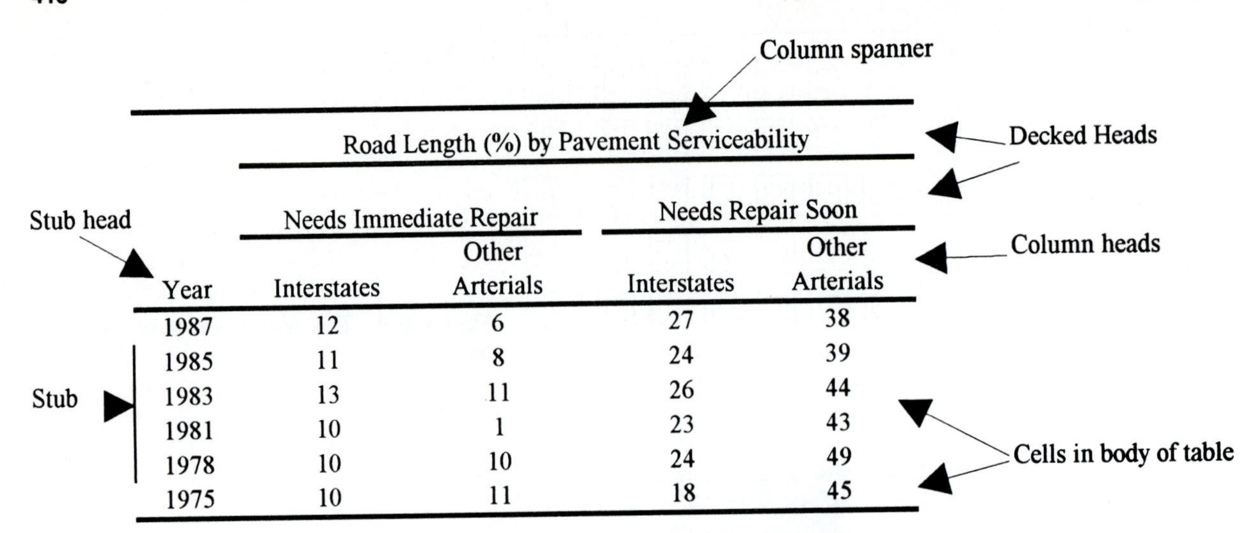

	Road Length (%) by Pavement Serviceability			
	Needs Immediate Repair		Needs Repair Soon	
Year	Interstates	Other Arterials	Interstates	Other Arterials
1987	12	6	27	38
1985	11	8	24	39
1983	13	11	26	44
1981	10	1	23	43
1978	10	10	24	49
1975	10	11	18	45

Figure D-6 Structural elements of a table.

comprehension by providing condensed information in meaningful form. A short, well-designed tabulation can replace a lengthy section of explanatory text loaded with statistics. Tables designed for oral presentations are usually most effective when they are simple. The design of tables for oral presentations is discussed in Appendix F. Tables included in written reports may be more complex, as the reader has the time to study and understand them. Reference tables generally provide material to support the text but are not needed in the text for clarity. These tables are generally placed in the appendices.

Structural Elements of a Table

Some general guidelines for the development of effective tables are described in this section. For reference, Figure D-6 shows the structural elements of a table. Formal tables are titled and numbered for reference. Tables are normally numbered in consecutive order, starting from the beginning of each article, chapter, or book. Arabic numerals (Borland, 1989; Michaelson, 1987; White, 1988) are used to identify tables occurring in the main body of a work. The use of suffix letters to numerals (1A, 1B) can cause confusion. Tables appearing in appendices are identified consecutively in each appendix (A-1, A-2, B-1). Double numbering (1-1, 1-2, 1-3) can be used to locate tables in their respective sections of a report. Double numbering is appropriate when sections contain numerous tables or graphics. Roman numerals are rarely used today. Position tables as close to the textural reference as possible but after the initial reference to them. The title should briefly explain the content of the table and should be pertinent to the text. If the data are not original, reference the source. Data items can also be referenced in the body of the table, as footnotes to specific data items, in columns, or in the title of the table. A subtitle contains any additional explanatory or descriptive information required to make the main title clear. Use the same type face as the title, but use either a smaller size or lighter weight.

Column and row headings should be brief. Vertical column headings and subheadings usually contain the dependent variables. Horizontal row headings (shown as the stub in Figure D-6) generally contain the independent variables. Define any abbreviations used in the headings as footnotes to the table.

The field (or body) contains the tabulation of numerical or verbal data referred to by the stub and headings. Each unit of the field where a horizontal row intersects a vertical column is called a *cell.* Whenever possible, data items in columns should have a uniform degree of accuracy. Fractions are usually converted to decimals and numbers are aligned by decimal point. Plus or minus signs are placed immediately to the left of numbers; missing entries are identified by a dash. Verbal entries that require multiple lines may be right-justified unless the spacing between words is noticeably large. Lines (rules) or grids are used to separate certain columns or lines, or to frame the table in a box. This may help to distinguish the headings, separating heading area from the data in the field, and improve the readability of the table. Rules should be thin lines; heavy lines may draw attention away from the table and make it difficult to read. Rules work best for very small or very large tables.

Each table should stand alone (be understandable without reference to text). All references and notes are handled separately, without reference to the text or other tables. Reference numbering starts at the upper left, with a new series for each table (e.g., a, 1, or *). Number tabular elements from left to right across the table. Footnotes should be the same width as the table. If possible, place each note on a separate line. Footnotes titled "Note" qualify, explain, or provide information relating to the table as a whole. They are always placed first in the sequence of footnotes. Footnotes titled "Source" cite the source of the information and are always last in the sequence of footnotes.

Organization

Readers are accustomed to accepting information from left to right. However, by definition, tables are structured around vertical columns. Data in vertical columns appear to be easier to read than horizontal comparisons. The critical problem is to make the left-to-right relationship clear (White, 1988). Organize the data to make interrelationships as visible as possible. Present data so that comparisons are possible both within a table and among tables. Organize columns of data to support the purpose of the table (e.g., to show similarities or differences, statistical trends, or interactions). For example, a table presenting data on high accident locations may be ordered from high to low based on the accident experience. If users must extract data from a table based on a name, place the entries in alphabetical order based on the major street. When designing reference tables, consider the ways in which readers may want to use these tables.

Avoid large gaps between columns. Gaps tend to confuse the reader because the eye must move across empty areas to locate the next item in a row. Using generous spacing between the lines can help improve horizontal tracking, which makes it easier to read across a table. If a table is large or complex, place it on a separate sheet. Small, simple tables should be placed directly in the text. Make sure that the table supplements and complements, but does not duplicate, information in the text. All tables should be referred to in the text.

GRAPHS

A *graph* is a visual representation of numeric information and is useful as a presentation device and in the interpretation and analysis of data. It is important to choose the most appropriate type of graph for each case. The most common types of graphs and charts are described below (Borland, 1989). When appropriate, the names used correspond to the names used in common microcomputer software packages.

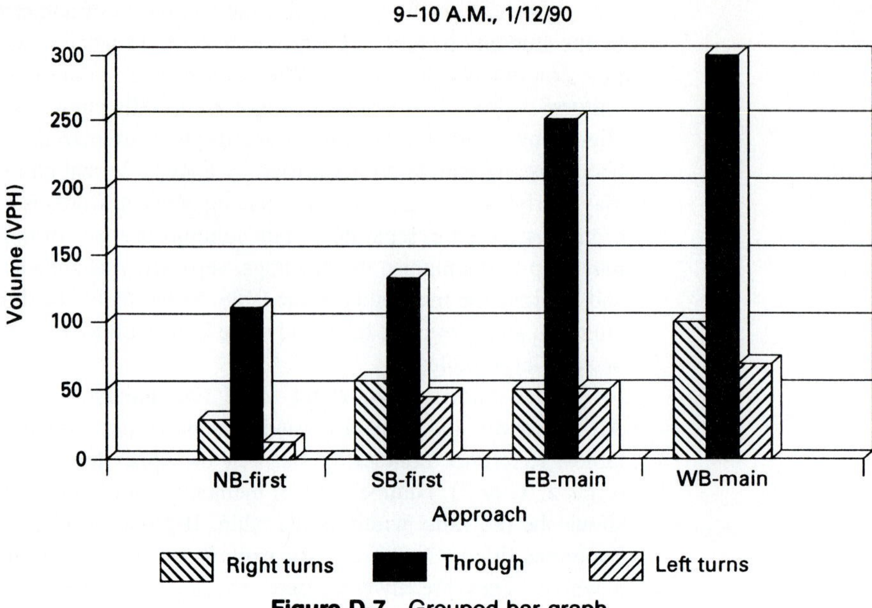

Figure D-7 Grouped bar graph.

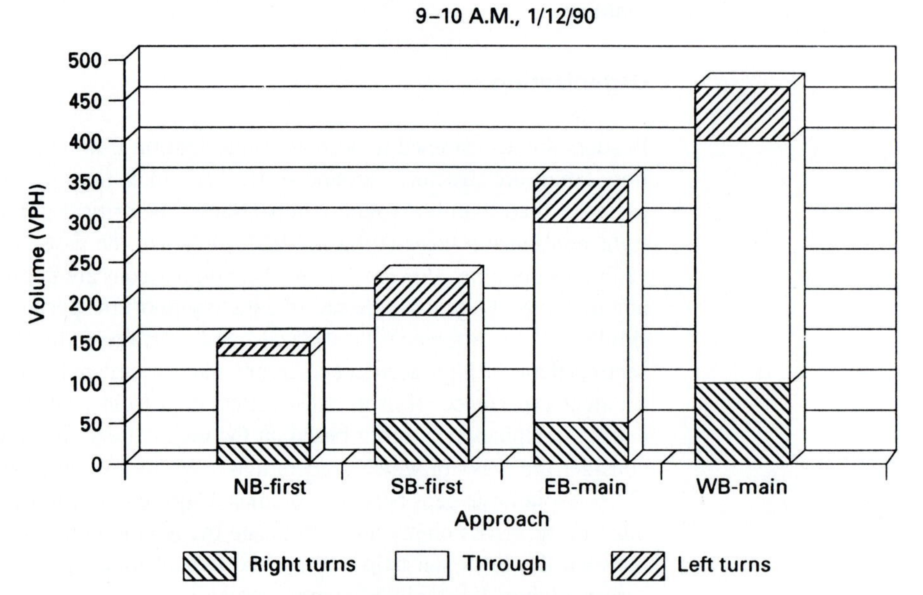

Figure D-8 Stacked bar graph.

Bar Graphs

A *bar graph* uses vertical or horizontal bars to represent each value in a series. Bar graphs are typically used to compare the values of different items at specific points in time. The bar graph shown in Figure D-7 presents vehicle counts by movement for four intersection approaches. By contrast, "stacked" bar graphs display the values of a series in a vertical column. The graph in Figure D-8 shows the same turning movements in a stacked format. The total height of the bar corresponds to the total approach volume for the time period named.

In the *line graph* shown in Figure D-9, each value in a series is connected with a line. Line graphs are used to display time series and trend data. While the use of a line graph implies continuous data, they are commonly used to show trends in time series. Be certain that line graphs of noncontinuous data do not imply data continuity.

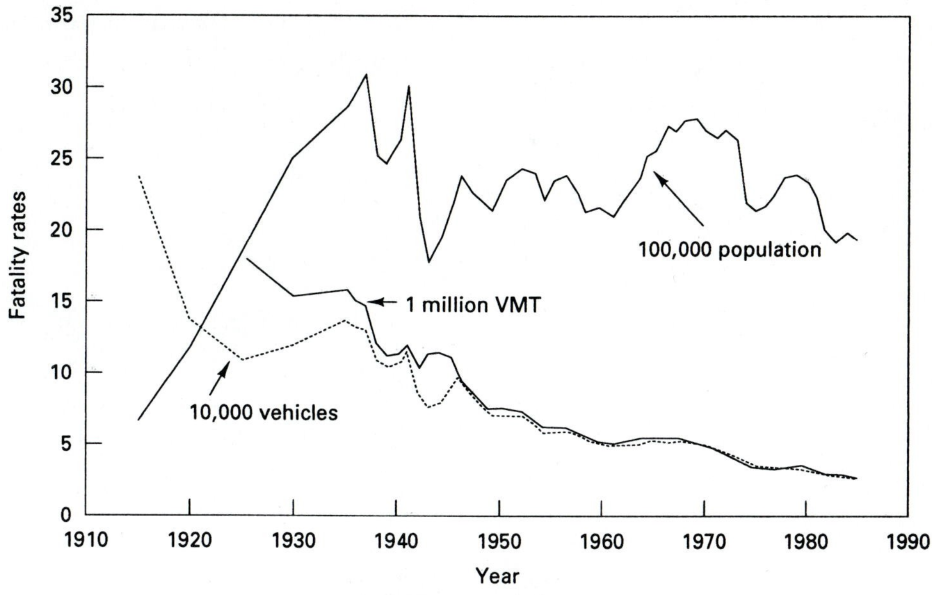

Figure D-9 Line graph.

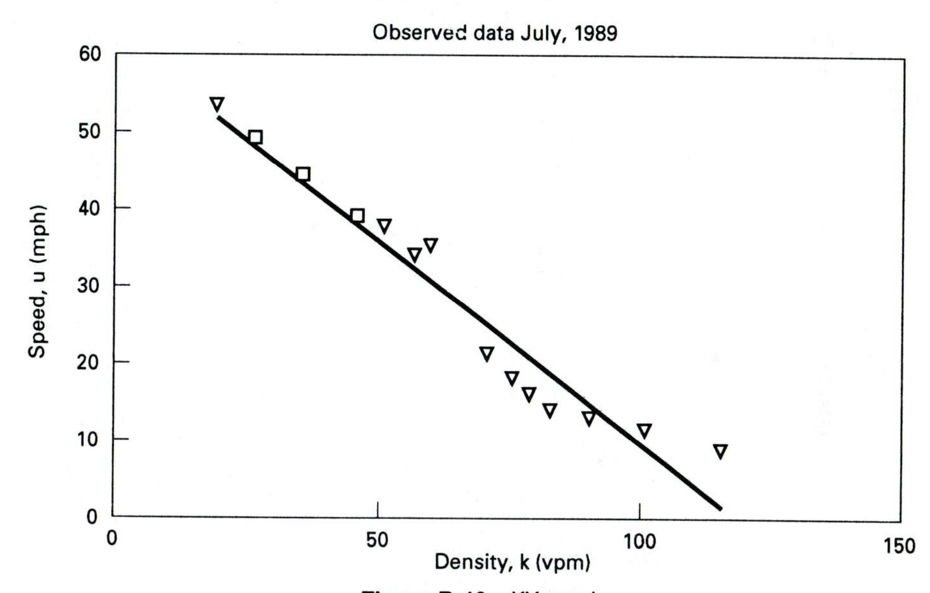

Figure D-10 *XY* graph.

Most spreadsheet programs require data for line graphs to have evenly spaced x values (e.g., one year or one month). If x values are not part of a discrete, uniform series, an *XY* graph should be used. Show data points only if they are actual measured values. Current spreadsheet software also allows line graphs to be merged with bar and other graphs for interesting effects.

XY Graphs

An *XY graph* differs from a line graph in that it is a plot of X (horizontal axis) versus Y (vertical axis). *XY* graphs demonstrate relationships among two or more series of data. Plots may be made of data points only, as a line (with no data points shown), or as a *best fit* line overlaid on the points. Examples of these three types are shown in Figure D-10.

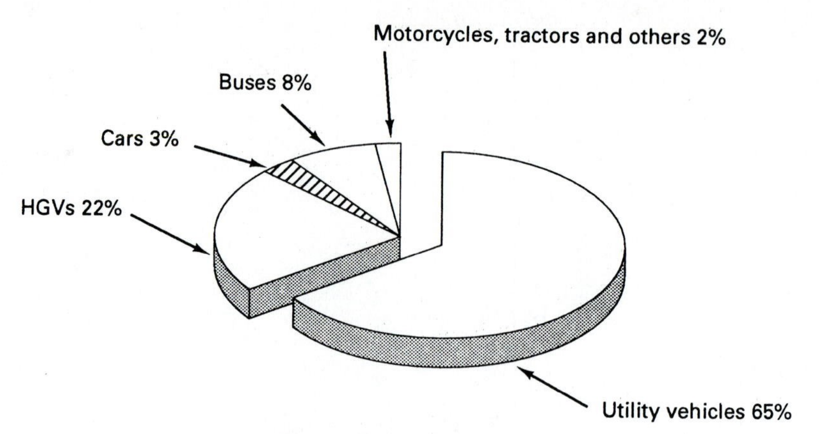

Figure D-11 Pie chart.

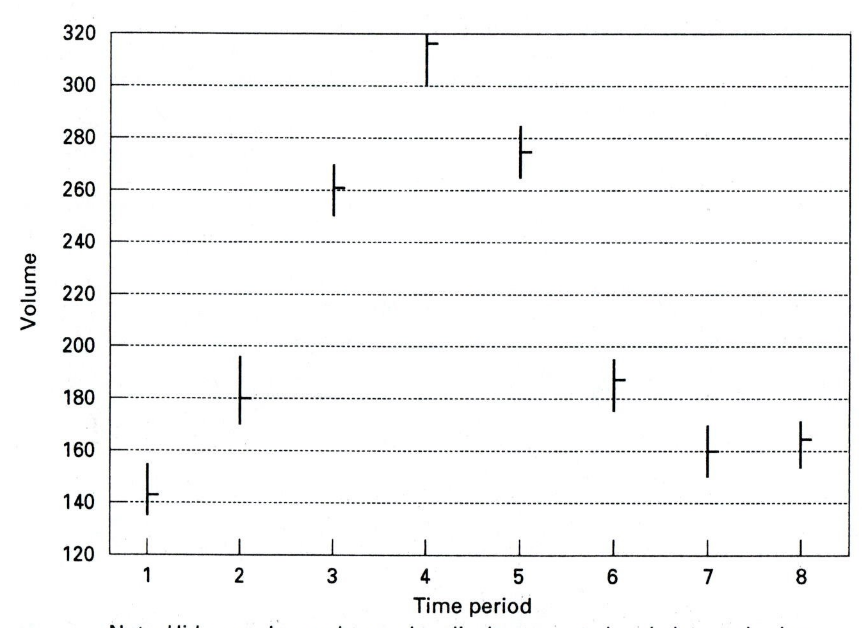

Note: Hi-Lo graphs can be used to display expected and observed values.

Figure D-12 High–low graph. Tick marks indicate observed values.

Pie Charts

Pie charts represent a single series and use a circle or ''pie'' to show the relative contribution of each value to the whole (see Figure D-11). Pie charts cannot display negative numbers. Pie charts do not work well when there are a large number of items making up the pie; the chart becomes too cluttered to be easily understood and a bar graph would be more effective in this instance. Extracting a slice from the pie or changing the color or size of a slice can change the meaning associated with it.

High–Low Graphs

High–low graphs use two data series to create a set of vertical lines, one for each pair of values. They can be used to illustrate the high, low, and mean values of any series of data collected. The tick marks indicate on Figure D-12 indicate observed values.

Area Graph

An *area chart* presents each series as a filled-in area. The *X*-axis represents the number of data points; the *Y*-axis indicates the number of series or accumulated series points be-

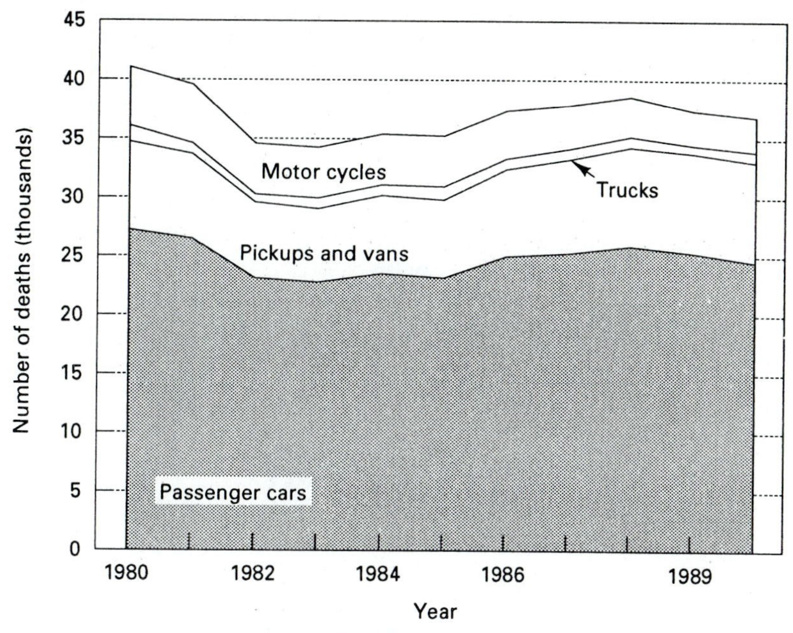

Figure D-13 Area graph.

ing plotted. It is similar in concept to a stacked bar graph; data series are stacked rather than overlaid. An example is shown in Figure D-13.

Pictograms

A *pictogram* integrates pictograms or icons into various types of graphs, including bar graphs, line graphs, and pie charts. Figure D-14 is an example of the use of pictorial symbols to illustrate sign legibility distances. Pictograms that are too complex may be difficult to interpret.

Statistical Maps

Statistical maps are designed to show the locations of a variable collected over a geographical area. The map shown in Figure D-15 illustrates this type of graphical presentation. Maps are included with many graphics packages.

Organization Charts

An *organization chart* is shown in Figure D-16. Chart elements are connected by lines that trace the flow of authority within the organization. The divisions of any department are placed below the department; all elements on the same level have a comparable level of authority. Elements in an organization chart need not be enclosed by a rectangles or other geometric shape. Enclosed items work well on simple charts. However, on large complex charts they make individual text items more difficult to focus on and may require smaller text to make the chart a reasonable size.

Flow Diagrams

Flow diagrams are useful for representing processes or procedures. Geometric shapes (boxes, circles, triangles) denote specific types of activities. Arrowed lines (vectors) trace possible actions. For example, a flow diagram might be used to illustrate the

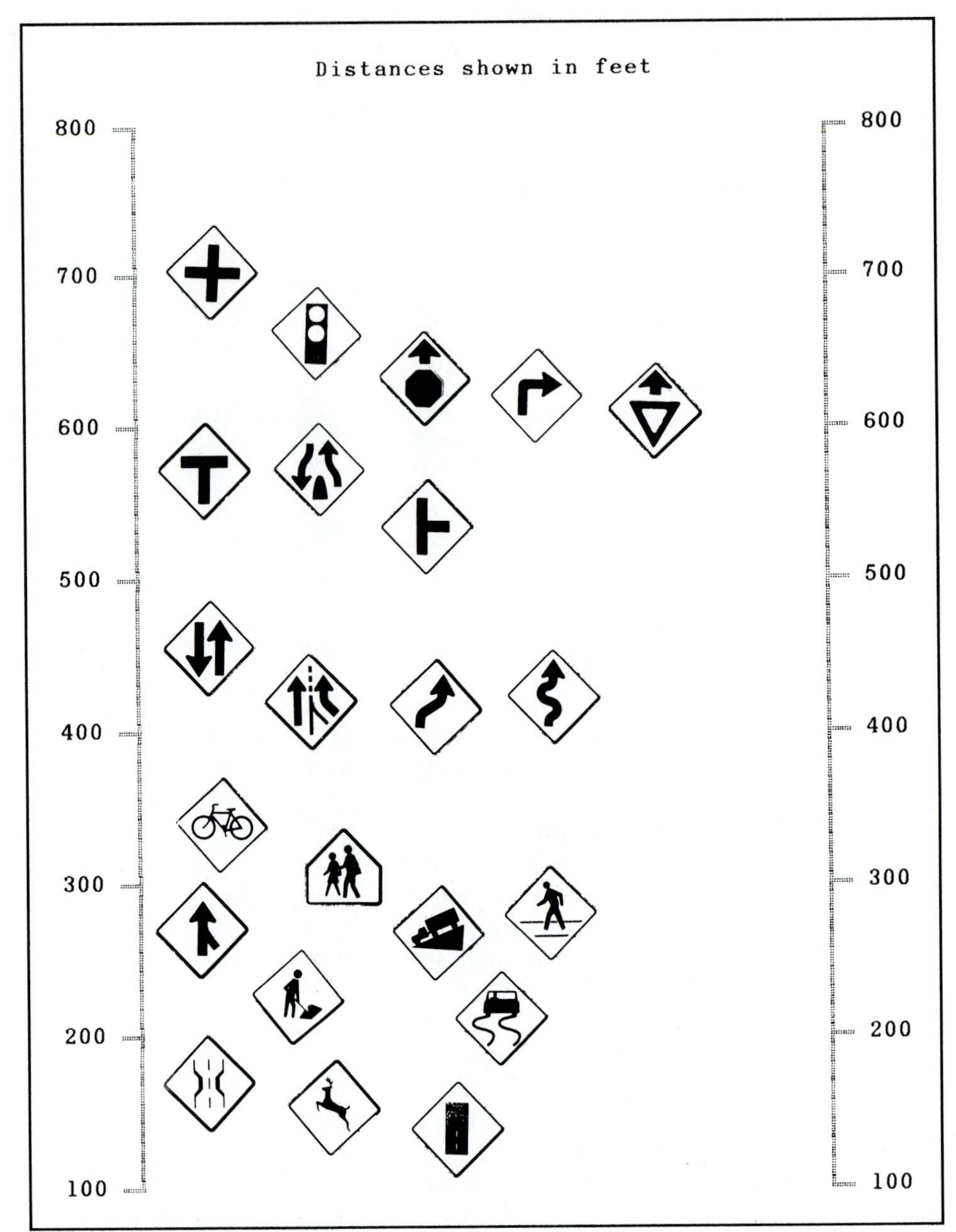

Figure D-14 Pictogram. *Source:* J. F. Paniati, "Legibility and comprehension of traffic sign symbols," Research Report, Federal Highway Administration, McLean, VA.

required steps in gaining approval for a new project, or, as shown in Figure D-17, a decision-making process.

Project Progress Charts

Project progress charts illustrate the time schedule for a study or project. The *X*-axis is the time span of the project in appropriate increments. Each task or set of tasks is represented as a line or bar spanning the time period over which that task should be accomplished. Progress charts are used as planning tools and as a method of checking progress as a project proceeds.

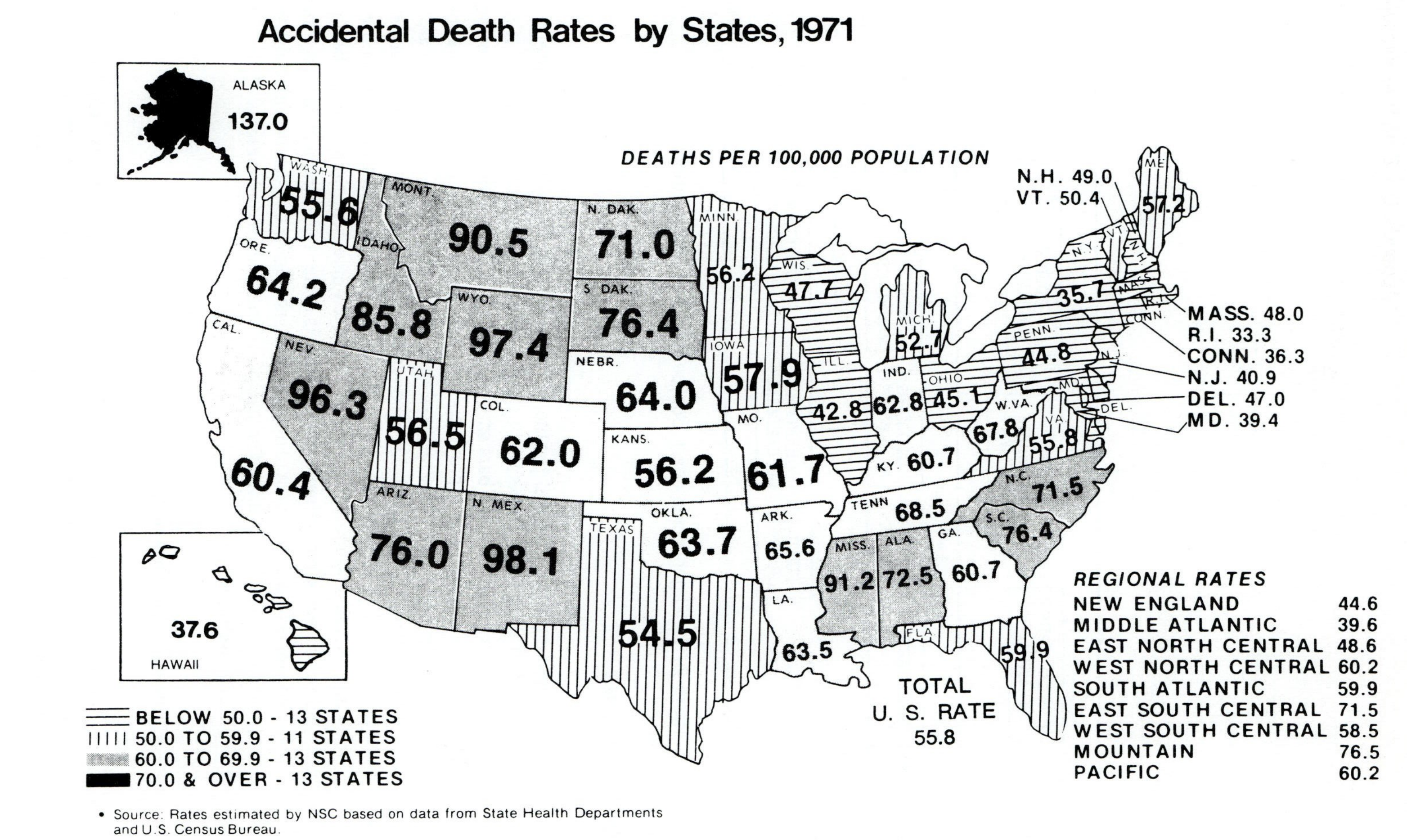
Accidental Death Rates by States, 1971
DEATHS PER 100,000 POPULATION
ALASKA
137.0
WASH.
55.6
MONT.
90.5
N. DAK.
71.0
MINN.
56.2
ORE.
64.2
IDAHO
85.8
WYO.
97.4
S. DAK.
76.4
WIS.
47.7
MICH.
52.7
CAL.
60.4
NEV.
96.3
UTAH
56.5
COL.
62.0
NEBR.
64.0
IOWA
57.9
ILL.
42.8
IND.
62.8
OHIO
45.1
KANS.
56.2
MO.
61.7
KY.
60.7
W.VA.
67.8
VA.
55.8
ARIZ.
76.0
N. MEX.
98.1
OKLA.
63.7
ARK.
65.6
TENN.
68.5
N.C.
71.5
S.C.
76.4
TEXAS
54.5
MISS.
91.2
ALA.
72.5
GA.
60.7
LA.
63.5
FLA.
59.9
PENN.
44.8
N.Y.
35.7
ME.
57.2
N.H. 49.0
VT. 50.4
MASS. 48.0
R.I. 33.3
CONN. 36.3
N.J. 40.9
DEL. 47.0
MD. 39.4
37.6
HAWAII
TOTAL
U. S. RATE
55.8
BELOW 50.0 - 13 STATES
50.0 TO 59.9 - 11 STATES
60.0 TO 69.9 - 13 STATES
70.0 & OVER - 13 STATES
REGIONAL RATES
NEW ENGLAND 44.6
MIDDLE ATLANTIC 39.6
EAST NORTH CENTRAL 48.6
WEST NORTH CENTRAL 60.2
SOUTH ATLANTIC 59.9
EAST SOUTH CENTRAL 71.5
WEST SOUTH CENTRAL 58.5
MOUNTAIN 76.5
PACIFIC 60.2
• Source: Rates estimated by NSC based on data from State Health Departments and U.S. Census Bureau.

Figure D-15 Statistical map. *Source:* Box and Oppenlander, 1976.

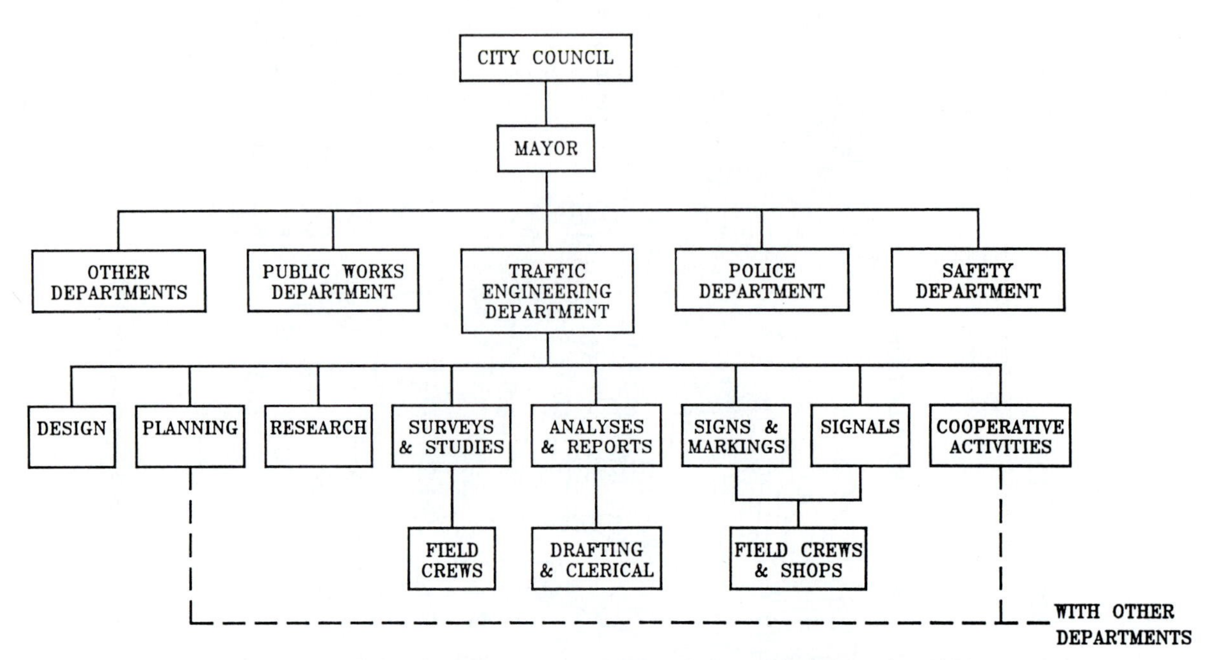

Figure D-16 Organization chart. *Source:* Box and Oppenlander, 1976.

DESIGN CONSIDERATIONS

Graphs must be designed for the intended viewer and with the purpose of the graph in mind. If the graph will be used to make a slide, it should be simplified. If possible, limit the number of curves or bars shown to two for visual aids. Graphs showing trends need not be as large as graph provided with data points to be used in calculations.

The scale (or aspect ratio) of the graph also affects the way data are perceived. A ratio of height to width of 3:4 is common. Designers may need to experiment, changing the scale of a graph to determine the most effective aspect ratio for its use. Too great a range of values causes the graph to appear compressed. Some spreadsheet packages will allow the use of more than one vertical axis scale. Label them clearly to avoid confusion. Type fonts (e.g., Helvetica, Times Roman) should be consistent with those used on tables and in the text if possible. The use of different sizes and weights of the same font type works well.

Engineering Drawings

Engineering drawings of roads, intersections, signs, markings, and other traffic features are often included as illustrations in technical reports. Photoreduction of drawings is often necessary to fit the standard page size of the report. Letter sizes and line widths on the original drawings should be selected so that the reduced copy is legible and readable. Fine lines and light-stroke letters tend to disappear during reproduction. Remember to show the scale of the drawing accurately on both the original and reduced versions. Bar scales work well for most illustrations. Clearly label any drawing that is not prepared to any specified scale as "not to scale." If an original drawing has been produced using CAD, it is fairly easy to control the scale of the plotted graphic.

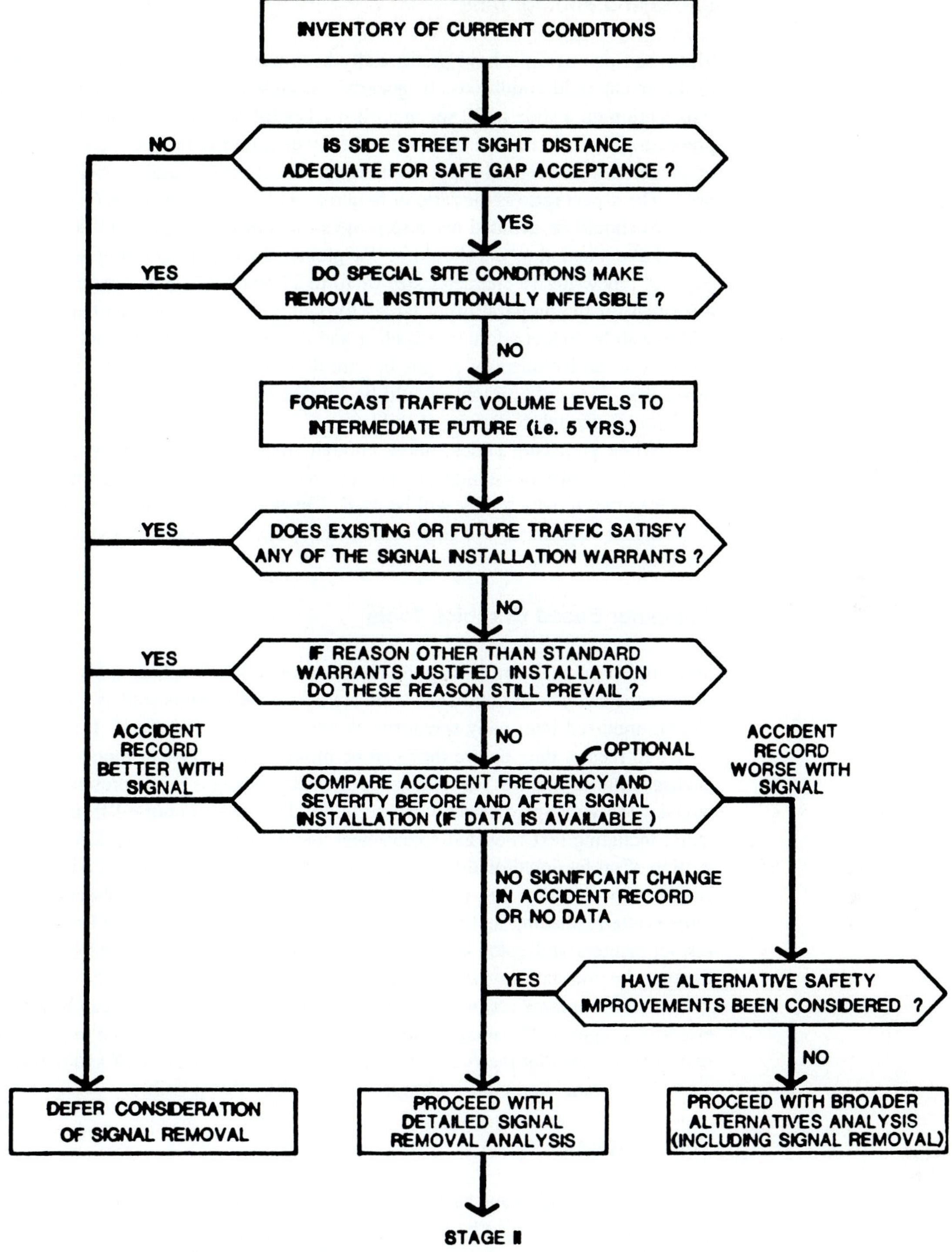

Figure D-17 Flow diagram. *Source:* J. L. Kay, L. G. Neudorff, and F. Wagner, "Criteria for Removing Traffic Signals and Users' Guide, (FHWA-IP-80-12) Alexandria, VA; JHK Associates Inc. and Wagner-McGee Associates, 1980.

Slides and Photographs

Slides and photographs have traditionally been used to strengthen oral presentations and to document field conditions. In general, an audience should be able to assimilate the information on a slide in 30 seconds. If it takes longer, the slide should be simplified by presenting the information on a series of slides rather than a single slide. Slides should be prepared from original copy. Commonly, aspect ratios ranging from 1:1.3 to 1:1.5 are used. The aspect ratio is the ratio of height to width. Vertical slides (aspect ratio = 0.66 to 0.75) should be avoided because projection equipment may need to be readjusted.

Letters and numbers should be 2% greater than the larger dimension of the original copy. A typewriter (a character size of 10 to 12 characters per inch or 10 to 12 point) is too small for the preparation of tables for slides. Be consistent on the use of type fonts within a slide. Select a font that is clear and readable. For clarity, captions and details do not need to be included on graphs but are described by the speaker. Tables should have fewer than 15 words; graphs should be limited to two curves if possible. Lines should wide (bold) enough to ensure visibility.

When preparing slides, allow enough time before the presentation to check the slides and remake those that are not effective. Display the slides in an environment similar to that in which they will be used. Check to see that all text is legible, that there is sufficient contrast to see important details, and that the slides communicate the desired message.

Computer-Based Graphics Tools

Spreadsheet graphics are among the most common data graphing tools. Spreadsheets emphasize the analysis of data. Data from statistical software as well as text (or ASCII) files can be imported into many spreadsheets with a special command. It is usually not necessary to reenter data before they can be plotted. A number of report graphics software packages are available. The emphasis of these packages is on presentation rather than analysis (as with spreadsheets). Some allow data to be imported from various file formats, including text files and spreadsheet and database managers. CAD allows the generation of professional-quality line drawings for slides and reports. CAD drawings can be generated with either pen plotters or laser printers capable of full-page graphics. With laser printers, shading and patterns can be used to enhance simple line drawings. Color ink jet printers and color laser printers are becoming more affordable and their use is likely to be common in the near future.

Paint programs create and manipulate graphics images that can be interesting and effective graphics. The images created by paint programs are stored in a bit-mapped format. This means that the files cannot usually be directly imported and manipulated with a CAD program. Some utility programs can make this possible.

Video Capture

In recent years, videotape has been used increasingly to log roadway inventories. Images from videotapes can also be captured and used to make slides for presentation. The images are captured and annotated using a software and hardware system installed in a microcomputer. The slide layout is similar for photographic images and video images. Video images, however, are generally of lower resolution (quality) than slides generated using a 35mm camera.

Scanned images and captured video frames are being used increasingly in presentation graphics alone or as a background for other computer-generated graphics. CCD (charge-coupled device) video images can be captured using a video capture card and software. A digitized image is a bit-mapped file and, like paint program files, cannot be imported directly into most CAD programs. Some CAD programs may allow the use of bit-mapped files as background or ''reference'' layers. Bit-mapped images can be manipulated only using PAINT software or other software capable of working with the file formats (video capture cards commonly come with a program capable of annotating these captured images). There is also a variety of file formats used by various software packages. The file formats are not directly compatible; however, conversion programs are readily available that will allow files to be transferred among applications software.

Captured video images and scanned images are also difficult to print. While they may be captured at fairly high resolutions and in color depending on the computer system, most common laser printers produce only 300 to 400 dots per inch (dpi) in black and white. The quality of paper output is generally disappointing. Slides, however, can be generated in color directly from a color computer display monitor using hardware produced by several different manufacturers. The resolution of these slides, while less than that of a slide produced with a 35mm camera, is adequate in most cases.

USE OF COLOR

Color is a very powerful tool in illustrations. Color influences the viewer's perception of the illustration. How people interpret color depends on their background, the current use of color in advertising, the part of the country they live in, and most important, what is currently in style. Colors can induce feelings (sometimes very strong ones) in the viewer that are not related to the illustration or its subject. These feelings can support or weaken the intent of the presenter.

Color is normally used on illustrations to (Turnbull and Baird, 1975):

- Attract attention
- Produce psychological effects
- Develop associations
- Build retention
- Create an aesthetically pleasing product

There are some general rules for using color in pleasing ways. A color wheel (which shows the 12 basic hues or colors) is a very useful tool in choosing colors. It can be purchased in any art supply store. The primary colors are red, yellow, and blue. The other (secondary) colors are produced from these three primary colors. Cool colors (blue and predominately blue) are relaxing and appear to recede on the page. Warm colors (red, or red and yellow) are stimulating and advance to the foreground. Green, brown, and red-purple lie between the warms and cools and are therefore relatively neutral.

An important use of color is to attract attention, primarily through contrast. Color should be applied to the elements of greatest significance. A bright color used with black is the most effective. Complementing warm and cool colors can also be used sparingly. Blue appears to be the most popular color. The colors used in an illustration should fit the overall mood. Too much color can detract from rather than improve the readers understanding of the graphic.

GRAPHICS CHECKLIST

It takes time and experience to produce the most effective graphics. Slides (or overheads) often make or break a presentation. Before finalizing illustrations, go over the checklist below.

1. Are the illustrations suitable for the intended audience?
2. Are the illustrations legible?

3. Is the message clear?
4. Is the layout neat and properly aligned?

REFERENCES

BORLAND INTERNATIONAL INC. (1989). *User's Guide for Quattro Pro,* Borland, Scotts Valley, CA.

BOX, P. (1976). *Manual of Traffic Engineering Studies,* 4th ed., Institute of Transportation Engineers, Washington, DC.

INSURANCE INSTITUTE FOR HIGHWAY SAFETY (1991). *Facts,* IIHS, Washington, DC.

KAY, J. L., L. E. NEUDORFF, AND F. WAGNER (1980). *Criteria for Removing Traffic Signals and Users Guide,* FHWA-IP-80 12, JHK and Associates Inc., and WagnerMcGee Associates, Alexandria, VA.

MICHAELSON, H. (1987). *How to Write and Publish Engineering Papers and Reports,* 2nd ed., iSi Press.

NATIONAL SAFETY COUNCIL (1986). *Accident Facts,* NSC, Chicago.

PANIATI, J. F. *Legibility and Comprehension of Traffic Sign Symbols,* Research Report, Federal Highway Administration, McLean VA.

TURNBULL, A., AND R. N. BAIRD (1975). *The Graphics of Communication: Typography, Layout and Design,* Holt, Rinehart and Winston, San Francisco.

WHITE, J. (1988). *Graphic Design for the Electronic Age,* Watson-Guptill Publications, New York.

E

Written Reports

H. Douglas Robertson, Ph.D., P.E.

INTRODUCTION

The results of many well-planned and well-executed engineering studies have been misunderstood, misconstrued, or disregarded because of poorly written reports. Considerable resources are wasted if the study findings, conclusions, and recommendations are not clearly conveyed to those charged with acting on the study's results.

With careful attention to a few writing principles, reports of engineering studies can be powerful tools to inform and aid in the decision-making process. Regardless of the size or importance of the study, the same fundamental principles apply.

1. Write to the level of the intended audience.
2. Use clear, concise language, not technical jargon.
3. Present findings and conclusions in a logical sequence.
4. Clearly show how the findings support the conclusions and recommendations.
5. Use figures and tables to portray the most important results.
6. Get to the point.

These principles are amplified in the sections that follow. Remember, good engineering must be accompanied by good reporting.

STYLE OF THE REPORT

Target the Audience

Adapting the language, style, and presentation of a report to the intended audience is not easy. The writer must first know the audience before writing to them effectively. This is true not only of readers known personally, but also for audiences when little information is available concerning their knowledge of and interest in the subject.

Many transportation engineering reports are developed for engineers, planners, or public officials with some knowledge of or experience with transportation issues. In these reports, less effort is required to define commonly used transportation engineering terms. If the report is prepared for the general public or a special-interest group, however, define terms that may be unfamiliar to them either in the text or in a glossary. In all cases, define acronyms the first time they appear in the report.

Start with a Summary

All reports should begin with a brief summary of the purpose, important findings, conclusions, and recommendations to serve readers who do not have time to read the entire report. The summary may range in length from a paragraph in a short report to an executive summary of several pages in a longer or complex report. A one-page summary is usually most effective.

How Long Should the Report Be?

The purpose of the report generally dictates its length. The merits of a report are not measured by its thickness. Clarity, conciseness, and completeness should be the guiding principles. The body of the report should summarize the methods employed, the data collected, the analyses performed, and the results or findings. Detailed supporting data should be placed in appendices. Generally, the shorter the report, the more likely it is that it will be read.

Make the Report Attractive

A report should be attractive to the reader. The appearance should be neat, the tone authoritative and professional, and the content accurate. It should be free of spelling and grammatical errors, which tend to detract the reader from the report's content. Break lengthy sections of text by using headings and subheadings liberally and by inserting tables, figures, photographs, or lists to portray results or present data. Make the report a convenient size (e.g., 8 ½ by 11 inches). Where possible, design tables and figures so they can be read without turning the page sideways. In addition, a standard page size is easier and cheaper to reproduce.

ORGANIZATION OF THE REPORT

The components of a transportation engineering report are determined by the report purpose, length, and complexity. In general, shorter reports have fewer components. The following is a list of components, any or all of which may be included in a given report:

- Letter of transmittal
- Title page
- Copyright notice

- Disclaimers
- Forward or preface
- Acknowledgments
- Table of contents
- List of tables
- List of figures
- Summary or executive summary
- Body of the report
- List of references or bibliography
- Appendix
- Glossary
- Index

The sequence presented follows normal conventions but may be altered to fit a given situation. Each component is discussed briefly below.

Letter of Transmittal

A letter addressed to the person or agency for whom the report is prepared is often attached to the front of the report. The representative of the organization that conducted the study and prepared the report signs the letter. This letter is usually a short statement that the signer is pleased to submit the report in the number of copies required. It may also briefly explain how and why the study was conducted and its single most important outcome. The principal purpose of the letter is to document the submission of the report. Copies of the transmittal letter may be included in copies of the report.

Title Page

The title page includes the name of the report, the date the report was prepared, the authors' names, and the sponsors' names.

Copyright Notice

If the report is copyrighted, this notice appears on the title page or the page that immediately follows. Authors can obtain information and an official copyright application form from the Register of Copyrights, Copyright Office, Library of Congress, Washington, D.C. 20025.

Disclaimers

In general, a disclaimer alerts the reader that the results, findings, conclusions, and/or recommendations presented in the report are those of the author(s) and do not necessarily reflect the opinions, views, or policies of the sponsoring agency or organization. Disclaimers most frequently appear in research reports. Disclaimers may also include explicit information about the report that might otherwise be assumed incorrectly.

Foreword or Preface

A foreword or preface provides the history leading to the report and, if applicable, its relationship to other reports. A foreword may be written by the author(s) of the report but is commonly written by a representative of the sponsoring agency or organization. A preface is generally written by the author(s).

Acknowledgments

This section recognizes the persons and agencies who assisted or contributed to the study. Acknowledgments may also appear in the letter of transmittal.

Table of Contents

Major divisions, sections, or chapters of the report are listed in order of appearance along with the page number beginning the division. Subdivisions of a section or chapter may also be listed in the table of contents.

List of Tables

The number and title of each table and the page number where it is located are presented in order of appearance in a list of tables. The inclusion of this list depends on the length of the report and the significance of the tables as stand-alone presentations.

List of Figures

Figures are listed in a similar fashion to the tables. If tables are listed, figures should also be listed, and vice versa. The list of tables precedes the list of figures.

Summary or Executive Summary

This section briefly summarizes the purpose of the study, major findings, conclusions, and recommendations. It is designed for those who do not have time to read the entire report and as a means of refreshing a reader's mind at a later time. Because it is almost certain to be read, it is the most important section of the report. The summary must not overstate or understate the findings or cause the reader to take conclusions or recommendations out of context. A typical summary ranges from one to five pages. However, an executive summary for a complex, multivolume report may be 25 to 30 pages long and be published in a separate volume.

Body of the Report

The body is the heart of the report and is supported by all of the other components in this list. The organization and content of the body are discussed in a later section.

List of References or Bibliography

When material from another report or book is used, it should be properly acknowledged. Authors should cite the document in a list of references and refer to the list at the places in the text where the material is used. Place the list either at the end of the report or at the end of each chapter. Another option is to cite each reference in a footnote on the page where the material is first used. Examples of reference listings can be found at the ends of the chapters in this manual.

A bibliography is a list of books and reports that contain materials that are useful to readers who want to pursue the subject matter further. Reports describing transportation engineering studies do not usually contain bibliographies unless the author wants to encourage further study or explanation.

Appendix

Detailed material that supports but is not essential to the body of the report should be placed in an appendix. Appropriate appendix materials include supporting data, detailed explanations of methodologies or procedures, derivations of formulas, conversion factors, lists of symbols, data collection formats, data collection protocols, and checklists. Appendices are an effective means of fully documenting and supporting the results of a study without cluttering the body of the report.

Glossary

If the report is written for nontechnical readers or if new and unfamiliar terms are introduced, a glossary or list of definitions should be included in the report. If few new terms are used, footnote definitions may suffice.

Index

Transportation engineering reports seldom contain indices unless the report is voluminous or intended for frequent reference. The index is more exact and detailed than the table of contents. The index lists all major subjects alphabetically along with the pages where each subject is addressed in the report.

BODY OF THE REPORT

The body of the report is composed of a series of sections or chapters. In a transportation engineering report, the body contains answers to the following questions:

1. What was the objective or purpose of the study?
2. Why was the study necessary?
3. When, where, and how was the study conducted?
4. What were the findings?
5. What conclusions were drawn?
6. What recommendations were made?

Organizing the Body

The first step in effectively arranging the body of the report is to construct an outline. The outline should list the chapter, section, and subsection headings and may contain notes or topics to be covered under each heading. The outline helps the writer organize the body in a logical, comprehensive, and complete manner, and makes the task of writing easier. There are a number of ways to organize and present the material in the body of the report. Some of them are described briefly in the following paragraphs.

Sequential Logical Statements Proceeding from Problem to Solution

The problem is stated in the opening chapter or section. The recommended solution in logical steps are developed in succeeding chapters.

Cause and Effect

The report begins with a description of the causal factors surrounding a problem. Then the report addresses the resulting effects, as indicated by the appropriate measures of effectiveness (MOEs). For example, the report of a congested route study might begin with a description of traffic volumes, vehicle classification, occupancy, and roadway geometrics. The effects of these factors would then be described in terms of measured speed, delay, level of service, and accidents.

Time Sequence

Previous conditions are addressed, followed by present conditions, followed by projected future conditions.

Problems in Order of Importance

Each problem and proposed solution (if known) is presented in order of priority beginning with the most severe.

Location

Each intersection or roadway section (link) is described in turn. The order of presentation may be based on type of road, size of facility, area of the city, magnitude of the problem, or other logical scheme.

Order of Audience Interest or Familiarity

Treating the more popular subject first, regardless of its importance, may be a way to gain the readers' interest and lead them into less familiar subjects.

Grouping Similar Subjects Together

In this approach, chapters are grouped by subject area. For example, planning chapters may be followed by design chapters, which in turn, are followed by operations and safety chapters.

Tips on Body Content

Format

The body of a typical transportation engineering study report generally includes:

- Purpose or objectives of the study
- Background (i.e., what led to the study or why was the study needed)
- Scope of the study (i.e., what limits were placed on the study)
- Methods used
- Data collected (e.g., type, amount, when, where, etc.)
- Analyses performed
- Findings
- Conclusions
- Recommendations

Other subjects that may be appropriate for a transportation engineering study report include:

- Alternatives developed or examined
- Selection of alternatives, traffic control devices, or routes
- Evaluation results
- Cost analysis or financial impact
- Environmental impact
- Traffic impact
- Implementation plans or recommendations

Use Simple, Clear, and Concise Language

Remember, a transportation engineering study report is a presentation of technical facts and their implications, not literary art. On the other hand, write the report to hold the reader's attention. Do not use complicated wording in an effort to "sound important." Most readers prefer simple, common language.

Use a Good Style Manual

The elements of style in writing technical reports vary from one manual to the next. Pick a good manual, adopt a set of conventions, and maintain uniformity in spelling, grammar, punctuation, and format throughout the report. Several good technical writing style manuals are included in the references at the end of this chapter. Agencies such as U.S. Department of Transportation and the Transportation Research Board also maintain technical writing style manuals.

Page Numbering

Number every page except for the title page and the letter of transmittal. The summary or the first page of text can be "Page 1." Advance pages are numbered with lowercase Roman numerals. Appendices may continue the numbering of the main body or can be numbered A-1, A-2, and so on. Place page numbers at the top or bottom center of each page.

Break Up Lengthy Narratives of Pure Text

Headings and subheadings lead the reader through the author's train of thought and presentation of facts. Make ample use of summary tables and figures. Photographs may also be useful in describing certain situations or conditions. See Appendix D for specific guidance on constructing tables and figures.

Explain the Methodology Used

The credibility of the results of a transportation engineering study often hinges on the type and amount of data collected and the methods used to analyze those data. Therefore, the report should clearly, yet concisely, provide the reader with sufficient information about the data and methodologies employed to establish an acceptable level of confidence in the findings. Explain the application of standard methods by referencing a text or manual on the subject. Special methods require more explanation that may be placed in an appendix to avoid cluttering the body of the report. The sources of data and information used in the study should be clearly identified. Authors should also specify the amount, place, time, and conditions under which field data were collected.

Relationship of Findings to Conclusions and Recommendations

This relationship is similar to the construction of a house, as illustrated in Figure E-1. The findings are the foundation on which the conclusions rest. The conclusions, in turn, support the recommendations.

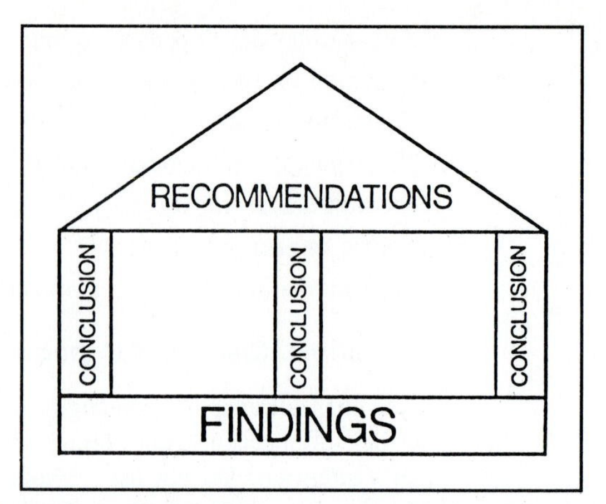

Figure E-1 Relationship of findings to conclusions and recommendations.

Reviews and Editing

A good technical writer, a panel of reviewers, or an experienced editor should review the report, if possible. This simple practice often turns a dull, listless recitation of facts and figures into an interesting, understandable, and perhaps even entertaining piece of writing. The goal for any report is that it be read and understood correctly. A report that readers ignore or misunderstand has no value and the work and effort to produce it will have been wasted.

PRODUCTION OF THE REPORT

The bound final report represents the product of the work, effort, and expense involved in conducting the study. It should therefore reflect the quality of the overall effort. Given the availability and affordability of word processors, desktop publishers, laser printers, and copiers, high-quality report reproductions are possible at a reasonable cost.

Duplicating

Three means of duplicating reports are photocopying, printing, and mimeographing. Photocopying is the most common method of producing transportation engineering reports, particularly those with few pages and a small number of copies required. Almost all reports are typed on word processors. Readily available software packages can provide editing, spell checking, and thesaurus assistance and can produce high-quality graphics to depict data displays, charts, and diagrams. Letter-quality and laser printers produce high-quality masters for use in photocopying. Copy centers usually provide fast and convenient services, including color copying.

Printing is generally cost-effective for large reports or when agencies need many copies. Printing requires a high-quality "camera-ready" master copy, but produces a very high-quality product that can include the use of color. Mimeographing is practically obsolete. Although inexpensive, it requires a stencil of the master copy and produces a low-quality product. If the equipment is readily available, however, it may provide an organization with copies suitable for internal use.

Binding

Reports may be bound by staples; metal fasteners; tape; plastic, metal ring, or velo binders; glue; or thread. The type of binding chosen depends on the thickness of the report, the expected amount of handling, and the cost. Plastic spiral ring binders are inexpensive and convenient but do not stack well. Velo binders provide a neat, compact binding method for small reports but may hold pages too tightly for frequently used documents. Staples and tape are the least expensive binders and work well on reports less than $\frac{1}{2}$-inch thick. However, staples tend to pull out of thick reports that are used frequently. Large reports requiring many copies and a degree of permanence should be hard-bound with glue or tape. The method of binding should be commensurate with the quality of the product it holds together.

REFERENCES

EDITORIAL STAFF OF THE UNIVERSITY OF CHICAGO PRESS (1982). *The Chicago Manual of Style,* 13th ed., University of Chicago Press, Chicago.

HODGES, J. C., M. E. WHITTEN, W. B. HORNER, S. S. WEBB, AND R. K. MILLER (1990). *Harbrace College Handbook,* 11th ed., Harcourt Brace Jovanovich, San Diego.

KRAMER, M. G., J. W. PRESLEY, AND D. C. RIGG (1985). *Prentice-Hall Workbook for Writers,* 4th ed., Prentice Hall, Englewood Cliffs, NJ.

MICHAELSON, H. B. (1987). *How to Write and Publish Engineering Papers and Reports,* 2nd ed., iSi Press.

F

Presentations

Donna C. Nelson, Ph.D., P.E.

INTRODUCTION

Oral reports and presentations are an important means of communication for traffic engineering professionals. Presentations range from simple progress statements to extensive reports on traffic studies. All effective communications must fit the requirements and experience of the audience. This is especially important in oral presentations. Define all unfamiliar terms. Arrange and present the main points of the presentation so that they are easily understood and allow a logical conclusion made. If a listener cannot follow the presentation with at least a reasonable degree of understanding, it makes little difference how thorough the investigation was. There is little chance for a listener, once lost, to catch up. This appendix presents some general guidelines that help to make a good presentation. Many of these are obvious but are often forgotten in the rush of preparing material for presentation.

PURPOSE AND SCOPE

Effective oral presentations are not condensed versions of written reports, nor are they speeches read aloud. Each presentation should be prepared for the specific conditions under which it is to be given and adapted to the presenter, the audience, and the presentation environment. The following five basic characteristics distinguish effective oral presentations from written reports:

1. Specific audience
2. Limited scope
3. Personal presentation
4. Need for instant understanding
5. Limited time for presentation

Audience

A number of people with widely different technical backgrounds may read a written report. Therefore, the writer must try to adapt the writing to a broad level of reader understanding and interests. The oral report, however, is usually prepared for a specific audience. A speaker must orient the report to the particular interests and levels of understanding of the audience. Begin the presentation talk with a foundation that the listener will understand and is interested in. Do not talk down to the audience; be careful to avoid the trap of condescension. Remember that the listeners have given up their time to come to hear your presentation. Do not waste their time.

Scope

Readers may read the entire written report or only certain portions. However, an oral presentation is usually prepared with a specified time limit in mind. The scope of the material must be appropriate to this limit, which often includes a question-and-answer period. The presentation should include a clear explanation of the subject and the general conclusions. Use brief summaries of substantiating data when they are critical for understanding the report. A brief description of the procedure is usually all that is required. If a written report is available, the talk is probably best limited to the most important parts of the written report, concentrating on those items that are of greatest interest to the audience. It is better to present less and cover it thoroughly than to discuss too much and have the listener feel that you did a superficial job.

Presentation

The medium of the oral report is the speaker. Therefore, the delivery style of the speaker is critically important. Posture, gestures, eye contact, and facial expression, voice projection, enunciation, pronunciation, and the degree to which the speaking style seems relaxed, conversational, and expressive all affect the attitude and receptivity of the audience. Four styles of presentation delivery include extemporaneous, impromptu, manuscript, and memorized techniques (Michaelson, 1987). The *extemporaneous* talk is usually preferred for the oral presentation of a technical report. In this method, the outline of the speech is carefully prepared but not committed to memory. The *impromptu* (off-the-cuff) method is obviously inappropriate for the presentation of the report; however, it is an appropriate style for the question-and-answer session. Prepared and read *manuscripts* or *memorized* methods tend to be inflexible and sound artificial. Practice, however, will improve the quality of any presentation. Oral presentations frequently include questions from the audience. Questions may be asked during the presentation or be deferred to a question-and-answer period that follows the formal presentation. While each method has its advantages, the most satisfactory arrangement is to defer questioning until the formal presentation has been completed. Potential questions are often answered during the remainder of the talk.

The presentation should be made in an appropriate location. Ideally, the space needs to be comfortable and sized for the number attending. It should contain all the necessary equipment for the presentation, usually a lectern with a light, a chart stand, a black (or white) board, and projection equipment with easily adjusted lighting. The space needs to be reasonably quiet and free of distractions.

Instant Understanding

If some portion of a written report is not immediately clear, the reader may reread it, refer back or ahead, or even consult other sources. However, the spoken word only lasts during the moment of presentation. The speaker must be exceedingly careful to be as clear as possible at all times, paying particular attention to audience reaction. Instant understanding is facilitated through voice, language, and the use of transitions and summary statements.

Proper use of voice includes adequate projection and distinct enunciation to permit even those persons seated farthest away to hear and understand the material presented. The speaker should use pauses freely to break up the flow of ideas into meaningful thought units. Also, speaking with sufficient forcefulness and using a variety of voice inflections avoids monotony and gives life and meaning to words. A speaker must be sensitive to the rate of delivery and must pace the various remarks for understanding, variety, and emphasis. Practice is essential to master all of these techniques.

The audience must understand the vocabulary used. If the readers of a written report encounter an unfamiliar word, they can refer to a dictionary. In an oral report, the speaker must define unfamiliar terms for the listeners. Obviously, the problem varies from one situation to another, but the speaker should be sure that the listeners understand the terminology. A broad vocabulary permits an expression of thoughts with clarity and precision in a vigorous and colorful fashion. Transitions and summaries help to guide the listener through the development of the presentation topic. After a section of the report is developed, summarize the main points briefly before moving on to the next section. Also, show the relationships among various sections.

Time

Organize and rehearse the presentation for the established time limit. The average rate of speaking is 100 to 150 words per minute. Timing can be checked by practicing with audiotape, videotape, or with a live audience. If available, videotape is the most effective. It gives the presenter the opportunity to see how he or she is coming across and how to improve his or her presentation skills. Simply speaking aloud helps to check the organization of the talk; however, when practicing in an ''empty room'' there is a tendency to speak much faster than on tape or to an audience. Rehearsal will help the speaker achieve a relaxed, at-ease posture, and a smooth, confidant delivery.

ORGANIZATION OF THE PRESENTATION

A well-conceived talk, carefully tailored to fit the audience, will fail if it is difficult to follow. A confused reader of a report can regress as needed to wade through an obscure passage; a confused listener is likely to be lost forever. Technical reports are organized around three major divisions: the introduction, the body, and the conclusion.

Introduction

The introduction prepares the audience for the body of the presentation. The introduction should motivate the listener, catch the interest, and inform about the content of the report. Motivation is accomplished by dwelling briefly on the importance of the subject and by establishing a distinct impression that something worthwhile is to be offered.

Listeners can be informed about the topic of the presentation by:

1. Identifying and defining the general subject
2. Providing the necessary background
3. Giving a preview of the main divisions to be covered in the presentation

A brief introduction sets the stage and provides prospective on the problem, its importance, and the state of knowledge about it. During the introduction, lights should be on and eye contact established.

Body

The body is generally organized in terms of the steps that were involved in the study. However, the report does not always need to follow the entire sequence. Sometimes the talk may be limited to analyzing the causes of a problem or explaining and evaluating the solution to a problem. The discussion of each phase should close with a statement that gives the subconclusions for that phase.

Conclusion

The conclusion normally fulfills the following three main functions:

1. Summary of the various subconclusions presented during the presentation
2. General conclusions that are drawn from the subconclusions
3. Recommendations that arise from the general conclusions

The conclusion section summarizes the material and may provide a transition to the question-and-answer period. It should flow logically and clearly from the material presented in the body. Surprise endings may be dramatic, but are rarely useful.

VISUAL AIDS

Simple, legible visual aids are essential to most transportation reports. The basic rule for designing visuals is to make them bold, simple, and colorful (Berthouex and Hindle, 1981). For most subjects, visual aids serve two purposes: They increase the understanding of the audience and help to guide the speaker through main points of the talk. Coordination of text and slides is very important. Each visual should make a contribution to the important points of the presentation. If the speaker does not refer to the visual, the listeners will assume that the speaker has forgotten something. A speaker should not assume that the audience intuitively grasps a visual's meaning; rather, one should succinctly point out the salient features, defining the meaning of any nonstandard phrases or symbols.

A common mistake is to show too much material on one visual. An audience should be able to read and understand a visual in less than 30 seconds. For complex items the best solution is to break down the material into a number of visuals with lettering and numbers as large as possible. For example, tables should have no more than a total of 10 or 15 words and numbers (20 is an absolute maximum recommended by some). Limit the length of listings or bulleted lists to 20 characters per line and eight lines per page. Limit the number of curves and bars on a graph to two. Captions and details may be simplified, and in some cases omitted, from graphs and illustrations used for slides. Details are described during the presentation.

The use of a greater number of simple visuals appears to be more effective than a few complex slides. Audiences may get bored staring at a slide they have assimilated, particularly if the speaker has moved on to another topic. Try out slides and other visual aids in a situation similar to that in which they will be presented, preferably before a test audience. Look not only for the content and clarity of the visuals, but the organization as well. A wide variety of visual "media" are available for use in presentations. Appropriate use depends on the nature of the presentation, the presentation environment, the expectations of the audience, and the skills of the presenter.

Slides and Viewgraphs

Unusually formatted or cropped slides, such as square, wide, or vertical images, tend to alert the audience. They make a strong impact when used in moderation and can be used very effectively to emphasize a point. Be careful: vertical slides will require readjustment of projection equipment. Use them only when you are sure that the setup does not require readjustment during the presentation. A standard slide tray holds 80 slides. For most presentations, time, not space, will be the limiting factor.

Computer-Aided Presentation

Computer-aided presentation tools are becoming widely available. While they can be used to create effective presentations, there are a number of pitfalls as well. Computer-aided presentation tools are those in which the computer is used as an integral part of a presentation. Visual displays are generated by the computer and are projected to the audience using specialized equipment. Projection techniques are discussed below. A major drawback is that computer equipment must be available at the presentation location. Borrowing computer equipment can be a risky venture, due to incompatibilities that may arise from video display cards, operating system versions, and other hardware/system configurations. Computer presentations that require the speaker to enter data into the computer are usually best when limited to very small interested groups or for demonstrating equipment or software. If the use of computer-generated output is an important part of the presentation, use a computer-based slide show if at all possible.

Computer-Based Slide Shows

Specialized software can be used to create and execute computer-driven slide shows. In general, software recalls and projects graphics or text file images to the screen. These slide shows can be pretimed or user timed. These images are then projected to the audience using projection equipment that is compatible with the computer system and software being used. These computer slide shows can usually be converted to videotape for easier showing and to avoid moving computer equipment. The creation of computer graphics is discussed in Appendix D.

Animated Presentations

Animated, computer-based presentations are being used in the business world. Although they are impressive, at present they are time consuming to create. Their best use is in situations where the presentation is going to be used repeatedly, or for very important presentations where cost is no factor. The rapidly improving capabilities of three-dimensional modeling systems and multimedia software/hardware systems suggest that these technologies will play an increasing role in the future. Some current CAD programs allow designs to be displayed in three dimensions and can give the user the feeling of "walking" through a project. Prototypes of new, virtual reality systems allow users to feel as though they are part of the system. The use is fitted with a special helmet and with sensors that allow the system to react to the user. An example is a system that allows a bicyclist to view a CAD-generated road environment, with the perspective and sensation of speed changing with the rider's rate of pedaling and head position. New multimedia systems that integrate video clips, laser disk, and other computer-generated graphics are becoming less expensive and more powerful.

Computer-Compatible Projection Equipment

Computer-generated images are projected in a variety of ways: by making slides, viewgraphs, or videos from computer images or by direct projection. Standard 35mm slides are made by directly photographing the screen, by photographing hard-copy images from a printer or plotter, or from special equipment that creates the slide directly from a computer file. The latter is usually the best and also the most costly method. Viewgraphs can be made from hard-copy images or from special equipment. Images can be projected directly from the computer screen using two types of technology; an LCD display device fitted to a low-wattage overhead projector or a three-color projector. LCD display equipment is the least expensive. Color LCD projectors are new on the market; however, the colors projected tend to appear faded. Three color projectors project very high-quality color graphics. Although quite expensive, these projectors are excellent when the presentation requires using graphics, CAD, or GIS software as part of a presentation. They can also be used to project videotapes.

Video Presentations

Videotapes can also be integrated into a presentation. Short video sequences of an accident scene or other situation may make a positive contribution to a presentation. Comprehensive video presentations should be professionally produced; videotapes produced by amateurs are almost always of poor quality. These presentations, usually intended as background information or public information tools, must be used repeatedly to justify their expense.

Presentation Boards, Flipcharts, and Chalkboards

Chalkboards and white boards are effective for discussions or presentations that involve audience participation. Presentation boards and flipcharts on easels are typically of two types: prepared media, and materials developed during the presentation or discussion. This proven, inexpensive technology can be quite effective for presentation before small groups. These methods become more limited before larger groups because of visibility. Choose colors for their visibility and contrast. Some colors, such as yellow, will fade under certain lighting conditions. Use broad-tipped pens to increase legibility.

If materials are developed during the presentation, they must be printed legibly. Lettering should be large enough to allow everyone in the audience to read. The presenter should also have adequate spelling skills. The general format and organization of material to be developed during a discussion should be planned in advance. Although these techniques are not high-tech, they also do not suffer from equipment failure and incompatibilities.

ANSWERING QUESTIONS

Often the most difficult part of a presentation is fielding questions. If the audience is hostile or upset, this time can be particularly challenging. It is extremely important to anticipate questions and objections that may arise during and after a presentation. Responses should be rehearsed so that the speaker appears to know the topic well and can remain cool and thoughtful even in confrontational situations. The following guidelines suggested by Berthouex and Hindle (1981) provide a few thoughts on handling a question-and-answer session.

1. Be brief; do not say more than is required. A brief answer can be unequivocal. It carries a punch. Many questions can be answered in one word: yes or no. By simply saying yes or no you save time and your answer will not be misunderstood. It may not be necessary to respond to editorial comments, even if you disagree. Listen carefully to determine if confirmation or clarification is expected; avoid giving a nonanswer to a nonquestion. If the questioner wants to know more, he or she will ask you to explain. Do not give a lengthy explanation of the problem and then say, "The answer to your question is yes." First answer yes and then add supporting details if, and only if, they are truly important. Go quickly to the point.
2. Do not use an opening line of nonsense patter to gain time to think of your answer. For example, do not start with, "Let me just think out loud for a minute" or "That is a very difficult question." Pause briefly before starting to speak if you need a moment to organize your ideas. Repeating the question clearly will allow you to gain some time and confirm that you understand the question. Do not assume that the questioner will deduce the answer correctly from a long discussion of pros and cons. Be explicit. If your answer must be long or complex, make a brief summary statement.
3. Do not bluff. If you do not know the answer to a question, the following choices are available: "I am sorry, I cannot answer your question" or "If you please, I would like to consult my notes (references, partners) before I answer that question." "I cannot answer that now. I will have the answer this afternoon (tomorrow, next week)."
4. Don't be evasive. Answer the question you are asked, or do not answer. Do not answer a different question. The questioner will not be impressed or fooled. Avoid hedging. Sometimes the answer is not simple; it must be qualified or have limits set. In these cases, try to make the limits quantitative and make the qualifications precise.
5. Be specific. Use precise words. Favor citing a number over using a descriptive phrase. Never assume that you and the listeners have the same scale of reference for imprecise words (words such as *deep, high, warm, small*). Cite data or numerical facts whenever possible.
6. Do not try to answer a question that you do not understand or one that is imprecise. If the question is unclear, your answer will be unclear. Even worse, it may be wrong. Do not guess at what the questioner has in mind, ask for clarification. You may rephrase the question slightly, but do not change it. Ask for confirmation that your interpretation is correct. If you are answering questions from a group and receive a comment far off the point, or far wrong, decide if it is best to assume that others also recognize this and move on as politely as possible. Do not let people put words into your mouth; however, do not be argumentative or insulting. Above all, remain calm. If you are asked a hostile question, restate it using positive phrasing. Avoid single or double negatives. It may be helpful to restate a negatively phrased question or one that uses angry words or emotionally loaded terminology into one that is clear, precise, and contains less inflammatory language.
7. Never use slang and avoid jargon. Slang is imprecise and it can give the impression that the speaker is poorly educated. Jargon is vocabulary known to a group of specialists but generally unfamiliar to outsiders. Jargon can be often be replaced by a few more common words.

PREPARATION AND PLANNING CHECKLIST

Preparation and planning will help ensure a quality presentation. During the process of planning and giving a presentation, the following items will help assure success.

Organization

Well before the presentation date, consider the following:

- Does the introduction explain why the subject is significant?
- Are the major points covered without excessive detail?
- Will the audience understand all the terminology? Is the talk simply arranged in a logical sequence? Are the visuals simple and visible?

Rehearsal

- Rehearse the presentation until you feel comfortable with the material.
- Rehearse with the visuals you intend to use.
- Identify and practice fielding possible questions.

At Presentation Time

On the day of the presentation, arrive early enough to check the following:

- Does the sound equipment function; do you know how to use it?
- Is the room equipped with the necessary projection equipment; is it in working order?
- Is the temperature of the room neither too warm nor too cold?
- How is the lighting adjusted?
- Is the seating arrangement right for your needs?
- Will everyone in the audience be able to see your visual aids?

It is too late to correct any deficiencies or learn how to use the equipment when walking to the podium. Rest assured that the first time you fail to confirm that the equipment is in place and working, it will not be.

REFERENCES

About the Presentation of "Your Paper," American Society of Civil Engineers, New York.

BERTHOUEX, P. M., AND D. HINDLE (1981). Oral communication: some guidelines on answering questions, *Engineering Education,* December, pp. 243–244.

MICHAELSON, H. B. (1986). *How to Write and Publish Engineering Papers and Reports,* 2nd ed., iSi Press.

G

Useful Forms for Various Transportation Studies

Commonly used data collection forms are contained in this appendix. Each form may be removed or photocopied for reproduction as printed forms. The forms are identified as to source in the manual by chapter name, number, and page on which they appear. The details of use are described in the text near the page on which the form appears or is first mentioned.

Before going to the site, do not forget . . .

____ 1. To check the data collection equipment. Make sure that it records, stores, and/or produces output as you require. Make sure that the equipment is calibrated properly.

____ 2. To label the equipment as needed (e.g., turning movement counters can be labeled with the approaches being watched).

____ 3. To bring the data collection equipment. Also, bring spares of equipment such as stopwatches that are small and unreliable.

____ 4. An accurate watch set to the correct time.

____ 5. The correct sizes of spare batteries for all the equipment.

____ 6. An abundance of forms. Also bring a clean copy of the form from which more copies can be made if needed.

____ 7. Paper for taking notes.

____ 8. Plenty of pens.

____ 9. Clipboards or other writing surfaces.

____ 10. A letter from the landowner (if private property will be used) and/or from a responsible agent of the highway authority giving permission to collect data. Contact names and telephone numbers are sometimes adequate.

____ 11. A few business cards of the engineer supervising the study. A contact name and telephone number are sometimes adequate.

____ 12. A short, simple answer to the question, "What are you doing here?"

____ 13. The telephone number where the supervising engineer can be reached on the day of the study, in case questions arise.

____ 14. A map showing the site or directions to the site, if it is unfamiliar.

____ 15. Folding chairs.

____ 16. For a long study, a cooler or insulated container with beverages.

____ 17. A hat, sun visor, or sunglasses.

____ 18. Sunburn protection.

____ 19. Extra cold-weather clothes, such as a sweatshirt and gloves.

____ 20. Extra warm-weather clothes, such as a T-shirt and shorts.

Figure G-1 Refer to Figure 1-1 on page 4.

METRIC CONVERSION FACTORS

Approximate Conversions to Metric Measures				
Symbol	When you know	Multiply by	To Find	Symbol
LENGTH				
in	inches	25.4	millimetres	mm
in	inches	*2.54	centimeters	cm
ft	feet	0.3	meters	m
yd	yards	0.9	meters	m
mi	miles	1.6	kilometers	km
AREA				
in^2	square inches	6.5	square centimeters	cm^2
ft^2	square feet	0.09	square meters	m^2
yd^2	square yards	0.8	square meters	m^2
mi^2	square miles	2.6	square kilometers	km^2
	acres	0.4	hectares (100,000m^2)	ha
MASS (weight)				
oz	ounces	28.4	grams	g
lb	pounds	0.46	kilograms	kg
lb	pounds	4.448	Newtons	N
	short tons (2000 lb)	0.9	metric tons	
VOLUME				
fl oz	fluid ounces	29.6	milliliters	ml
pt	pints	0.47	liters	l
qt	quarts	0.95	liters	l
gal	gallons	3.8	liters	l
ft^3	cubic feet	0.03	cubic meters	m^3
yd^3	cubic yards	0.76	cubic meters	m^3
TEMPERATURE (exact)				
**F	Fahrenheit temperature	5/9 (after subtracting 32°)	Celsius temperature	**C

Approximate Conversions from Metric Measures				
Symbol	When you know	Multiply by	To Find	Symbol
LENGTH				
mm	millimetres	0.04	inches	in
cm	centimeters	0.4	inches	in
m	meters	3.3	feet	ft
m	meters	1.1	yards	yd
km	kilometers	0.6	miles	mi
AREA				
cm^2	square centimeters	0.16	square inches	in^2
m^2	square meters	10.8	square feet	ft^2
m^2	square meters	1.2	square yards	yd^2
km^2	square kilometers	0.4	square miles	mi^2
ha	hectares (100,000m^2)	2.5	acres	
MASS (weight)				
g	grams	0.035	ounces	oz
kg	kilograms	2.2	pounds	lb
N	Newtons	0.225	pounds	lb
	metric tons	1.1	short tons	
VOLUME				
ml	milliliters	0.03	fluid ounces	fl oz
l	liters	2.1	pints	pt
l	liters	1.06	quarts	qt
l	liters	0.26	gallons	gal
m^3	cubic meters	35.3	cubic feet	ft^3
m^3	cubic meters	1.3	cubic yards	yd^3
TEMPERATURE (exact)				
**C	Celsius temperature	9/5 (then add 32°)	Fahrenheit temperature	**F

Figure G-2 Metric conversion factors.

VEHICLE TURNING MOVEMENT COUNT

FOUR-APPROACH FIELD SHEET

Time ____________ to ____________

N/S Street ____________________ Date ____________ Day ________

E/W Street ____________________ Weather ____________________

P = passenger cars, stationwagons, motorcycles, pick-up trucks. Observer ____________________

T = other trucks. (Record any school bus as SB; other buses as B).

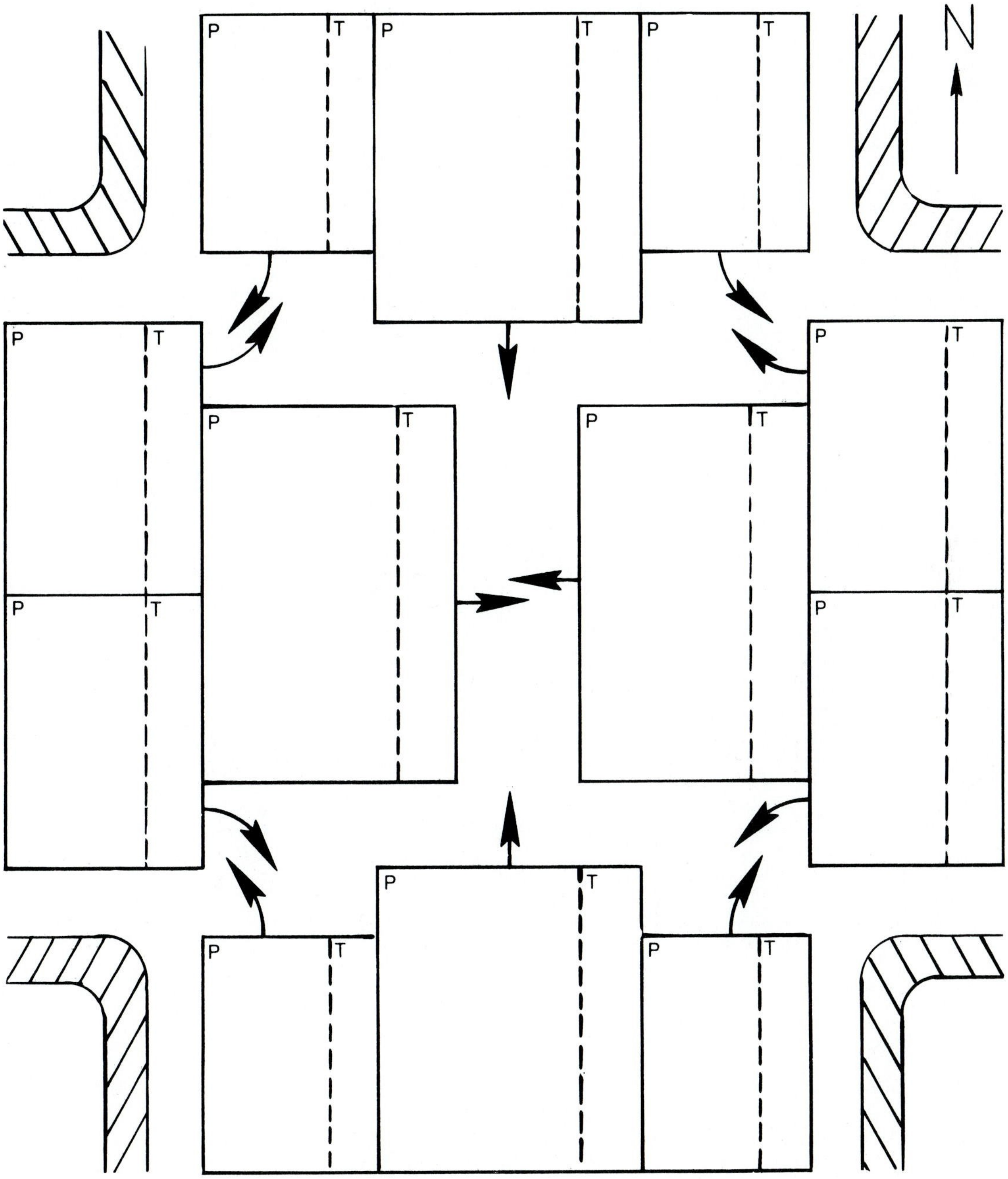

P T P T P T

N

P T P T P T

P T P T

P T P T P T

Figure G-3 Refer to Figure 2-1 on page 8.

TABULAR SUMMARY OF VEHICLE COUNTS

Observer __________ Date __________ Day __________ City __________

INTERSECTION OF __________ AND __________

R = Right turn
S = Straight
L = Left turn

TIME BEGINS	from NORTH				from SOUTH				TOTAL North South	from EAST				from WEST				TOTAL East West	TOTAL ALL
	R	S	L	Total	R	S	L	Total		R	S	L	Total	R	S	L	Total		

Figure G-4 Refer to Figure 2-3 on page 10.

DATA COLLECTION FORM

Site ID # ______________ Route ______________ County ____________ State ______________

Mileposts ________________________ Expanded Mileposts ________________________

Date ________________ Day ______________ Weather ________________________

Start Time ________________ End Time _________ Extraction Date ________________

TSR Unit ________________ Diskette # ______________ Directory ________________

Data Collected: Raw Vehicles _________ TSR Stats __________ Speed Stats ______________

Length Stats __________ Counts _________ Stream Stats ________________

Speed Bins ___________ Length Bins ________________________

Mat configuration:

Lane ________

Direction ________

Lane ________

Direction ________

Lane ________

Direction ________

Lane ________

Direction ________

Noted Problems ____

____

____

Site Characteristics beginning at (Milepost or location) ________________________

Lane Width ____________________ Shoulder Type ________________ & Width ___________

Median Type ________________________ & Width ________________ Bridges ___________

Intersections (4-leg): Signalized ___________________________ Unsignalized ______________

Intersecting Roadways: Right Side ___________ Sig. _________ Unsig. ________________

Left Side ____________ Sig. _________ Unsig. ________________

Driveways: Right Side ________________ Left Side __________

Horizontal Curvature ____

____

Terrain ____

Development Type ____

Other Characteristics ____

____

Figure G-5 Refer to Chapter 3.

DATA COLLECTION LOG

Site ID #	Route	County	State	Mileposts	Date Collected	Deployment		Removal	
						Driving Time	Setup	Driving Time	Retrieval

Figure G-6 Refer to Chapter 3.

DATA COLLECTION SUMMARY

Site ID Number	Area Type	Speed Limit	Mile Posts	No. of Lanes	No. of Driveways	No. of 3-Leg Intersections Sig. Unsig.	No. of 4-Leg Intersections Sig. Unsig.	Terrain	Horizontal Curvature	Other

Figure G-7 Refer to Chapter 3.

TRAVEL-TIME AND DELAY STUDY
AVERAGE VEHICLE METHOD
FIELD SHEET

DATE ______________ WEATHER ______________________ TRIP NO. ______________

ROUTE ______________________ DIRECTION ______________________

TRIP STARTED AT ______________ AT ______________________
(LOCATION) MILAGE)

TRIP ENDED AT ______________ AT ______________________
(LOCATION) (MILEAGE)

CONTROL POINTS			STOPS OR SLOWS		
LOCATION	TIME		LOCATION	SEC Delay	

TRIP LENGTH	TRIP TIME	TRAVEL SPEED
RUNNING TIME	STOPPED TIME	RUNNING SPEED

SYMBOLS OF DELAY CAUSE: S-TRAFFIC SIGNALS SS-STOP SIGN LT-LEFT TURNS

PK-PARKED CARS DP-DOUBLE PARKING T-GENERAL

PED-PEDESTRIANS BP-BUS PASSENGERS LOADING OR UNLOADING

COMMENTS ____

RECORDER ______________________

Figure G-8 Refer to Figure 4-1 on page 57.

TRAVEL-TIME AND DELAY STUDY
MOVING VEHICLE METHOD
FIELD SHEET

ROUTE________________________________DATE____________________

START POINT________________________END POINT__________________

WEATHER____

RUN	START TIME	FINISH TIME	TRAVEL TIME	VEHICLES MET	VEHICLES O.TAKING	VEHICLES PASSED
__BOUND						
1						
2						
3						
4						
5						
6						
7						
8						
TOTAL						
AVERAGE						
__BOUND						
1						
2						
3						
4						
5						
6						
7						
8						
TOTAL						
AVERAGE						

COMMENTS ____

RECORDER(S)______________________________

Figure G-9 Refer to Figure 4-6 on page 63.

INTERSECTION STOPPED - DELAY FIELD SHEET

Intersection ________________ Time ____________ To ______________________

Date ________ Day ______________________

Weather ______________________

Observer(s) ______________________

Street ________________ ______ Bound Traffic Lane ______________

	Number of Stopped Vehicles; V_s							
Sec / Min								
Totals								

$\Sigma V_s =$ ____

Volume, $V =$ ____

$\text{Stopped Delay} = \dfrac{\Sigma V_s \times I}{V} =$ ______ sec/veh

Figure G-10 Refer to Figure 5-1 on page 71.

INTERSECTION STOPPED - DELAY FIELD SHEET

Intersection ____________ Time ________ To ____________

Date ______ Day ____________

Weather ____________

Observer(s) ____________

Street ____________ ______ Bound Traffic Lane ____________

Sec / Min	Number of Stopped Vehicles; V_s							
Totals								

$\Sigma V_s =$ ____

Stopped Delay $= \frac{\Sigma V_s \times I}{V} =$ ______ sec/veh

Volume, V = ____

Figure G-10 Refer to Figure 5-1 on page 71.

FIELD SHEET - SATURATION FLOW STUDY

Location: ______________________________

__________ Bound Traffic Lane: ______________________________

Date: ______________ Time: ______________ City: ______________

Observers: ______________ Weather: ______________

Grade: ________ % Lane Width: ________ ft Area: ______________ Other: ______________

Obs No.	Time (seconds) between 4th vehicle and ...			
	7th veh.	8th veh.	9th veh.	10th veh.
1				
2				
3				
4				
5				
6				
7				
8				
9				
10				
11				
12				
13				
14				
15				
16				
17				
18				
19				
20				

Obs No.	Time (seconds) between 4th vehicle and ...			
	7th veh.	8th veh.	9th veh.	10th veh.
21				
22				
23				
24				
25				
26				
27				
28				
29				
30				
31				
32				
33				
34				
35				
36				
37				
38				
39				
40				
Column Sums:				
	(a)	(b)	(c)	(d)

$$\text{Mean Saturation Flow (vph)} = \frac{3600 * \text{Total Number of Observations}}{\frac{(a)}{3} + \frac{(b)}{4} + \frac{(c)}{5} + \frac{(d)}{6}}$$

Figure G-11 Refer to Figure 5-3 on page 76.

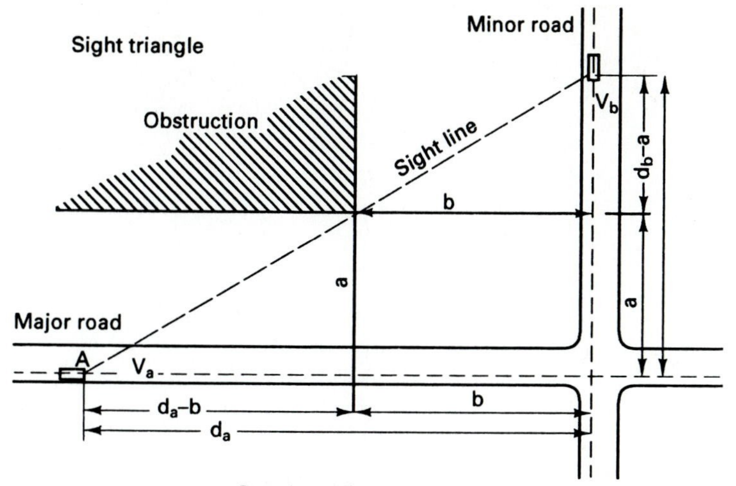

Case I and II
No control or yield control on minor road

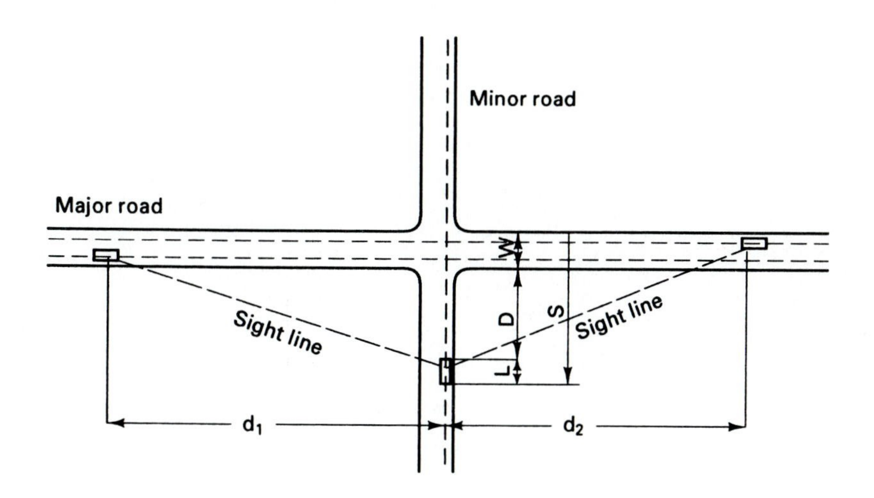

Case III
Stop control on minor road

Figure G-12 Refer to Figure 5-5 on page 84.

Origin-Destination Study
Field Sheet

Location ______________________ Station Number ______________________

Time: Begin: ______________________ Inbound ______________________

End: ______________________ Outbound ______________________

1	2	3	4	5
Origin	Destination	Route Used	Parking	Other
Indicate by block, street, zone, or other city		Streets, zones, or highway	Location and type	

Date: ______________________ Observer ______________________

Figure G-13 Refer to Figure 7-5 on page 115.

License Plate Study
Field Sheet

Location ______________________ Direction of Traffic ______________________

Time: Begin: ______________________ Station Number ______________________

End: ______________________ Weather ______________________

License Number	Time	Truck or Bus	Out-of State ?

License Number	Time	Truck or Bus	Out-of State ?

Date: ______________________ Observer ______________________

Figure G-14 Refer to Figure 7-8 on page 119.

PARKING INTERVIEW FORM

CITY____________________ DATE__________ FACILITY NO.__________

RECORDED BY____________________ TABULATED BY____________________

Time Parked	Time Leaving	Purpose 1. Work 2. Shop 3. Business 4. Other	Destination (building name, street address, etc.)		Office Use Only	
				Duration		Walking Distance
				hr	min	

Figure G-15 Refer to Figure 10-10 on page 186.

INTERSECTION TRAFFIC CONFLICTS SUMMARY

Location________________________________ Leg Number(s)________________

Day__________ Date______________ Observer(s)__________ Length of Recording Period____________

C = Conflict SC = Secondary Conflict

Count Start Time (Military)	Time Period	Approach Volume	Left-Turn Same Direction		Right-Turn Same Direction		Slow Vehicle		Lane Change		Opposing Left-Turn		Right-Turn From-Right		Left-Turn From-Right		Through From-Right		Right-Turn From-Left		Left-Turn From-Left		Through From-Left		Right-Turn On-Red		All Same Direction		All Through Cross-Traffic		Other
			C	SC	C	SC	C	SC	C	SC	C	SC	C	SC	C	SC	C	SC	C	SC	C	SC	C	SC	C	SC	C	SC	C	SC	
Total																															
C + SC																															
Daily Count																															
Rate Per 1,000 Veh																															

Figure G-16 Refer to Figure 12-15 on page 232.

ACTOR CODES	ACTION CODES	Name: Date: Time Period: Intersection: Direction (leg with actor 1): Weather:

Time	Actor 1	Action	Actor 2	Action	Comments

Figure G-17 Refer to Figure 12-16 on page 233.

CROSSWALK FIELD SHEET
PEDESTRIAN COUNT

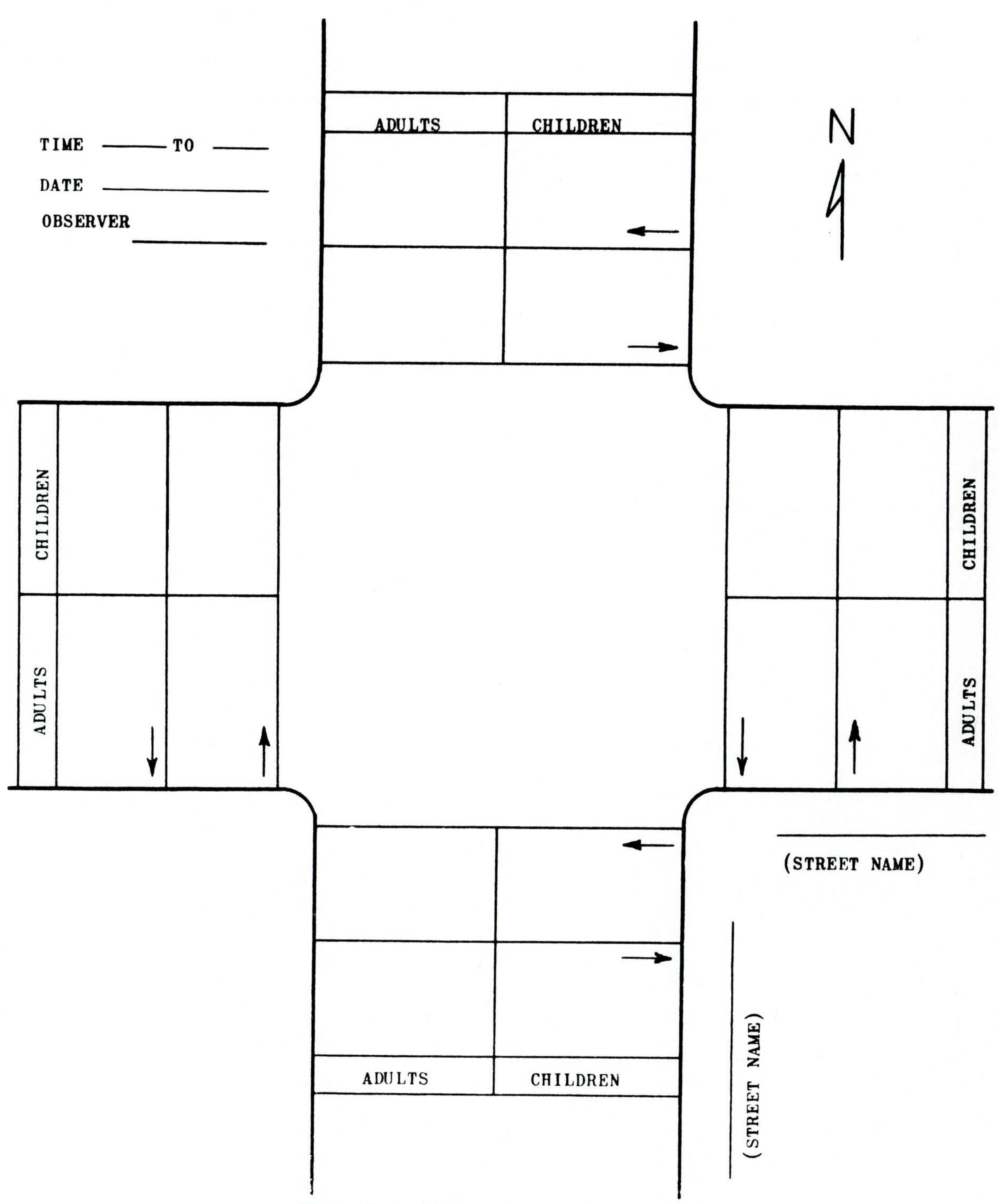

TIME ______ TO ______

DATE ____________

OBSERVER ____________

ADULTS | CHILDREN

CHILDREN | ADULTS

CHILDREN | ADULTS

ADULTS | CHILDREN

N

(STREET NAME)

(STREET NAME)

Figure G-18 Refer to Figure 13-1 on page 238.

CROSSWALK FIELD SHEET
PEDESTRIAN COUNT

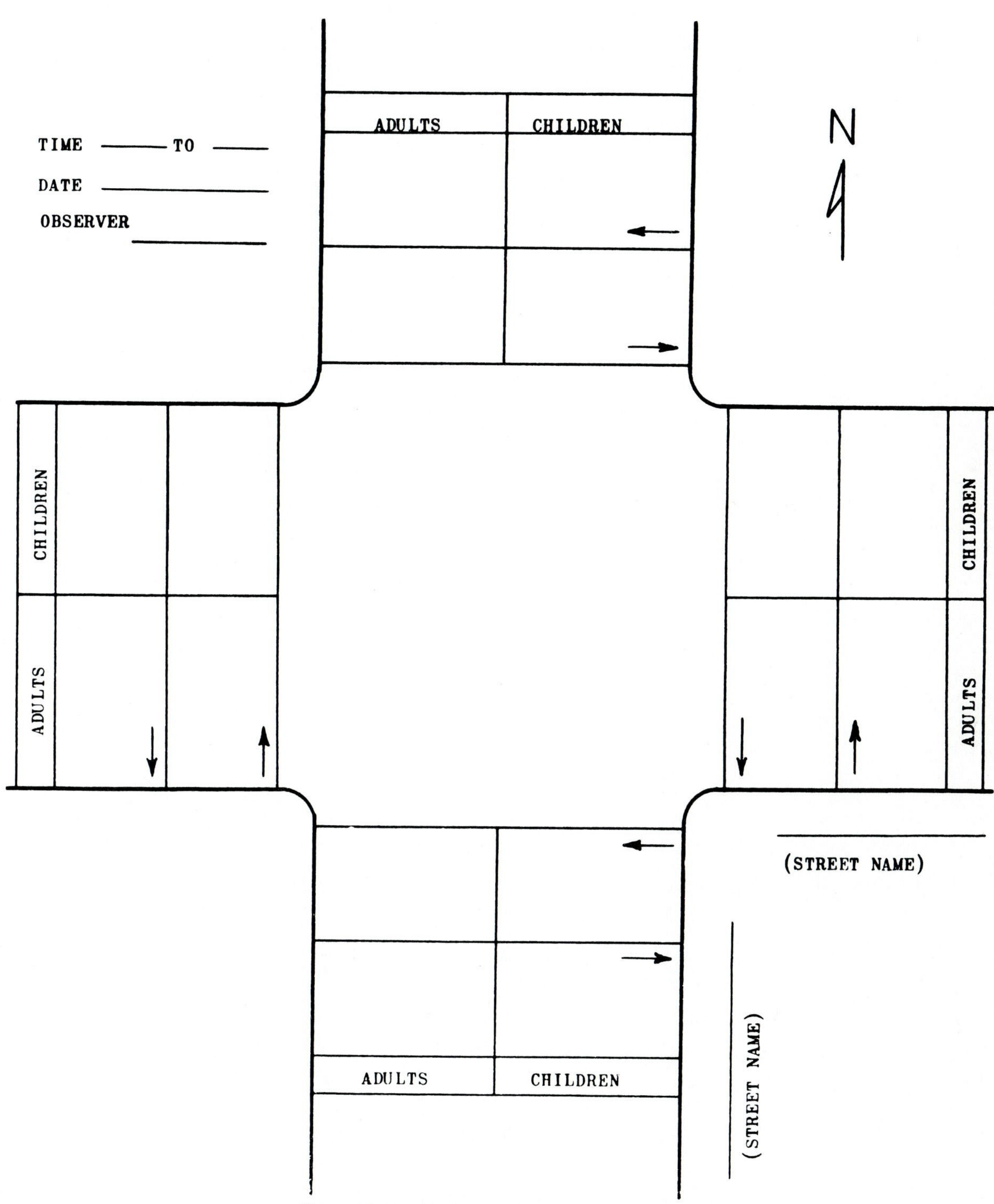

Figure G-18 Refer to Figure 13-1 on page 238.

GAP STUDY FIELD SHEET

Group Size Survey   Location______________________
Date______________   Time: From__________ To__________
Crossing Distance__________ ft   Walking Speed__________ ft/sec

No. of Rows	Tally	Total	Cumulative
1			
2			
3			
4			
5			
6			
7			

N = 0.85 x _____ = _____ or ____ rows   H = ____sec   R = ____sec

Minimum Acceptable Gap, G = W/S + (N-1)H + R = ________ sec

Available Gap Survey   Date______________
Time: From__________ To__________   Duration__________ min

Gap (sec)	Tally	Total
10		
11		
12		
13		
14		
15		
16		
17		
18		
19		
20		
21		
22		
23		
24		
25		
26		
27		
28		
29		
30		
31		
32		
33		
34		
35		
36		
37		
38		
39		
40		

Total Adequate Gaps __________

Figure G-19 Refer to Figure 13-5 on page 248.

Survey Date ___________

Location ______________________ Cross walk across _________________

End of Survey (to nearest minute) ________	Number of lanes 'N' ________
Start of Survey (to nearest minute) ________	Roadway Width 'W' ________
Total Survey time (Minutes) ________	Adequate gap time 'G' ________

Gap Size Seconds	Number of Gaps Tally	Total	Multiply by Gap size	Computations
8				
9				G = 1+ [W/35} + 2N
10				
11				G = ________sec
12				
13				
14				T = total survey
15				time X 60
16				
17				T = ________sec
18				
19				
20				D = [(T-t)/T] X 100
21				
22				D = ________%
23				
24				
25				
26				t = total time of all
27				gaps equal to or
28				greater than G
29				
30				
31				
32				
33				
34				
35				
36				
37				
38				
39				
40				
41				
42				
43				
Totals				

Figure G-20 Refer to Figure 14-9 on page 267.

Driver Observance of Stop Signs
Field Sheet

Location: ____________________

Time: ________ to ________ Weather: ____________________

Date: ________________ Recorder: ____________________

N S E W on ____________

Right	Straight	Left
	NON-STOPPING	
	PRACTICALLY STOPPED (0-3 MPH)	
	STOPPED BY TRAFFIC	
	VOLUNTARY FULL STOP	

N S E W on ____________

Right	Straight	Left
	VOLUNTARY FULL STOP	
	STOPPED BY TRAFFIC	
	PRACTICALLY STOPPED (0-3 MPH)	
	NON-STOPPING	

Notes: ____________________

Figure G-21 Refer to Figure 15-2 on page 289.

No Left Turn Compliance Study
Field Sheet

Location: ____

Direction of Travel: ____________ Area Type ____________________

Time: __________ to ________ Weather: ____________________

Date: ______________________ Observer ________________

	LEFT	RIGHT AND THROUGH	TOTAL
CARS			
TRUCKS			

	LEFT	RIGHT AND THROUGH.	TOTAL.
TOTAL CARS			
TOTAL TRUCKS			
TOTAL			

Notes: ____

____

Figure G-22 Refer to Figure 15-3 on page 290.

Driver Observance of Traffic Signals
Field Sheet

Location: ______________________________

Time: __________ to __________ Weather: ______________________

Date: ____________________ Recorder: ______________________

N S E W on ______________

RIGHT	STRAIGHT	LEFT
	GREEN	
	YELLOW AFTER GREEN	
	RED	
	JUMPED SIGNAL	

N S E W on ______________

RIGHT	STRAIGHT	LEFT
	GREEN	
	YELLOW AFTER GREEN	
	RED	
	JUMPED SIGNAL	

Notes: ______________________________

____

Figure G-23 Refer to Figure 15-4 on page 291.

Right Turn on Red Compliance Study
Field Sheet

Location: ______________________

Direction of Travel: ____________ Area Type ______________

Time: ________ to ________ Weather: ______________

Date: ______________________ Observer ______________

INDICATION	ACTION	CARS	TRUCKS	TOTAL CARS	TOTAL TRUCKS
ON GREEN	Turned on G or Y				
	Stopped on R Waited for G, Turned on G				
	Behind a Waiter (above) Turned on G				
	Attempted to turn on R, turned on G				
ON RED NO QUEUE	Full Stop				
	Stopped by Cross traffic				
	Stopped by Pedestrian Crossing				
	No Stop				
ON RED QUEUE	Full Stop				
	Stopped by Cross Traffic				
	Stopped by Pedestrian Crossing				
Totals					

Notes: ______________________

Figure G-24 Refer to Figure 15-5 on page 293.

Pedestrian Observance of Traffic Signals Field Sheet

Location:____________________________________

Time: _________ to _______ Weather: ____________________

Pedestrians crossing ____________________ in____________direction

STEPPED FROM CURB ON	CROSSED STRAIGHT (crosswalk)	Total
RED / WALK	NON-STOPPING	
YELLOW / FLASHING DON'T WALK	PRACTICALLY STOPPED (0-3 MPH)	
GREEN/ STEADY DON'T WALK	STOPPED BY TRAFFIC	
	CROSSED DIAGONALLY	
RED / WALK		
GREEN OR YELLOW/ DON'T WALK		
TOTAL		

Date: ________________ Recorder: ____________________

Notes:____________________________________

Figure G-25 Refer to Figure 15-6 on page 294.

BOARDING COUNT FIELD SHEET

ROUTE ______________________ BLOCK NUMBER ______________________

DAY ____________ DATE ____________ WEATHER ______________________

OBSERVER ______________________

Route Segment		Boarding Passengers					
From	To	Full Fare	Reduced Fare	Transfer	Full + Transfer	Reduced + Transfer	All Passes

Figure G-26 Refer to Figure 17-1 on page 314.

POINT CHECK FIELD SHEET

ROUTE (S) ______________________ BUS STOP NUMBER ______________

DAY __________ DATE __________ WEATHER ______________

☐ ARRIVING LOAD

☐ DEPARTING LOAD OBSERVER ______________

Route Number	Direction	Block Number	Vehicle Capacity	Arriving Time		Passengers
				Scheduled	Actual	

Figure G-27 Refer to Figure 17-2 on page 315.

RIDE CHECK FIELD SHEET

BLOCK NUMBER ____________________ ROUTE NUMBER ____________________

DAY __________ DATE __________ WEATHER ____________________

DIRECTION OF TRIP ____________________ OBSERVER ____________________

SCHEDULED START TIME ____________________

Location	Passengers			Time Check	Remarks
	On	Off	Load		

Figure G-28 Refer to Figure 17-5 on page 318.

PUBLIC TRANSPORTATION VEHICLE DELAY FIELD SHEET

Day:
Date:
Weather:
Route:
Direction:
Vehicle Type:

Observer:
Method:
Trip Number:
Trip Start Time:
Trip End Time:

(1) Location	(2) Time at control point	(3) Time slower than walking speed	(4) Stop time	(5) Time faster than walking speed	(6) Delay cause	(7) (5) – (3) Delay time (sec.)
					TOTAL DELAY TIME, SECONDS	

Symbols for delay cause: P = passenger loading, S = traffic signal, SS = stop sign, PK = parked cars, DP = double parked, PED = pedestrians, RT = right turns, LT = left turns, T = general congestion, KT = intentionally killed time, O = other (explain).

Remarks:

Figure G-29 Refer to Figure 17-6 on page 319.

LOG DATE ______________________

Truck Mileage Reading at Start of First Trip________(miles) Truck Mileage Reading at End of Last Trip __________(miles)							
Stop Number	Arrival and Departure Time at Each Stop		Location of Each Stop by Zone	Type of Business or Activity at each Stop	Type and Quantity(units) of Freight Handled at Each Stop		Load Factor for Truck or Trailer after Each Stop
Start of 1st Trip							
1st Stop							
2nd Stop							
3rd							
4th							
5th							
6th							
7th							
8th							
9th							
10th							
11th							
12th							

Figure G-30 Refer to Figure 18-7 on page 340.

LOADING DOCK OPERATIONS LOG FORM

LOCATION: ______________________ DATE: ____________

SIZE OF BLDG:____________SQ. FT. OCCUPIED SPACE__________

LAND USE______________________________

		TYPE OF VEHICLE			COMMODITY		
COMPANY NAME	TIME IN	VAN	SUT	T-T	DELIVERED	PICKED-UP	TIME OUT

Figure G-31 Refer to Figure 18-8 on page 341.

P.U.D. OPERATIONS SURVEY

A. LOCATION ID____

B. DATE____

C. TIME OF ARRIVAL______________ TIME OF DEPARTURE_________

D. PARKING MODE

Vehicle Parked In

1. Travel Lane_________________
2. Non-Travel Lane Illegally_______
3. Non-Travel Lane Legally_______
4. Other______________________

E. ACTIVITY RECORDED

1. Location____
Operation____
Commodity________ No. of Parcels____ Total Weight__________

NOTE: Continue below only if additional establishments are visited

2. Location____
Operation____
Commodity_____ No. of Parcels_______ Total Weight_________

Figure G-32 Refer to Figure 18-9 on page 342.

Index

A

B

C

D

E

F

G

H

I

L

M

N

O

P

Q